T0260643

August Weismann

August Weismann

Development, Heredity, and Evolution

Frederick B. Churchill

HARVARD UNIVERSITY PRESS

Cambridge, Massachusetts
London, England
2015

First printing

Frontispiece: This portrait of Weismann was painted by Otto Scholderer, a German painter
who practiced in London. Weismann knew both Otto and his brother Emil in Frankfurt.
Writing to Emil in 1907, Weismann commented that Otto visited him at his Freiburg
home on a daily basis while painting the portrait and that it "had from the beginning
excited the admiration of his family" ("Das Bild, das er damals von mir malte hat die
Bewunderung meiner Familie eregt"). Weismann to Emil Scholderer, 16 June 1907. Klaus
Sander, ed., *August Weismann (1834–1914) und die theoretische Biologie des 19. Jahrhunderts.
Urkunden, Berichte und Analysen.* Freiburger Universitätsblätter, vols. 87, 88. Freiburg:
Rombach Verlag, 1985.

Library of Congress Cataloging-in-Publication Data

Churchill, F. B. (Frederick B.)
  August Weismann : development, heredity, and evolution / Frederick B. Churchill.
    pages   cm
  Includes bibliographical references and index.
    ISBN 978-0-674-73689-4 (alk. paper)
    1. Weismann, August, 1834–1914.   2. Biologists—Germany—Biography.   I. Title.
QH31.W45C48 2015
570.92—dc23
[B]   2014035141

To Sandra Riddle Churchill

Contents

Color plates follow page 386.

Preface

Few contemporary historians and philosophers of science know of August Weismann. Still fewer life scientists know of his achievements in biology, although they may have heard his name in connection with the scientific turmoil over the processes of heredity and evolution in the period between the publication of *The Origin of Species* (1859) and the beginning of World War I (1914). This was a time when the scientific world was arguing with more opinion than exactitude over the operation and mechanisms of biological evolution. To promote Darwin's theory of natural selection or to assemble some combination of competing causes, be they physical, chemical, or organically more complex, were issues foremost in the minds of both zoologists and botanists. What follows is an intellectual biography of Weismann, which places him on the same level of importance in shaping the biology of evolution as historians have placed other frequently heralded luminaries in the world of biology.

Overall, the text of this biography is a chronological narrative that follows Weismann's life from birth to death. It is based to a large extent on Weismann's publications, correspondence, Tagesbücher, and other relevant university documents now housed in the Freiburg University Library. The analysis focuses on the development of the entire corpus of Weismann's work as manifested in his publications. It necessarily also examines the relevant publications of contemporary biologists with whom Weismann interacted. In general, this means that each chapter introduces the setting for important nonscientific and institutional events in Weismann's life during the period covered by that chapter and then moves on to the relevant biological developments during that same period. Occasionally the two are interwoven within the same chapter.

Included in the narrative are vignettes of Weismann's scientific contemporaries who interacted with his science during the period being covered. This is particularly the case with the careers of his exact contemporary Ernst Haeckel and his antagonists Rudolf Virchow, Herbert Spencer, Oscar Hertwig, George John Romanes, Theodor Eimer, and Richard Semon. The parallel and diverging careers of Weismann and Haeckel are so important (and are documented in already published correspondence) that they are treated in an early single section, which then makes it possible to refer to Haeckel throughout Weismann's life. There are points when August Gruber, Weismann's brother-in-law and long-term institutional colleague, becomes an important part of the history of Freiburg University. This chronological pattern, however, must be interrupted at key points when the surrounding study of life develops rapidly and initiates a change in the whole of biology as well as in Weismann's own career. Thus, Chapter 2 pictures the state of biology at the time Weismann enters the science; likewise, Chapters 9 through 11 compare in a general way the nature of heredity and the rise of nuclear cytology. This latter development, which demarcates a transitional period in Weismann's research career, terminates Part I and initiates Part II of the biography. The publication of Weismann's best-known work, *Das Keimplasma,* also begins a period when Weismann is criticized from many directions and responds in kind. The important cytological research of Theodor Boveri and the closely connected advent of Mendelian genetics are biological developments that challenged Weismann's germ-plasm theory, which the by then older Weismann had to accommodate.

Important to Weismann, and to most prominent German zoologists of the day, was the university institute. These housed both museums of natural history with material important for zoological research and scientific laboratories, providing the benches for the training of doctoral students and the space and equipment for the development of the early scientific careers of a small number of teaching and research assistants. Of course, it also served as the forum in which Weismann lectured to university students, many of whom were medically inclined, pursued his technical research, and spent most of his academic career. Unlike Haeckel, Weismann stuck to his last at his institute; and unlike much of Haeckel's writings, most of Weismann's published work addressed technical matters rather than popular problems about the role of the evolution of mankind. However, like Haeckel, most of Weismann's major publications were quickly translated and published in English, an indication of how important his work and ideas were for English and American biologists.

I have written this text with historians and philosophers of science particularly in mind. It focuses primarily on Weismann's ongoing research at his zoological institute in Freiburg, Germany. I also hope that biologists with an interest in an important period in the discipline's past will find an intellectual biography of this important nineteenth-century zoologist both interesting and valuable, particularly for the period between the publication of Darwin's *The Origin of Species* and World War I. There is to my knowledge no secondary source in English, German, or French that covers Weismann's scientific career in such a comprehensive fashion. The few surveys of biology that cover this period generally consider only Weismann's germ-plasm theory and then only in a cursory fashion. Yet his career and accomplishments were well known during his lifetime, and he focused attention on how the fields of embryology, heredity, and evolution must ultimately be understood as a comprehensive whole. Ernst Gaupp, a young contemporary of Weismann in Freiburg, provided a digest of his scientific work in his *August Weismann: Sein Leben und Sein Werk* (1917). Although he was cognizant of Weismann's uncompleted "Vita Propria," Gaupp's study was more a summary of Weismann's mature germ-plasm theory than a historical account of his career. The historical writings of Peter Bowler, the sweeping and detailed history of biology by Ernst Mayr, and the informative conference on the sesquicentennial of Weismann's birth edited by Klaus Sander all offer important but constrained historical reflections of certain aspects of Weismann's career but do not provide a historical analysis of its development. Much the same can be said of the chapters in Leonid I. Blacher's *The Problem of the Inheritance of Acquired Characters* (1982),[1] Reinhard Mocek's *Die Werdende Form: Eine Geschichte der Kausalen Morphologie* (1998), Lynn K. Nyhart's *Biology Takes Form* (1995), and Robert K. Richards's *The Tragic Sense of Life: Ernst Haeckel and the Struggle over Evolutionary Thought* (2008). Although the last three comment on Weismann within the context of their own valuable analyses of the period, they do not explore the historical development of Weismann's science. Nevertheless, Weismann looms as the most important of the German contributors to a synthesis of microscopic and evolutionary sciences at the end of the nineteenth century. It is my intention that this biography should fill this lacuna. Only with a fuller picture of Weismann's career and accomplishments can we understand the development of zoology both in Germany and in England between 1859 and 1914.

As this biography proceeds, a chronological map will unfold of the different personages and materials as they reacted with Weismann's separate accomplishments. One pattern that becomes clear is that Weismann tended to work from one biological problem to another. Often these problems were immediately related to one another, as when he moved from the study of hydromedusae (i.e., hydroids) to heredity theory; at other times, they were less directly connected, as when he moved from dipteran development to butterfly biogeography. Whatever he pursued was meticulously presented, and all his studies fed into his understanding of heredity, development, and evolution—and ultimately into his germ-plasm theory. Personal circumstances and accidental opportunities may have shaped his choice of study, but Weismann appeared always to have kept the bigger picture in mind. The same may be said of his choice of dissertation topics for his advanced zoology and occasional anatomy students, fifty-nine in all.[2] But in these respects, his research and institutional careers may not have differed much from later twentieth-century patterns.

August Weismann

I

EXPLORATIONS INTO NATURAL HISTORY AND DEVELOPMENT

1

Growing Up and Early Research

August Weismann was born in 1834 into what is commonly called the "Bildungsburgertum," or educated middle class, in the free city of Frankfurt on the Main, some thirty kilometers up river from where the Main flows into the Rhine. The Main valley in itself was highly agricultural, and heavily cultivated according to contemporary accounts. South across the Main and connected by a single bridge sat the much smaller town of Sachsenhausen, and visible to the north were the low-lying Taunus Mountains. At the time, Frankfurt, one of four free cities of the German realm, consisted of roughly 55,000 people. The city contained narrow, twisting streets and many partially timbered medieval buildings.[1] Its medieval walls with the exception of some towers had been torn down during the occupation of Napoleonic troops, and the city leaders later took advantage of the vacated space to turn it into green promenades. It was a quaint city run by its own diet, more commercial than industrial, and proud of its past. Frankfurt was also the birthplace of Johann Wolfgang Goethe, the most noteworthy of many illustrious citizens, who had died two years before Weismann's birth.

When Weismann was twelve, the city became the site of the Frankfurt National Assembly, which strove to find a representative solution for the disparate and competing states in the German realm. Although the city did not possess a traditional university, it sported the locally established and funded Städelsches Kunstinstitut, or art institute, with its large collection of paintings and drawings, and the similarly local Senckenberg Naturforschende Gesellschaft, or science society, which focused on physiology and anatomy. Both institutions were important to Weismann during his upbringing.

A mid-twentieth-century genealogist of the large Weismann family traced August Weismann's lineage, with some uncertainty, back to Valentin Weissmann [*sic*], a citizen ("bürgerlicher Einwohner") who lived in Weierburg in Nieder Östereich and who apparently was martyred for his Lutheran ("evangelisch") beliefs. According to the genealogy, his undeterred followers remained overwhelmingly Lutheran. A large number of the men appeared to serve as teachers in their careers. The Frankfurt stem of the family, however, began with a banker, Immanuel Friedrich Weismann. His first son, Jean (Johann) Konrad August Weismann (1804–1880)—who would be the father of the zoologist—became in 1831 a teacher and later a professor at the city's gymnasium. Jean married Elenore Elizabeth (Eliza) (1803–1850), daughter of Christian Leopold Lübbren, the mayor of and civic official in the northern city of Stade situated close to the mouth of the Elbe River, and of Marie Margarete (née Römhild). Jean and Elenore conceived four children: Leopold Friedrich August (1834–1914), Agathe Marie Julie (1836–1900), Julius Heinrich (1838–1863), and Therese Marie (1844–1857). August Weismann wrote of all three of his siblings, and he also produced two short autobiographical sketches, which were published for specific purposes. He also began but did not complete an autobiography, which he entitled *Vita Propria: Erinnerungen aus altester Zeit*.[2] It is to this latter work I turn to recount his boyhood years and early career up to the summer of 1871. At this point he, unfortunately for historians, ended his account. He terminated this more orthodox autobiography in 1913, roughly a year before his own death on November 5, 1914. When exactly he began or ended the text is not clear; needless to say, the reader must take care in interpreting this life story written at the age of seventy-nine, but the names and facts contained therein do appear to be accurate.

We learn of an idyllic childhood life as painted by the seventy-nine-year-old Weismann. The family first lived near the Eschenheimer Turm, the tall four-spired tower that had been spared during the Napoleonic occupation of Frankfurt. Gardens existed between the neighboring houses. The promenade fashioned in place of the city wall was easily accessible for play and walks, and the former moat provided a brook for additional play. The family soon moved to Mainzerstrasse, which followed the remains of the old fortifications. It was a new street, lined by handsome new houses, a standard setting for the German educated middle class of the period. Weismann remembers, from that period, the bugle sounds of the post, music played on the sides of the streets, a paver resetting street stones, his rocking horse in a sunny room, his father playing the piano ("perhaps some Bach"), and cherry trees behind their house. These

Figure 1.1. Weismann and his sister Agatha in 1839.

August Weismann (1834–1914) und die theoretiche Biologie des 19. Jahrhunderts. Urkunden, Berichte und Analysen. Freiburger Universitätsblätter, vols. 87, 88. Freiburg: Rombach Verlag, 1985. Helmut Risler, "August Weismanns Leben und Wirken nach Dokumenten aus seinem Nachlass," p. 24.

cheerful scenes allowed him to recall his siblings, Agatha and Julius, and walks with his parents in the neighboring fields before the city that were full of flowers and butterflies, which fascinated him.

Later, Weismann recalls moving to a different house near the Gallustor by the former fortifications on the west side of the city. It lay in the neighborhood of the large three-story house that served as the Prussian consulate. Still as a young boy, Weismann recalls hiding at the bottom of the steps of the consulate and watching King Frederick Wilhelm IV of Prussia, who was visiting Frankfurt, climb them. The king appeared to the inconspicuous youngster to be a slow moving, heavy, and unwell man. (Frederick Wilhelm IV was born in 1795 and ascended to the Prussian throne in 1840; so he may have been in his late forties at the time. Later he was declared insane and superseded by his brother, who became regent.)

On a cultural level, it is evident that his mother had a lasting influence on the young Weismann in many ways. She was a trained artist in both aquarelle and oil, and even before her marriage she was painting formal portraits. Apparently she had learned to copy old masters in the style of the Düsseldorf Academy, which had begun to emphasize realism and nationalism.[3] In turn, she encouraged her son to draw, and by the time he reached fourteen years of age, Weismann took art lessons given by a Jacob Becker at the local Städelsches Kunstinstitute. His mother also encouraged him to play the piano at an early age, which led him to take lessons from a Herr Heymann. He readily took to music. He recalls listening to Haydn's *Creation* and to Beethoven's "Pastoral" symphony, but at the age of 13 he opined that the cuckoo call in the Andante of the latter was unrealistic. By the age of 16 he claims to have understood all of Beethoven's symphonies except the Ninth. It is interesting that he barely mentions his siblings receiving the same attention, but later he mentions that Julius had a nice singing voice and should have been encouraged to develop it.

It happened that this music teacher was also a butterfly collector and interested his young pupil to develop his own collection and to learn not only the names of the specimens he caught but also about their larvae. When the family moved outside the city to Taunusstrasse, Weismann had at his disposal the outlying fields of clover and other flowers and more butterflies. He mentioned that his father was not entirely happy about his involvement with nature, considering it only a hobby for rich people, but that his mother continued to support his interest and his grandparents (presumably on the maternal side) gifted him Bertuch's book of natural history.[4] When the teachers at the higher levels of his school did not support his collecting of insects either, he turned to plants and put together a herbarium. Only after an older friend, a Dr. Bagge, at the time a candidate in theology and a good botanist, began taking him on collecting trips, did Weismann deeply pursue the subject. Together, with a druggist friend, they went on collecting trips to the southwest toward Mainz and southeast toward Aschaffenburg. Weismann felt the range of locations provided him with a great number of species, and he began to learn their scientific names, but regretfully, he later added, not their physiology.

Weismann also enjoyed traveling with the entire family when all the children were old enough to make such adventures by a combination of boat, carriage, train, and foot. He recounts traveling by boat, then by foot, up the Ahrtal, a trip to Königstein in the Taunus, and exploring the Odenwald from Miltenberg over Amorbach to the Rhine and back to Frankfurt. His later descriptions are not long or detailed, but what is noteworthy is that he remembers best the living nature about him. He recalls the names of butterflies, rock crystals, and flowers, to a point when his mother remarked at one point that he appeared to miss an unusually beautiful sunset for the sake of a meadow in bloom.[5]

Spring of 1848 came early in March. Having just entered his fifteenth year, Weismann was tuned into some of the political events sweeping Germany and the rest of Europe. When the "Vorparlament" met in Frankfurt, Weismann was aware that Friedrich Hecker, a political radical from Baden, stayed in the house next to his. He remembered Hecker speaking to a sympathetic crowd from the window. Adding to the same recollections, Weismann mentioned that he also heard "many" speeches in Paulskirke, the site that served as the seat of the Frankfurt National Assembly. He claimed that at the time he was an outright "Republikaner" who wore a "black-red-gold freedom cockade" and a "blood red scarf" and with his friends played at war in the hedges of the neighborhood. He added, interjecting his later convictions, that he gradually realized the "dark side" of the republic when he visited the barricades before the fighting really began. "It seemed to me that the arbitrariness of the foolish multitude was too uncertain a leader," and "I soon became a convinced supporter of Prussian authority." With these thoughts he finished his paragraph by asserting that "if we no longer want to remain a joke with other people and if we also want to be of value to the world, then we must become united into a solid whole with a functioning government on the top of which stands one man and that can be no other than the Prussian king."[6]

During the same spring of 1848 Weismann visited his maternal grandparents in Stade near the mouth of the Elbe. Whether this was an attempt on his parents' part to get him out of the turmoil of Frankfurt or not is unclear from Weismann's sixty-four-year retrospective. It was a long trip alone for a fifteen-year-old. He went by steamboat from Mainz to Köln, then by train to Hannover. On the ride he had struck up a friendship with the wife of a non-commissioned officer, and when they reached Hannover, she offered him the extra bed in her room at the inn where she was staying. He felt the situation was awkward (*genierlich*), but he assured his potential reader that nothing unusual transpired. He slept soundly, and they parted in a friendly way the next day. Weismann then went by train to Hamburg and by post to Stade. Travelling through sand dunes and heath and passing many small, colorful houses, he finally arrived at his destination situated on a winding stream, which led into the Elbe. The new setting and the large delta area of the Elbe clearly impressed him. Although there were a few steamships most of the shipping on the river still consisted of large sailing ships bound upstream to Hamburg or returning to the ocean. "It was a magnificent view to see such a ship, covered with full sails from bottom to top traveling up the Elbe; the swans gliding calmly and easily behind and often following one another."[7]

He describes a trip with his "dear and friendly grandmother" and his aunt Auguste by boat to Hamburg where he was again impressed by the ships in the harbor presenting a "veritable forest of masts." Unlike those in London or Marseille, he commented, the ships in Hamburg were not forced to lie together in a confined space. "All the masts, yardarms, ropes, etc., present a confusion of lines, and reach so high in the heaven, that that alone really impressed us."[8] When Weismann returned to Frankfurt he wrote a theme for school with the title "a day in Hamburg." His father, who was his German professor at the time, read the theme aloud praising it to his class.

On account of the death of his grandfather, Christian Leopold Lübbren, Weismann again visited Stade, this time with his father and younger brother Julius. He relates an adventure that he and Julius had in a boat with the plan to return to Stade by way of the winding stream leading from Stade to the Elbe. The tide in the Elbe was such that they could not complete their return and had to overnight in the half-open cabin of a small fishing vessel. Because the boat's occupants could only speak "plattdeutsch," it was hard for them to communicate. Nevertheless the ending of the story was a happy one. When they returned the next morning their father was waiting for them on the shore, but he did not say a word. Weismann surmised that he was so relieved that he could not reprimand them.

Weismann felt it important to indicate the living standard of both his grandfathers. His father's father, Immanuel Gottlieb Friedrich Weismann, moved to Frankfurt where he worked in a bank. He lost much of his estate when, during the Napoleonic wars, soldiers were quartered in his house. As a consequence the family, consisting of two successive wives, who bore him collectively nine children, had to go into debt in order to succeed in raising five of them to maturity.[9] Of the four surviving boys, Weismann's father, the oldest, and the next two became teachers, the fourth broke ranks and became a banker. Three of the four, including the banker, remained in Frankfurt; the fourth taught in Coburg. His mother's father, as mentioned above, was evidently well off and became the mayor and local official in Stade. He was able to leave Weismann's mother a modest sum upon his death, which in turn helped Weismann in his training as a zoologist when she passed away at the early age of forty-seven.

Weismann strongly felt the loss of his mother when, shortly after he had turned sixteen, she died. She has been "a true friend" not only to him but to Agatha and Julius, and she could hardly be replaced by his father, who tended to be strict and constantly scolding. As Weismann mentioned a number of times, his mother had opened up the world of music and fine arts to him and encouraged him in his collection of butterflies and plants. As the eldest child, Weismann felt that he received the brunt of his father's rebukes. Later he learned to appreciate his father more, and when Weismann lived elsewhere, he visited him frequently in Frankfurt.

Both his mother and younger sister Theresa died of typhus, the former in December 1850, the latter in 1857. Typhus had been a disease of long standing in Europe, documented as far back as the sixteenth century. Dr. G. Spiess, a medical friend close to the family and father to one of Weismann's closest friends at school and university, was called on to treat both mother and sister. He followed the older practice of first giving her "normal salt, which irritated the intestines even more—at any rate did not help, and then came the hunger treatment with oat gruel." His mother craved food while she wasted away from starvation; in retrospect, Weismann was outraged. By the time seven years later his sister Theresa contracted the disease, Weismann was a young physician and had become aware of more modern treatments, such as allowing the patient to drink milk as both a soothing and nutritional medicament. Alas, he was in Rostock when she came down with the disease, and by the time he returned to Frankfurt, Theresa was nearly gone. Fifty years later, Weismann still felt strong regret that he could do nothing to save her.

Soon thereafter, his other sister, Agatha, and his father also came down with typhus. However, this time Weismann could insist on supplementing Dr. Spiess's gruel with milk. Both patients survived, and Weismann wrote a bitter retrospective passage about how the new treatment was already known by the clinically more advanced physicians of the time.

Before turning to Weismann's scientific career, it makes sense to extend briefly this genealogical account rather than trying to place individuals in chronological sequence later on. While striving with the legal system—unsuccessfully as it turned out—to win his inheritance from his mother which was held by his aunt Dorotea Römhild on his mother's side, Weismann was introduced to the Gruber family in Genua. While his elaborate negotiations over what turned out to be a nonexistent last will and testament were coming to naught, the Gruber family, a large German family led by a successful merchant, Adolf Gruber and his wife Julie Schönleber, provided a substitute home for him in Genua. Among them he partook of family life and frequently played the piano for everyone; in fact, he became most attached to the eldest daughter, Marie (known as Mary) (1848–1886), who often vocally accompanied him. The two married in 1867. Over the course of twenty-one years, the couple lived in Freiburg, where Weismann habilitated and rose to be a full professor of zoology with his own institute. Together, they raised six children: Therese (b. 1868), Hedwig (b. 1870), Elise (b. 1871), Bertha (b. 1873), Meta (b. 1876), and Julius (b. 1879). Meta died early, but the rest of the children lived to raise families of their own. Julius became a professional pianist, composer, and conductor of the Freiburg symphony orchestra. The others we frequently hear of in Weismann's published letters.[10] His wife, Mary, died early at the family retreat in Lindenhof on the Bodensee.

Gymnasium and Göttingen

When he reached the primary level at the gymnasium, Weismann read Sophocles and Horace (presumably in the original Greek), and took German with his father. Later he claimed to have found all these studies stimulating. He struggled a bit with mathematics but turned down the offer by his mother to receive special assistance. In retrospect, he found that he "lost nothing thereby." Later after becoming a young doctor, he made an effort to study a book by the French mathematician Leonhard Euler on the algebraic solution to geometrical problems, but he neglected to report how his independent study went.[11]

By the time he was thirteen years old, Weismann was taking piano lessons with a Franz Messer,[12] and attending lectures on chemistry and physics with Rudolf Christian Boettger, a highly regarded teacher at the Physicalishen Verein in Frankfurt and later the director of the Senckenberg Naturforschende Gesellschaft.[13] Boettger's lectures in chemistry stimulated Weismann enough to consider professionally entering the field. His father asked for advice about this plan from Friedrich Wöhler, who had been born near Frankfurt and after his appointment in chemistry at Göttingen was an occasional visitor to his home city. Wöhler advised the young student that before becoming a chemist he should secure his living by first becoming a physician. The advice turned out to be prophetic, and the contact with Wöhler may have been one of the reasons that in the fall of 1852 Weismann attended Göttingen to pursue his medical doctorate.

Weismann took a course with Wöhler after matriculating, and the contact, although he did not say so, may well have shaped his enduring mechanistic belief about life. Limited finances were also a reason for his not pursuing botany or zoology, even if the latter had been independently available. Nevertheless, Weismann looked forward to studying the acceptable medical sciences, particularly those of anatomy and physiology. By the mid-nineteenth century, as the historian H-H. Eulner has pointed out, "histology, comparative anatomy and embryology were included in physiology and therefore part of the responsibility of the anatomists."[14] By the mid-nineteenth century the trend in medical anatomy was working toward comparative anatomy, which in itself might form a bridge to zoology. In addition to the presence of Wöhler, Jakob Henle, a former prosector for Johannes Müller and a renowned microscopical anatomist in his own right, had just been appointed professor of anatomy at the university. New opportunities would open up for the beginning medical student Weismann.

Throughout his four-year stay in Göttingen, encouraged by Alexander Spiess, childhood friend and the son of his family's doctor, Weismann roomed with the highly reputable professor of surgery Wilhelm Baum and his family. Wilhelm Baum was a professor of surgery at Göttingen between 1849 and 1875. At the time, Baum also lectured on ophthalmology.[15] Weismann mentions that Baum was a mild, thoroughly good man who was artistic, musical, and devoted to his students. Weismann remembered Frau Baum as an unusual woman, who had good intuitions about people, who also was highly musical and played well four handed on the piano. They had two daughters (ages eighteen and sixteen) and a son. So for four years Weismann felt at home and gradually, through the family, met many local professors.[16]

As for his friends, he mentioned Spiess, who was attending Göttingen at the same time, and Theodor Otto von Heusinger from Marburg, who later returned to his native city as a physician. As anticipated, Weismann heard the anatomy lectures of Henle, but he considered them dry and too confined to the refined details of human structure. Despite the fact that Henle had transformed his lectures into the more exciting examination of comparative anatomy when he taught at Heidelberg, he resisted without explanation such an expansion of the subject in Göttingen. Weismann, of course, heard Wöhler's lectures in chemistry, but after having taken Boettger's chemistry course in Frankfurt, he felt he learned nothing new at the university. In addition, Weismann discovered that the renowned Göttingen physiologist Rudolf Wagner gave engaging lectures but was ill, so left much uncovered; the famous *Handbuch* by Johannes Müller did not provide a substitute, as Weismann considered it by the early 1850s to be out of date, although it still contained many valuable ideas.[17]

In retrospect Weismann remembered the four years as a medical student more for the music he participated in than the sciences he pursued. Despite his disappointing lectures, Henle provided Weismann and his friends with opportunities to play quartets, the piano, and even sing at his home. Weismann mentioned that he particularly enjoyed playing Chopin's Nocturnes and Scherzos but not the "technically unattainable Études." Weismann also alluded to a four-year attachment he had with a "fine, clever girl." Both realized, however, that their relationship could not continue, for Weismann had to leave Göttingen and could not seriously consider marriage before he found his way in the world.[18] Weismann's inaugural dissertation at Göttingen concerned the formation of hippuric acid in humans.[19] An expanded version of his work as it applied to herbivores received a royal award from the medical faculty.

Rostock and Chemistry

After passing his doctor's examination with a First in July 1857, Weismann accepted a position as an assistant with Benjamin Theodor Thierfelder, who had just become professor of medicine at the city hospital in Rostock. His fee for service provided him with 100 Thaler and living expenses. In part because Weismann was not really interested in extending his clinical experiences and in part because the work often seemed foolish (*närrisch*), he allowed himself to be enticed by his "best friend" to join Professor Franz Schulze's chemical laboratory. Here, he was set up with a small laboratory room, which also served as his living quarters, where, with Schulze's help, he carried out a chemical examination of the fluctuation of salt in the seawater of the Ostsee.[20] This work won a prize from the faculty. Again in retrospect, he was most impressed by the professor of comparative and pathological anatomy Hermann Friedrich Stannius, author of an important text on vertebrate anatomy. The professor was friendly to Weismann and other advanced students.

As soon as the semester was completed, Weismann traveled with a friend from Frankfurt through the East Sea to Copenhagen. Fifty years later, he would still wax with romantic enthusiasm for the blue sea, the medieval fortifications of the royal city, and the starry night sky as the two traveled by boat and carriage through Flensburg and the straits between Schleswig and Holstein. At the time, both of these duchies belonged to Denmark.

Weismann claimed to have learned very little from his chemical research about either urea or seawater. In retrospect, he felt that he did not have the "patience and exactness of a chemist for experimenting."[21] Nevertheless, we can reasonably assume that after spending four years in Göttingen studying and writing a dissertation in chemistry and a year pursuing a chemical problem in Rostock, he had learned to think like a chemist. He read the classic chemical works by Fourcroy, Vauquelin, Wöhler, and Liebig, and he learned to juggle the atomic quantities of carbon, hydrogen, nitrogen, and oxygen in the production of urea and its products. He determined the specific weight of seawater with changes in temperature, wind, and barometric measurement and with an eye to the changes of salt concentration in different locations and under different physical conditions. His minor publications in both areas received encomiums from the respective faculties of two universities. Never again would he dabble in chemistry, but philosophically his chemical experience formed the bedrock to all of his later biology. This generalization includes his research on seasonal polymorphism and his examination of the changing distributions of variations over time as well as his cytological research, which eventually led to his theories of heredity.

Return to Frankfurt

Weismann left Rostock and traveled home to Frankfurt when his younger sister, Therese, came down with typhus. As mentioned earlier, his mother and sister had succumbed to typhus, and then Agatha and his father also had come down with the disease. Weismann, now on the scene, insisted on a milder regimen that included a milk diet. It was a treatment he had learned from Thierfelder in Rostock. What had been a lethal disease for his mother and younger sister became a serious but treatable disease for the rest of the family. At the same time that further research in Rostock left his thoughts, Weismann toyed with the possibility, encouraged by his father, of earning a living as a physician in Frankfurt. During the summer of 1858, he went through the motions of studying medicine under some of the great stars on the faculty of medicine in Vienna. He visited their lectures but interacted on a personal level only with Carl Ludwig, who at the time was professor of anatomy and physiology in the Austro-Hungarian capital, and with the chief physician at the general hospital whose daughter, Weismann mentioned, later became an opera singer in both Karlsruh and Munich.

Most memorable of his summer experiences was a lengthy overland mountain excursion, which started in the region of the Grossglockner, descended to the Dolomites, and ended after an unspecified transit in the Austrian Ötztal. Here he aspired to continue, but on account of the weather he was unable to climb Mount Similaun, which at 3,602 meters is the highest peak on this stretch of the Italian-Austrian border. The trip included acquiring rugged mountain boots, frequent glacier climbing, and many hours of ascents and descents with a personal guide, who at one point saved him from sliding out of control down a glacier. Weismann mentions one comical moment: when he was dozing on the mountainside, he was awakened by the peering, bearded face of a large mountain goat. The trip ended in a more relaxing way when he turned to the visual arts in the Munich galleries and museum. The heritage from his parents of many Rhineland travels and the tutelage from his mother in both music and art weighed more strongly on Weismann's seventy-nine-year memory than the scientific experiences gained at this time.[22]

2

The Age of Development

By the time Weismann was sixteen in 1850, some of the most important nineteenth-century scientific approaches to the understanding of life had already been initiated. Above all, there emerged a single-minded focus on embryological development throughout the animal kingdom, which was associated with the defining accomplishments of Karl Ernst Baer in the third and fourth decades of the century. These were paralleled in exquisite detail by Baer's contemporaries Christian Pander, Heinrich Rathke, and Johannes Müller. Continuing full swing after the publication of Darwin's *The Origin of Species* in 1859, descriptive embryologists, not yet steeped in experiments, assumed a dominant role in driving zoological and botanical research to the end of the century. Their successes invited the next two generations of students who aspired to become professionally trained anatomists and zoologists to follow their lead. Derived from descriptive embryology and born of the same desire to understand the microscopic details of the organism, the cell emerged as the basic unit of organic structure and function. Articulating the cell theory in twin publications, Matthias Schleiden and Theodor Schwann provided a visible and variable unit out of which descriptive and later experimental embryologists fashioned their own tapestries of whole organisms. Contemporaries such as Müller, Jacob Henle, and younger investigators such as Albert Kölliker, Robert Remak, and Rudolf Virchow seized upon the cell as the unit for understanding such diverse organic phenomena as development, reproduction, and disease as well as the contrasts between sexual and asexual generation. Little was envisioned about organic form without reference to the cell. At the same time, while investigating life throughout the world, many naturalists began studying in depth the alternating forms and processes of different generations of the same species. The meaning of this apparent violation of the classical formulation of sexual reproduction was not immediately obvious. Investigations on this score by Japetus Steenstrup, Richard Owen, Karl von Siebold, Rudolph Leuckart, and Thomas Henry Huxley did much to reveal the overall complexities of animal reproduction. The lion's share of work in all three of these areas of investigation was done after the Napoleonic period in Germany at the reformed state universities. Essential to all three—descriptive, cellular, and reproductive biology—was the continued development of the science of microscopy, about which we will have much to say in later chapters.

The Study of Development: Baer

It is historically curious that three medical students on the Baltic periphery of German cities initiated the earliest descriptive studies of development in the nineteenth century. Pander, Baer, and Rathke knew each other as students and followed in one another's footsteps as each advanced in his early career. They were in turn assisted by the Dorpat, later Königsberg, anatomist and physiologist Karl Burdach and by the Würzburg anatomist and physiologist Ignaz Döllinger.[1] The combination of both anatomy and physiology in the same Lehrstuhl may have simply spoken to the time and limitations of the medical faculties at small universities in the German states, but it also may have encouraged thinking in terms of embryology. Both Burdach and Döllinger wrote about development and considered it a necessary dimension to their belief in the unity of science and life itself. Their influence on Baer, Pander, and Rathke should not be underestimated. To capture the impact of descriptive embryology on zoology at the end of the fifth decade of the century, when Weismann began his own embryological studies, one cannot do better than focus on the achievements of Karl Ernst von Baer.

Baer is remembered for two outstanding accomplishments in embryology. The first entailed his successful demonstration of the mammalian egg in 1827, which had previously eluded many scientists; once realized, it unified the general picture of reproduction among all higher organisms.[2] The second encompassed a two-part work entitled *Über Entwickelungsgeschichte der Thiere: Beobachtung und Reflexion* that traditionally shaped the rise of modern descriptive embryology.[3] A careful examination of this substantial work of nearly 600 pages and seven plates of multiple figures (four of them in contrasting red, yellow, and black) helps explain why the book was justly elevated to the status of a revered exemplar during the second half of the century.

As a totem, the entire work possessed an awkward structure about which to navigate. Its organization and its interrupted publication resulted in two quite differently constructed parts. The first, appearing in 1828, was in turn separated by two distinct objectives. The second was cobbled together by the publisher in 1837 from a series of Baer's lectures describing additional observations on chick development and an extended detour by Baer on the development of reptiles, numerous mammals and their different eggs, and finally fish. An introductory letter credits his friend Pander with the initial inspiration for his embryological studies. Its casual construction notwithstanding, the totality of the work evinced both an observant and a thoughtful investigator who made great strides in the factual details of the development of not only the chick but also mammals including humans and other vertebrates. Although he focused on the chick, Baer's observations reflected the importance of comparative embryology. Equally significant was the fact that the second half of part one consisted of general principles of development that could easily be adjusted over time as the factual details changed with new investigations. As Jane Oppenheimer pointed out many years ago, the secondary title to Baer's *Entwickelungsgeschichte* stressed that the work contained both *"Beobachtung und Reflexion"*—observation and reflection.[4]

Turning first to the *Beobachtung* or the observation of details of animal development, we find that Baer focused initially on the twenty-one days of development of the "Hühnchen" or chick. He segmented this process into three unequal sections. It is clear that Baer must have used a microscope as well as the lens that had led to his discovery of the mammalian egg. This instrument is inferred from his fine analysis of the early germ formations: the stratification of germ layers, the primitive streak seen from various angles, the bending and pinching off of the embryo from the rest of the egg, and the formation of the neural tube and primitive gut as well as, by the end of the first day of observation, some of the somites. Baer continued identifying tissue layers, the aorta, the notochord, the cephalic region, the foregut, and the vessels stretching throughout the egg, and he noted the torsion of the body onto its left side to accommodate the future expansion of the heart and to allow for greater flexion in the cephalic region. "The turning of the embryo onto its left side," Baer rationalized, "is a very important moment in the formation of the fetus, for on it depend many changes, especially the metamorphosis of the heart in the interior." As a consequence, he explained, the yolk sack and umbilical vesicle (*Nabelblast*) also lay on the left side. The sequel of this asymmetry Baer also found in the lizards, snakes, all birds, and all mammals he examined, and he documented his personal observations by indicating that only two of more than a hundred chicks that he had studied had turned instead onto their right side.[5] Thus, his studies continued through the detailed examination of development and were embroidered with many explanations and comparisons with other species through the twenty-one days of chick development and 140 pages of compelling prose.

Baer's account of the twenty-one days of chick development was brilliantly descriptive and suggestively comparative. At that time, however, zoology was struggling with the relationship between the development of individual particulars and the general outline of taxonomy of the whole. It was essential that Baer instruct his readers on how to deal with both the particulars and generalities of zoology. It is within this context that Baer presented his readers with a lengthy discussion of the principles of embryology. It was the distilling of these principles into Scholia and Corollaries that made his *Entwickelungsgeschichte* the premier statement of the goals of embryology for the rest of the century. In resorting to six Scholia and associated Corollaries, Baer embraced a device used by Euclid and Newton, but his lawlike summary for embryologists appeared much more a lesson from his empirical observations than from deduced generalizations. What follows is an abbreviated account of these Scholia and Corollaries.[6]

Scholium I, "On the Certainty of Observation of the Embryo," dealt with Baer's recognition that young embryos are coarser in outline than older ones and they simply do not exist in miniature below the resolving power of the common microscope of the day.

Scholium II, "The Formation of the Individual in Relation to Its Surroundings," relates Baer's observation that during development the embryo increasingly separates itself from the rest of the egg and then sequentially incorporates into itself the remains of the blastoderm, other egg structures, and some of the extraembryonic membranes. The same self-promotion also applied to the production of the ovum and offspring so that "generation is here a direct relocation [*Verlängerung*] of growth past the boundaries of the individual[,] and propagation is nothing more than an extension of growth beyond oneself."[7] Here, Baer employed a metaphor about reproduction and heredity that would reemerge in zoological texts for the next fifty years.

Scholium III, "Internal Development of the Individual," is where Baer points out that development is a process of differentiation from the homogeneous to the heterogeneous. This pattern in itself was not surprising, but Baer recognized its simple empirical onset by describing the initial formative process as consisting of the appearance of two plates: the animal and vegetative. He further observed that each plate then divided to produce four germ layers: the skin, the muscle, the vessel, and the mucous layers. Baer found that these structures emerge through the transformation rather than new formations. "Each organ is thus a modified part of a more general organ, and in this respect one can say that each organ is already contained in the fundamental organs."[8]

Scholium IV, a much longer and more complex section, presents the generalized "Scheme That Vertebrate Development Follows." As the germ layers become distinct, Baer noted that they grow into a double tube, each with its internal and external cylinder. These tubes are positioned above and below the longitudinal axis indicated by the notochord.

Scholium V, the longest of the scholia, addresses the taxonomic "Relationships of the Forms Which the Individual Assumes in the Different Stages of Development." This was clearly the climax of the Scholia, and this and Scholium VI were the only ones translated into English by T. H. Huxley.[9] Baer first set out to refute the widely held law of parallelism, which derived its name from the postulated parallel between the hierarchical scale of taxonomy and the teleological progression of development. He argued not against a limited correlation between the two, but denied the existence of a universal, uniserial hierarchy of all organic forms that could be traced to embryogenesis. Instead, Baer generalized the relationship between the type and grade of development in terms of four different laws of development. (1) The more general characters of a group of organisms appear earlier in development than the more specialized characters. (2) The less general forms develop from the more general forms. (3) Every embryo of a given animal form rather than passing through other forms becomes separated from them. (4) The embryo of a higher form never resembles the adult of any other form, only its embryo.

Scholium VI concluded the first volume of the *Entwickelungsgeschichte* by generalizing Baer's results: "The History of Development of the Individual." Baer asserted in italic type that this "is the history of its increasing individuality in all respects."[10] Only within the context of the preceding five Scholia and the body of empirical studies presented in the first 140 pages could this aphoristic generalization exert its full impact.

Finally, Baer closed his text with a reflective passage that reminded the reader that cosmic design—that is, a teleological perspective—was still very much a part of the study of life. As deftly translated by Huxley twenty-five years later, this key passage concluded:

there is one fundamental thought which runs through all forms and grades of animal development, and regulates all their peculiar relations. It is the same thought which collected the masses scattered through space into spheres, and united them into systems of suns; it is that which called forth into living forms the dust weathered from the surface of the metallic planet. But this thought is nothing less than Life itself, and the words and syllables in which it is expressed are the multitudinous forms of the Living.[11]

Many readers have noted how the concluding paragraph in Darwin's *The Origin of Species* is reminiscent of—though not metaphysically equivalent to—this passage from Baer.

Baer's fame did not rest exclusively on the first part of the *Entwickelungsgeschichte,* but there is no need to survey the second part beyond that which has already been discussed. Baer's interest in tracing the development of the fertilized egg and the separate organ systems of the chick became the model, providing the organizational themes for later texts on descriptive embryology. From a more general perspective, Baer's extraordinary efforts to juxtapose the development of birds, reptiles, mammals—including humans—and some invertebrates established the importance of comparative embryology for anatomy, taxonomy, and eventually for the study of phylogenies. Above all, Baer's understanding of early development in terms of the dynamics of germ layer development provided his contemporaries and later generations with both the empirical grounds for establishing homologies and a powerful tool for focusing on the mechanical processes of development. Rathke, followed by Johannes Müller and Robert Remak, Carl Bergmann, Rudolf Leuckart, Wilhelm His, and many others, carried the descriptive tradition well into the 1860s and beyond.

A recent biographer of Baer, Boris Raikov, has delved into the wealth of manuscripts left in St. Petersburg where Baer lived out his long life after moving from Könnigsberg in 1834.[12] From the archives, Raikov has determined that when Baer took up the position of prosector then professor of anatomy at the Prussian university in Könnigsberg, he also took up the task of rebuilding its zoological museum. Early in his academic career, Baer gave popular lectures based on his teaching human anatomy with an eye to comparisons with other vertebrates and his maintaining the museum with its growing collection of specimens. He lectured on chick development twice in 1821 and on reproduction in 1822. Shortly thereafter, he published the first volume of a proposed but never completed collection of more formal lectures on physical anthropology, which comprised a standard introductory course on the description of adult human anatomical systems and organs.[13] Baer never published the proposed second volume, but a manuscript of its contents shows that Baer intended to discuss embryology along with anatomy and physiology. Above all, the manuscript reveals how immersed at the time Baer was in the Naturphilosophie of Friedrich Schelling and Lorenz Oken, which urged an understanding of the transcendental laws of nature through the developmental processes. Later, Baer would repudiate the implied philosophical system, but he pursued the empirical details of animal development—in part because he could not accept the recent mechanical and preformationist conclusions about development of the Swiss and French chemists Jean Louis Prevost and Jean Baptist Dumas.[14]

Of immediate value to Baer's exploration of the chick was the fact that he drew upon Pander's earlier dissertation on the development of the same organism. Both Pander and Baer depended upon a discovery by Döllinger about how to open the chick egg to reveal the living embryo inside. In need of an exemplary of invertebrate development, Baer could peruse Rathke's recently published examination of the embryology of the crayfish.[15] Besides promoting the onset of descriptive embryology, Pander and Rathke, like Baer, pointedly described the earliest development of the embryo in terms of germ layers. Both argued that there were three basic germ layers, although Baer argued for four. Three or four layers notwithstanding, the implied mutual mechanical interaction of germ layers served the same function in the formation of primitive embryonic structures. Pander's account was brief but described the appearance of these layers: "Actually a unique metamorphosis begins in each of these layers and each hurries toward its goal; although each is not yet independent enough to indicate what it truly is; it still needs the help of its sister travelers, and therefore, although already designated for different ends, all three influence each other collectively until each has reached an appropriate level."[16] Not only did Pander identify interacting germ layers, but his experience and words strongly suggested that each layer was "designated for different ends," that there was a unique specificity in their actions and in their developmental destiny. This specificity challenged the research of embryologists for the rest of the century.

In her detailed account of the rise and fall of the germ layer doctrine, the embryologist and historian Jane Oppenheimer demonstrated how the germ layers became the principal element in a mechanical explanation of development during the third through sixth decades of the century. Thereafter, descriptive generalization was transformed into an inflexible causal biological principle as its apparently valid generality joined forces with the theoretical controversies over evolution theory incited by the publication of *The Origin of Species*. By the 1870s, experienced embryologists such as T. H. Huxley and Edwin Ray Lankester in England and above all Ernst Haeckel in Jena advocated the strong taxonomic connection between the specific destiny of germ layer formation and evolution.[17] Oppenheimer's conclusion, which only concerns us for the very end of the century, comes with the maturation of experimental embryology during the first third of the twentieth century, which demonstrated the complete falsity in the claim of a lawlike specificity of the germ layers.

The Rise of Cellular Biology

At the very end of the century and in an all too brief overview of zoology and its successes during the past hundred years, Weismann, then sixty-five and a renowned and experienced zoologist, commented on the consequences of the improvement of the microscope. "This goes for the discovery, which forms the basis of our examination of the fine structure of living forms for the discovery of the cell by Schleiden and Schwann." He recounted, "it goes, however, even more for the science, which bases itself on cells, for histology or the doctrine of tissues, that are bound together for the different functions of the higher animal body in particular ways and organized cellular masses. Elsewhere it was also first the microscope, which allowed us to recognize the lower and smallest animal world as single celled organisms."[18] Alas, this comment is brief, part of a frustratingly short summary. Not surprisingly, it reflects the three areas of Weismann's own interests but distorts the history of cell theory to the point of ignoring the contentious issues over cells between the forties and sixties, in which Weismann in a small way became also involved.

In a comprehensive historical survey of cell theory, Henry Harris recovered the complicated narrative about the recognition of and discussions of that "fine structure" of organisms, which preceded and followed the well-known monographs by Schleiden and Schwann at the end of the 1830s. Harris brought together the various strands in a way that is less unidirectional, providing a picture of a research frontier in the throes of debate and flux.[19] In retrospect, it may seem a tedious and excessively elaborate story, but Weismann's early and later works were unavoidably involved. Harris provides a historically so-phisticated picture of the many microscopists who were examining the fine structure of both botanical and zoological materials. From seventeenth- and eighteenth-century observations by Anton van Leeuwenhoek to Lazzaro Spallanzani on what we now call "microorganisms," and from the detailed identification of fission in the cells of the plant meristem by Barthélemy Charles Dumortier and Hugo von Mohl, and from the casual illustration of bodies within cells by Robert Hooke and the 1833 paper by Robert Brown, who with a compound microscope identified what he called the "nucleus," cells became increasingly identified as central to the study of histology. The Czech histologists at the German university in Breslau—Jan Evangelista Purkyné and his students, particularly Gabriel Gustav Valentin—worked in the 1830s with achromatic compound microscopes to describe cells in great detail; Purkyné in particular generalized the structures to include both animals and plants. Johannes Müller and his students and associates in Berlin were zeroing in on the same structure.

What becomes evident from Harris's valuable account is that when Matthias Schleiden and finally Theodor Schwann generalized these and their own observations in 1838–1839, their "cell theory" reflected a broad research frontier, not an isolated discovery. What made their theory so attractive was that it not only provided descriptions of structures but embraced a developmental conception of growth, differentiation, and pathological abnormalities in cells. Schwann's monograph in particular captured the contemporary efforts to generalize upon the fundamental building blocks of all life. It also explicitly accepted endogenous, exogenous, and binary fission as different modes of cell reproduction. Harris writes repeatedly of "Schwann's mistake" of embracing multiple modes of cell reproduction, and it took until the mid-1850s, following the work of Franz Unger, Robert Remak, and Rudolf Virchow, to establish that only binary fission of a cell into two daughter cells provided the only pattern of reproduction. As such, however, their conclusion turned out to be too simplistic. The revised cell became doctrine, and cytologists refocused their energies on the importance of the nucleus, the nucleolus, and other cellular structures and processes. By this time, Weismann had matriculated in Göttingen, had listened to Henle's excessively descriptive lectures in anatomy, "ohne alle leitenden Gedanken" ("without any guiding ideas"),[20] and had written his dissertation under the guidance of the physical chemist Friedrich Wöhler in the philosophical faculty.

The Alternation of Generation and Other Modes of Reproduction

During this thirty-year period between 1830 and 1860, another issue focused the attention of biologists. It was more a collection of phenomena that resulted in a new perspective than a set of overall guiding principles or a single theory about fine structure. This research revolved around the different modes of reproduction in individual organisms. The Canadian biologist and historian John Farley has written cogently and in detail about the subject in the first half of his comprehensive survey of the concepts of gametes and spores during nearly one and a half centuries of biological investigations.[21] The questions that particularly arose at this time were initially framed not only by advancing microscopic techniques and discoveries but in reaction to eighteenth-century animalculist and ovist debates over which sex and in which anatomical structures a preformed offspring appeared.

By the early nineteenth century, the issue had evolved into questioning the nature of the action of the sperm and pollen on the ovum and ovule of animals and plants. Did these donated entities convey a chemical or physical stimulus to begin the process of egg development? Was there instead a material penetration and maybe a physical union of male and female "gametes"? By the 1840s, an expanded version of these questions asked how sexual products were to be interpreted in terms of the cell theory and recent discoveries of eggs throughout the animal kingdom. And finally there arose the additional question of how the characters inherent in the donor gametes transmitted to the recipient's gametes explained heredity from each. Must the spermatozoa or pollen grain first mature? And how were they able to join and penetrate the ovum or ovule? The multiplicity entailed in the contribution of a single male's donation was also a puzzling matter. It is possible to formulate these questions from the perspective of just forty years later, but the ambiguities of the time, the uncertainties of different living materials in each investigation, and the generalizations that were needed to resolve these questions for both animals and plants make it harder to look back with understanding from our perch in the twenty-first century.

The basic biological information emerging by the mid-1850s was readily available. In the first three decades of the nineteenth century, it was common to believe that the spermatozoa were parasites of the sperm. In fact, Baer, who coined the designation "spermatozoa," did so in the belief that these were parasites in the semen. Farley points out that others, such as Gustav Valentin, Fredrich Gerber, Christian Ehrenberg, and Felix Pouchet, held similar views.[22]

In his substantial *Grundzüge der wissenschaftlichen Botanik* of 1842, Schleiden argued that pollen grains were simply cells that entered the embryo sack and developed therefrom. His theory needed to be revised at the end of the decade by Wilhelm Hofmeister, but even he maintained that fecundation was a matter of chemical excitement.[23] The mammalian ovum discovered by Baer affirmed the belief that the egg was universal to both invertebrates and vertebrates alike. A reaction to the positive conclusion to the universality of sexual reproduction was introduced by Adelbert von Chamisso with a casual description of an alternation of generation.[24] In 1842, the Dane Japetus Steestrup dramatically publicized the alternation of generations in many organisms, and within three years his ideas had become widely disseminated in both German and English.[25]

The process of alternating generations was further recognized in other organisms and became tangled up with an increasing recognition that parthenogenesis occurred in an array of organisms, notoriously including aphids and hymenoptera. The studies by Karl Th. E. von Siebold and Rudolf Leuckart on bees in the 1850s helped sort out the difference between alternating generations and parthenogenesis by using the microscope to differentiate the physical facts of fertilization and asexual reproduction. In a widely read overview of nearly 300 pages on "Reproduction," Leuckart argued there was no basic difference between asexually reproductive cells such as buds and spores, and sexually produced ova.[26] The thirty-year period that began with a triumphant succession of discoveries related to sexual reproduction ended with the general recognition of a spectrum of reproductive modes that ranged throughout the living world.[27] As Weismann began his first zoological explorations into reproduction and development, it was to the new science of histology that he turned.

The Development of Histology

There can be no understanding of the zoology practiced by Weismann and many of his contemporaries without recognizing the importance of the microscope and the general application of Schwann's cell theory. This may seem a trivial comment from our perspective of 150 years later, but it was not for the zoologists working in the 1840s and 1850s and particularly not for those wanting to understand the fine structure of organs and tissues, their origins and development, and the criteria for normal and pathological. Weismann was not unique in seeking answers to questions about structure and form from the perspective of his medicine or from the need to understand elementary organic parts. The microscope was the principal tool of use, and histology was the young science to exploit this instrument. Together they promoted a framework for organizing and describing tissues, organs, and the composite nature of any organism. Weismann had already been introduced to histology through his Göttingen teacher and musical friend Jacob Henle. When he returned to science again in the winter of 1859–1860, it must have seemed natural for Weismann to pick up what he had already explored. When he acquired a microscope to pursue histology is not clear, but he likely used one of the best German instruments. The reigning German textbook in histology at the time was Albert Kölliker's *Handbuch der Gewebelehre des Menschen für Ärzte und Studierenden* (1852), which came out as a digest of his much larger survey of contemporary histology.[28]

Besides a brief history of recent histology, Kölliker promised to follow the principles of the science of histology laid down by Bichat in 1801. Since that time, however, Kölliker recognized that histology had fallen into the danger of becoming "lost in minutiae." Only with the advent of Schwann's cell theory could histology again advance on account of a "law of cell genesis, or of a molecular theory." Now the histologist could examine the structure and development of these morphological elements to understand the functions and composition of cells, the cell nucleus, and the form of the organism—its structure and development. With the details collected, Kölliker hoped that histology would attain the "rank of a science" by studying the fully formed structures and their morphological elements, that is, their cells. In turn this would allow the histologist to look for more facts, to look at the actions and the elementary parts—their nuclei and greater elements—and to examine their development and to find their associated laws. Ultimately, Kölliker held, the facts, their generalizations, and the comparative studies would lead to a single law about the overall form of all organs and then all organisms. It was an ambitious challenge to future histologists who could now discern a path through the thicket of details by focusing on cells and their development.

Weismann's Research and Papers in Histology

We have followed the beginning of Weismann's career from his childhood in Frankfurt to his studies in Göttingen, to his first employment as a chemist in Rostock, and to his return to Frankfurt. During the summer of 1859, Weismann served with the government as a physician in the Franco-Italian war.

The conflict, however, soon resolved itself after the bloody battle of Solferino. Participation in the military gave him the opportunity to visit the Italian medical facilities in Verona and Bozan. Returning to Frankfurt thereafter, Weismann felt that he "did not know what to do there other than to practice medicine." "At the same time," he reflected,

> I continually searched for pure scientific work, and having come from the university, I was somewhat versed in histology, and also since there was available at that time a "comparative" histology text, I thought of entering zoology by means of histology. I turned to Henle, my teacher at Göttingen, for a work theme. Earlier he had suggested to me such a theme with the words that he had considered it for 14 days but could not recommend any for me.[29]

This time, however, Henle suggested that the tissue-like structure of the human placenta provided the opportunity to adjudicate between the "Haupt-Matadoren der Histologie," that is, between Henle and Virchow. Weismann attended to their differences, unfortunately for him deciding in favor of Virchow, but the comparison gave him the opportunity to examine the structure of muscles in general as an area of concentration.

He was successful in examining in microscopic detail the heart muscles of "worms" of many sorts as well as snails, fish, and insects, among other organisms. By producing six papers on the subject, he not only made a scientific name for himself with his work on the cellular structure of muscles, but also convinced others that he really wanted to become a zoologist.[30] The most important result of his work in histology was the demonstration that the muscle structure of coelenterates, echinoderms, mollusks, and a great variety of worms was composed of simple cells whereas the muscle structure of arthropods and vertebrates possessed muscle tissues composed of a variety of muscle bundles. Structurally, the muscles of these latter two groups were similar, but in their origin and in the histological nature of their muscle types they were very different. This was because all vertebrates possessed muscles of both types, both the simple cellular type and the primitive bundle type. The arthropods, on the other hand, lacked tissues with the cellular type structure. The muscles of the lower classes lacked primitive bundle type tissues. Insects revealed a mixture of both types of muscles, which Weismann felt must be related to their need to chew their way through their pupae while hatching from the larval to the adult stage.[31]

Although Weismann did not pursue further this style of histological study, his experience prepared him for future opportunities. The outcome revealed some important of features that would characterize Weismann's later microscopic investigations. Weismann realized the importance of studying the cells in their details to pursue an understanding of the tissues and organs of the organisms as a whole. In surveying many animal types, Weismann recognized the importance of comparative examinations within the spectrum of the animal kingdom.

Thus, the details gathered from the microscope would lead to generalizations about the gross level of organisms. Finally, he consciously recognized that important comparative generalizations could be extracted from the microscopic structure of animal tissues.

3

Becoming an Embryologist

Insect Development

The task of converting from a practicing physician with an MD degree and some research experience in human histology into an academic zoologist required Weismann's return to the university, where he would become familiar with the entire taxonomic and anatomical range of vertebrates and invertebrates and produce a substantial piece of independent zoological investigation. In Germany at the beginning of the 1860s, the most direct way of achieving this goal was becoming associated with a thriving institute that focused on the entire animal kingdom, but this type of position was not common at the time. It required becoming an expert in the focus of zoological research— taxonomy, anatomy, and embryology—and publishing independent research results in a substantial form. Achieving this, however, would be only part of the undertaking. In addition, one had to find an opening, likely in a medical faculty, where one would be expected to offer a range of courses for medical students. This might mean teaching comparative anatomy, which included both vertebrates and invertebrates. This was the trajectory Weismann followed. After he had worked in Göttingen with the prominent histologist Jacob Henle, it comes as no surprise that Weismann next turned to Rudolf Leuckart's institute in Giessen.

Rudolf Leuckart

In a persuasive chapter of his innovative work on *The Strategy of Life* (1982), Timothy Lenoir placed the methodological and philosophical dimensions of

Rudolf Leuckart's zoology within the contemporary struggle of the fifth and sixth decades of the nineteenth century to locate morphology within the context of a mechanical understanding of life. This was the time when the internationally famous chemist Justus von Liebig was transforming the chemistry of life's processes into the study of specific organic reactions that could be identified and experimented upon in the laboratory. Liebig's chemistry successfully championed a mechanical understanding of basic organic processes but stumbled in explaining more general features of life, such as the form, development, and taxonomy of organisms and their physical and chemical attributes. To do so, Lenoir explains, Liebig had recourse to a "Lebenskraft," or life force, which he did not consider superadded in the construction of life but rather as a potential capacity inherent in the complex of the physics and chemistry of matter that became life. Further, according to Lenoir, Hermann Lotze, a professor of philosophy at Göttingen, refined Liebig's point of view by insisting that the purposeful, organized character of life "is, therefore, nothing other than a particular direction and combination of pure mechanical processes corresponding to a natural purpose."[1]

In moving from an examination of Liebig's and Lotze's understanding of life to the sympathetic and parallel articulations of these ideas in the works of Carl Bergmann and Rudolph Leuckart, Lenoir points out that there were "close personal contacts between the four men." The latter three were associated with one another as students in Göttingen; Bergmann soon became a colleague of Lotze at that same university, and Leuckart became Liebig's colleague in Giessen.[2] Lenoir continues his narrative with an examination of Bergmann and Leuckart's "monumental synthetic collaborative work entitled *Übersicht des Tierreichs* of 1852," as well as of some of their individually authored pieces. In his detailed analysis of their major publications in the mid-1850s and early 1860s, Lenoir argues that both Bergmann and Leuckart found ways of directly attributing the origin of an organism's apparently purposeful anatomy to the material of the egg. Acted upon by the stimulation of fertilization and the external environment, the complex molecules of the egg turned into a new living organism. It was a view of transmission and development based on the catalytic processes promoted by Liebig's chemistry. This was not a view of heredity promoting a mixture of the material compositions of the zygotes of two parents. Lenoir concludes that the morphological processes were seen by the group as evolutionarily progressive but not Darwinian in nature.

Leuckart had arrived in Giessen from Göttingen in 1850 to become the colleague of Liebig, the anatomist and embryologist Theodor Bischoff, and

the botanist Alexander Braun. In retrospect, it was a formidable group of scientific colleagues, with Liebig and Bischoff being senior enough to ensure the
thriving of a community of like-minded scholars. Klaus Wunderlich has provided a brief picture of the Zoological Institute that Leuckart inherited in the
philosophical faculty at Giessen: "No instrument, not even a knife or twee-
zers, no collection, or as well as none, no money," as Leuckart explained to
his former student Carl Claus in 1863.[3] In a few years, the institute was relocated to newer quarters on the top floor of "the New Anatomy Building."
Leuckart assembled a suitable collection of animals for teaching purposes, and
employed a curator to assist him. He created a laboratory course that emphasized hands-on experience, and he acquired the necessary instruments for dissecting and examining specimens in detail. Leuckart's reputation grew rapidly with his encouraging teaching style, his laboratory practicum, and his
long list of informative as well as challenging publications. Leuckart's most
recent bibliography lists over a hundred articles published during his tenure
at Giessen between 1850 and 1868. These covered the gamut from descriptions
of specific taxonomic items to a defense of morphology in response to criticism
by the physiologist Carl Ludwig, from discussions of the processes of metamorphosis and alternation of generations to the discovery of the micropyle
through the chorion of insect eggs, and from a detailed examination of parthenogenesis in bee colonies to a lengthy article on reproduction in general. A
major textbook coauthored in 1852 with Bergmann provided a comparative
survey of the anatomy and physiology throughout the animal kingdom.[4]

It is not surprising that, after being a former chemistry student at Göttingen
and a frustrated chemist and physician in Rostock and Frankfurt, Weismann
migrated in the winter of 1861 to Leuckart's institute in Giessen. With its
geographical setting on the Lahn River flowing through Hessen, with its regional proximity to Weismann's native Frankfurt, with its reputation in the
natural sciences, and, above all, with Leuckart's Institute of Zoology, Giessen
was perhaps the best choice of universities he could attend to establish his credentials in zoology. Leuckart confirmed Weismann's immediate involvement
at his institute in a letter to his student Carl Claus that winter. "I have, however, many distractions," Leuckart must have complained, "but here at the
institute a beginning zoologist, Herr Dr. Weissmann [*sic*] at the moment in-
vestigates and studies living sea urchins with archduke Stephan from early to
late and scarcely needs to come to me."[5]

Weismann's deep absorption in zoology is off-handedly confirmed in Leuckart's letter. Also relevant is the presence of Archduke Stephan at the institute,

which may help explain Weismann's move to Schloss Schaumburg a year later to pursue an independent line of zoological research. In retrospect, Weismann made it clear that the study of zoology had always been his utmost desire.

Schloss Schaumburg

Archduke Stephan of Austria had been one of the casualties of the aborted revolution in Vienna of 1848. He had aspired to become king of his homeland of Hungary, but with the collapse of the anti-Habsburg movement, he became persona non grata in the Austrian empire. As a consequence, he had taken up residence at Schloss Schaumburg atop a high wooded hill overlooking the village of Balduinstein, which lay on the Lahn some sixty air kilometers west of and down river from Giessen (between Diez and Bad Ems). The archduke had recently inherited the castle and its surrounding forests from his mother. Between 1850 and 1855, he had added a neogothic extension to the castle, complete with four towers, the tallest of which (42 meters) overwhelmed the twelfth-century building. He had included in his expansion an art gallery, a library, a mineral collection, and an uncompleted "Rittersaal" ("knight's hall," a banquet room) situated between the twin west towers. Weismann described the addition as "im gotischen Burgerstil" ("in bourgeois gothic").[6] A railroad along the Lahn valley was completed in 1862, the year Weismann arrived, which soon made the castle readily accessible, principally to the nobility who continued to visit long after the archduke had passed the castle on to his heir in 1867. Elegant despite its mixture of architectural styles, the castle was, and still is, perched above the Lahn River, surrounded by extensive forests, and it served as a place of repose for the archduke.

Whether Leuckart had suggested that Weismann pursue research at Schloss Schaumburg or whether the possibility arose as Weismann and the archduke examined organisms together at the institute is not clear, but Weismann accepted the position of the archduke's personal physician the following year. The location offered an ideal setting for an in-depth study of insect development. A proposed contract, written in December 1861, offered Weismann the job for a year and a half starting January 1, 1862.[7] Above all, he was offered an annual wage of 1,200 usw (*und so weiter* = and so forth) for the first five years' employment, should he stay that long, and an additional 400 florins (fl.) for his annual expenses. One of the provisions stipulated that an additional 1,600 fl. would be added to his salary should he wish to stay beyond the five-year period. Serving as the personal physician to the archduke, Weismann would be available to attend to the personnel of the estate and military attendants living in the village of Balduinstein. He was assigned three rooms in an older tower for his lodging and personal use. Windows provided him a view of the surrounding forests and a short stretch of the Lahn. As Weismann explained fifty years later, "That was thus my world for the near future, except with one of the teachers[,] Hoffmann[,] and when necessary with the old honorable Archduke[,] I had no intellectual interaction, but this apparent loneliness was for me not a concern. Forests, mountains, here and there a small village surrounded by woods, appealed more to me than they would have repelled me."[8]

Weismann's own recollections, written a half century later, confirmed the setting and most of the contractual terms. In addition he left us some insight into his reasons for accepting such an unusual position on the verge of his entering academia. He used his earlier experience as a physician to his advantage, and as an aspiring academic zoologist, he saw in the year at Schloss Schaumburg an opportunity that enhanced his knowledge of natural history and a chance to pursue embryology without the distractions of living in a city or university town. He took with him from Frankfurt his microscope made in Wetzlar (Albert Kölliker does not mention the make in his *Manual* of 1852), a dissecting knife, scissors, preparation needles, a dish of wax, and a dissecting scope.[9] That was all he needed besides his personal effects to plunge into the embryology of the order Diptera.

Although the initial offer of employment was for a year and a half, Weismann remained in service with the archduke for only a year.[10] Upon his own request, he was released at the end of 1862. He had seldom been called upon to perform medical duties—twice he mentioned treating the archduke's soldiers for minor injuries or physical conditions. As he had anticipated, he had a productive year in insect natural history and embryology (as will become clear in the next section). It was also a year of personal growth and tragedy, in which he managed to accommodate to the royalty at the castle. He made a favorable impression on the archduke, his wife, and his two young sons who went to school on the premises. The archduke appeared to Weismann as a knowledgeable and affable man, but he was more interested in hunting than "intellectual conversation." His many requests "robbed" Weismann "of much working time." However, the archduke enjoyed hunting with a savvy Bohemian guide, who Weismann learned was also knowledgeable about the bird-life of the area.[11] From a contemporary manuscript with entries starting with the third week of May and lasting through the first week of September, we learn that Weismann was infatuated with "the queen" and believed that she also found him interesting. At the same time, he also recognized that being over fifty she was far too old to encourage him with personal favors.

Weismann's interest in younger women at this time of his life, however, was real, and he became attached to an Emma Amtmanns, the second daughter of a local family. Perhaps no more than fourteen years old, she was attractive and a proficient singer, and the two enjoyed making music together. In later life, Weismann saw no difficulty in justifying this platonic attachment: "I find it very natural, for one gladly interacts for an extended time with a clever and educated girl, particularly if one otherwise lives completely alone."[12] The archduke, too, was eager to promote a more lasting affair between the two, perhaps viewing it as a way to keep Weismann in his employ.

During the year at Schloss Schaumburg, Weismann also learned directly from his younger brother Julius that he would be traveling to Veracruz, Mexico, to pursue his budding career in textile merchandising. It was a natural move for Julius at the time; the American Civil War was making it difficult to continue his trade from Manchester, England, where he was employed. Weismann warned him about the dangers of yellow fever in Mexico, and he tried to persuade his father to prevent Julius from going. Again while visiting the family in Frankfurt, Weismann urged his brother to reconsider, but all to no avail. At the age of twenty-six, capable and ambitious, Julius shipped to Veracruz; within eight days of his arrival, he contracted yellow fever and died. The way in which Weismann recounted the events fifty years later indicated how deeply and painfully Julius's death was etched in his memory. There can be no surprise that he named his only son after his deceased brother.[13]

General Understanding of Eggs and Larval Development

At the end of the century, Weismann recognized how central technology had been in advancing study of animal development.

> Not many things compare with embryology, an area of research that first took root in this century. It appears almost unbelievable that for most of the series of forms, which complete their development in the egg, an equally great number of species may be followed down to the smallest detail, as [may be done] with the structure of the whole animal. Indeed this would only have been possible, when research developed completely new aids, above all, those of the staining and slicing technique, which

permitted, even the smallest animal bodies to be laid out in an entire series of the finest slices and be displayed before the microscopical eye. The indispensable condition for this method was the perfection of the microscope, for without this the far greatest portion of zoological work of the century is unthinkable.[14]

A cursory glance at the major German, English, or French nineteenth-century journals and texts in zoology confirms Weismann's assessment. Technology, as Weismann appreciated firsthand, unquestionably made the explosion of detailed embryological studies possible. After the publication of *The Origin of Species,* the microscopical side of biology only accelerated. Weismann was right. During the last four decades of the century, meticulous and exhaustive studies of the development of a myriad of different organisms played a big role in unraveling the phylogenies demanded by evolution theory. Embryology moreover had its own principles to ratify and expand upon. The cell theory of the late 1830s became an essential tool for understanding and organizing the tapestry of animal and plant tissues and ultimately for framing a comprehensive microscopic picture of their relationships. The refinement of the cell to the level of the nucleus and other subcellular and even subnuclear particles added further dimensions relevant to the developmental process. The doctrine of the specificity of germ layers, recognized initially by Baer and Oken, became a central theme for developing possible relationships between organisms even before the impact of Darwin. Finally, the untangling from one another the processes of embryogenesis, alternation of generations, metamorphosis, parthenogenesis, gamete maturation, pedogenesis, and heterogenesis became an essential goal of microscopists if they were even to communicate with one another about the particulars of development. Weismann's zoology mentor, Rudolf Leuckart, had become one of the most influential voices in articulating the complexities of generation of all types.

Because, as we have noted, the position of physician to the archduke was hardly taxing on his time, Weismann enjoyed the full opportunity to hone his skills as a microscopist by focusing his attention on dipteran development from egg to adult, and to interpret his findings to the world of zoology in a series of significant studies. Released from service at Schloss Schaumburg at the end of 1862, Weismann became a resident of Freiburg and by April published his Habilitationsschrift on an important aspect of his recent dipteran research. The expanded account of his research led to a major set of articles on dipteran development, which he promptly republished as his first book in zoology.[15] The subject was not accidental. The work he had done under Henle in Göttingen gave an important introduction to the use of the microscope, and Leuckart's encyclopedic knowledge of zoology in general and interest in insect embryology in particular gave Weismann the confidence and credentials he needed to pursue the time-consuming and exacting study of the eggs and larvae of the midge, *Chironomus nigro-viridis* Macq. (Weismann appeared to be somewhat uncertain of the species.) As holometabolic insects, these dipterans go through embryonic, larval, and pupal phases before becoming adults. Their eggs are merely 0.24–0.28 millimeters long and 0.096–0.099 millimeters broad; the adults are the well-known nonbiting midges that swarm above water during the warm days of spring and summer. Weismann also examined in detail the development of the common house fly, *Musca vomitoria,* whose eggs at 1.41–1.49 millimeters are somewhat larger.

The Understanding of Insect Development, 1860–1900

Before detailing Weismann's investigations, however, a few general words about comparative embryology and insect development, as it was understood at the end of the second half of the century, will be helpful. Cambridge zoologist Francis Maitland Balfour's two-volume textbook on comparative embryology was much admired both in the English-speaking world and in its German translation. This extensive work established the modern standard for comparative embryology.[16] Balfour succinctly pointed out why this comparative approach was important and why his *Treatise,* in particular, was timely and valuable. He made clear that the comparison of embryos and their parts was essential in any evolutionary interpretation of phylogenies. At the same time he was adamant that the study of development also made clear the "origin and homologies" of germ layers, their role in producing primary tissues, and the relationship of these germ layers to adult organs ("Organology" as he called it). Balfour emphasized both of these attributes in an address before an audience of anatomists and physiologists at the 1880 meeting of the British Association for the Advancement of Science. For him, the construction of phylogenies and an enhanced interpretation of germ layers justified the comparative approach.[17]

Starting with suggestions made by Darwin and Fritz Müller and emphasized by Ernst Haeckel, Edwin Ray Lankester, and many others, Balfour's first objective of the comparative approach, the construction of phylogenies, was widely discussed both at the time and later from an historical perspective. The second objective, an increasing understanding of the germ layers and their derivatives, is often taken for granted and rarely examined in detail. If one looks at the immediate successor of Balfour's *Treatise,* the multivolume German work on invertebrate embryology by Eugen Korschelt and Karl Heider, one finds these two objectives repeated but with a different emphasis. The chief problems for comparative embryologists, the German authors insisted, "consist in the investigation of the formation of the germ-layers, the origin of organs, and the development of the general form of the body. Its purpose is the recognition of the laws of development, the determination of the homologies of organs, and the deduction of the ancestral history of the Metazoa."[18] Descriptions of the formation, origin, and homologies of germ layers, organs, and the laws of development comprised the nuts and bolts of comparative embryology. Their determinations embraced the bulk of the comparative studies of development through the second half of the century. Only then was descriptive embryology overtaken and transformed, but not eliminated, by experimental embryology and its set of mechanical quests.[19]

To continue our understanding of insect development from the perspective of the 1880s to 1900, we should add that Balfour included the insects among the Tracheata, a major invertebrate taxon of arthropods at the time. This taxon also included the myriapods (such as centipedes and millepedes) and arachnids (such as spiders, scorpions, and ticks), and it is clear from Balfour's account that even in 1880 the formation of germ layers in both the insects and myriapods was of central concern but had only been "imperfectly worked out." Ten years later, Korschelt and Heider would write their corresponding chapter based on their own recent work. Over three quarters of their cited studies had been published since 1880—that is, since the appearance of the first volume of Balfour's decade-old textbook. This concentration of works at the turn of the century was to be expected in a rapidly advancing field of science. Nevertheless, despite the heavy emphasis on recent literature in this rapidly expanding field of zoology, out of the ninety-eight citations for the section on insect embryogenesis, four of the citations were of Weismann's by then classic studies of the 1860s—two of which came from his Schaumburg period. Weismann's works were not only remembered a quarter of a century later by Korschelt and Heider but were examined with respect, if not with full acceptance, as the starting point for their own presentation of early insect development.

So what was Korschelt and Heider's basic message about insect development in the 1890s? With a few exceptions, they pointed out that the insect egg is rich with yolk. It is elongated, and at the time of the union of the male and female gametes during fertilization, it is surrounded by peripheral proto-plasm, a vitelline membrane surrounding the yolk, and an outer shell or cho-rion, through which the sperm penetrates to the interior through a micro-pyle. (Incidentally this physical penetration had been identified by Leuckart in 1854.) When united, the nuclei of the sperm and egg migrate to the cen-tral region of the egg; the position of the future embryo, fore and aft, dorsal and ventral, may be easily determined externally by the curvature of the egg itself. Because some investigators still believed that numerous cleavage nuclei arose through a process of "free nuclear formation" in the yolk, the authors and their English editor found it important to stress that "every nucleus orig-inates from a pre-existing nucleus either by mitotic or amitotic division."[20]

As we will see, Weismann at first held that "free nuclear formation" took place within early cleavage, but he certainly did not believe this beyond 1883. The blastoderm, the germinal disc representing the earliest embryonic struc-ture, forms after many of the cleavage nuclei migrate to the surface of the vitelline membrane. (The editor uses the term "cleavage-cells," but warns that these are not true cells but nuclei surrounded by yolk islands!) The cleavage cells that remain in the yolk participate in the organization of the yolk bodies into yolk cells. Korschelt and Heider also pointed out that certain cleavage nuclei become isolated "pole cells" at the posterior end of the egg. In Diptera, the cleavage nuclei do so prior to the time when the blastoderm has fully formed, and these cells become the rudiments of the genital glands.[21]

Continuing their examination, these embryologists described how a germ band forms longitudinally on the ventral side of the egg through a process of cellular growth, invagination, and a differentiation of cells into an outer layer, soon known as the ectoderm, and into two inner layers, which eventually form the mesoderm and endoderm. As the sides of this blastoderm fold in upon the center of the band, they become known as amniotic folds. Their closure, formed by a meeting of their edges, creates a hollow space known as the am-niotic cavity. The two coauthors reported, "The germ-band, after its develop-ment, thus appears covered by a double cellular envelope derived from the amniotic fold."[22]

Weismann Views Dipteran Eggs and Larval Development

The rough description of early insect development to the point of the appear-ance of the embryonic band suffices to provide an outline for what a microscopist might also have seen—less clearly—while looking at a developing dipteran egg in circa 1860. There are, however, many assumptions about early cell formations, the appearance of nuclei, the ordering of cells into germ layers, and the pole cells that were poorly understood thirty years earlier. Above all, the difference between early insect development and the more familiar germ layer formation of vertebrates was investigated by Baer, Oken, Rathke, and Geoffroy St. Hilaire. While Weismann performed his initial studies at Schloss Schaumburg, the relationship between nuclei in the yolk and the blastoderm, the significance of the pole cells, and the formation the early germ band attracted his attention. He organized his first paper around these poorly understood but observable phenomena. In an excellent historical account of Weismann's initial studies, Klaus Sander examined the issues in roughly the same order in which Weismann pursued them in the early 1860s.[23] The prime theoretical concerns for Weismann embraced the earliest period of embryogenesis, the nature of cell formation, and the significance of the novel organization of germ layers in arthropods as exemplified in Diptera. To examine these issues, Weismann chose the eggs of the gnat *Chironomus nigro-virdis,* which may be easily found on the still surface of small ponds or in strands attached to floating vegetation. A single egg's dimensions are a meager 0.24–0.28 millimeters in length and 0.096–0.099 millimeters in breadth. For Weismann, this small size was compensated for by the fact that the chorion of the egg is transparent.[24]

Weismann devoted the second part of his monograph to examining the late larval and pupal stages with an eye to the transition of structures from the hatched larva to the adult. Despite its more opaque chorion, he chose the common house fly, *Musca vomitoria,* to serve as his exemplar of larval growth and pupal transformation. *Musca's* larvae were abundant and easy to find on carrion. They increase in size from 0.3 to 2.0 centimeters in length during a gestation period of between twelve and fourteen days, and in the summer the pupation period lasts an additional ten to fourteen days.[25] Weismann's focus on the structural minutiae of embryonic and larval development enhanced his understanding. When we recognize that his mentor, Rudolf Leuckart, had written about the early development of insects and had just completed an important study on their alternation of generation and parthenogenesis, the choice of Weismann's research topic is fully understandable.[26] In commenting on the latter of these works, Weismann regretted that Leuckart had not dealt sufficiently with the earliest stages.

Leuckart, however, had not worked in a vacuum. The starting point for his and Weismann's studies was the inaugural dissertation of 1842 by Albert

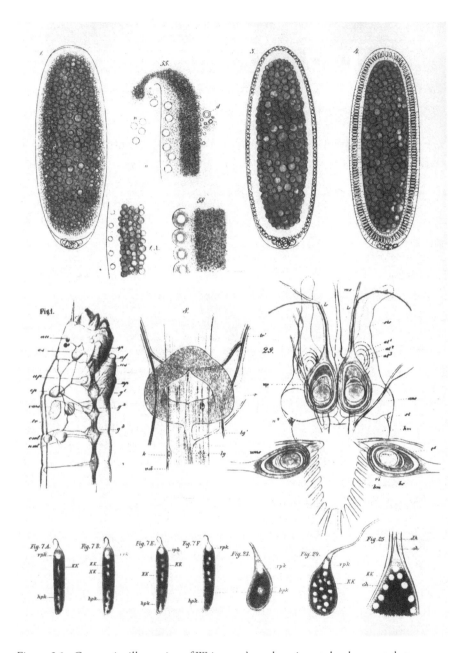

Figure 3.1. Composite illustration of Weismann's work on insect development that was assembled by Klaus Sander to accompany his insightful paper illustrating Weismann's contribution to insect development. Note particularly the pole cells in the upper figures, "Weismann's ring" in fig. 8, and the imaginal disks in fig. 29.

Klaus Sander, ed. *August Weismann (1834–1914) und die theoretische Biologie des 19. Jahrhunderts. Urkunden, Berichte und Analysen*. Freiburger Universitätsblätter, vols. 87, 88. Freiburg: Rombach Verlag, 1985. Klaus Sander, "August Weismann's Untersuchungen zur Insektenentwicklung 1862–1882," p. 45.

Kölliker devoted to the early generation of insects.[27] Shortly thereafter, Kölliker had became Ordinarius for physiology and comparative anatomy at the University of Würzburg, and as we will see, Weismann was to interact with him on a number of occasions throughout their overlapping careers. Also important for the context of Weismann's research were the works of Thomas Henry Huxley on the reproduction of aphids (1858), Edouard Claparède's earliest examination of arthropod (araignée) development, which appeared while Weismann was finishing his own studies of *Chironomus* and *Musca,* and, above all, G. Zaddach's early 1850s publication on the development of the egg of phryganeides or caddis flies. Insect development was thus not simply a project drawn from an empty hat, but it was still fragmented and imperfectly understood.

As for the contents of his studies, we can clearly see that Weismann struggled to understand the formation of the blastoderm or single-celled epithelium of the egg. Where did the many cells come from? Were they derived from the nuclear "flecks" in the yolk? How were they connected with the germinal vesicle, which is evident before fertilization but thereafter disappears into the yolk? A number of caveats are necessary when asking such questions, for they ring in our ears with modern overtones. In 1860, fertilization was not yet demonstrated as the outcome of a union between a male and female pronucleus. That was not to come until fifteen to twenty years later. A modern understanding of the embryonic relationship between walls, contents, and nuclei was also a product of a later period. In the early 1860s, Weismann had to wrestle with such questions while focusing on the totality of embryogenesis and pupation. He agreed with Kölliker that the majority of embryologists favored Remak's and Virchow's doctrine, expressed in the phrase of the mid-1850s "Omni cellule e cellule" ("All cells come from cells"), which in this instance dictated that after the germinal vesicle vanished into the yolk, it must fractionate and produce numerous nuclei or fragments. These, in turn, necessitated that in a given time these structures were propelled to the surface of the egg to form the walls of the blastoderm. "Pertinent here is the question of the continuity of all organic elements of form (cell or nuclei)," Weismann said, "which stands with the continuity of all living beings in the most exact connection." The reference here to organic evolution was significant in the context of his having recently read the Bronn edition of Darwin's *Origin* at Schloss Schaumburg.[28]

Remak's and Virchow's phrase about all cells coming from cells notwithstanding, and despite his own immediate enthusiasm over Darwin's theory of evolution, Weismann felt that the two biological principles of cellular continuity and biological evolution failed to explain the origin of the blastoderm cells. In opposition to the findings of many of his peers, Weismann felt that a continuation between the germinal vesicle on the one hand and the blastoderm cells on the other had not been demonstrated. Although he would have been looking for them, the eggs of *Chironomus* did not reveal to Weismann the necessary nuclei or their fragments in the yolk to explain the rapid multiplication of nuclei at the egg's surface. For him, the appearance of unexplained nuclear fragments on the surface of the yolk suggested that with Diptera and perhaps all arthropods, there was a different sort of embryonic cleavage than that which occurred with nonarthropod eggs.

On the other hand, a majority of commentators had argued intermediate positions. According to Kölliker, they felt that the Keimbläschen (i.e., the germinal vesicle) vanished upon fertilization, so the first cleavage nuclei must be new formations in the yolk. Müller, Leydig, and Gegenbaur each took a different tack claiming that the Keimbläschen does not disappear and arguing in turn that the cleavage nuclei could be derived directly from it.[29] Kölliker, Rathke, Zaddach, Huxley, and Leuckart were unable to confirm a continuity between the two, and Claparède simply concluded that there must be continuity based solely on "Omni cellule e cellule." To complicate these matters, Zaddach and Leuckart felt the "Kernflecken" were really cells in their own right, but they avoided the question of their continuity with the Keimbläschen.

Because of his resolute view concerning the lack of visual proof of a continuity and after his examination of the appearance of nuclei in the yolk and the cells as the peripheral layer of protoplasm—and, moreover, because he did not find a proliferation of cell formations in the yolk—Weismann was inclined to argue that "we will simply say, that the cells of the blastoderm in insects (probably in the arthropods in general) arise through free cell formation." He added, in justifying this radical reversal of Remak and Virchow, that with the sudden appearance of peripheral cells, "we are reminded of the theory of cell formation, which the discoverers of the cell Schleiden and Schwann announced more than twenty years ago to be generally valid, [but] . . . now again needs a few modifications in order to comply with the facts."[30] Weismann's reminder provided not only a solution to the lack of continuity puzzle but reminds us in the twenty-first century that the continuity of cell formation in the 1860s, despite the claims of Remak and Virchow, was still a contentious issue. Weismann's particular solution, however, did not last long. Sander has pointed out that it was only two years later that another Leuckart disciple, Elie Metchnikoff with further observations on Diptera, corrected Weismann's error about the free formation of the peripheral cells.[31]

Pole Cells

It was in the context of observing the peripheral cells that Weismann identified a cluster of cells at the posterior pole of the egg. He distinguished these from the polar bodies formed during egg maturation, but he designated them for their posterior location as "Polzellen." Their function "puzzled" him, so he simply commented "that they are truly cells, which later join the cells of the blastoderm (in *Chironomus*) to be sure, [but] their meaning here also remains completely unclear."[32] His identification was not elaborate, and his observations about their joining with the blastoderm cells remained ambiguous. Again, it was Metchnikoff who corrected Weismann and traced the polar cells to the larval rudiment of the gonads, where they multiplied and led to the production of the larval germ cells.[33] Weismann closed his examination of the embryonic development of Diptera with short descriptions of the developing germ band and the germinal envelopes later designated as the amniotic folds. He adamantly disagreed with Zaddach's identification of them as germinal folds or later as organ formations that played the same role as the germ layers in vertebrate development.[34] Finally, Weismann ended the first part of his study with what for its day was a striking example of insect embryogenesis, an elaborate and prescient description of the rotation of the embryo of *Chironomus* from the dorsal to the ventral side of the egg. He went no further in the embryogenesis.

The second part of Weismann's study focused on the development of the larval and pupal stages of Diptera, or the process that transforms the hatched embryo into the very different form of the imago or adult. The general pattern of this complex "metamorphosis" was well known by the nineteenth century and remains a phenomenon of speculation and excitement to grade school science classes today. What actually happens on a cellular and functional level began to be investigated with the technical tools Weismann had mentioned only at the beginning of the second half of the nineteenth century. In the 1870s, both V. Graber and John Lubbock followed Weismann's example.[35]

We have already seen that Weismann examined only the earliest embryological changes that eventuated in the next metamorphic stage, that of the self-feeding and mildly active larva. The larva's increase in size and its substantial but not radically contrasting changes in form might easily be seen as simply an extension of the embryological processes, except for the fact that it was ex-ovo, self-feeding, and developmentally constructed in a manner to produce the inactive pupal stage. The real question left for the embryologist was how did the larval form produce the adult of the insect with its wings, stabilizers, antennae, elaborately articulated legs, and other adult structures—both external and internal? Only after they had established the "genetic connection" between the parts of larval and adult stages, Weismann insisted, could they really understand the morphological and functioning parts and thence understand the whole life cycle. At this moment in the history of biology, Weismann explained, "an embryological history of the insects in this sense was impossible before the founding of a scientific histology." Now, however, Weismann could justify his research: "It appears well timed that the investigation should be attempted, and I believe that it has produced much that encourages further research advances along this beaten path."[36] Technological advances in microscopy were beginning to open up new questions in biology as well as in medicine. We will discuss this aspect of the science in detail in a later chapter.

Weismann's achievement in solving the problem of understanding the connection between larval and adult parts was first presented to academia in his Habilitationsschrift, which he submitted to the medical faculty in Freiburg upon his release from his duties at Schloss Schaumburg. It was formally published with the major scientific institution of his former native city-state of Frankfurt, the Senckenbergische Naturforschende Gesellschaft (Senckenberg Scientific Society). Another version of his solution appeared as the second part of his monographic article on the development of Diptera.[37] Weismann, however, was not the first to be engaged in finding an answer to this radical metamorphic change in many insects. As he pointed out, Jan Swammerdam in the seventeenth century had described that the "adult parts," such as the antennae, proboscis, wings, and legs, could be recognized under the skin of the fully grown larvae.[38] Weismann also quoted a passage from Hermann Burmeister's five-volume *Handbuch der Entomologie* (1832–1855), in which the author identified traces of wings under the skin and alterations in the alimentary canal, the forelegs, and proboscis in the last stages of butterfly caterpillars. Burmeister, born in the Baltic seaport of Stralsund, was at the time working on the publication of his *Handbuch* and Ordinarius at Halle.[39]

Louis Agassiz on Insect Development

A more complete account than those earlier observations was written by the Swiss émigré to America, Louis Agassiz, whose study of the classification of insects was presented in 1849 to the American Association for the Advancement of Science. Weismann briefly examined this "American" monograph and found that Agassiz was promoting embryology as the basis for insect classification.[40] Studying both the caterpillars and pupae of butterflies and the maggots and "cased pupae" of flies, Agassiz had focused on the adult appendages such as legs, wings, eyes, and antennae, which he had found in miniature within the advanced larva after he had removed the epidermis surrounding the hardened casing, specifically within the pupae of Lepidoptera and Diptera. In both orders, Agassiz found a gradual appearance or a "transformation" into adult parts lying just under the skin before and after the final larval molt. There were, of course, differences in the metamorphosis of Lepidoptera and Diptera. The caterpillar of the former molts its skin a half-dozen times or more; the maggot of the latter molts only when it changes into a pupa. Referring to the Lepidoptera, Agassiz had claimed that "it is plain that the Lepidopteron arises from the larva with most of its perfect features, only developed in a less finished manner, while the changes which the animal undergoes during the pupa state only consist, as it were, in the last perfecting of the final development already introduced during the last period of the larval life." He added with reference to the more transparent Diptera pupa that "parts of the mouth, had lost their larval appearance, and assumed the character which they exhibit in the perfect insect."[41] It is evident from Agassiz's account and penciled sketches that he must have used a magnifying glass or dissecting microscope, but he certainly did not avail himself of the more powerful microscope used in the new field of histology. His real goal, as mentioned, was to clarify the taxonomic relationship between the insects (with their superficially wormlike larval stage) and the annelids.

Contrary to Weismann's comment, Agassiz had consulted the first volume of Burmeister's *Handbuch der Entomologie* but had found Burmeister not inquisitive enough about the development of adult structures in the advanced larval and early pupal stages. More important for both Burmeister and Agassiz, however, was the fact that they described the appearance of such structures as though they arose from a mere transformation of the larval parts into the adult parts late in the larval stages and early during pupation.[42] Weismann's examination of Diptera with their nearly transparent larval skin must have begun with this "transformation" process in mind.

Discovery of Imaginal Disks: "Umbildung" and "Neubildung"

Upon careful microscopic investigation of the transformation process, Weismann found instead something quite different.[43] He noticed on the front three segments of the maturing larva disklike bodies ("Scheiben") distributed evenly on both sides in two rows, one ventral the other dorsal—in all adding up to twelve. He began referring to them because of their future involvement in imago development as "imaginal disks." (These *Imaginalscheiben* comprised the *Untere* and *Obere Prothoracalscheiben,* the *Untere* and *Obere Mesothracalscheiben,* and the *Untere and Obere Metathrocalscheiben.*) The disks lay under the external chitinous layer, that is, under the "Hypodermis," as Weismann named the most external layer of the epidermis. Lying in the subcutaneous layer, "they appeared as light, round or oval small dishes and consisted of a uniform mass of very small, grain-like cells." He explained that as the larva grew larger, these disks grew proportionately even greater; they first consisted of an inner and outer layer of cells, then their internal contents developed cleavages. Some of the cell masses became wrinkled, and others enclosed a spiraled cylinder of cells. Watching as they continued to develop, Weismann described the imaginal disks of each segment eventually joining to one another, and in a brief initial account he added that the ventral rows of disks of the second and third thoracic segments possessed the "Anlagen" of the three pairs of legs, and the dorsal rows contained the "Anlagen" for the balancers, the wings, and appeared attached to respiratory spiracles. The foremost segment, forming the functional head of the larva, comprised the seat of two opposing pairs of imaginal disks. The foremost, the "Stirnscheiben" or frontal disks, possessed potential antennae, while the latter, the "Augenscheiben," consisted of the precursors for the eyes.[44]

Weismann's discovery was made on the water-dwelling larvae of a species of the notorious black fly (*Kriebellmücken,* the family of Simuliidae). The more detailed examination of the disks came from *Musca vomitoria,* the same species of house fly studied by Agassiz. This common domestic pest allowed Weismann to identify the relationship of the disks to the other developing organs of the larva and then to name and examine in greater detail five of six pairs of disks, two on each of the first three thoracic segments. He described and sketched them as semicircular, and throughout the process of larval development all the disks continued to grow in diameter. For example, those disks of the dorsal second segment (i.e., "Mesothoralscheiben") grew from 0.071 to 0.13 millimeters while their component cells increased in diameter from 0.008 to 0.010 millimeters. He further pointed out that all the imaginal disks in the larva existed only on the first four larval segments, with the greater number appearing in the first three that he had so carefully examined.

With his discovery of imaginal disks, which implied the precursor, not the miniatures, of specific adult structures, Weismann reasoned that he was dealing with a process akin to an alternation of generations. In other words, he claimed that identifiable adult structures of the head region and the thoracic appendages arose not through the "transformation" of molting but through a "metagenesis," as he began calling it. This counterintuitive process, however, applied only to those specialized adult appendages. Most of the internal organs (e.g., the neural, alimentary, and respiratory systems) merely underwent a transformation or metamorphosis from corresponding larval parts to the early pupal counterparts. In contrast, the formation of antennae, eyes, mouthparts, complex articulated legs, wings, and balancers was altogether different. These products clearly arose from the imaginal disks, which must contain their molecular or cellular precursors. Some of these disks were even identifiable as defined clusters of cells in the advanced embryo. These remained unchanged until the pupal stage, at which time they grew and developed into the specified morphological structures of the adult.

Weismann likened the process to the metagenesis in the life cycle of starfish, where the adult is completely different from its free-swimming and self-feeding larva and only arises as an outgrowth from the latter. To be sure, when most of the internal larval systems of dipterans degenerate during pupation, their replaced systems arise through a direct metamorphosis of the same material. The dipteran larvae also contain provisional organs, each of which represented, as Weismann put it, a "neuesindividum." It resulted in a metagenesis akin to the alternation of generations in echinoderms. As Weismann suggested, the "Scheiben" or imaginal disks contained nuclei in an amorphic "Grundsubstanz," and when they are triggered during pupation, the nuclei, which merely represent the structure of the adult, take on the form of that later structure. In other words, such adult structures are anticipated or predetermined—not preformed. His words make this transition clear:

> A surprising similarity, however, to the development of the body of the echinoderm strikes us in the imaginal disk. As the body of the echinoderm initially places indifferent groups of cells at many points along the circumference of the larval gut and [which] then gradually grow together

into a mass, so arise in different places inside the larval body of the fly indifferent groups of cells,—here also in genetic connection with the larval organs—which then differentiate during the course of development into parts of the adult body and grow together into a general whole.[45]

Weismann could only partially apply his vision of the role of the imaginal disks in the pupation of the muscid flies to other insects. Crane flies went through a functionally active larva to pupa transition and so did not possess as many inactive imaginal disks. The articulated legs of butterflies appeared to be transformed directly from certain larval legs rather than arising from a separate disk. In both cases, metamorphosis rather than metagenesis took place.[46] There was also flexibility in Weismann's system, which rendered the imaginal disks dependent upon the functional demands of individual life cycles. This made the imaginal disks less valuable for classification even within the class of insects itself.

"Umbildung" versus "Neubildung"

The imaginal disks allowed Weismann to consider two contrasting modes of development: transformation through metamorphosis and the rise of new formation through metagenesis—an "Umbildung" and a contrasting "Neubildung." In the context of dipteran development Umbildung took place through metamorphic processes even where "the products of degeneration do not disperse, but remain together and preserve the form of the organ in its whole even when no single histological element is any longer present." In contrast, the imaginal disks, "concern the complete 'Neubildungen,' that is all those parts which in the larva are not yet present, or at least not in a fully constructed and functioning form."[47] One can easily imagine how there might be a blurring of lines in particular circumstances, but this contrast for Weismann was absolute and not an expression of a spectrum of developmental possibilities. Within this contrast lay a more extended organic principle that was to emerge repeatedly throughout Weismann's career and that colored his discussions not only of development but of heredity and evolution.

One final and important aspect of Weismann's research was how it reflected on the germ layer doctrine championed a generation earlier by Baer and Oken and which soon would be championed by Ernst Haeckel and reexamined by the Hertwig brothers. The germ layer doctrine was first fashioned from studies of vertebrates, extensively but hardly exclusively, from the chick embryo. As embryologists widened their scope to the detailed study of other types, particularly arthropods, a comparison of early development became inevitable. According to Weismann's retrospective account, Baer, Pander, Reichert, and Remak had all in one way or another homologized the articulated limbs of vertebrates with those of arthropods. To imply by this a functional analogy might be acceptable, Weismann felt, but to assume a morphological homology contravened the studies he had just completed. The claim of homology between parts of different organisms, Weismann insisted, implied that their earliest formations must also be homologous. The issue was important for Weismann because the most recent studies of dipterans, a study Weismann often referenced in his own account, were done on crane flies and applied to the whole insect class by Zaddach in 1854. The latter had claimed that the germinal band in crane flies developed from two germinal layers that were the homologues of the middle and upper germ layers in vertebrate development. To accept this, however, Weismann insisted that, "in my mind the following embryonic parts may only be considered homologues with those that have arisen in the same way and out of which arise morphologically equivalent parts." Weismann's examination of *Chironomus* with its amniotic folds had revealed a very different mode of early development. When he got to Freiburg a year later, he went on to investigate the development of crane flies, of which he found plenty in the warm September and October on the river Dreisam flowing through the city center. Their earliest development being similar to that of *Chironomus* confirmed his concerns about homologizing them with vertebrates as urged by Zaddach.[48]

Both 1861 and 1862 were fruitful years for Weismann. He had refashioned his credentials from physician to zoologist, he had delved deeply into the cellular events of early insect embryology, he had recognized and named the imaginal disks as an important key for understanding insect metamorphosis, and he had assembled the material for three important papers. During the same period, Weismann had proved to himself the value of working in an uninterrupted fashion surrounded by an ideal setting, and by the age of twenty-eight he had resolved a number of professional and social questions that inevitably confront a talented and ambitious young man in his late twenties.

4

Studies in Descent, Part I

1863–1876

After leaving Schloss Schaumburg, Weismann had a number of reasons for going to Freiburg and becoming attached to the Anatomy and Physiology Department of the medical faculty. At the time, Freiburg, situated at the southern end of the Schwarzwald, served as access to the western end of the Bodensee and the Swiss city of Constance, the French border, and the lower Rhine as it flowed northward along the Baden frontier and into the German Rhineland. Above all, Weismann was attracted to the town's proximity to the mountains and the chance to work unhindered in a modest university setting. He indicated in his *Vita Propria* written over forty-five years later that he had accomplished most of the embryological research he had planned under the employment of Erzherzog Stephan. The duke, however, turned out to be less interested in his microscopical work than he had let on in Giessen. ("The duke also made the impression of a well-educated man. I never saw him take particular joy in intellectual conversation, but I never noticed him taking part in an intellectual conversation."[1]) There is some indication that Leuckart may have had some influence on Weismann's choice of Freiburg as a place to begin an academic career. More pertinent, however, was Weismann's travels to Genua. His rambling account written over forty-five years later indicates that this trip was part of an effort to secure his mother's estate, as she unfortunately had failed to make clear that her son was to be the beneficiary. In the process of working through the legal morass, Weismann made the acquaintance of August Gruber, a successful German merchant with a large family, who had abodes both on the Bodensee near Lindau and in the hills above Genua. Weis-

Figure 4.1. August Weismann and Mary Gruber in 1866. They were married in June of the following year.

Klaus Sander, ed. *August Weismann (1834–1914) und die theoretische Biologie des 19. Jahrhunderts. Urkunden, Berichte und Analysen.* Freiburger Universitätsblätter, vols. 87, 88. Freiburg: Rombach Verlag, 1985. Helmut Risler, "August Weismanns Leben und Wirken nach Dokumenten aus seinem Nachlass," p. 29.

mann entertained the Gruber family with his piano playing; in the process, he became attracted to and eventually would marry Gruber's eldest daughter, Mary.

The 1850s had introduced an upswing in the growth of industry in Freiburg, with an emphasis on textiles, metal works, and railroad construction. The city, which overall was Roman Catholic in orientation, was going through cultural turmoil in the form of a struggle between ultramontane and local aspirations along with a strong influx of immigrant workers.[2] As Weismann indicated, the university was small, "scarcely 300 students." According to his later recollections, he felt, "Not that I particularly wanted to attain much, but peaceful work time appeared to me the most important and I should find it there."[3] It helped, to be sure, that the current professor teaching zoology, Professor Otto Funke, was more a physiologist and thus wanted Weismann to teach zoology and manage the zoological collection.[4] Weismann habilitated at the university in April 1863 while he lived in "Hozmarktplatz," then Müsterplatz. It was here in the center of the city that he received the news that his younger brother had died in Veracruz only a few days after arriving in Mexico to advance his business career, and Weismann would later emphasize how deeply he felt this loss.

While first in Freiburg, Weismann received a yearly stipend of 2,000 fl. For his early efforts, he gained the opportunity to submit much of this dipteran work to the university as a Habilitationsschrift, thereby initiating his academic life in zoology.[5] He supplemented this innovative work with further professional publications on dipteran development placed in the prestigious *Zeitschrift für wissenschaftlishe Zoologie* (1863–1866). A university career seemed assured until in 1864 he experienced the first devastating shock: an illness that impaired his eyesight, which would strongly influence his long career. For the next ten years, this budding insect anatomist had to relinquish his career as a microscopist. He would return to the microscope between 1874 and 1884 but only as a supplement to his broader research. Throughout the sixty years of his university research, his students and assistants at the institute generally served as "his eyes" for any sustained microscopic work.[6]

Friendship with Haeckel

One of the delightful coincidences that the history of science keeps tossing into our laps is that the two most influential evolutionary biologists of the immediate post-Darwinian period were born within thirty days of each other. As earlier pointed out, Weismann was born in Frankfurt on January 17, 1832; his professional colleague and long-term friend, Ernst Haeckel, was born on February 16, 1832. Such a coincidence tempts us to overexplain their parallel careers. After all, the parents of both were members of the nascent nation's Bildungsburgertum and were evangelical in commitment. Both men had been encouraged as youths to botanize and collect insects. Both studied medicine (although Haeckel never practiced it) before turning to zoology. Both read Darwin's *The Origin of Species* within a year of the appearance of the first German translation, and it became a formative point in their careers between their commitment to academic zoology and their habilitation. Both were convinced by Darwin's work that the common descent of organisms provided the only framework for explaining the phenomena of the living world. Both found in Darwin the opportunity to promote mechanistic explanations in biology and thereafter vigorously opposed the introduction of teleological causes into their science. They belonged as well to the cohort of German academic biologists who had freed the discipline of zoology from a subordinate relationship with anatomy and physiology in the medical faculties; they helped to reshape it as an independent science, complete with Lehrstühle, institutes, and museums within the philosophical faculties. As young men, both had expressed concern over the rise of Prussian influence in German affairs, but both were ardent nationalists, who celebrated the unification of Germany and paid homage to the retired Bismarck on his eightieth birthday. Both were productive investigators and prolific writers. Both established programs for training doctoral and academic zoologists at peripheral universities—Weismann at Freiburg and Haeckel at Jena—and both remained with their universities for over half a century, during a time when their programs became two of the foremost zoology programs in Germany and when their universities rose to national stature and gained international reputations in science and medicine. Finally, both independently tangled with Rudolf Virchow at plenary sessions of the Versammlung on a matter that later extended into acerbic exchanges in print. Many, if not most, of these parallels were shared to one extent or another by other academic zoologists who came of age at about the same time. Nevertheless, in the last quarter of the century, Haeckel and Weismann's names, above all others, were associated with the Darwinian revolution that had captivated not only academic but public discourse.

At this point, however, the parallels begin to break down. It is evident that the two friends and colleagues had different personalities, different styles in their pursuit of nature, and different relationships with their students and other colleagues. Haeckel was open, irrepressible in his enthusiasm, and heroic in his ambitions. He was effusively grateful to his parents, was expressively romantic in his letters to both of his intended brides, and he became deeply depressed when his first wife died two years after their marriage. Only in writing his foremost book, *Die Generelle Morphologie,* did he find salvation. Haeckel loved the forms and colors of nature and expressed his feelings in rich and daring illustrations. He developed into an accomplished watercolorist. He lavishly documented his many travels not only with vivid descriptions but also with landscape paintings, many of which he published toward the end of his life. Therein one can sense Haeckel's inner romantic feelings radiating from the forms and colors he chose to capture with his brush. One only has to view the way he portrayed his trees, twisted, gnarled, with crooked branches extending into a brilliant sky, to sense how much he had moved beyond a naturalistic representation to an expression of inner feelings. This strong desire to communicate the intuitive penetrates his scientific illustrations as well.

For his part, Weismann was more formal and cautious. He appears to have been intense in concentration, whether climbing in the Alps or peering through the microscope. He suffered through much of his career from a debilitating sensitivity of his eyes, as he explained to Haeckel, of which a "great sensitivity of the conjunctiva has turned out to be the chief cause of the periodic increased sensitivity of the retina."[7] This illness not only curtailed his research but influenced his social activities, his reading, and his research. Although Weismann was loyal, he was not emotionally indebted to his father. He always attributed his love of nature, music, and art to his mother, who died when Weismann was sixteen; she could sketch excellent portraits and painted commercially acceptable copies of oil masterpieces.[8]

As for Weismann's art, there exist a carefully executed drawing of a Frankfurter house and barnyard done when he was ten and many illustrations for his scientific works that were turned by others into colored lithographs. His art provided measured representations rather than expressed inner feelings, and his scientific illustrations were intended to serve as demonstrations for his textual arguments.

On the piano, however, Weismann excelled.[9] He played with musicians in Frankfurt during his brief career as a physician, he regularly participated in musical evenings with colleagues throughout his life, and he had a music room specially built onto his house. He passed his love of music to his children—particularly to his son, Julius, who would become a professional pianist, composer, and conductor. Weismann was fond of the classical repertoire from Bach to Beethoven but disliked the "moderns," such as Wagner and Brahms.

When Weismann's first wife, Mary Gruber, died in 1886, he was despondent but did not descend into depression. According to his great-grandson Helmut Risler, his wife's family was somewhat offended by his apparent emotional detachment after her death.[10] It is likely that Mary's death may well have brought on the return of his retinal sensitivity that ensued later in the year and curtailed further intensive microscopical research for the rest of his life.

These personal contrasts help us infer how their personal engagement with evolutionary biology may have differed between Weismann and Haeckel, but we can push deeper into the matter. The best place to start is with their correspondence with one another. Fortunately, sixty-four letters, representing a span of forty-nine years, have been preserved (though internal evidence indicates that there may have been additional correspondence that has disappeared). From what remains of their letters, we can work out that Weismann and Haeckel met for the first time in 1863 or 1864—perhaps in Jena, or at the Versammlung in Stettin (1863) or Giessen (1864), but that record still remains unclear.[11] Haeckel would have been a young *außerordentlicher Professor* who was about to or had just propelled himself to the attention of the community of scientists and physicians by a challenging address in Stettin on Darwin. Weismann would have just earned his Venia legendi from Freiburg and the right to teach zoology in the medical faculty. Early on in their careers, they suggested ways whereby they might meet during vacations. Personal affairs, the onset of Weismann's retinal sensitivity, Haeckel's compulsions to travel and collect marine invertebrates, and Weismann's more limited scientific travels and need to relax at the family estate on the Bodensee all seemed to conspire to keep them and their wives from developing a close social relationship. Nevertheless, their paths occasionally crossed at congresses, and toward the end of their lives they faithfully exchanged birthday greetings on major milestones: the sixtieth, seventieth, seventy-fifth, and eightieth birthdays.

Overall, their correspondence provides us with a tangible sense of how they interacted with one another, how each supported the other's ambitions and endeavors, and how they agreed to disagree about the pursuit of their common goal to investigate and promote "the Darwinian theory."[12] The correspondence starts out when Weismann complained to his friend how even a visit to the renowned Hamburg aquarium strained his eyes and how the doctors had prescribed rest and sunbathing for his "miserable" ("jammervoll") eyes but to no avail.[13] Haeckel responded with a long and sympathetic letter, in which he suggested to Weismann various Darwinian projects that could be pursued without irritating his eyes. His suggestions were accompanied by a telling passage that is worth repeating in full:

I consider it as one of the most widely held and grievous errors of the dominant zoological trend that people generally believe that the best has already been done simply with good observations, and that analytical research should be the single goal. To the contrary I hold the synthetic utilization of the empirically acquired material and the contemplative penetration of the same, which alone can lead us to the recognition of general laws, as the far higher and more noble goal; and how few know at all of this goal![14]

Should Weismann have missed his point, Haeckel went on to berate histology as being particularly guilty of the dearth "of a thoughtful and correct comparative reflection." He exhorted his friend, "What a desert of nonsense, contradictions, obtuseness [exists] in all of our histological textbooks!"[15] "The contemplative penetration" of the empirical material—how close this comes to Haeckel's beautiful, enveloping trees or the illustrations of his half-mysterious, pulsating medusae!

Weismann was grateful for the suggestions but did not take up Haeckel's proposed Darwinian projects. He did, however, respond to Haeckel's gratuitous comments on the limitations of "empirically acquired material" and the desert of histological facts. He felt that the zoologist must read—at least a couple of hours a day—to keep informed, and that as important as intuition may be for a complete understanding, facts are needed, too. "In summarizing," he continued,

> I also believe that it is timely to bring in sense and understanding, where they still fail, but with all such tasks it is a question of surveying a large quantity of facts in their entirety. For that, however, the earlier accumulated store [of facts] is, at least for me, not sufficient. The facts, which the philosopher needs to speculate, he carries around within himself; the scientist, however, even with the most prodigious memory, can never know whether that, which immediately fails him, will provide the key to the whole. No! I have realized that I must still wait! But one should not give up.[16]

Where Haeckel had recommended that his friend read philosophy—that is, Kant, Kuno Fischer, and J. S. Mill—Weismann rejected the first and expressed his sarcastic delight that the second had "translated Kant into German." He cheerfully confessed that Mill's *Inductive Logic* had remained untouched on his must-read list for a long time. There is no record of whether he ever satisfied the implied resolution to read Mill's work, but Weismann already had a feel for the English empirical tradition.

As their early correspondence grew, they both became concerned about more mundane events. The success of Prussia in defeating Austria pleased neither one, despite the fact they both supported the new direction in which Bismarck was pulling the German nations. They swapped news about their respective new brides, exchanged reprints, and commented on zoology positions at other universities. In the period between 1865 and 1870, two themes dominated their correspondence. On his side, Haeckel was concerned about the professional reception of his *Generelle Morphologie;* he feared that a negative response might hurt their common Darwinian goals, and he alluded to other drawbacks ("Schattenseiten") of the book.[17] He seemed, however, more at ease with himself and his zoology when two years later his *Naturliche Schöpfungsgeschichte,* written for a more general audience, appeared. Weismann complimented Haeckel on some details of the former work, but confessed that its two large volumes had to be read aloud to him and this exercise made for very slow going.[18] For his part, Weismann was deeply concerned about his ailing eyes and what the affliction might mean for his career. He read the *Naturliche Schöpfungsgeschichte* with interest, but it also pained him, for he recognized that with his forced inactivity he was being left behind, and he confided to Haeckel that it might even force him to forego future research and his academic career.[19] He kept Haeckel informed about his two-year leave of absence from teaching duties and his reoriented research plans. Haeckel sympathized, encouraged, and offered him a publishing platform in his *Jenaische Zeitschrift.* After Darwin published *The Descent of Man,* Haeckel was prepared to fight for the "Catarrhinen-Theorie" and hoped that his friend would join him "as a valiant companion in arms" for an impending battle of "giant dimensions."[20]

During his recuperation, Weismann did not remain idle. His leave of absence from 1870 to 1872 provided him with a respite from lectures and academic affairs. He used the time wisely by steeping himself in the mechanics of species formation, which in turn involved him in the shortcomings of Mortiz Wagner's migration theory. Wagner first presented his ideas as an alternative to Darwin's mechanism of natural selection, but Weismann thought through the problem in a way that many had not, and he studied the phenomenon of polymorphism in butterflies in the mountains of Sardinia to test Wagner's claims. His research with butterflies will be the subject of a later section of this chapter; relevant here is Weismann's education of Haeckel not only on the contrast between Darwin and Wagner but on how they might complement each other.[21] This exchange came during the period when Haeckel was writing his important monograph on calcisponges.[22]

Both Haeckel's *Kalkschwämme* and Weismann's *Einfluss der Isolirung auf die Artbildung* appeared in 1872, and these works could not be a clearer sign of how their research interests in evolution theory were beginning to diverge. The recent embryological work of Alexander Kowalevsky, Elie Metchnikoff, and Nicolaus Kleinenberg had redirected Haeckel's attention to the ubiquity and evolutionary significance of the primitive germ layers. Their advances in microscopic techniques allowed exquisite studies that embraced a large range of invertebrates. For Haeckel, the calcisponges were not only aesthetically appealing marine invertebrates that had to be organized in a modern form— that is, in an evolutionary taxonomy—but they presented him with a bilaminar organism in which the two germ layers could be easily identified and which formed the core to his phylogeny. The well-illustrated, three-volume monograph allowed Haeckel to utilize for the first time primitive germ layers as an empirical demonstration of the mechanical bond between evolution and embryology. The germ layers guaranteed and specified analytically what he had formally demonstrated synthetically eight years earlier. "In my *General Morphology*," Haeckel explained both in the original and an English abstract of his work, "I sought to demonstrate synthetically that all the phenomena of the organic world of forms can be explained and understood only by the monistic philosophy; and now this demonstration is furnished *analytically* by the morphology of the Calcispongiae."[23] His earlier arguments had been intuited and formalized whereas his taxonomy of calcisponges provided him with a factual base. It was only then that he triumphantly declared the import of what he now called the "Biogenetische Grundgesetz" (biogenetic law). From and after 1872, this expression became the hallmark of Haeckel's phylogenetic studies and encouraged him to construct what he felt must have been the original metazoan—a bilaminar, invaginated, gastrula-like organism that he baptized "the Gastraea."[24] It was for him, even at this time, a natural set of steps to go from sponges to gastraea, from gastraea to man, and from man to a monistic universe. "The general results, which emerge from the monograph lying before us," Haeckel confidently proclaimed, "are purely philosophical in nature and permit themselves to be summarized in the sentence: 'the biogenie of Calcisponges is a coherent proof for the truth of monism.' "[25]

Weismann showed a keen interest in Haeckel's reports on his work in progress and loaned him specimens from his collection in Freiburg. Because the published volumes turned out to be very expensive, Haeckel could only advance his friend a copy of the final signatures of the General Part, the section in which he drew conclusions about the germ layers and detailed his biogenetic law.[26] Weismann was particularly impressed by Haeckel's discussion of the coelenterate body and the conclusions he drew from comparative embryology. "I do not need to say," he added to his friend, "that I immediately read through the latter (i.e., the general part of the *Kalkschwämme*) and with the greatest interest and am now very excited to get a hold of the entire work and to learn in detail the proofs to your conclusions."[27] There are no recorded comments on Weismann's impressions of these proofs or Haeckel's extension from the sponges to monism; their conversation quickly moved to the publication of Haeckel's monographs on his gastraea theory.

The first eight sections of Haeckel's monograph came out in early 1874. For his part Haeckel felt that he had progressed beyond that which Kowalevsky and Kleinenberg had inferred about the homology of the germ layers throughout the metazoans, but he was eager to learn Weismann's reactions—particularly because of the latter's detailed work on the germ layers in Diptera.[28] He did not have long to wait. His colleague responded with enthusiasm for the general notion of the homology of the primitive germ layers, but he also recognized that there was opposition to Haeckel's bold hypothesis. He cautioned that "the discoverable facts must teach us whether it is right or not. The idea is very attractive and it would really be a pretty undertaking to make the opposition to it plausible." Weismann continued, "I would be eager once again to examine the development of insects—if the eyes would endure it, which yes, even now want to be protected."[29] Weismann did not feel the development of Diptera to be the hurdle that others, such has Gegenbaur, had maintained. Apparently Gegenbaur had argued that the limbs were derived partially from the neurolemma and partially from the trachea. Weismann's response was that these structures were both ectodermal, thus preserving the primitive germ layer homology. Weismann was also opposed to Gegenbaur's hydrostatic origins of the trachea. This provisional support pleased Haeckel, and like an appreciative brother he immediately invited Weismann to review two of John Lubbock's books on insect development for the *Jenaer Literaturzeitung*.[30] Although they dealt with development and phylogeny and were deferential to Haeckel, neither of Lubbock's books had anything to do with germ layers or even cellular anatomy beyond early cleavage. For the moment, Haeckel's phylogeny based on germ layer taxonomy was in the vanguard.

A year later, sections nine through twelve of the gastraea theory appeared in print. Haeckel had gathered much of his material on a collecting trip to Corsica and Sardinia accompanied by the Hertwig brothers during the long spring vacation in 1875. Stationed in Ajaccio, a seaport on the western side of Corsica, they had found ample material for their separate researches despite the fact that Haeckel found the fishermen "lazy" and "ill-informed" about the variety of sea life.[31] The invertebrate eggs Haeckel studied complemented his first publication, and four more empirical sections were shipped off to his publisher in August of the same year. These consisted of a lengthy discussion of palingenesis and cenogenesis followed by a detailed taxonomy of egg types, cleavage patterns, and a discussion of the phylogenetic significance of the first five stages of development from monerula and cytula, through morula and blastula, to the all-important gastrula. Each stage in Haeckel's taxonomy had its corresponding hypothetical primitive organism, with the gastrula being reflective of his now well-publicized gastraea. This step in the ontogenetic/ phylogenetic pathways was important because it represented the evolutionary starting point of all metazoans. In a "Nachschrift" (postscript), Haeckel once again mentioned the critical importance of the embryological work of Franz Eilhard Schulze, a professor of anatomy in Rostock and a participant of the recent North Sea Expedition: "The communication by F. E. Schulze confirms not only the correctness of my conception, but it significantly strengthens it at the same time, by showing that the typical formation of the gastrula through the invagination of the blastula with the lower metazoans also occurs with sponges."[32]

Though Elie Metschnikoff and Oscar Schmidt were skeptical of Haeckel's endeavor, Weismann waxed even more supportive. "I am convinced," he assured his friend, that "with it [i.e., gastraea] you have now taken an enormous step forward." He wasted no time in addressing Haeckel's phylogenetic insight, revealed in the context of the latter's insistence that arthropods as a group, despite important differences in the formation in their primitive germ layers, comprised a single lineage. "It is surely not conceivable," Weismann explained, "that *a* portion of the arthropods should have another origin than the *other* portion! So when one portion of them is known to have the gastrula form, it must be accepted that it has been lost in the other portion."[33] The same reasoning applied to his insects with their superficial cleavage, partially invaginated germ band, and a frequent overgrowth by the peripheral blastoderm.[34] But here was the catch that was to plague the entire enterprise of having embryology be the final arbiter for phylogenetic determinations: how was one to distinguish a primary attribute from a secondary adaptation?

Other than suggesting that paleontological evidence might decide, Haeckel never found a satisfactory answer. Nevertheless, he employed two neologisms that emphasized the contrast and that became central in the debate over the connection between the theory of descent and development.[35] "Palingenesis" referred to the embryological development of the lineage traits in phylogeny; "cenogenesis" referred to new traits which were modifications through adaptations in the original evolutionary pathway. For Haeckel, "palingenesis" was the recapitulation of the established past or the repetition of the ancestral lineage; "cenogenesis" was a "falsification" of this phylogenetic past, a deviation of development away from a true recapitulation. From the outset, the two concepts caused confusion. Many zoologists pointed out that the opposition of these terms rendered the biogenetic law untestable. Some changed their meaning and even their etymological roots; others, such as Gegenbaur, insisted that adult anatomy not embryology should be the prime indicator of homology and relatedness; still others soon recognized that Haeckel's nomenclature and the fundamentals of his phylogeny were too simplistic to serve the complex and varied patterns of animal development. Weismann, however, appeared to jump on the Haeckelian bandwagon. "Moreover, your catch words of cenogen[etic] and palingen[etic] development appear to me very useful and happily chosen," he complimented his friend; "it is always good for such a complicated concept to have a simple expression."[36]

Again, Haeckel was grateful for Weismann's support, and after receiving and examining the second segment of his friend's study on seasonal dimorphism in butterflies, he recognized, maybe even confessed, that there might be another way to pursue the Darwinian program. Weismann's field studies provided a level of analysis on species change that Haeckel's phylogenetic taxonomy simply could not.

Two years later, Haeckel was giving Weismann brotherly advice about publishing in the radical Darwinian journal *Kosmos* and plotting to recommend his friend to John Murray of the Challenger Expedition as the most plausible person to write up the collections of Phyllopoda and *Daphnia*. At the time, however, Charles Wyville Thomson was still in charge of the project, and Haeckel confided that the British doyen was against farming out material to foreigners—particularly to Germans![37]

After a period between 1876 and 1878, when no less than a dozen more letters were exchanged between Jena and Freiburg, the correspondence between the two friends altered in character. Several years would go by without even a birthday greeting. Nearly four and a half years would elapse before Weismann asked Haeckel if he might spare some specimens of Obelia and other hydromedusae and thanked him for an interesting but unspecified gift.

Another year would pass before he was in touch with Haeckel again. The lacunae in their correspondence unquestionably reflect the vagaries of preservation of Haeckel's letters; there are no letters from Haeckel to Weismann between 1879 and 1894. The contents of what remains indicate a shift in the relationship between the two evolutionary comrades in arms. It is clear that during the fifteen years following 1878 the two exchanged publications. Weismann asked for the architectural plans of the new institute in Jena and invited Haeckel to stay at his home in Freiburg, where the Naturforscher Versammlung was about to assemble. There is some indication that they swapped information about positions in zoology, discussed family and academic gossip.

Nevertheless, their personal, professional, and research styles were diverging. In the 1880s and 1890s, they both were leading busy, productive, and occasionally heartrending lives. Haeckel made a semester-long trip to the tropics of Ceylon (1881), which was enormously exciting for him from both a natural historical and a social perspective. This adventure was followed by shorter trips to Russia, North Africa, and again to Corsica. His lust for traveling and collecting came to a climax in 1900 when he voyaged to the Malay Archipelago and set up a field laboratory in Java. As for the specifics of his research, Haeckel pursued his organization and publication of Challenger material and published his comprehensive *Systematische Phylogenie* between 1894 and 1896. A new institute and a new house, baptized "Villa Medusa," the first edition of his immensely successful *Welträtsel,* and the first ten installments of his *Kunstformen der Natur,* with its hundred magnificent full-spread illustrations, rounded out the century for the popular Jena professor.

During those same years, Weismann immersed himself in cytological research. At first, he was attentive to germ cell production and migration among hydromedusae; then, armed with the newest cytological equipment, he and his students focused on polar bodies and maturation division. As we shall see, this led to a new perspective on the process of heredity and to his developing and defending his germ-plasm theory. Tragedy struck in 1877 when his youngest daughter, Meta, died, and again in 1886 when Mary, after bearing him six children, died at the family "compound" in Lindau on the Bodensee. He was left in the position of running the family home and the new institute he had just established. He traveled, too, but only within Europe. Three times he pursued research at the newly founded Zoological Station in Naples, and he visited other marine stations in Marseilles and Bretagne. Once, he traveled to Constantinople with his new wife since 1895, but the trip was purely recreational without professional intent. Further, he visited Great Britain three times to deliver lectures. For the most part, his vacations were spent at his family's Bodensee retreat, where he spent hours writing, collecting freshwater fauna, and making excursions into the Alps. After Mary's death, Weismann's eye affliction returned with a vengeance and grew increasingly worse. He became dependent upon his students for microscopical research and upon family members and hired domestics to read aloud professional and recreational works to him in the evenings.

Within these personal parameters, both zoologists continued to share the goal of promoting a mechanistic evolutionary biology. Haeckel continued to be enthusiastic about "the fundamental biogenetic law" and ignored the ambiguities in contrasting palingenetic and cenogenetic development. He incorporated the biogenetic law, elaborated repeatedly on the opposing processes—variously identified as palingenesis, recapitulation, and heredity, on the one hand, and cenogenesis, adaptation, and variation, on the other—into his popular works and then, as we have noted, forged the mixture into the foundation of his monistic worldview. Despite the centrality of these concepts to his work, he added little to their clarification after presenting and developing his gastraea theory. His soul was constantly energized by his collecting trips, by taxonomy, by the construction of theoretical phylogenies, and with the endlessly fascinating diversity of life and landscapes captured with his paintbrush. Weismann stuck to a narrower last: he mastered the new techniques of cytology and instinctively recognized that only at the microscopic level would evolution, development, and heredity be brought together. The steps in his professional career are, of course, the subject of this and later chapters.

The careers of Haeckel and Weismann thus diverged with the different objects they investigated and the research approaches they selected. It has sometimes been said that Haeckel provided nothing new to science after his *Generelle Morphologie.* There may be something to this, but I believe the claim leaves out the inspirational value of his later research, his interest in the evolution of man, and his popular writings. As we have seen, his fundamental biogenetic law and its associated concepts were developed in the 1870s, and his popularizations of monism as the natural extension of his evolutionary beliefs were the products of the 1890s. Equally important, Haeckel's continued study of lower invertebrates provided a sumptuous library of taxonomies, which from the point of view of later generations may have come up wanting but were very much an expression of a side of embryology and natural history of the period.

It was perhaps symbolic of their changing professional influence that Haeckel delivered his third and final plenary address to the 1882 Naturforscher Versammlung just one year after Weismann presented the first of his three plenary addresses. (He delivered the other two in 1885 and 1888.) Haeckel's professional star had reached its apogee just as Weismann's was gaining its full momentum. Haeckel reinforced this parting of scientific trajectories a decade later by riveting attention in a provocative address in Altenburg on his status as the titular head and major promoter of a monistic worldview.[38] What he had foreshadowed in his monograph of calcisponges became a second career. By this time, Weismann had just put the final touches on the book *Das Keimplasma,* which became the hallmark though not the high mark of his career. (Haeckel's address was given on October 9, 1892, and Weismann signed the preface to *Das Keimplasma* in the middle of May 1892.)

More, however, was to happen between the two friends. In 1893, Haeckel wrote a scathing critique of Weismann's research and the direction in which he had taken biology. Embedded in an essay entitled "On the Phylogeny of Australian Fauna," the attack came in an interlude placed between a brief set of published comments on Darwinism and Australian fauna and a lengthy discussion of the phylogeny of Australian monotremes and marsupials.[39] It was disguised as a manifesto defending what Haeckel had long called "progressive heredity," that is, his formal way of describing the inheritance of acquired characters. Darwin, after all, had considered such a process as a supplement to his mechanism of natural selection. Ever since his *Generelle Morphologie,* however, Haeckel had insisted that "progressive heredity" was an essential element in evolution.[40] Throughout the 1880s, as Weismann marshaled his cytological research against such inheritance and emphasized that natural selection was sufficient to drive evolution by itself, Haeckel had remained more or less above the fray. However, by the 1889 edition of his *Natürliche Schöpfungsgeschichte* and the 1891 edition of his *Anthropogenie,* Haeckel expressed his disapproval of Weismann's "neo-Darwinism."[41] *Das Keimplasma,* subtitled by Weismann "Eine Theorie der Vererbung," could only have been construed by Haeckel as an affront to his own evolutionary beliefs. So Haeckel marshaled his complaints and fired back.

Accordingly, Haeckel maintained that (1) the continuity of the germ-plasm was neither empirically nor theoretically acceptable; even the new cytological research on fertilization and cleavage could not be construed as its proof. (2) Weismann's separation of the individual into soma and germ-plasm was likewise neither empirically nor theoretically valid because internal physiological "correlations" continued to exist between the two. (3) Phylogenetically there was no sharp divide between single-celled Protists (i.e., protozoa and algae) and multicellular histones (i.e., metazoans and metaphyta); there was a spectrum of developmental and reproductive modes between the two. (4) Weismann's hypothesis of the immortality of Protists and germ-plasm was ill conceived, for all cells might be considered immortal because of the continuity of material throughout the living world. Besides, individual cells—that is, the Protists—may die just as do their collectives, the histones. (5) "Progressive Vererbung"—that is, evolutionary heredity—is indirectly proven through anatomy and morphology because adaptation to the environment is everywhere evident. (6) Finally, "progressive Vererbung" is provable through experiments and artificial breeding, which produce irrevocable and useful facts. From Haeckel's perspective, these six propositions were self-evident. They pointed to a conclusion that to support them was to support the inheritance of acquired characters and a mechanistic theory of evolution. To deny them was to reopen the door to a teleological theory, which by his account must ultimately be based on creation theory! It was a real slap at his friend to place him and his ideas alongside Carl Nägeli, Albert von Kölliker, and Karl Ernst von Baer as throwbacks to a "transcendent creation."

Contentious disputes in the Wilhelminian academy, however, did not always end in lasting enmity. (Haeckel was known to have emotionally and intellectually "disinherited" many of his students and colleagues, such as Oscar Hertwig, Anton Dohrn, Wilhelm Roux, Miklucho-Maclay, and Hans Driesch.) A few months later in mid-January 1894, Haeckel sent Weismann well wishes for his sixtieth birthday and for his past accomplishments. The congratulations broke the ice, for he even expressed regret about their differing views of heredity:

In the meantime I hope that because of this our old collegial friendly relationship will not be ruined. It is true I consider the problem of progressive inheritance as one of the most important in all of biology and am therefore so strongly convinced in the "inheritance of acquired characters," that it seems to me without it—and without the connected "phyletic adaptation"—the doctrine of descent loses its causal explanatory value. Nevertheless . . . I gladly admit that this is basically a philosophical article of faith and that it provides as little a convincing proof as your opposing opinion. Gradually will the continuing discussion of this surely bring us a few steps closer to the truth.[42]

Haeckel's words of appeasement had their intended results. In a return letter, Weismann insisted that what was important was pursuing the truth, not who ended up being right or wrong. Acknowledging that he had always valued Haeckel's work and insights, he continued:

> I would be exceedingly sorry, if the differences in certain points of our scientific convictions—as certainly exist—should become transposed into personal ill feeling. No! I also regret that we for the moment do not agree on everything, but that has never prevented me from recognizing the bounty of observations and thoughts, which you have brought to science, with gratitude and pleasure. We work, not so much to be right with our momentary conceptions, but to come somewhat closer to the truth, and it is definitely not said that a path, which finally turns out to be false, will therefore have been also useless. I believe: to the contrary, if it is false, then we stand in the future on the other [path] all the more firmly and certainly.[43]

The exchange brought them both back to a regular, though not intense, correspondence and mutual respect. Landmark birthday greetings were swapped, reprints and photographs were traded, professional advice was sought, and Weismann longingly admired Haeckel's continued travels. When in 1908 the conservative Catholic Kepler-Bund attacked Haeckel for misrepresenting many illustrations in his popular texts, Weismann, along with many other zoologists, issued the so-called Leipzig Declaration to defend their professional colleague. Their eightieth birthdays were the occasion for their last exchange of letters. Haeckel's has not survived, but he must have again alluded to their different views about heredity as Weismann sent a full four-sided letter in return. In it, he discussed their common perspective won over a fifty-year period of research in evolutionary biology and marveled at the distance their science had come. "He who has even had only a small part in this enormous progress of knowledge, can look back on his life with pride; just as you can too."[44] If there is a moral to be read into this personal relationship, it is that a disagreement, even at the most fundamental level of science, does not have to be personally vindictive. If there is a reason for narrating the interaction between Haeckel and Weismann, it is to establish how important each was for the other, and to anticipate the importance Haeckel's biogenetic law might have on the development of Weismann's ideas.

Moritz Wagner and Evolution by Isolation

There was a silver lining to the dark tragedy of being forced to forsake his microscopical work until his eyes improved. At a time when he despaired about his future in science, Weismann applied to the university senate and indirectly to the Baden cultural minister for a total of five semesters of leave from his teaching responsibilities and from the charge of director of the zoological collection in the medical faculty.[45] Nevertheless, he neither abandoned his goal of becoming a university professor nor his commitment to explore the depths of evolutionary biology. Others had faith in him as well. As we have seen, Haeckel suggested ways for him to continue his involvement in zoology. His colleagues in Freiburg not only supported his leave requests but the Baden cultural minister sanctioned the senate's recommendations that he be placed on a permanent (*planmässig*) status as außerordentlicher Professor in 1867. A year later the same parties elevated him to *ordentlicher Professor* (full professor) and made him director of a newly established zoological institute—again within the medical faculty. In the same year as he became a regular faculty member of the university, he felt secure enough of his future to marry Mary Gruber, and his future father-in-law August Gruber, shrewd businessman that he was, must have felt confident enough to sanction the union.

During the two and a half years of his leave, the state of Baden became engaged in the unification of Italy, in war with France, and in framing the German empire—events that energized Weismann's German loyalties. For a time, however, these sentiments came into conflict with the natural anti-Prussian feelings that came of his Frankfurt heritage.

There was another silver lining aspect to Weismann's time off. By leaving descriptive, laboratory embryology behind, Weismann could turn to field problems in evolution. He prepared for the transition by critiquing Darwinian theory for his colleagues in his inaugural address as a full professor.[46] Four years later, he published an extended critical review of Moritz Wagner's isolation theory of evolution, which included a favorable examination of Franz Hilgendorf's studies of fossil snails.[47] (I will comment later on both responses.)

At the same time, Weismann introduced his readers to his own studies of butterflies. Just prior to his marriage and later during his leaves of absence, Weismann also arranged to make two field trips: the first to Corsica, the second to Sardinia. Both allowed him to investigate the fascinating and complex problem of polymorphism and seasonal dimorphism in butterflies and to amass material to sustain his judgment of Wagner's ideas. In short, Weismann spent his years away from the microscope wisely. He was to demonstrate that he was every bit a field and museum naturalist as he had earlier shown that he was a descriptive embryologist focused on the histological and cellular levels of development. As a field naturalist, he was not collecting, classifying, and establishing phylogenetic lineages as Haeckel did. Nor was he concerned about morphological changes over vast periods of time, as was the chronological scope embraced by Haeckel's gastraea theory. Instead, Weismann focused on defined problems associated with geographic and temporal distributions after the most recent ice age. He was interested in the causes of variational adaptations to changing climates in the field, and he focused on manipulating larvae and pupae at his institute with varied and controlled temperatures.

This last feature of his science is worth emphasizing, for Weismann is rarely viewed as an experimental zoologist by later generations of biologists and historians. At the time, he began many experimental series on the influence of abnormally high and low temperatures on caterpillars and pupae. His experiments soon developed into observations on the development and evolution of caterpillar markings, which reflected Haeckel's biogenetic law on the species level, which also led to a short essay on the metamorphosis in the Central American salamander Axolotl. His extensive research, focused on the patterns and mechanisms of evolution, collectively appeared as his *Studien zur Descendenztheorie* of 1875–1876. This turned out to be a large publication capped by the important essay "On the Mechanical Conception of Nature," which now provides the historian with the only metaphysically focused essay Weismann ventured.[48]

Justification of Darwin's Theory

It was likely both an august and festive occasion at the university auditorium when on Wednesday, July 8, 1868, Weismann presented his inaugural address on Darwin's theory.[49] Descent was of great topical interest in German academic and educated circles of the day. By 1868, however, a talk on the subject was hardly novel. The second edition of Darwin's *The Origin of Species* had been quickly translated into German by the paleontologist Heinrich Georg Bronn, but the translation was defective in style and encumbered by critical editorial comments and an editor's concluding chapter that questioned Darwin's mechanistic thrust.[50] The Bronn translation was extensively reworked and updated in 1867 by the Leipzig zoologist J. Victor Carus in accordance with the *Origin*'s fourth and later editions. In the process, Carus mitigated the earlier metaphysical confusion, and his edited translation of *Origin* and of many of Darwin's other books soon would become the authorized German translations well into the twentieth century.

In his famous Stetin address of 1863, Haeckel had thrown down the gauntlet before his professional colleagues by urging them to heed the metaphysical message of Darwin's theory. By the time of Weismann's inaugural address, many noted biologists and physiologists, such as Gustav Jäger (1862), Carl Claus (1863), Friedrich Rolle (1863), Fritz Müller (1864), Carl Nägeli (1865), Albert von Kölliker (1864), Ludwig Büchner (1868), and Ludwig Rütimeyer (1868), had already or were about to become engaged at both the professional and popular levels with lengthy treatises and general lectures on Darwin's theory. These represented just the tip of the iceberg.[51] Dozens of German monographs and articles had already propounded a range of negative and positive reactions to Darwin's great work. They had variously promoted materialistic and extraphysical theories of transmutation, recognized Darwin's mechanism of natural selection, repeated with anecdotal evidence the importance of Lamarck's strong support of the inheritance of acquired characters, and urged the consideration of internal perfecting forces; some even insisted on the traditional creation story. It made sense that in Freiburg a general, educated audience would be interested in what their newly elevated professor of zoology would say about Darwin's theory, though whether Weismann pleased or disappointed them is unknown. It is doubtful that his talk aroused passions— as was Weismann's way, it was too thoughtful and too provisional to be strongly partisan. However, his address was explicit about his support of Darwin's ideas, and he did not try to overwhelm his audience with a catalog of erudite information but rather organized his thoughts around a few general points.

In brief, Weismann structured his address around two arguments. The first, reminiscent of Darwin's approach, was an argument by exclusion. Comparative anatomy, embryology, and the study of larval forms, the paleontological record, and zoological evidence collectively affirmed a theory of descent, for combined they embraced more of nature's phenomena than did the orthodox creation story. But was this conclusive proof of descent? On this count, Weismann, in a curious pre-Popperian way, was more temperate than Haeckel and perhaps more subtle than Darwin: "As we thus have seen, a scientific theory to be sure is never proven, except perhaps, if it is false, it is refuted, and it is therefore a question of whether facts can be introduced, which stand in inextricable opposition to one of the two hypotheses and so will eliminate it."[52]

The more central issue was laid out in the second part of the address: how effective was Darwin's theory in explaining evolution? Here, Weismann revealed the extent to which he had absorbed the Darwinian worldview, but he was hardly precise about the details. Gradualism rather than saltation defined the rate he envisioned for organic change. He considered the category "species" to be a conventional demarcation along a continuum of change. However, Weismann recognized that a species had to adapt to changing conditions over the duration of hundreds of generations rather than during a single lifetime. Adapt and evolve into new species it must, or go extinct. He recognized that variability and heredity were the two basic processes that needed to be understood scientifically to round out the Darwinian explanation, but zoology could not yet explain either one in even a rough hypothetical way. For Weismann, variability was simply the capacity of the body to respond to either external or internal influences. Heredity (or "Vererbungsfähigkeit"), as he described it, simply designated the ability to transmit a developmental direction ("Uebertragung einer Entwicklungsrichtung") from one generation to the next. Each organism then must have a specific nature or a "chemical and physical constitution" ("chemischen und physikalischen") that reflected the morphological attributes of the individual and which could be altered and passed on. The developmental direction was both inherited and acquired. Ultimately, "species are the result of natural selection and the quality of variations of their ancestors."[53]

This quite vague Darwinian scenario presented all kinds of problems as biologists began to unpack it over the next forty years. Adaptation had to be understood as both an individual and a collective phenomenon. Weismann's nominalistic species concept may have been an attempt to free biology from the static typology of the creationist explanation, but it produced its own uncertainties about the unit of change. Weismann's invocation of an organism's capacities to inherit and vary was at best a delaying tactic for processes not yet specified; at the time, heredity and variation commonly were envisioned as distinct and even antithetical activities of life rather than as complementary and overlapping phenomena. Weismann was confident that "as soon as one looks at the capacity to inherit as the transmission of a direction of development, the explanation for variability cannot be hard."[54] He also saw the external environment serving in a double capacity: first as a modifier of inheritance, then as the conditions for the struggle for existence and selection. As he recognized, this double capacity brought into focus a parallel—as it turned out, a confusing one—between the development of the individual and the evolution of the species. In 1868, Weismann was hardly an ultraselectionist, for in addition to the above-mentioned obscurities, he recognized that the small constant details of the butterflies he was studying appeared to have no utility.

Moritz Wagner and Isolation

One message, however, that emerged from Weismann's address was his unqualified rejection of a creationist explanation or of any nonphysical, non-chemical, internal directing force. This still left a spectrum of physical and chemical mechanisms to be sorted out. It was prescient of him when he wrote Haeckel two and a half years earlier that the "accumulated store [of facts] is, at least for me, not sufficient. . . . No! I have realized that I must still wait! But one should not give up."[55]

The printed version of his inaugural address closed with a seven-page addendum in which Weismann commented on a sixty-two-page pamphlet by the journalist, explorer, and accomplished amateur naturalist Moritz Wagner. Wagner (1813–1887) was the younger brother of the Göttingen anatomist and physiologist Rudolf Wagner (1805–1864). In 1864, he attained the status of Honorary Professor of Geography and Ethnography and became the director of the ethnological collection in Munich.[56] Wagner's newly appeared work touched off a series of exchanges, which included Darwin and Haeckel among others, on one of the most intricate and profound subjects facing the future development of evolutionary theory: the role of isolating factors in the evolutionary process—were they essential?[57] How were they to be construed? Were there differences among geographical, ecological, and behavioral isolating factors? Such questions continued to be discussed and heatedly debated well into the twentieth century and often, though not always, seemed to boil down to whether evolution by means of natural selection could occur sympatrically or must be allopatric—as Ernst Mayr defined the problem in 1959. Even when biologists became clearer about the contrast between monophyletic evolution and speciation through divergence, and even with the development of modern classical and population genetics, the necessary role of isolation continued to be a contentious issue.

In Wagner's original pamphlet and in the later expanded, emended, and somewhat differently oriented version, the historian finds a very suggestive, well-informed, natural-historical theory that suffered from its author's inability to think clearly about the implications of his ideas. In 1868, Wagner's principal focus was on the migration of "pioneers" of an established species across a natural boundary into a region where they became isolated from the rest of the species. He drew upon many examples of closely allied species collected during his travels, particularly in the Andes, and from Darwin's experience of related, isolated species on the Galapagos Archipelago. In a new region, Wagner's argument continued, these pioneers would be subjected to different living conditions, which might be nutritional, climatic, or ecological or simply the fact that as pioneers their existence was unique. Over many generations, the new conditions would change their physiological makeup, and they would begin exhibiting new traits. Natural selection would then push a new trait until a new variety and finally a new species would come into existence. In his own words, Wagner explained,

> The migration theory of the organism and natural selection stand in an intimate relationship to one another. The geographical division of forms would not be explainable without Darwin's theory. On the other hand, natural selection without migration of the organism, without a long isolation of a few individuals from the home territory of the original species, will not be effective. Both phenomena are narrowly correlated with one another.[58]

Wagner may have intended his theory to supplement Darwin's *The Origin of Species,* but at times he claimed more. He felt he could explain how external causes experienced through isolation enhanced individual variability and created the circumstances for variations to appear before natural selection operated on them.[59] What Wagner may not have anticipated was Darwin's objection, when he conveyed to Darwin a copy of his original talk in Munich prior to the publication of his *Darwin'sche Theorie.* Darwin was grateful for Wagner's many examples of geographical barriers and depiction of pioneer migrations, but he informed the author that he had already thought long and hard about the role isolation played in evolution and had eventually come to the conclusion that most evolutionary change must have taken place gradually through incremental changes in large, even pan-continental, populations.[60]

Having spent four summers collecting closely related species of butterflies at different elevations in the mountains near Freiburg, Weismann had also thought about the problem of isolation and evolution.[61] Like Darwin, he was moderately impressed by Wagner's discussion; and like Darwin he could not accept the dominant role that Wagner had accorded to migration and isolation. There were degrees of isolation, different sizes of pioneer populations, and varying intensities of natural selection to consider. Like Darwin, Weismann envisioned intermediate forms, which if eliminated by competition, would leave distinct species without prior isolation. Polymorphism among butterflies and examples of transcontinental species suggested how new forms could come into being without migration and how an established species on either side of an ocean barrier could remain stable. Drawing upon his recent experience in the Black Forest, he also recognized that the floral differences at different altitudes made it possible for two closely related species of butterflies, such as *Erebia medusa* and *E. stygne,* to inhabit overlapping ranges at different months of the year—thus providing an example of environmental rather than geographic isolation. As brief as he was, Weismann offered constructive counterexamples, which suggested that the problem was more complicated than Wagner had envisioned. His remarks also revealed how strongly Weismann himself was wedded to a broad geographical process of species change, which a century later would be described as sympatry, and how unconcerned he was about the difference between phyletic evolution and speciation. This and his other notions relating to evolving species will be considered in a moment.

Wagner would not take the criticism graciously. He had his reputation as a naturalist to defend and a mechanism for evolution to promote. He made matters worse, however, by doubling his efforts to emphasize the mechanism of isolation at the expense of natural selection.[62] After praising Darwin's contribution to the theory of descent, Wagner made an about-face: "I entertain the equally deep conviction that . . . the 'natural selection' of new species through the variations of favored individuals in the sense conceived by Darwin is an error."[63] The operation of that which he now called his "Migrationsgesetz" (migration law) and under which he subsumed a mix of physical separation of a few pioneers, the physiological and morphological production of new variations in response to new living conditions, the inheritance of these traits over many generations, and a continued isolation of the new and expanding colony, would, he envisioned, produce a new species more rapidly than would the action of natural selection. "Vererbungskraft" and "Variationskraft" (the forces of heredity and variation), not natural selection, fashioned the new species. According to Wagner, only in the evolution of humans, who possessed the capacity to manage artificial selection in their agriculture and to hurdle natural barriers during their travels, did the migration theory cease to operate in species formation.

Wagner made a major concession in the new version of his theory, in response to a criticism by Haeckel. He restricted his migration theory to higher plants and animals, where sexual reproduction dominated. After all, migration and isolation from interbreeding made no sense with asexually propagating species, among which he included "bryozoans, coelenterates, infusorians, foraminiferans, radiolarians, etc." As a result, the universality of his explanation disappeared. For the addendum to Weismann's published address, Wagner reserved strong, and in places justified, criticism. He pointed out that Weismann had not proved his counterclaims of sympatric evolution; his reliance on the freshwater snail *Planorbis* as an exemplar could be legitimately questioned, and some of Weismann's butterfly examples, he felt, were equally problematic.[64] However, it did not improve his standing as a scientist to claim that Weismann placed "authority over truth."

Given the novelty of his ideas and his unquestioned experience in natural history, Wagner's longer essay had to be confronted. Its arrival on the scene prompted Weismann to correspond with Haeckel about the migration theory and convinced them both of the one-sidedness of its author's new presentation. Weismann purposefully set out on his butterfly-collecting trip to Sardinia to examine the finer details of isolation. The resulting monograph, "On the Influence of Isolation on Species Formation," not only provided his formal response to Wagner but represented his first major publication in six years.[65] This substantial contribution to evolution theory helped to identify Weismann as a serious, careful discussant of the factors involved in the process. Darwin responded warmly to a complimentary copy, and despite the fact that he read German slowly, he appears to have read it carefully.[66] He disagreed with Weismann on certain details, but he found the discussion of "amixie"—that is, a lack of interbreeding—"to throw an important light on an obscure problem" and reinforced Weismann's conclusion with his own recollection: "I rejoice to think that I formerly said as emphatically as I could, that neither isolation nor time by themselves do anything for the modification of species." Darwin also took genuine pleasure in Weismann's use of sexual selection, which he had recently championed in *The Descent of Man* (Weismann had read J. V. Carus's German edition). His major disagreement concerned the Freiburg professor's claim that species alternated between periods of constancy and variability—an idea Darwin said he had also pursued but with little conviction.

After his return from Sardinia, Weismann had planned on his publication presenting both negative and positive messages. The former turned out to be his sharp rebuttal of Wagner's 1870 essay. He was obviously disturbed by Wagner's style, which he interpreted as "slanted" ("schief") and rife with "uncivility" ("Ungezogenheit").[67] Such polemics alone, however, would not have induced him into print. He found isolation to be a fascinating biological problem unto itself, and Weismann was convinced, after his experiences in the Black Forest and Sardinia, that Wagner's migration theory could be decisively refuted by examples of sexual dimorphism in butterflies.

Planorbis *and* Vanessa

Despite Wagner's criticism, Weismann returned to Franz Hilgendorf's studies of fossil shells of the freshwater snail *Planorbis* as his primary exhibit of sympatric evolution and periodic variability. A convinced Darwinian himself, Hilgendorf had excavated in the gravel quarries on the sides of a Miocene meteorite crater in the Schwabian Alps. Over a period of 14 million years, this impact crater had turned into a freshwater lake, gradually filled with sediment, and partially drained, until in modern times it formed an area of farmland encircled by banks of beach gravel. Named after the farming village in the area, the Steinheim Basin (Steinheimer Becken) provided a continuous sequence over time of the evolving fossil snail *Planorbis multiformis,* a pulmonate gastropod found nowhere else in the world. Hilgendorf described nineteen "varieties" in ten identifiable geological zones. Furthermore, he sketched out a phylogeny based on the morphological similarities of the shells and their position in the geological column, with the realization that all of these varieties, because they were air breathing, must have lived close to what was then the shoreline of the lake during their lifetime. The study, which Hilgendorf carried out for many years, provided him and Weismann with a prime example of a simple Darwinian phylogeny based on gradual change, a lack of a clear demarcation between species and varieties, a simple branching lineage, which stressed phyletic change over speciation, and a record of nonprogressive evolution. He did not doubt that the snails had evolved solely within the Steinheim Basin. The fundamental issue—the matter of sympatric evolution, upon which Wagner had attacked Weismann—appeared vindicated by *Planorbis.*[68]

Second, dimorphism, a subject Weismann had studied intensely during these years away from the microscope, also reinforced his rejection of Wagner's migration theory. Whether the dimorphism was of secondary sexual characters within a single species of butterflies living in the same locale or of the different forms of caterpillars found in the same brood, Weismann had a strong argument in support of sympatric evolution. "According to Wagner's isolation theory sexual dimorphism should not exist at all."[69] This 1872 claim seemed irreproachable as long as one assumed, as he and Darwin did, that there was no significant difference between varieties and species, or between the development of larvae and the evolution of species.

Weismann's third major argument against Wagner concerned the most cosmopolitan butterfly of all, *Vanessa cardui* or the painted lady. Indigenous throughout the world, distinguished by only the smallest regional differences, this active butterfly was recognized by most lepidopterists at the time to be a single species. For this worldwide species, Wagner had claimed that there could only be partial isolation separating the regional forms. This struck Weismann as a circular argument that ended up discounting isolation because of the lack of significant morphological difference rather than claiming the regional forms were different species because of cases of obvious, oceanwide separation.[70] The negative message seemed to be clear: isolation itself was ill-defined and often problematic.

Weismann's positive message, embracing two thirds of the text, hardly mentioned Wagner. Instead, he provided an elaborate discussion of how isolation, seen as the root cause of amixie, or the preclusion of interbreeding, must be considered with other factors in mind, such as the size of the primary and secondary populations, the degree of isolation over time, the rate of migration, the production of indifferent "morphological" characters, the contrast of living conditions on either side of a natural barrier, the implications of sexual selection for colony divergence, and, of course, the action of natural selection. Evolution, he argued, was a complex process of variables, not a simple mechanical sequence. Such complexities drew Weismann out of his own understanding of evolution. He agreed that to a limited extent migration of a few individuals across a physical barrier could, on the one hand, lead to change in the newly established colony under certain conditions. The migration had to occur, he insisted, when the species was in a state of variability. If a new colony was then established by only a few individuals, if there was little or no reinforcement by immigration across the barrier, and if the living conditions on either side of the barrier were the same, then such a colony would contain only a small percentage of the spectrum of varying traits of the original species. Thus, through amixie these traits might dominate in the new colony. Given these circumstances, however, Weismann insisted that nothing new had been evolved, only the proportion of traits had altered, and he could consider these traits only indifferent, "morphological" traits. That is, they possessed no additional utility. As mentioned previously, Weismann had also recognized the possibility that sexual selection might bring about evolutionary change, but in the described situation he insisted this would produce only morphological change within the already available range of traits, which might be functional for mating but might at the same time be disadvantageous for the individual. Weismann felt such a factor had no need of isolation to operate.

On the other hand, when the migration took place across a barrier into an area where different living conditions prevailed, and by this Weismann meant not only climatic conditions but differences in flora and fauna, the new colony might thrive with new opportunities. Under these conditions, additional "reinforcements" from the original stock would at first also find a place in the new region, for there would be little intraspecific competition. As the colony filled its new territory, however, the reinforcement rate would be reduced, and the colony would begin to feel the full force of natural selection.[71] Such a situation often occurred in an archipelago, such as the Galapagos Islands, but for Weismann it also happened in the broad expanses of a continent where complete breeding isolation—that is, amixie—was not present. In brief, isolation was neither necessary nor sufficient.

The positive side to his monograph provided a nuanced discussion of Hilgendorf's presentation of *Planorbis* and an attempt on Weismann's part to reconstruct 14 million years of isolated evolution in the Steinheim Basin. His discussion moved on to Weismann's own experiences on Corsica and Sardinia with various species of tortoiseshell butterflies of the genus *Vanessa*. All of them appeared to be derived from species on the mainland but had contrasting histories on their isolated islands. *V. urticae* was broken up into nine clearly identifiable varieties (or species). (*Vanessa urticae*, which is now *Nymphalis* or *Anglais urticae*, is a tortoiseshell, which is common throughout Europe and much of Asia.) Finally, the similarities and dissimilarities between Arctic and Alpine species of butterflies provided a similar demonstration of how the isolation of populations after the Ice Age might or might not coincide with divergence of forms. Isolation, in and of itself, was evidently not the decisive factor of evolutionary change. Much more was involved in the evolutionary process.[72]

Trait Transmission

The most novel aspect of Weismann's analysis consisted of a hypothetical model to illustrate the dynamics of trait transmission in both open and closed populations. Weismann chose to imagine a trans-European land snail population divided into many small semi-isolated zones and set himself the problem of estimating how rapidly an internally induced variation might travel through interbreeding from one end of the population to the next. He started with the simplest situation of uniformity in all living conditions throughout the area. Because, according to him, natural selection operated only when conditions changed, this simplest of situations eliminated the need to consider its involvement. Lacking the conception and vocabulary of a hereditary unit, Weismann instead asked, with perhaps a crude hydraulic conception in mind, how rapidly does a new trait in the "blood" move from one zone to the next until it passed from one end of the snail's territory to the other, given the assumption that one-tenth of the new blood was passed along to each generation? He then asked the same question after imposing certain new conditions, and the action of natural selection favored the new trait to a significant degree—he suggested 50 percent.

He then explored what this might mean in an isolated colony and a string of isolated colonies, such as found on an archipelago. He also recognized that the currents and countercurrents of "blood" flow and inbreeding would have a bearing on the outcome, so he manipulated his hypothetical model through sixteen pages. Even to ask such a question, to focus on the flow of single traits instead of on a general change in the whole organism, was to penetrate the problem from a very different angle compared with Wagner or Haeckel. Weismann admitted that his quantitative estimates were only hypothetical estimates, with no empirical ratification other than the common belief that the traits of an offspring would reflect each parent in a 1:1 ratio. Nevertheless, what he was proposing was quite extraordinary given the contemporary discussion about the role of isolation in evolution. He concluded that small, isolated colonies could provide an important condition for the expression of indifferent morphological characters, and more importantly they could provide a venue for the action of natural selection and evolution when conditions changed. Even more important for him was the argument that in an open, continental population, natural selection introduced by a spectrum of different and changing living conditions would, albeit more slowly and with the production of intermediate forms, drive an evolutionary change sympatrically.

Weismann was never again to use such a model, and he would soon drop the traditional metaphor of "blood" and its dilution to elucidate the hereditary process. It is fair to ask, however, what such a model suggested about his understanding of evolution and heredity at this stage in his career. The common metaphor of blood carried with it all the connotations of blending heredity. The blood became increasingly diluted by one-half with each generation. Weismann's crude hypothetical calculations were just that, such that by the end of ten generations along the geographic range of his postulated snails, the end product contained 1/1,024 of "the blood" of the initial progenitor. Introducing natural selection into the equation, Weismann clearly recognized that the composition of the blood would change and that the species would evolve. Darwin, too, had assumed a blending theory of heredity even as he developed his provisional theory of pangenesis that helped him mentally negotiate between varying traits of the individual and the evolution of species.

It took more than a material, quasi-particulate hypothesis of heredity to overcome the supposition of blending inheritance. The possibilities, however, were there. As Weismann explained a year later in a review of nine evolutionary works that appeared in 1872,

> the changes thus follow completely determined directions, but not all individuals change in the same direction, rather they deviate in different directions from one another, so that one might anticipate the final formation of more new species, which however in this case does not happen, but rather a blending of different new characters [develops] into a single new species.[73]

Not only did Weismann seem to conceive of the new species as being composed of many new and different characters, but he envisioned a deeper cause than simply their hydrodynamic intermixing; "this lay far deeper, namely in the physical constitution of the species."[74] It would take him a dozen years to realize fully how such a change might be mechanically possible.

Curiously, Weismann minimized through neglect the subject of the operation of the direct influence of the environment on the production of traits. For the most part, he wrote of living conditions that either assisted in the stabilization of a population by remaining unchanged or, if changing, ushered in the operation of natural selection. He maintained throughout his monograph that a species tended to exhibit little variability during most of its existence. During intermittent periods, however, he insisted that variability increased. This was a bold assertion, with the fossils of the Steinheim Basin serving as his best evidence. It was a belief shared by many of his contemporaries. A year later, however, Weismann was to recognize that further research on *Planorbis* by the American paleontologist Alpheus Hyatt and the Würzburger geologist Fridolin von Sandberger had brought Hilgendorf's results into question. He could sarcastically report that "Hr. Wagner may be pleased, for the time being I am unable to exercise my battle elephants (namely, the Stein-heimer snails) against him."[75]

Weismann's entire monograph, with its focus on the meaning and action of isolation, bears witness to the fact that Wagner, despite their testy disagree-ments, had an important influence on shaping Weismann's own investigations of the evolutionary process. Despite his rejection of Wagner's "slanted" and "uncivil" promotion of isolation and his dismissal of Wagner's underestima-tion of the role of natural selection, Weismann was forced to think about evo-lution in terms of the dynamics of population movements and expansion, the influence of changing conditions, and the quantitative impact of natural se-lection over time. His first full monograph on evolution found Weismann ex-tending his interest in polymorphism in butterflies. In this work, he would turn to the fruits of the English tradition of natural history.

Weismann's Studies of Descent: Seasonal Dimorphism

Where his inaugural address of 1868 and his monograph on the *Influence of Isolation on the Formation of Species* of 1872 had served as platforms for Weis-mann to enter the discussion over the mechanisms of species change, the first of his *Studien* provided a detailed, ninety-page discussion of his eight years of research on the phenomenon of seasonal dimorphism in butterflies. This latter monograph set his research style apart from Haeckel's phylogenetic tra-dition, while at the same time suggesting how closely bound Weismann re-mained to the theoretical framework of Haeckel's discussion of evolution and development.

A number of things emerge from the study, which significantly appeared first in the *Annali del Museo Civico di Storia Naturale di Genova* and then later in the year was published by Wilhelm Engelmann in Leipzig as the first study in his *Studien zur Descendenztheorie* of 1875–1876. By this time, Genoa had in a way become Weismann's second home. It was where his in-laws lived, whom Weismann admired, and it was where Mary Weismann found support for her growing family as her husband traveled to Sardinia, Corsica, and later Naples—first in pursuit of butterflies and then to find hydroid polyps. That his study of seasonal dimorphism appeared as a publication by the local mu-seum there reflects its strong natural historical orientation.

More than being a descriptive study of the geographical distribution of mul-tiple forms of the same species, this monograph reveals Weismann examining contrasting "winter" and "summer" forms of *Vanessa levana/prorsa* and *Pieris*

napi/bryoniae after carefully planned manipulation of the ambient temperature of the larvae and pupae. The experiments were not complicated, to be
sure. They required rearing boxes, appropriate plants for the larvae, ice chests,
and an incubator. The important point was that Weismann had a clearly
defined problem where artificial manipulation became a logical and doable
complement to the geographical information gathered by collecting, studying
museum samples, and corresponding with dedicated lepidopterists.

Finally, as a zoologist Weismann had a strong incentive to associate the patterns of seasonal dimorphism that he had found with the current academic
deliberations over polymorphism, parthenogenesis, and the alternation of generations. Such discussions, as we already saw in Chapter 3, had been central
to the discussions of development and propagation during the previous thirty
years. They forced Weismann to generalize the results of his investigations of
seasonal dimorphism. Today, we can find some of his conclusions understandable and "modern," but in the context of the terms of the conversation of his
time, they probably appeared quite abstruse and irrelevant. This is the hidden
dilemma for any creative scientist: to engage in the contemporary conversation and yet to find ways to reshape it.

Alfred Russel Wallace and Seasonal Dimorphism

Let us begin with a familiar, well-studied figure in English evolutionary
thought, Alfred Russel Wallace. In 1864, Wallace, who was residing in England after eight years of collecting and writing in the East Indies, read a paper
before the same society that had earlier secured his place alongside Darwin
as the codiscoverer of natural selection. Five years later, it would be his brilliant study in natural history with the disarming title *The Malayan Papilionidae, or Swallow-Tailed Butterflies, as Illustrative of the Theory of Natural Selection* that would again bring him to the attention of naturalists. The first
published version was complete with plates, tables, and other scholarly attach-
ments of a seasoned naturalist.[76] What Wallace dealt with was not only an
order of insects that would occupy much of Weismann's scientific research
throughout his career but did so in a way that embraced many of the problems Weismann would repeatedly wrestle with. Lepidoptera, that is, butterflies and moths, were an attractive group for zoologists to work with. By midcentury, there existed a rich corpus of descriptive literature assembled by a
hundred years of field collectors who were interested in not only the classification of specimens throughout the world but their ranges, habits, and life

cycles as well as the native plants that sustained their caterpillars. For Wallace, the wings of butterflies, in contrast to their other body parts, were not only structures of beauty but provided the ideal focus for research into evolution. In an elegant passage, Wallace endeavored to excite his readers to the possibilities of "an endless variety of pattern, colouring, and texture." Wallace declared,

> The scales, with which they are more or less completely covered, imitate the rich hues and delicate surfaces of satin of velvet, glitter with metallic luster, or glow with the changeable tints of the opal. This delicately painted surface acts as a register of the minutest differences of organization . . . they enable us to perceive changes that would otherwise be uncertain and difficult of observation, and exhibit to us on an enlarged scale the effects of the climatic and other physical conditions which influence more or less profoundly the organization of every living thing.

As Wallace knew all too well, his perceptive words echoed a comment of his former collecting companion in the Amazon, Walter Henry Bates, who believed that the wings of butterflies "serve as a tablet on which Nature writes the story of the modifications of species."[77]

There is much to be gleaned from Wallace's paper: his analysis of family rankings, his discussion of the term "species," his invocation of "the laws" of geographic distribution, and (in the 1870/1973 version) his selective use of Darwin's concept of sexual selection. Central to the paper, however, was Wallace's discussion of variations among the Malaysian swallowtails. Here, he focused on what he called six "laws and modes of variation."[78] To us, these six may appear somewhat of a hodgepodge of taxonomic, developmental, genetic, and ecological matters that had yet to be sorted out. To Wallace, Weismann, and many of their contemporaries, they entailed a necessary first step in trying to examine systematically the confusing range of variations upon which selection or the conditions of life might act. Wallace identified the first of these modes as "simple variability," an expression recognizing that progeny range in a spectrum of forms—sometimes greatly, sometimes only a little—from the parents. As he explained it, "specific form is to some extent unstable," and with sexual reproduction the offspring in some way reflected the differing traits of both parents. The second type of variation he placed under the heading of "polymorphism or dimorphism," terms he invented to distinguish the phenomenon from simple varieties and the physical differences between good species.

Wallace conceived of both polymorphism and dimorphism as a reflection of distinct forms that interbreed "without producing an intermediate race." He included as illustrative examples cases of Batesian mimicry, "alternate or seasonal dimorphism," and associated human characters, that is, cases that might appear to us as examples of linked Mendelian traits. Finally, Wallace pointed out that increased knowledge of variations and geographical ranges made the matter of identifying distinct species more difficult rather than easier. Again, drawing upon Bates's work in the Amazon, Wallace saw that where there once appeared to be distinct species of Lepidoptera, they now were "in reality most intricately combined in a tangled web of affinities."[79] Wallace's multiple distinctions should alert historians as to how confusing and tenuous the phenomenon of variations must have seemed before the rise of classic genetics.

Wallace was more experienced than Weismann with this "tangled web of affinities" as revealed through his study of biogeography, but as an embryologist turned lepidopterist, Weismann was better attuned to the morphological differences in the variations that his English contemporary had attempted to catalogue. In early 1871, he applied to Haeckel for a copy of Wallace's essay on mimicry, to which his friend responded that it had already reappeared in a recently published German translation of Wallace's essays.[80] That Weismann was interested in mimicry at all is in itself significant, for the subject was emerging as one of the few defining natural phenomena known in the nineteenth century that historians have identified as unambiguously favoring the Darwinian mechanism of natural selection.[81] The new reprint of Wallace's essays in both its English and German editions contained not only the less technical essay on mimicry but the more substantial essay on Malaysian Papilionidae. In fact, Weismann shortly thereafter employed Wallace's term "sexual dimorphism" in his 1872 essay on *Isolirung*.[82]

Weismann opened his study on "The Seasonal Dimorphism of Butterflies" by again acknowledging Wallace's expression. Although he noted that its "heterogenous composition may arouse the horror of the philologist," Weismann found it so "concise" and "intelligible" that he proposed to use it himself in its Germanized form.[83] It is curious that Weismann would single out the "philological horror" of the expression at this point, for in 1872 he had no qualms about using the expression "sexuelle Dimorphismus."[84] Today zoologists use the term "polyphenism," which captures the notion of possible, multiple phenotypes derivable from the same genome. This expression would be anachronistic in Weismann's day, for it is laden with connotations from classic and postclassic genetics.

As we see him embark on his own investigations, it becomes immediately clear that Weismann was concerned about the underlying causes of the dimorphic phenomena. Might it be simply a chemical response to the different temperature of the spring and summer broods? Was it the response of the different "pressures" of natural selection presented by the two seasons—a cause he considered to be "indirect"? Or might there be some direct influence of the seasonal temperature on the larvae or even pupae? These were questions that were formed in such a way as to invite experimentation, and in 1866 and 1873 Weismann began controlling the ambient temperature at both the larval and pupal stages of his locally raised butterflies. He worked with the common *Pieris napi* and became actively involved with the investigations of the West Virginia lepidopterist William H. Edwards, who had examined in detail the polymorphism in the North American Ajax or zebra swallowtail (*Papilio ajax,* forms *telamonides, walshii,* and *marcellus*). Above all, however, Weismann studied the common European map butterfly, *Araschnia levana* (Linnaeus 1758), whose spring and summer adult forms appear so different that their taxonomic identity was not recognized before the 1830s. (Weismann used the older genus designation *Vanessa,* but *Araschnia* was also in use at the time.)[85]

"Map butterfly" is the descriptively appropriate common English name for *A. levana,* reflecting the complex maplike pattern of the underside of both wings. (The common German and French names Landkärtchen and "Carte géographique," respectively, connote the same representation.) It is a stunning perspective of an overall handsome butterfly, but unfortunately the underside is rarely portrayed in classical guides. However, the commonly illustrated upper surface of the wings clearly demonstrates the contrast between the two forms involved. *A. levana* designates the winter generation—that is, the generation that hatches from its pupae, takes flight, and propagates in the spring. Its wings' dominant color ranges from yellow to orange and is checkered with black spots and partial bands. *A. prorsa* designates the summer generation— sometimes two generations. This is a larger butterfly, with somewhat more pointed wings, and with the upper wing surface dominated by a black background with a dramatic cream-white band that extends from the middle of the costal or front edge of the forward wing to the middle of the inner margin of the hind wing. Although the male is somewhat smaller than the female in both forms, the two sexes display roughly the same seasonal change in color and pattern. Rarely in the wild a third, intermediate form appears, *A. porima,* which upon close inspection appears to be a blend between the spring and summer patterns.

In his *Vita Propria,* written over forty years after the fact, Weismann maintained that in 1864 he began collecting butterflies and their larvae on the Schlossberg, then in the Mooswald northwest of Freiburg, on the moor near Hinterzarten near Titisee, and in the high Schwarzwald. He further made it clear that this activity was both a renewal of a boyhood passion and a surrogate for the microscopical research he was unable to pursue. (There is no reason to doubt his word, for Weismann appears to have been referring to notes he had earlier kept.) Within two years, the rekindling of a childhood hobby had become an earnest exploration into the cause of the seasonal dimorphism of *A. levana.* He went on to explain that he subjected the pupae of the winter generation (i.e., *levana*) to cold temperature and got a partial reversion of the ensuing summer generation (*prorsa*) to the rare intermediate form (*porima*). Interestingly, suspecting a problem, he reasoned that he needed to increase the cold; so he negotiated with "Herrn Sommer," the proprietor of the Zähringerhof in the center of Freiburg, to keep a box of pupae in the kitchen's icebox. He admitted that these makeshift conditions were far from ideal, for he could not control the temperature, but what else could he do at this stage of his career before he had a zoological institute of his own with its own equipment?

Nevertheless, even under these less than ideal experimental conditions (how long Weismann continued to use the icebox of the Zähringerhof is uncertain), he succeeded in producing *levana* when there should have been *prorsa.* After he was promoted to full professor and the director the Zoological Institute within the medical faculty in the middle of 1868, Weismann mentioned controlled refrigeration the next year. In all, Weismann systematically recorded twelve sets of cold and heat experiments that carried into the spring of 1873 on *A. levana/prorsa* and four sets of experiments on the sulfur butterflies *Pieris rapae* and *P. napi/bryoniae.* He generally used collected larvae but occasionally carried the generations on through the year by successfully breeding some of the offspring. He collected and experimented on broods of up to seventy individuals, and he often divided his broods into controlled lots and occasionally used other species of the same genus as further controls. He described with fine detail the types of variations he obtained, which overall varied along a spectrum from *levana* through *porima* to *prorsa.* Although the variables were complex—after all, they included the natural generations, the degree of temperature manipulation, the stage of development, and the length of time of the artificial exposure—his basic findings consistently showed that it was possible to direct the larvae and pupae of the summer form toward or into the spring *levana* form but not the other way around.

Only after his early success did Weismann discover that the Austrian naturalist Georg Dorfmeister had been raising certain species since 1845 with the hope of discovering a relationship between wing color and larva and pupa rearing temperatures. Dorfmeister, however, had not published his work until 1864 and then only in the local natural history society of Steiermark.[86] Although he did not say so, Weismann would have learned nothing from Dorfmeister's account even though he acknowledged its existence. There was indeed a difference between old-style butterfly collecting and scientific experimentation, between descriptions and problem solving. In 1880, Dorfmeister was more systematic in the way he approached his experiments. He admitted that, in contrast to Weismann's monograph, he had been more interested in the production of variations and in the stage in development when adult color and markings were fixed.

William Henry Edwards

Far more valuable to Weismann was the work of the American lepidopterist William Henry Edwards. W. Conner Sorensen has provided an excellent chapter-length discussion of Edwards's work on polymorphism in American butterflies in general and his exchange of ideas with Weismann over a five-year period in particular.[87] Born in the Catskills in 1822 and educated at Williams College and at a law school in New York City, Edwards swung between a successful business career and the study of butterflies. His entrepreneurial father migrated from partnering in "the straw goods" business in New York to the family tannery business in the Catskills, then back to banking and insurance in Brooklyn. He was evidently very successful and encouraged young Edwards to explore the rapidly changing mining industry in the trans-Appalachian counties of Virginia.[88] The newly constructed Chesapeake-Ohio Railroad had opened up new commercial possibilities, so Edwards founded the Ohio and Kanawha Coal Company in Coalburgh toward the end of the American Civil War (i.e., shortly after West Virginia became recognized as a new state independent of the confederate state of Virginia).

In the meantime, the natural history and taxidermy, which had been encouraged in Edwards's youth, was put to good use when in 1846 he voyaged with an uncle up the Amazon. Together they explored the delta islands and traveled the broad river from Pará to the ingress of the Rio Negro and somewhat beyond. It was the experience of a lifetime, which Edwards turned into a readable travelogue the following year—full of description of life along the river, native activities, and general observations on natural history. It is clear that at this time Edwards concentrated on collecting plants, shells, and birds, and he appears to have acquired some detailed knowledge of the latter. This future dean of North American butterflies, however, barely mentioned Lepidoptera. At the end of his account, Edwards included some advice to future explorers and collectors about useful medicines, clothing, and rudimentary collecting materials for other explorers.[89]

Edwards's book, published both in New York and London, found two readers in Wallace and Bates on the eve of their departure for the same region. They even visited Edwards, who by chance happened to be in London at the time. He appears to have written letters of introductions to further their explorations. Edwards was to correspond with both later, as well as with Darwin, Raphael Meldola, and on a number of occasions with Weismann.

In the 1850s, after his return from the Amazon and England, Edwards began collecting butterflies in earnest; he became well known in American entomological circles, and by the outbreak of the Civil War he was publishing accounts of new species of Lepidoptera. By 1867, ensconced in his own coal-mining business in Kanawha County, Edwards became a regular contributor to major entomological journals and published the first installment of his life-long project *The Butterflies of North America*.[90] Besides this elegantly illustrated three-volume work, one of his major accomplishments was developing a simple method for enticing gravid female butterflies to lay their eggs in captivity. The technique turned out to be as essential as it was simple for establishing the genetic connection between different forms within polymorphic species. He took a "powder keg" with open ends, placed one end over the food plant of the caterpillar, and netted the opposite end. By placing the impregnated female in the keg, he found that she readily laid her eggs on the food plant, and from there Edwards could observe the sequential instar stages, the chrysalis, and the subsequent generation, and he inadvertently set the stage for easily altering the conditions at any point in the cycle. In Sorensen's assessment, "Edwards is regarded as the most innovative nineteenth-century American lepidopterist in terms of breeding butterflies for the study of their evolutionary development."[91]

It was probably the appearance of the earliest issues of Edwards's *Butterflies of North America* that first caught Weismann's attention. He recognized in Edwards a careful and knowledgeable worker who was wrestling with vicariant North American species at a time when Weismann had debunked Moritz Wagner's migration theory. He sent Edwards a copy of his *Einfluss der*

Isolirung along with a request for specimens of certain species of North American butterflies.[92] This was the beginning of fruitful collaboration by correspondence and exchanges, which lasted at least through 1876.[93] The most convincing confirmation of the application of Edwards's breeding technique involved the zebra swallowtail. Edwards demonstrated that what had been assumed to be three different species were in reality three seasonal forms of one species. Over a two-year period, he went on to show that there were two spring forms—*Papilio walshii,* the earliest to appear, and *P. telamonides,* which appears a few weeks later. *Telamonides* produced the summer form *P. marcellus* and occasionally an overwintering chrysalis that emerged as *telamonides* the following spring.[94] To complicate things, *walshii,* when carefully monitored, revealed the capacity to produce all three varieties in the summer. Although *marcellus* continued to produce more *marcellus* for two further summer generations, approximately 10 percent of each of its three summer broods remained at the chrysalis stage and overwintered to emerge the following spring as *walshii* and *telamonides.* Not only had Edwards shown the lineal relationship between the three forms, but he had presented a far more complex cycle than the European *Araschnia.* Darwin, Wallace, and Bates, each of whom must have received a personal copy of issue no. 9 of Edwards's *Butterflies of North America,* responded to Edwards's findings with enthusiasm. Bates thought it opened up a long vista of consequences. The significance would have been in Weismann's mind, as well. In a letter to Coalburgh, written some time after their first communication, Weismann explained his interpretation of the new events:

> The case of *Walshii* and *telamonides* is indeed very singular and not easy to explain. Nevertheless, I should believe that the ordinary warmth of the room in the winter is the cause which prevents the chrysalids acquiring the perfect winter form *Walshii.* The case of ajax is more complicated than the other cases of seasonal-dimorphism. It seems now to me possible that not the form *Walshii* is the primary, but *telamonides.* It seems *telamonides* results from all generations. This primary form could have been changed by summer heat into *Marcellus,* by winter cold into *Walshii.* But this would pre-suppose that *telamonides* has originated in the south and there resided at the time of the great glaciers.[95]

Weismann repeated the same scenario in his 1874–1875 monograph on seasonal dimorphism with *Araschnia* and *Pieris,* but in the more elaborate sequence discovered by Edwards. He must also have urged Edwards to carry out similar temperature experiments and enjoined other lepidopterists that "experiments ought to be able to decide this [primacy]."[96] This may have been the first time Edwards learned of the experiments in Freiburg, and he seems to have taken the first opportunity to perform his own in Coalburgh (1875–1876). Although these efforts were visited by a number of mechanical problems caused by the uneven exposure of pupae to ice, he succeeded in forcing the summer broods to produce variations of *telamonides* rather than *marcellus*. He concluded, in accord with Weismann's hypothesis, that *telamonides* must be the "primary form" of the species and the other two derived through temperature variations at pupation—*walshii* to cooler, *marcellus* to warmer.

A somewhat similar exchange passed between Coalburgh and Freiburg in 1876 when Weismann learned of Edwards's experimental work with *Phycoides tharos marcia,* the common pearl crescent. Here, the primary and spring brood had darker wing markings than the summer brood, *morpheus.* The results were less predictable, however, and there appeared to be a sexual difference in the response to temperature manipulation. "This proves, as it appears to me," Weismann wrote, "that the males are changed or affected more strongly by the heat of summer than the females. The secondary form has a stronger constitution in the males than in the females," and in a sign of collegial confidence, he continued, "As I read your letter, it at once occurred to me whether in the spring there would not appear some males which were not pure *Marcia,* but were of the summer form, or nearly resembling it; but when I reached the conclusion of your letter I found that you especially mentioned that this was so!"[97]

These episodes, which connect Weismann's monograph and Edwards's publications, reveal not only the scientifically rewarding communication between the two but also a developing community of avocational lepidopterists in general, whose familiarity with the particulars of individual species and whose experience at finding and raising specimens—in part for taxonomic reasons, in part to serve a regional inventory, and in part for commercial purposes—provided logistical support to the professional zoologists, whose central concern was to fashion generalizations about the phenomena of nature. Field experience, laboratory experiments, and museum studies came together to form a bond firmly soldered by common enthusiasm, goodwill, and a mutual respect between investigators of different callings, objectives, and nations.

From his perspective, Weismann saw that an explanation must embrace a large variety of cyclical patterns. There was the straightforward case of the seasonal dimorphism of *Araschnia levana/prorsa;* there were the examples brought to his attention by Edwards, where the zebra swallowtail (*Papilio telamonides*) appeared to segregate imperfectly into an early spring form (*walshii*) and a summer form (*marcellus*). Then there was the case of the pearl crescent (*Phycoides tharos*), in which seasonal dimorphism was partially associated with sexual dimorphism. Weismann was also familiar with the small copper, *Polyommatus* (now *Lycaena*) *phlaeas,* which is seasonally dimorphic in Italy but not in Germany, despite the fact that two broods a year occurred in both. Lepidopterists began to recognize that polymorphism was a common occurrence among butterflies, and the challenge was to explain not only the cyclical events but to unite them in a general theory about cyclical changes and the evolution of species. After all, if seasonal dimorphism might be seen as a splitting up of a species into two different forms associated with different seasons, might not the same process operate with respect to species evolution at large? This was the basic question Weismann always bore in mind.

The question might seem at first simple, but sorting out the many variables and demonstrating their actions as part of the search for unambiguous explanations was not the work of a single lifetime. Nonetheless, Weismann's understanding of science prompted him to push as far as possible into this shrouded world of partially framed questions and hypothesized explanations, to clarify the first and to chisel the second into testable theories. His first task was to distinguish between the direct chemical and physical actions of temperature on color and markings, on the one hand, and indirect influences, on the other. The map butterfly provided a clear case in point. As mentioned earlier, both female and male winter *levana* show a general yellow to orange background and a distinctive black-checkered pattern whereas the male and female summer *prorsa* show a strongly contrasting black background and a cream-white mid-wing band across the dorsal surface of the front and hind wings. The artificially reared individuals revealed an intermediate pattern but always in the form of an overlay of one pattern on the other, not a blending of one color and marking into the other. One could not speak of a bleeding of a chemically induced color from one form to the next.

5

Studies in Descent, Part II

1873–1876

It makes logical sense that Weismann devoted the second part of his *Studien* to "The Final Causes of Transformation" (Plate 1). His focus had deliberately swung from polymorphism to the actual mechanisms of evolutionary change. This, after all, was the primary biological problem of the day, and it was a subject that he had thought and written about, though less directly, in the first part of the *Studien* and earlier in his examination of Wagner's migration theory and Hilgendorf's work on *Planorbis*. What is not immediately clear is why he chose such different material and subjects for the pursuit of his goal: that is, why he chose the larval forms and markings of sphinx moths and a vertebrate he had previously not worked with; why he showed a concern for phyletic forces, then wrote a general statement about mechanical conceptions in biology. The logic must be found both in the sequence of his research objects and in the contemporary philosophical debates of the time.

The social framework that provided him the opportunities, inspiration, and support, however, may be seen in the administrative details of Weismann's career, the growth of his profession, and above all his family network. So before digging once more into the details of his research activities, it is worth briefly viewing this larger picture. In May 1873, Weismann had completed his seasonal dimorphism experiments with butterflies and began another long-term study, turning this time to the daphnids of the Bodensee. With the full endorsement of the medical faculty, he had become a full professor for zoology in the philosophical faculty (*ordentlicher Professor der Zoologie*). At the same time, he remained a member of the *ärztlichen Vorprüfungskommission,* that is, the preliminary commission for the testing of prospective physicians, a position he held for the rest of his career. This transfer in Freiburg was important for wresting zoology with its many unique scientific issues from the medical sciences, which were appropriately focused on vertebrate anatomy and morphology. By the summer of 1874, after ten years of protecting his eyes, Weismann was again able to use a microscope. He put this to good use examining the germ layers and germ cells of the daphnids he was collecting. (These subjects will be the focus of Chapter 6.)

During the same summer semester, Weismann taught for the first time a "short summer course" of one hour per week on evolution.[1] The course served as a trial run for the full winter semester course, first taught in 1875–1876 and yearly thereafter for the rest of Weismann's teaching career. Ultimately this course led to his two-volume *Vorträge über Descendenztheorie* (1902), which will be the subject of a much later chapter. In 1875 and 1876, he was deeply immersed in the investigation of the daphnids of the Bodensee and was assembling a series of essays, which, along with his previous essay on "Seasonal Dimorphism," collectively became his *Studien zur Descendenztheorie*.[2] It is to these four later *Studien* we turn in this chapter, focusing on the essays that occupied him during this interlude in his life.

Weismann and the Nature of Phyletic Series

If a temperature-dependent chemical action appeared improbable, then what would indirect influences entail? Weismann, of course, thought of natural selection, but there was a problem. All his experiences with diurnal Lepidoptera indicated that the upper surfaces of the wings were not consistently exposed while the butterfly was at rest, and when the butterfly was in flight, the fine details of markings and color seemed unlikely candidates to give one specimen an advantage over another. Although he appreciated such a possibility in other circumstances, he was disinclined to see the operation of Darwin's sexual selection in fashioning the fine winter and summer details that seasonal dimorphism appeared to enlist; "in my mind," he opined, "Darwin ascribes too much power to sexual selection when he attributes the formation of secondary sexual characters to the sole action of this agency."[3] This was as far as Weismann could go at this stage of his career; he repeatedly wrote in this monograph of the "physical constitution of the species," a built-in feature of the organism that made every species, every variant, and every polymorphic regularity a unique form.[4] When he thought in evolutionary terms, he thought of one form replacing another. The numerous intermediate forms that can be produced artificially, he added, "appear to me to furnish a further proof of the gradual character of the transformation. Ancestral intermediate forms can only occur where they have once had a former existence in the phyletic series."[5]

A "phyletic series" was always in the background of Weismann's thinking. It came to the fore when he distinguished between primary and secondary forms of a polymorphic cycle. It was shared with Edwards and soon after his readers when he associated cold weather forms—let us say, *Araschnia levana* or *Papilio telamonides*—with the first to appear phylogenetically—*levana* in the northern climes with the receding of the glaciers, *telamonides* as a southern form that split into *walshii* in the north and *marcellus* in the south as temperate climes reappeared in North America. It became central in his deliberations when he communicated with "my friend," the commercial lepidopterist Otto Staudinger, about the differences between the Arctic and Alpine forms of *Pieris napi* var. *bryoniae*, both of which he considered primary, and the warmer regional form *P. napi* and the summer form *P. napi* var. *napeae*.[6] The retreat of the glaciers in the northern hemisphere and the retreat of the cold weather forms of butterflies into the north or their retreat into the high mountains were long-term changes that became very much a part of his biogeographical and evolutionary perspective.

If Weismann's monograph on seasonal dimorphism had ended at this point, it might from a modern perspective be viewed as a classic in early experimental natural history. Science, however, rarely marches forward in a unilinear fashion that is convenient for later historians. The second half of the monograph embraced a contemporary concern about modes of cyclical heredity and their origins. The subject was central in the context of Darwin's *The Origin of Species* and was important to Haeckel, Claus, Leuckart, Kölliker and Lubbock, and Darwin. Weismann was no exception.

In Weismann's day, this meant participating in a contemporary conversation about the alternation of generations and metamorphosis as well as one about descent. In Chapter 2, we reviewed the onset of the first half of this conversation, which began in 1819 when Adalbert Chamisso coined the expression *Generationswechsel* to describe what he interpreted as an alternation between budding and sexual forms of reproduction in pelagic tunicates. It had taken center stage with the publication of Steenstrup's book on the alternation of generations, and had become more complex with Owen's investigations of parthenogenesis and Leuckart's comprehensive 300-page survey of different modes of reproduction published in 1853. An additional twenty years of studies in natural history had only further revealed the incredible diversity in patterns of reproduction in the animal world, but they had also complicated these considerations by adding an evolutionary backdrop as an essential context. What inevitably took place was a conversation about development and evolution.

When Leuckart completed his monumental survey on reproduction, he devoted only the last forty of the 300 pages to discussing the forms of asexual reproduction and polymorphism. As we saw in Chapter 2, the focus of his monograph had been on surveying the efficiency of various modes of sexual reproduction. One of Leuckart's goals in the last forty pages had been to examine and contrast the sexual and asexual forms of reproduction; as he did so, he arrived at the conclusion that the physiological mechanisms of adaptation to the diverse conditions of life rendered it possible to imagine a transition in reality between these two reproductive modes. A continuum existed between forms of budding and germ production on the one hand, and the alternation of generations, polymorphism, and sexual reproduction on the other. (To the anguish of his mentor Rudolph Wagner, this move appeared to threaten the essential significance of sexuality and the two sexes.) This continuum was established by recent discoveries of parthenogenesis, particularly but not exclusively by himself and von Siebold, among termites, gall wasps, scale insects, bagworm moths, honeybees, other Hymenoptera, *Daphnia, Limnadia,* arachnids, mites, and possibly among certain gastropods. "The history of generation," Leuckart pointed out, "shows in a clear way how the role which the sperm plays in fertilization has become even more restricted by the advance of science. Initially the sperm was seen as a young germ, which used the egg merely as a cradle and the yolk for sustenance; later it was seen as an element which had to a certain extent an equal status with the egg, which joined the egg and only with the union produced the germ; now the egg has become the germ even though something more is perhaps needed for development."[7]

Moreover, on the eve of the publication of *The Origin of Species,* parthenogenesis appeared to be the key to understanding the relationship between the alternation of generations and sexual forms of reproduction. As early as 1851, Leuckart had been prepared to contrast the alternation of generations, or that which he called "metagenesis," on the one hand, with a sequence of changing sexual generations, which he called "heterogenesis," on the other. The contrast was not always clear, yet these two categories lived on in name and often in spirit into the 1860s and 1870s, with, of course, the theory of descent by then making the conversation more complex.[8]

One strategy to clarify the difference in the post-Darwinian period was seized by the Nestor of German histology. Albert Kölliker belonged to the generation who had already achieved professorial status by the time Darwin published *The Origin of Species,* by which time he had become head of both the Institute for Physiology and Comparative Anatomy and the Institute for Anatomy at Würzburg.[9] In his early years, Kölliker had spent much of his career in descriptive anatomy and the embryology of both invertebrates and vertebrates. As an experienced investigator of marine organisms, he had a deep appreciation for the comparative approach to morphology and extensive experience with the process of reproduction. In 1848, he had also joined Karl Theodor von Siebold in founding the *Zeitschrift für wissenschaftliche Zoologie,* with the intention of establishing the up-and-coming field of zoology on scientific principles. As Nyhart has pointed out, the prospectus for the new journal placed a premium on works that sought to generate the laws of nature by connecting different kinds of phenomena. In the context of this chapter, the same prospectus spoke of the "cultivation" of "morphology, comparative anatomy, and histology . . . bordering . . . the physical laws of life phenomena," but it made no mention of experimentation. Comparative techniques were the operative—and in their own way profound—method of proceeding.[10]

In 1864, Kölliker presented an address to the Physikalische und Medizinische Gesellschaft von Würzburg on Darwin's theory.[11] Up to a point, Kölliker was complimentary of Darwin's accumulation and organization of a large body of zoological information, but he recited many of the standard criticisms of *The Origin of Species* that had already become commonplace, such as the lack of transitional forms, the lack of transitions in the fossil record, no demonstrable production of a new species through natural selection, the absence of useful variations within a species, the continued presence of the simplest forms despite a general evolution to higher forms, and the absence of demonstrated sterility between varieties. These and many other like-minded criticisms have revealed to modern students of Darwin how great the gulf was between Darwin's achievement and the scientific framework of so many of his contemporaries. Above all, it was ironic that Kölliker objected to Darwin's implied teleology by insisting that there must be a purposeful production of useful characters in the organism for the Englishman's system to work at all. Darwin, Kölliker argued, could not sustain descent on the basis of accidents alone.

Kölliker, however, was not a man to deny descent itself. By way of contrast, he offered an alternative explanation, which he called his "theory of heterogenous reproduction" ("Theorie der heterogenen Zeugung"). It was a physical explanation of descent through an identity with heterogenesis—the substantial change between one and a following sexual generation. In both cases, a general law of development, heterogenesis, was directly converted into an explanation for phylogeny and when under unspecified circumstances a fertilized egg proceeded to develop into a higher form by means of parthenogenesis. This produced a different form—not by "saltation" (not his terminology) but "only very gradually and slowly" would a higher, more evolved species emerge.

Kölliker's informal theory rested on a twenty-year bonanza of zoological and botanical research that had shown the enormous range of possible reproductive modes and varied forms in species of both kingdoms. The classical example of varied and complex reproduction for zoologists was the hydrozoans and scyphozoans; yet tremadodes, echinoderms, and arthropods all had recently revealed life's inventiveness in producing a range of contrasting forms within a single species. Even sexual dimorphism and insect castes emphasized Kölliker's point that varied reproduction could produce different forms. With reference to the hydroid polyps, one of the founders of the new *wissenschaftliche Zoologie* could wax both expansive and romantic:

> When one surveys the entire row of facts about these organisms, one cannot resist the thought that a creative act has proceeded here and possibly still runs on in the manner that hydroid polyps bring forth simpler and higher medusae or as I designate it heterogenetic reproduction: now from the simple polyps, which reproduce themselves directly through eggs, up to the medusae, which form themselves equally directly out of eggs, there exists an almost continual row of intermediate forms of reproduction.[12]

Kölliker's theory of heterogenous reproduction had little impact on what followed, and to Weismann's skeptical mind it seemed suspiciously supportive of an internal motive force in the evolutionary process. Weismann rejected both Kölliker's theory of heterogenous reproduction and Nägeli's later perfecting principle (*Vervollkommnungsprinzip*) for invoking a directional internal force.

Haeckel on Reproduction

Haeckel's *Generelle Morphologie,* published two years later, appeared to offer a more congenial way of bringing the many modes of reproduction into an evolutionary framework. After all, Haeckel depended on sorting out the various levels of individuality—cells, germ layers, organs, bions (i.e., the conventional organism), corms or colonies—and their various modes of reproduction, including sexual, asexual, metamorphosis, and alternation of generations, to provide a mechanism for the phylogenies of evolution.[13] In all, he listed and described at least twenty-two such patterns of bion multiplication among both plants and animals. He gave each its distinctive Germanized name and, as was often his wont, a corresponding Latinized version.[14] On the progression of reproductive modes hung a demonstration of the physiological division of labor and diversity of life. At the time, Haeckel did not echo what Kölliker had attempted to do, but later (1877) he was to take a swipe at Kölliker's theory of heterogenesis in the context of berating physiologists in general for their lack of understanding and attention to evolution theory. His attention to the issue may be seen as a general reflection of how in German zoological circles it was important to sort out the varied patterns of reproduction for the benefit of evolution theory. For Haeckel, such a classificatory exercise was central to his life's endeavor. "The objective consideration of Strophogenesis and its comparison with Metagenesis is extremely important and instructive, particularly for an understanding of the parallel between ontogeny and phylogeny."[15]

Without reciting the long list of Haeckel's distinctions, it is worth noting that Haeckel drew an interesting comparison between what he called "Generationswechsel" or "Metagenesis," on the one hand, and "Generationsfolge" or "Strophogenesis," on the other. The first category denoted that which Leuckart and others had considered the orderly alternation of sexual and asexual generations. Its formal cognate, "metagenesis," used both in German and English literature, implied a contrast in the forms of the asexual and sexual generations. "Generationsfolge" simply referred to a sequence of different levels of forms in a single individual from the cleavage of the egg, through other orders of individuality, ontogeny, and metamorphosis to the sexual generation at the end of the cycle. Strophogenesis appears to be a word of Haeckel's own devising; it was not used in Haeckel's *Natürliche Schöpfungsgeschichte,* and it appears to have passed quickly into oblivion.[16] He did not pursue Leuckart's cycle of heterogenesis—perhaps Kölliker's use of the expression had irrevocably spoiled it for him.

Carl Claus, a close contemporary of Haeckel and Weismann and like the latter a student of Leuckart, had become in 1872 a full professor of zoology and comparative anatomy at the University of Göttingen and director of its

Zoologisch-Zootomisches Institute. He excelled in the comparative morphology of marine invertebrates. As Nyhart has pointed out, Claus had been an immediate convert to Darwinism with a strong orientation toward understanding development in its functional context rather than in a formal, descriptive manner. This functionalism, an extension of his teacher's scientific approach to zoology, soon led him to become a severe critic of Haeckel's gastraea theory.[17] In the second edition of his famous textbook on the foundations of zoology, Claus reinforced and refined a more conventional taxonomy of reproductive modes.[18] It was an evolutionary taxonomy of reproduction premised on the rise of sex through multiple stages, which started with the simplest form of direct division ("Theilung") and moved on to budding, germ production, and the appearance of pseudo-ovae. First with the emergence of hermaphroditic sexual reproduction and the later appearance of separate male and female individuals, sexual reproduction finally arrived on the biological scene. Part of Claus's lesson included paragraphs on parthenogenesis, different forms of embryological development, metamorphosis, and the alternation of generations, all of which, when combined with the multifaceted world of sexual reproduction, seemed to sort themselves into two basic types: (1) metagenesis, or the cyclical alternation of generations in a number of specified patterns of sexual and asexual individuals, and (2) heterogenesis, or the production of a sexual form from another, contrasting sexual form or from an alternation of a sexual and parthenogenetic reproduction. Claus essentially ended up where his mentor had twenty years earlier, but where Leuckart had rationalized the continuity through his beliefs in the power of mechanistic adaptations, Claus as an evolutionist with strong leanings toward the mechanism of natural selection understood the complexities in the animal kingdom well enough to recognize that these two kinds of cycles were not distinct types but could be separated by whether parthenogenesis, "concubito sine lucina," involved the development of a nonfertilized true egg or a germ with the appearance of an egg (a pseudo-ova).

Heterogenesis and alternation of generations obviously exist in close relationship, but distinguish themselves by the asexual and sexual propagation of the intermediate generations. Because the boundary between germ and egg cell is obliterated by parthenogenesis, both developmental forms are not sharply separated from each other; thus, for example, the propagative manner of aphids can be located as well in heterogenesis (the viviparous aphids are a special generation of parthenogenetic females) as in the alternation of generations (the viviparous aphids are asexual, self-reproductive nurses).[19]

Complex! Uncertain! Distinctions that vanished under further scrutiny! Reproduction in its many forms, spread across a world of apparently limitless biological diversity, continued to frustrate the organizing skills of even the most careful naturalist in the immediate post-Darwinian age. By the fourth edition (1880) of his *Grundzüge der Zoologie,* Claus was clearer about his identification of reproduction modes, but he remained equally adamant that the distinction between metagenesis and heterogenesis in certain cases is problematic.

Weismann and Cyclical Generation

Fresh from his collection of and experimentation with the seasonal dimorphism of Lepidoptera, Weismann could not help but become involved in the conversation on reproduction as the status of cyclical generation was one of its central themes. And, after all, he too had been a student of Leuckart. Weismann's contribution, however, appears at first to be somewhat out of place. The seasonal, geographic, and climatic polymorphism that lepidopterists were reporting was almost exclusively the features of wing color and markings. These could be significant as attributes for species identification, but except for a few rare cases of parthenogenesis (e.g., *Solenobia,* the bagworm moth) such attributes did not appear to affect the basic morphology or physiology of these insects. In the bigger picture, were these attributes really meaningful? (In 1874, Weismann considered such wing variations as functionally insignificant although taxonomically important.) As a natural historian, Weismann considered the seasonal changes in butterfly markings to be closely correlated with temperature changes and the overall climate. As a trained embryologist he seemed equally at home in discussing seasonal dimorphism in ontogenetic terms and in associating such phenomena directly with the contemporary enthusiasm to link reproductive modes to an evolutionary process. At the outset, he complimented Haeckel's more recent analysis of progressive and regressive metagenesis, but he preferred to return to Leuckart and Claus's distinction of metagenesis and heterogenesis.[20] Briefly, for them metagenesis entailed an alternation of generations that has arisen when different ontogenetic stages acquired the ability to propagate. Heterogenesis denoted an alternation of radically dissimilar sexual generations. (See Plates 4 and 5.)

Even as he agreed with Claus that a sharp distinction could not be drawn between these two series of reproduction, Weismann was not at all clear how seasonal dimorphism fitted into the spectrum. Wallace, who had defined seasonal dimorphism in 1865, had considered it a form of the alternation of generations, but his use of that expression was very different from Steenstrup's and his continental commentators. Haeckel had at least tried to associate the two modes of alternation of generations with contrasting phylogenetic series of progression toward or retrogression from a bisexual state. Although appreciating his friend's evolutionary inferences, Weismann returned to Leuckart and Claus's descriptive terminology with the intention of reexamining it in terms of the phylogenetic origins of the two sequences. Appealing to specific examples, Weismann recognized that there might be three quite different phylogenetic origins of metagenesis: (1) when the larvae of sexual adults acquire the capacity to produce further larvae by means of asexual reproduction, such as *Cecidomyia;* (2) when some of the sexual adults regress to asexual producing larvae, such as trematodes; and (3) when in colonial polymorphism asexual and sexual forms develop independently. Weismann recognized that all of these phylogenetic scenarios were at best a guess because they all involved a conjectured evolution of a nurse generation.

Nevertheless, none of these scenarios seemed to match the phenomenon of seasonal dimorphism of butterflies; so, contrary to Wallace's earlier suggestion, Weismann felt he could rule them out as an exemplar of metagenesis.[21] First, as in the case of the recently celebrated *Cecidomyia* larvae, the larval stage might have acquired the "nursing" capacity to produce additional larvae asexually. This seemed to Weismann a secondary adaptation because closely related species do not have this capacity. In short, an adaptation in metamorphosis had turned the entire ontogeny into metagenesis. Second, in the cases of hydromedusae and trematodes, the universality of "nursing" stages suggested that this asexual mode of reproduction originated phylogenetically before or became adaptive when a later phylogenetic stage in the metamorphosis process assumed the sexual mode of production. Finally, a physiological division of functions in a colonial organism, such as hydromedusae, might become metagenetic through a divergence of asexual and sexual individuals. In this case, metamorphosis and metagenesis would phylogenetically emerge together.

Weismann, however, wished to extend the concept of heterogenesis to include alternation of sexual and parthenogenetic reproduction, as appeared in the case of the summer multiplication of aphids and the winter eggs of the daphnid *Leptodora hyalina*. In these and many other examples, parthenogenesis must be construed as a sexual form of reproduction—or in Owen's description, repeated by Leuckart, a "lucina sine concubitu."[22] Would heterogenesis be a more appropriate ascription for seasonal dimorphism? If it were, Weismann felt a significant conclusion followed, for "a knowledge of its mode of origination must therefore throw light on the nature of the origin of heterogenesis."[23]

Both seasonal dimorphism and heterogenesis, to be sure, were understood to involve a cyclical change in sexual generation correlated with alternating external conditions, but here again, there was a problem. On the one hand, Weismann had argued that the changes associated with the butterflies he had studied appeared small and nonfunctional but correlated with climate. His experiments had demonstrated that the influence of temperature was asymmetrical—the southern and summer forms could be partially or completely converted to northern and winter forms, but not the other way around. The changes were predictable and seemed to be directly caused by his manipulations. On the other hand, Claus had argued that the changes in heterogenesis, exemplified by Leuckart's *Ascaris nigrovenosa,* were substantial and adapted "the organism to important varying conditions of life."[24] A conclusion, by the way, with which Weismann concurred. This signified, Claus continued, that they were indirectly caused by the external conditions and explainable by natural selection. (Again, a conclusion to Weismann's liking.) Those insignificant changes in color and markings on butterfly wings, however, were in Weismann's mind not functionally important, not indirectly caused by natural selection.

> When I raise a distinction in the nature of the changes between seasonal dimorphism and the remaining known cases of heterogenesis, this must be taken as referring only to the biological or physiological result of the change in the transformed organism itself. In seasonal dimorphism only insignificant characters become prominently changed, characters which are without importance for the welfare of the species; while in true heterogenesis we are compelled to admit that useful changes, or adaptations, have occurred.[25]

Changes in seasonal dimorphism must be the products of the direct action of the conditions of life. It was at this juncture that Leuckart's two students could not agree. Weismann's studies had convinced him that changes in butterfly dimorphism were different from the adaptational changes traditionally associated with heterogenesis. He was focused on the appearance of a kind of variation that did not fit Claus's Darwinism. Did this mean there were no lessons to be extrapolated from the former that spoke to the origins of the latter?

Weismann believed there were. He felt that his work, emphasizing the origins (i.e., the phylogeny) of the contrasting generations at least forced a readjustment in the Leuckart/Claus functional definitions. By focusing on the changing ontogenies rather than on a formal contrast of the morphology of reproductive modes, as had many systematists, or on the adaptations caused by the changing conditions of life, as had Leuckart and Claus, Weismann maintained that he could join such apparently different cycles as found in seasonal dimorphism in butterflies, the summer and winter eggs of the daphnid *Leptodora hyalina,* the well-known cases of asexual generation among many Aphididae, and other invertebrate cycles under the same developmental umbrella. The development of parthenogenesis was a particularly telling but not essential stage in the heterogenetic process. Insignificant but taxonomically important generational changes could now be included as well. "A separation of the modes of cyclical propagation according to their genesis appears to me—especially if practicable—not alone to be the greater value, but the only correct one."[26] In this respect Weismann was really following in the footsteps of Fritz Müller and Haeckel.[27]

Weismann closed his long monograph in a dismally uncertain way. He left metagenesis and heterogenesis behind as though he had never discussed them at all. He tried to explain in the simplest of causal speculations how seasonal dimorphism in butterflies actually came about. He had already eliminated natural selection by declaring the color and marking changes as nonfunctional. He was aware that Darwin invoked sexual selection to explain such attributes, but he could not be satisfied that it applied to seasonal differences. (Weismann was later to change his mind about the role of sexual selection in influencing wing colors and markings.[28]) This left the influence of the changing climate.

As mentioned earlier, Weismann's experiments revealed an asymmetrical influence of temperature on the winter (primary) and summer (secondary) forms; the secondary form could be experimentally converted at the pupal stage to the primary but not vice versa. So temperature changes might explain reversion, but they could not in themselves be the agent of a permanent change to a new species. Something internal to the organism, some "internal constitution," he felt, must also be involved, and when it is affected, new patterns of color and markings appear on the wings; the older pattern, however, must remain intact in a latent state.[29] At this point, Weismann understandably began to spin his wheels. He had neither the information nor the vocabulary to articulate a fuller explanation. He resorted to his vision of amixia, forged at the time of his confrontation of Moritz Wagner's migration theory; he spoke of "innate tendencies," which react to "external factors" and "represent the life of the individual as well as that of the species."[30] He maintained that "the direction of development transferred by heredity is to be regarded as the ultimate ground both of the similarity and dissimilarity to the ancestors." Weismann closed his account with further acknowledgment of his orientation: "I can agree with Haeckel's mode of expression, and with him trace the transformation of species to the two factors of heredity and adaptation."[31]

Haeckel, whose gastraea theory had been highly praised earlier in the year by Weismann, read his monograph and answered in kind. Haeckel's words are revealing about their contrasting research styles. "That is the right way," he confessed to his friend,

> to make the theory of descent plausible to a very large circle, particularly among them the stubborn systematists. I am convinced that you work wonders with it and more than I do, for example, through many phylogenetic speculations. I can only implore you to proceed unswervingly on this highest agreeable path. The richest material lies in masses all around us and precisely the insects appear to me to be especially suitable for it.[32]

Caterpillar Markings and Phyletic Parallelism

Weismann published two complementary monographs in 1876. He continued to use Lepidoptera as the object of study, but in contrast to his examination of polymorphism in adult butterflies, Weismann refocused his attention on the larval stages of hawk moths instead. The consequence of this reorientation was that the study of larval stages propelled him into the center of the contemporary morphological discussions of the relationship between ontogeny and phylogeny.

Weismann almost never used Haeckel's language, but he felt certain that a recapitulation of antecedent ontogenies was an undeniable feature of all phylogenetic lineages. Nonetheless, in a passage at the end of the third monograph (one to be examined in the next section) he placed quotes around Haeckel's expression "fundamental biogenetic law" and attributed the phrase to both Fritz Müller and Haeckel. His own informal definition leaves the impression that the law was hardly structural and comprehensive. "The ontogeny comprises the phylogeny," he reported, but "more or less compressed and more or less modified." Haeckel would not have disagreed but might have been disappointed that his friend, after having praised his gastraea theory in 1876, did not stress more the "fundamental" aspects of his invention.[33]

Weismann turned again to the Lepidoptera as organisms on which to study the factors of evolution. Unlike in his previous work on adult butterflies, he dealt exclusively with the caterpillars of hawk moths, for their sizes and ease of rearing made them a suitable subject to study. They alone provided an unparalleled opportunity to study surface markings during development and to examine inductively the interface between development and evolution. Ultimately Weismann had to limit his studies to the larvae of the globally diverse family of Sphingidae, or sphinx moths. They provided him with complex and contrasting markings at subsequent larval stages and reflected the total life cycle of the caterpillars and their conditions of life.

In their own way the delicately patterned, horned sphinx larvae provided the same booklike library for the attentive reader that Bates and Wallace had praised in the opened wings of the imagoes. Unfortunately, the full sequence of larval stages of most species was not well known at the time. The classical literature on caterpillars, studies by such renowned naturalists as the Nürenberger miniaturist Rösel von Rösenhof, Jacob Hübner, and two English civil servants in the East Indian Company, Thomas Horsfield and Frederick Moore, provided much information and many illustrations.[34] In addition, Weismann's contemporary, the commercial lepidopterist Otto Staudinger, placed his incomparable Dresden collection at Weismann's disposal. Professor Gerstäcker of Berlin, Wolfensberger in Zürich, Riggenbach-Stählin in Basel, and Edwards in Coalburgh also provided material. Neither books nor collections, however, systematically covered the full life cycle of Lepidoptera in the way Weismann needed. Even the vocabulary for describing caterpillar markings had to be fashioned.[35] Furthermore, only a few of the species could be raised from eggs, so many of the earliest stages remained undocumented.

One cannot overestimate the tedious business Weismann must have experienced in trying to work out even a half-complete metamorphosis of the five to seven often contrasting larval stages of a large number of sphingids by combing the fragmented literature, studying incomplete illustrations, and examining those specimens he could get his hands on. More often than not, Weismann could not establish empirically the full developmental patterns he was after. Much of what he eventually presented in two lengthy monographs had to be arrived at deductively through his conviction that recapitulation in the form of terminal additions, backward transference, modification, and possible elimination of markings provided a fair mechanical picture of the phylogenetic past. Nevertheless, the effort was unavoidable if Weismann was to arrive at any conclusions about the course of phylogeny and the factors of evolutionary change—which was his real goal.

There are indications that he started his caterpillar studies in 1869.[36] They suggest that this project had begun long before his studies of seasonal dimorphism had been completed, before Leopold Würtemburger had documented through the paleontology of ammonites the backward transference and condensation of traits, and even before Haeckel had made his "fundamental biogenetic law" an enduring object of his personal biology. How many of Weismann's assumptions were derived from these colleagues and how much was due to their shared evolutionary and mechanical convictions is hard to say. Whatever the balance of influence may have been, there can be no doubt that Weismann was working at the time on the forefront of developmental and evolutionary theory, and that his research was uniquely his own.[37]

Weismann's overall strategy was to establish the phylogenetic origins and evolution of the superficial markings found on sphingid larvae. Were they also nonfunctional, as he had considered the trivial traits on butterfly wings? Because he was dealing with larval traits, they could not be attributed to sexual selection, but were they the product of natural selection or the laws of growth? There still remained the option of considering them the product of an "innate power of development," a vital phyletic force of some kind. Weismann's monographs in part two, however, are to be read as responses to the teleological theories being trumpeted by Kölliker, Nägeli, von Hartmann, and others. The only way to eliminate those nonmechanical options, however, was to bring the argument down to the nitty-gritty of specific phylogenies, and this meant distinguishing between apparent and blood relationships: "For this reason alone, larvae and pupae would have an important bearing upon the establishment of systematic groups."[38]

For ninety pages, Weismann presented his data. He had investigated ten genera and at least forty-four species from around the world. He described, as far as the record permitted, the development of markings in four to six stages or moltings with the hope of determining on the basis of the recapitulation of the markings the ancestor of a given genus. Such a project was relatively straightforward because in the larvae of sphinx moths, indeed in those of most moths and butterflies, the caterpillar at all its stages possesses a standard external structure of a head and fourteen additional segments which provide reliable reference points during development as well as during the disruption of ecdysis or shedding of the outer cuticular layer. Counting the head as the first segment, each of the successive three segments possesses a pair of clawed legs that correspond to the three thoracic segments and six legs of the future imago. Segments four and five are suspended, allowing the front of the caterpillar to twist and turn while traveling from leaf to leaf or in defense. Segments six through nine (excepting in looper moth larvae, inchworms) are supported by eight stubby prolegs that firmly clasp the surface on which the larva feeds or travels; segments ten and eleven are again suspended. Segment twelve is the terminal segment, which includes an anus and an anal proleg. With the sphingids, segment twelve also contains a dorsally located caudal horn used for defense or concealment. In its general form, the caterpillar conforms to its most important function in the life cycle of the organism: to eat and grow, to eat more and grow more, and to eat yet again and grow even more.

As mentioned previously, Weismann's efforts to track the surface markings on each segment of each stage of the many individuals of each species rarely provided a full sequence. Piecing together the multiple sources of his data—the investigation at his own institute, the descriptions and illustrations by other lepidopterists, and the museum specimens made available to him—he was able to draw up some generalizations about the phylogeny of the various markings. While doing so, he first identified four different elements of surface markings to be tracked: (1) the presence of a background color but the absence of any markings, (2) the presence of longitudinal stripes, which were composed of either a simple subdorsal line or in combination with a dorsal or spiracular line, (3) the expression of oblique stripes, and (4) the appearance of paired eye-spots or ring-spots, which might extend into rows.[39]

Three principles appear to have dictated Weismann's interpretation of these surface markings. The first he considered "as almost self-evident," for it consisted of the straightforward claim that developmental changes proceed from the simple to the complex. The second principle maintained that new characters most often, but not always, appear in the last stage of development. The third held that new characters are transferred during the evolution to earlier stages of development and in so doing force a compression and eventual disappearance of phylogenetically and ontogenetically older characters.[40] Logically it is possible that the last two principles could conflict with the first, as would become apparent in cases of degeneration of structures or parasitic development, but this was not an issue with sphingid larvae. Progressive complexity, terminal additions, and condensation drove the recapitulation process as he understood it. At one point, he was tempted to call these principles "laws." They appear to be conclusions drawn from his experience as an embryologist, his understanding of phylogeny, and what Weismann often referred to as the laws of growth, or Darwin's law of correlation. In reality, as he himself recognized, they were not laws in a profound sense but regularities of development that were useful to determine the affinities and phylogenetic lineages of his larvae.

When hatched, the ideal larva was small, naked of markings, and possessed of a solid green or brown coloring. As it grew and molted, it first developed longitudinal stripes and then oblique stripes. Eye-spots and ring-spots appeared at later stages and were associated with but not necessarily derived from either the longitudinal or oblique stripes. The color, the details of the markings, and the temporal placement and extension of all the markings were stage and species specific. Only a few species exhibited even a partially complete spectrum in their ontogeny; depending upon the complexity of their last developmental stage, they were considered phylogenetically the youngest. Weismann's mapping of lineages seems to be akin to the strategy of modern dendrologists, who must match up ring patterns in a large collection of living trees and preserved logs to achieve a chronology larger than offered by any single cross section.

For example, he was successful enough to demonstrate the very different development of the markings in the genera *Chaerocampa* and *Deilephila* on the basis of their adult morphology though they had been traditionally lumped into a single genus. "The developmental history of the caterpillars shows that there is a wide division between the two groups of species."[41] In brief, he determined that the "ring spots of *Deilephila* first originated on the segment bearing the caudal horn, and were gradually transferred as secondary spots to the preceding anterior segments"; with *Chaerocampa,* the "the eye-spots proceed from the subdorsal line, but they appear on two of the front segments and are transferred to the *posterior* segments."[42] Such a conclusion could only be drawn by examination of the minutiae of larval development—spot by spot, stage by stage, and species by species.

A reform in lepidopteran taxonomy based on development, however, was only a minor consequence of his investigations. More important was the strong functional component. On the basis of the larval feeding behavior and predation experiments, Weismann argued that given markings have utility at certain developmental stages. The youngest larvae blend naturally with the ground vegetation upon which they feed; they are protected by their solid green or

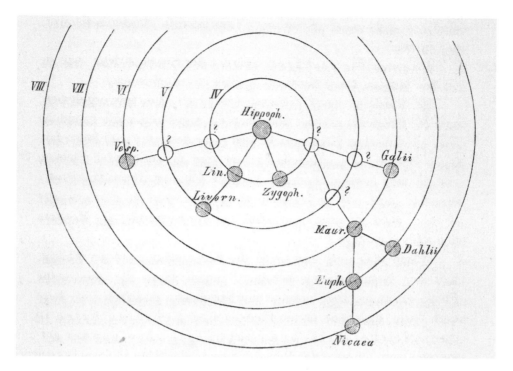

Figure 5.1. Hypothetical presentation of the possible phylogeny of the genus of *Deilephila*, drawn by Weismann while writing on the ontogeny and phylogeny of markings in Sphinx caterpillars. Weismann, *Studien*, Pt. I., 1875, p. 117.

brown coloration. As the larvae grow in size and as they feed upon taller grasses, the longitudinal stripes conceal them from predation. As the more advanced larvae of many species relocate their feeding to broadleaf plants, oblique stripes that mimic the veins of leaves protect them again. Weismann went to considerable lengths in discussing the utility of ring-spots and eye-spots. He argued that such markings would alarm or provide a warning about an unpleasant taste to predators; in very simple experiments, he tested a series of larvae with and without rings or eye-spots in pens with insect-loving wall lizards. He concluded, although the data were not presented in the monograph, that the rings and eye-spots did deter predation.

From a modern perspective, Weismann's presentation seems tedious, excessively provisional, and casually documented. He anticipated theoretical objections that sent him into lateral discussions, and in the narrative of his argument one can easily lose track of the central theme. He explored variability and polymorphism in larvae; he continually weighed the pros and cons of utility; he posed the possibility of changes in function of given features during development; he appeared to endorse the operation of natural selection and mechanics of growth throughout the larval stages; and he argued that older markings often lose their utility in their backward transference to earlier stages. Throughout, he identified recapitulation as the primary description of the linkage between ontogeny and phylogeny, but he failed to adopt Haeckel's "fundamental biogenetic law." Finally, he spoke vaguely about a constrained though unspecified constitution of the body that produced variations:

> I long ago insisted that it should not be forgotten that natural selection is, in the first place, dependent upon the variations which an organism offers to this agency, and that, although the number of possible variations may be very great for each species, yet this number is by no means to be considered as literally infinite. For every species there may be *impossible* variations. For this reason I am of the opinion that the physical nature of each species is of no less importance in the production of new characters than natural selection, which must always, in the first place, operate upon the results on this physical nature, *i.e.,* upon the variations presented, and can thus call new ones into existence.[43]

This "opinion," emphasizing the importance of variations, was to play an increasingly important role as his research proceeded.

In many respects, I find the nautical metaphor of "warping" a ship across a submerged coral reef an apt description of Weismann's narrative style of argumentation. The anchor is set fore, and the sailing ship is winched ahead along the most plausible channel. The anchor is then reset, perhaps at a different angle, and the forward winching is continued. Weismann faced a formidable reef of objections having to do with parallel evolution, imperfect morphological analyses, polymorphism, and the uncertain utility of many markings. There were, however, two general results to his warping through the reef of scientific discourse. First, the minute details of growth and development of sphingid markings, in his judgment, reinforced the importance of external conditions—both direct and indirect—as mechanistic "impulses" on the transformation process. Second, his study precluded the possibility of innate vital forces. At this point, his was an argument by exclusion—every feature he identified might reasonably be explained in recognizable mechanistic terms, but he could not eliminate the possible action of a phyletic force.[44]

His next monograph was more of a theoretical document that dealt with the taxonomic problem that Lepidoptera and other insects presented with the incongruities between a classification system based on the developing larvae on the one hand, and on the fixed imago form on the other. At the time, most classification decisions were determined by the markings and colors of the adult wings, and, at the subordinal level—that is, at the point when the distinction between butterflies (Rhopalocera) and moths (Heterocera) was made—on the form of the adult antennae. After increased investigation of insect development, it was recognized that the holometabolous insects, that is, insects that undergo a radical reorganization during a "quiescent" pupal stage, including Lepidoptera, Hymenoptera, and Coleoptera, produce two very different forms for the same organism. How should this difference be figured into the taxonomic calculation? The initial paradox, though occasional, seemed clear. For example, as Weismann pointed out, if one produced a classification of larvae of the subfamily Nymphalinae, one would end up with one set of taxonomic affinities; if one produced a classification on the basis of imagoes, one ended up with another.

The paradox was actually more complex and had deeper ramifications. It turned out that within the subfamily Nymphalinae the two genera *Apatura* and *Nymphalis* contain larvae that are closer in form to larvae of genera in the subfamily Satyrinae than to the larvae of the other genera of the Nymphalinae—this despite the fact that their imagoes are closer to the imagoes of the other genera of Nymphalinae. For Weismann, the issue was more profound: "It cannot therefore be well denied," he maintained, "that in this case the larvae show different relationships to the imagines." Does one raise these two genera to the level of a third subfamily?[45] The problem was one of the weighting of attributes, which is the bugaboo feature of a Linnaean-based system. In the post-*Origin* period, the problems in such a system were magnified by the general acceptance of evolution theory. Weismann and his contemporaries desired a "natural system" that reflected the phylogenetic past, and Weismann explained his general frustration with taxonomists who disregarded this dimension in a rambling passage that in retrospect seems again like warping his ship across the reefs of real and potential adversaries:

If the "natural" system is the expression of the genetic relationship of living forms, the question arises in this and in similar cases as to whether the more credence is to be attached to the larvae or to the imagines—or, in more scientific phraseology, which of the two inherited classes of characters have been the most distinctly and completely preserved, and which of these, through its form-relationship, admits of the most distinct recognition of the blood-relationship, or, inversely, which has diverged the most widely from the ancestral form? The decision in single instances cannot but be difficult, and appears indeed at first sight impossible; nevertheless this will be arrived at in most cases as soon as the ontogeny of the larvae, and therewith a portion of the phylogeny of this stage, can be accurately ascertained.[46]

As he proceeded from one case of incongruity in form relationship to another, he began to distinguish between two kinds. First, the larval stage, or inversely the imago stage, of two "systematic groups" might appear more closely related than the other stage. Second, as demonstrated in the case of the genera *Apatura* and *Nymphalis,* the divergence at the larval stage might be such as to force the taxonomic creation of a higher-level systematic group. Moreover, it appeared to him that both kinds of incongruity could easily be identified at the varietal and species levels (he provides many examples), although systematic groups based on larval divergence at the higher levels tended to be lower than the systematic groups based on the imagoes. Thus, many butterfly and moth groups diverged on the family level when the calculation was based on their larvae, but they diverged on the subordinal level of Rhapolocera and Heterocera when based on the form relationship of the imagoes, particularly with regard to the antennae. As a system group, the genus appeared to Weismann to reveal fewer incongruities than either the lower levels of species and varieties or the higher levels of families and above—an observation he elaborated upon for pages.[47]

Weismann might have been making the taxonomic point that a hierarchical classification based exclusively on embryology or adult anatomy risked invalidation by the other—an issue that would emerge in the next decade. The bottom line, however, for his examination of incongruities between embryo and adult, between one systematic group and another, was that by this means Weismann felt he could draw conclusions about the causative factors in the evolutionary process. While Haeckel, Carl Gegenbaur, Anton Dohrn, and others focused more on the phylogenies produced by evolution, Weismann appears more concerned about the mechanism of evolution itself. His point was that the direct effects of the conditions of life and the indirect effects of natural selection could explain the range of incongruities; the phyletic force could not. His argument against the phyletic force had now become more than simply an argument of exclusion, that is, that his mechanistic forces were sufficient to produce the effects he found. It was an argument about the impossibility of a teleological phyletic force to explain the contradictory genetic relationships. "The assumption that there is a transforming power innate in the organism," he insisted at one point, "indeed agrees quite well with the phenomenon of congruence, *but not with that of incongruence*."[48]

So as not to appear parochial in his concentration on Lepidoptera, Weismann next turned his attention to the ordinal level of insects and examined the form and "blood" relationships between Hymenoptera, Diptera, and Coleoptera—three other holometabolous orders of insects. His message was the same, but in these cases he dealt with major structural differences in larvae, which within the same order might be grub, caterpillar, wormlike, or maggotlike in form. He was addressing an adult morphology of what might at first blush seem fundamental to the organism, if one uncritically assumed that larvae recapitulated the past. Weismann noted that his conclusions were anticipated sixteen years earlier at Schloss Schaumberg where he had examined the dipteran family Muscidae. Then, he had observed that the larval head started out looking like a "well-developed" grub's head "with antennae and three pairs of jaws," but that during development it transformed into a reduced wormlike "suctorial-head armed with hooks and lying within the body."[49] At first, neither he nor others had drawn the obvious conclusion that the maggotlike larva is not an ancient form but represents a recent adaptation.

So by the mid-1870s, Weismann had repeatedly documented from the varietal to the ordinal level what other zoologists were also beginning to realize, that the conditions of life, not the historical past, shaped the larval form.[50] The point of the exercise was not to challenge Haeckel's fundamental biogenetic law, which was not even mentioned in these two lepidopteran monographs, but to refute through embryological studies a vital phyletic force.[51] The *Studien* were Weismann's way of exploring the mechanism of evolution.

The Transformation of Axolotls (1875) and the Mechanical View of Nature (1876)

In 1872 when Weismann was still an *außerordentlicher Professor* on the medical faculty, Albert von Kölliker delivered to his institute five specimens of the Mexican axolotl. The common name "axolotl" is of Aztec origin, perhaps initially referring to the local god of deformation and death, Xolotl, and his mythic transformation to a "water dog." In the wild, the axolotl matures in the form of a branchiate or gilled salamander and may continue to live and breed in that neotenic state; its numbers have been vastly diminished in the two southern lakes of what remains of the basin of shallow lakes surrounding the Aztec island capital city of Tenochtitlán. Today, the axolotl is kept as an aquarium pet and has long since earned its status as an important organism for genetics and developmental studies because of its large size, low maintenance requirements, and uninhibited reproduction.

The name axolotl is now reserved for the branchiate form of *Ambystoma mexicanum,* which is easily confused with its sympatric congener, the tiger salamander *Ambystoma tigrinum.* In the nineteenth century, it was also confused at times with North American species, some of which also reproduce periodically in the branchiate state. There also exist related forms with slightly different life histories in other Mexican lakes. All of this has created a tangle of identities and multiple generic and species names. Its scientific identification and the origins of laboratory stocks are still a matter of discussion.

Weismann referred to the land form of axolotl as "amblystoma," a popular name that Louis Agassiz had turned into a formal genus. Earlier axolotl had been identified with the generic names of "Siredon," "Axolotl," "Ambystoma," and at least twenty other variants. Tschudi's 1839 designation "Ambystoma" has survived in modern times under the rules of the International Commission on Zoological Nomenclature of 1963 and 1974. There have also been nine other specific names, including the proposal by Robert Wiedersheim (1879) to rename the axolotl "Weismanni" in honor of his brother-in-law.[52]

Nomenclature has its interest for the historian as well as the biologist. When Weismann acquired his specimens, the axolotl was well known in European zoological circles. Auguste Duméril, a professor at the Museum of Natural History in Paris and a specialist in fish, amphibians, and reptiles, had announced a decade earlier that he had managed to transform the lifelong water-dwelling "branchiate" axolotl into the lung-dependent, land-dwelling salamander known then as "Amblystoma." Duméril achieved the transformation on thirty specimens by reducing the amount of water in which his colony lived. It is easy to imagine what an impact his experiments had. The transformation was spectacular and interpreted by some as a leap from one species to another in a single generation.[53]

Duméril must have been a generous man, for he sent specimens to colleagues throughout Europe. In fact, it appears likely, as modern researchers explain, that today's widely dispersed laboratory stocks of *A. mexicanum* stem from Duméril's original colony. It also seems likely that this was originally a mixed stock that included not only *mexicanum* but *tigrinum*. Furthermore, since the former breeds naturally and in captivity only in its branchiate form and the latter breeds only in its pulmonate form, the results were initially confusing.[54]

Kölliker was either a direct or secondhand beneficiary of Duméril's largesse. There appear to be no published accounts of Kölliker's ensuing experiments, but Weismann reported later that he had succeeded in transforming merely one of one hundred axolotl into *Ambystoma*. For Kölliker, nevertheless, this sudden transformation of a single axolotl confirmed for him Duméril's general results and lent dramatic support for his own theory of evolution by heterogenesis.[55] Weismann, who would not buy into Kölliker's conclusions, had also commented in 1872 on Duméril's results in his critique of Wagner's theory of isolation.[56] Where the latter had argued that it was the isolation in Paris that had caused the transformation of the axolotls, Weismann had insisted that the change in conditions from Mexico to Paris had acted directly on the axolotl's organization, causing it "to reach a higher stage of development, such as many of its allies have already attained."[57] As we will see in a moment, his phrasing stuck in his craw when he came to reconsider the phenomenon three years later.

Sometime in the summer of 1872, Kölliker must have delivered specimens from his line of axolotl to his younger colleague in Freiburg. After successfully breeding them, Weismann tried his own transformation experiments on five specimens the following year. His efforts turned out to be a failure.[58] By then, however, Weismann had many irons in the fire. His eyes had improved to the extent that he could assume the responsibilities for moving his institute and official duties to the philosophical faculty. He returned to his formal teaching duties for the summer semester and then spent the summer vacation at Lindenhof with his family. It was also an opportunity for him to stake out his next research project, which concentrated on the daphnid fauna of the Bodensee. In September, he traveled to Wiesbaden for the Versammlung der deutscher Naturforscher und Ärzte, after which the winter semester (1873–1874) began. All the while, he was organizing his butterfly and caterpillar material to go into his *Studien*. It was under these circumstances that Weismann recognized that his axolotls needed more attention than he could devote to them if they were to be coaxed to transform into *Ambystoma*.

His problem was solved when "a lady living here (Freiburg), Fräulein Marie von Chauvin, undertook to rear a number of my larvae of the following year [1874] which had just hatched."[59] Chauvin evidently possessed just the right combination of "care in treatment and delicacy of observation" to compel five of twelve young larvae to metamorphose into *Ambystoma* after they had been forced to live in shallow water for six months. In a gracious move, Weismann published von Chauvin's written account of her experiments at the outset of his general discussion of the subject. At the same time, he pointed out that the curator of his zoological museum, a Herr Gehrig, had maintained a control brood of axolotl in deep water, none of which metamorphosed. The cooperative interplay between Weismann, von Chauvin, and Herr Gehrig showed the advantages of Weismann having been promoted to a full professor and director of the Zoological Institute and museum in the philosophical faculty. It was a pattern of respect, enthusiasm for the subject, and duty that characterized a well-run institute in Wilhelmine Germany. For whatever reason, von Chauvin's results were far more consistent than either Kölliker or Duméril had achieved. She went on to publish five papers of her own on the subject. With von Chauvin's success with a small number of axolotls, Duméril's earlier experimental results, and a number of contemporary discussions by de Saussure on the wild population of axolotls in Mexico (1868) and by Edward Drinker Cope on North American siredons, and experiments by the English ornithologist W. B. Tegetmeier (1870) and by Seidlitz to examine polymorphism, Weismann felt he had learned enough about axolotl to add an essay on the subject to his *Studien*—after all, he too was deeply immersed in the evolutionary implications of polymorphism, metamorphosis, phyletic parallelism in development, and Haeckel's consolidation and popularization of his biogenetic law.

First, by the time Weismann wrote his essay, the anatomical details of the difference between axolotl and *Ambystoma* had become well known. Durémil and Kölliker had pointed out that the degeneration of the gills was accompanied by changes in the branchial arches and the hyoid bone. They further mentioned that the dorsal crest disappeared and the tail became altered in form. The skin changed, and cutaneous glands degenerated as well. Kölliker had noted the difference in the size and shape of the pupils and eyelids, and that the toes lost their webbing and became narrower. Durémil had pointed out that the setting of the teeth and lower jaw were significantly different, and the anterior portion of the body became less concave. Weismann corroborated most of these details himself and added others. It would be surprising if he did not also consult with his brother-in-law about the anatomical changes, even before the latter arrived in Freiburg as prosector at the Anatomical Institute; there is no doubt that he did afterward.[60] The bottom line for Weismann, although evidently not for Kölliker, was that the anatomical differences between axolotl and *Ambystoma* were far too complex to manifest a heteroge-netic leap from one form to the other.

Then there was the question of its sterility once *Ambystoma* had appeared. This fact caused much discussion and comparative studies of various popu-lations even within the lakes of the central valley of Mexico. Duméril had reported that none of his thirty *Ambystoma* appeared capable of reproduction, and Armand de Quatrefages confirmed that the spermatozoa in the males were imperfect. Over the ensuing decade, the *Ambystoma* of Duméril's colony at the Natural History Museum, with one exception reported in 1876, simply did not, and apparently could not, reproduce. By the end of the decade, the matter became more complicated and less well documented, but Weismann followed it carefully and asserted that the best one could say was that *Ambys-toma* was "relatively sterile."[61] The point was crucial for him, for the transfor-mation of axolotl to *Ambystoma* could not be blithely co-opted as evidence of a phyletic force driving evolution to a more advanced level: "Mere reversion forms may die off without propagating themselves; but a new form called forth by the action of a phyletic vital force should not be sterile, because this is the precise 'aim' which the vital force had in view. The conception of a vital force comprises that of teleology."[62]

In the end, Weismann recognized that he could not explain the relative sterility of *Ambystoma* either. In fact, as he delved further into the problems of relative sterility, of earlier environmental conditions of the lakes of central Mexico, and the relation between axolotl/*Ambystoma* and other species that either remained in a perennibranchiate state or transformed into the salamander stage, the more complex and uncertain he became about a general explana-tion. Three conclusions emerged from his vacillations. First, the changes in the conditions of the lakes had obviously been enormous over the past 500 years and must be far better understood if one was to tease out a full explana-tion of the evolutionary pathway axolotl took in arriving at its current state. Here, as earlier with the rise of seasonal dimorphism in butterflies, Weismann showed a keen sense for the changing conditions of life during the evolution of any particular organism.[63] Second, Weismann revealed some concern about interpreting *Ambystoma* as the older phylogenetic stage because it is not reca-pitulated in the ontogeny of today's axolotl. Would he be contravening the biogenetic law of Müller and Haeckel? Not at all, he concluded. Rather, this was a reversion not an addition, that is, an elimination of a more advanced adult form of urodelas, not that form's compression and modification. "The only trace that remains to us of its former developmental status is the 'ten-dency,' more or less retained in each individual, to again ascend to the salamander stage under favourable conditions."[64] The focus of Weismann's future career would be working out what this "tendency" really meant.

Third, and probably the most important consequence for Weismann, was a deeper understanding of the meaning "reversion." It is clear that Weismann used the term in two ways. *Ambystoma* "reverted" through an evolutionary change to a larval form, and axolotl "reverted" to *Ambystoma* when the conditions compelled it to. The latter meaning was the focus of this essay, but he could not avoid the first use as well.[65] Reversion was a phenomenon that was often identified as "atavism," but this term failed to capture fully the essence for which he was groping. As he recognized, the dilemma was highlighted in his own misconceived passage of 1872 when he had written his critique of Wagner:

> Why should not a sudden change in all the conditions of life (transference from Mexico to Paris) have a direct action on the organization of the Axolotl, causing it suddenly to reach a higher stage of development, such as many of its allies have already attained. . . . In all probability we have here to do with the direct action of changed conditions of life.[66]

When Weismann reflected on these words three years later and then on the broadening perspective that von Chauvin's experiments had provided him, he wrote, "It is precisely on this last point that there lies the new feature furnished by these experiments."

The array of anatomical differences between axolotl and *Ambystoma* forced him to reconsider what reversion really entailed. It could not be an orthogenetic evolutionary advancement paralleling other salamandrians, as Haeckel and others believed, nor could it be a species-level leap propelled by a phyletic force, as Kölliker insisted. Weismann urged a third possibility: the way of Darwin and Wallace, who considered adaptation to be a gradual accumulation of traits over many generations:

> These naturalists thereby designate a gradual bodily transformation appearing in the course of generations in correspondence with the new requirements of altered conditions of life or, in other words, the action of natural selection, and not the result of a suddenly and direct acting transforming cause exerted but once on a generation.[67]

Axolotl's "reversion" to *Ambystoma,* unlike Triton's "reversion" to a reproductive larva, entailed the appearance of formerly adaptive traits of an "organically higher" form that because of changing conditions had ceased to remain viable. These new adaptive traits could not be recapitulated in axolotl's ontogeny but must somehow remain in a latent state. It took little imagination on Weismann's part to speculate that his seasonally dimorphic butterflies might represent the alternation of the older and newer adaptive forms in an alternation of generations.

It is interesting that Weismann should have made so much of axolotl, for so little new information came out of his and von Chauvin's experiments. As he pointed out, being able to observe the anatomical details of the specimens that she must have sent him clearly had their impact. The full answer may lie in the fact that he was not at this time in his career submerged in the details of microscopic embryology but had time to think deeply about the evolution of polymorphism and phyletic parallelisms of whole organisms such as butterflies, caterpillars, and salamanders.

"On the Mechanical Conception of Nature"

It is not easy to capture Weismann's mid-nineteenth-century narrative essays in a fair but encapsulated form. As alluded to previously, the warping of his argument across the reefs of contrary opinion could be laborious, but his ship did advance. At this point in his career, he showed a command of the pertinent literature; he argued with and against it, and he had a sense of how to draw upon the biological evidence then available. He recognized the value of simple experiments and the limitations to Haeckel's more single-minded program of producing phylogenies on the basis of the biogenetic law, but he did not turn against the latter. After all, he was a committed mechanist and shared more immediate and common concerns about the proliferation of non-mechanistic ways of interpreting evolution. If one must identify a moment of epiphany in the development of a scientist's career, the metamorphosis of the Mexican axolotl may have provided Weismann with the final push toward adopting an exclusively adaptational explanation of evolution.

Despite its portentous title, Weismann's final essay appears more of a summary to the *Studien* than an independent research project. Its title, "On the Mechanical Conception of Nature," may seem to promise a scholarly philosophical examination of an issue of real contemporary concern. After all, the 1870s were a time of the rise of neo-Kantianism on the continent, of Huxley's Lay Sermons, of Tyndall's presidential address at the British Association for the Advancement of Science, and, above all, of the publication of the much-expanded second edition of Frederick Alfred Lange's *History of Materialism*. In this context, Weismann revealed the influence of his Göttingen chemistry professor and dissertation director Friedrich Wöhler, who had first convinced him that organic compounds were not the product of a vital force but of chemical processes similar to inorganic chemical reactions. If there need not be a physiological vital force, need there be a special ontogenetic or phyletic force? Weismann's essay, however, ended up being less of an overview of the subtleties in the philosophy of science than a rejoinder to Edward von Hartmann's recent interpretation of Darwin's natural selection in his *Wahrheit und Irrthum in Darwinismus*.[68]

As earlier mentioned, von Hartmann's book was not as much an attack on evolution as it was a disagreement about Darwin's mechanistic spin. Central to the argument was von Hartmann's identification of three key principles in Darwin's theory, none of which at the time could be demonstrated to exist in the classic sense of mechanical, that is, in terms of matter in motion. The Darwinian theory of evolution by natural selection, for him, could only work if a teleological force was included in the mix. The three key principles for von Hartmann, in a different order from the way he presented them, were (1) variability ("die Variabilität"), (2) inheritance ("die Vererbung"), and (3) the struggle for existence ("der Kampf um's Dasein"). The last of these principles actually morphed into a principle about the viability of the organism in all of its complexity to survive under local conditions. He began referring to this process as the law of correlation ("Gesetz der Correlation").[69] In his critique, Weismann first resorted to empirical examples that brought into question von Hartmann's teleological interpretation of these principles. "Let us keep to the facts," he insisted at one point.[70] Whereas von Hartmann had postulated an alternative between unlimited variability or a directing phyletic force, Weismann denied this "unlimited" characterization and provided, among a number of examples, the case of some pierid butterflies of the tropics that failed to mimic heliconides as so many other butterflies, including other pierids, had done. If there had been a directing phyletic force, his argument continued, mimicry must have been its strategy for survival, but the nonmimicking pierids found other ways of adapting. In short, Weismann accused von Hartmann of failing to understand variability as simply "fluctuations in the type of the organism."[71]

Whereas von Hartmann had argued that it took similar variations in hundreds of individuals at the same time if heredity were to provide the critical mass for an evolutionary advance, Weismann turned to examples of polar animals that over a period of thousands of generations slowly developed white coloration. He spoke in terms of the percentage of favorable and less favorable traits, and in the cases of mimicry and protective coloration he insisted that "in all these it is always *improvement* that is concerned, and not the question 'to be or not to be' with which we have to deal."[72] So it was either Weismann's form of a gradual accumulation of variations through heredity or an original preestablished harmony as von Hartmann seemed to believe. Documentation through "facts" seemed to elude Weismann as much as von Hartmann, but it is more pertinent to note that during his discussion of heredity Weismann guardedly endorsed Haeckel's parallel between the mechanistic production of germ cells and the differentiation of embryonic cells, on the one hand, and the mechanistic reproduction of individuals and speciation, on the other. The analogy had been newly reinforced in a famous illustration in Haeckel's *Perigenesis der Plastidule,* in which Haeckel had emphasized that heredity was simply an "overgrowth" of molecular motions from one generation of cells to the next, and from one species to the next.[73]

Finally, whereas von Hartmann had considered a vital force necessary to understand the successful struggle for existence of the extraordinary correlation of parts within every organism, Weismann countered that correlation really represented a deep-seated relationship among heredity, available variations, and living conditions. For him, there was no need for a phyletic force. Empirical cases, however, might arbitrate between the two positions, so Weismann found it necessary to argue from what he considered was the contrast between his and von Hartmann's understanding of species. The philosopher assumed a heterogeneous leap from one species to the next in the evolutionary process; it required a phyletic force to guarantee a new harmony of parts. The zoologist was persuaded that there existed a mechanical continuum between varieties and species. The organism was simply a mosaic of traits, the parts of which might change gradually and independently of one another, as so palpably occurred with his larvae and imagoes, and even with inorganic crystals. Weismann insisted that in both the organic and inorganic realms there existed an internal harmony ensured by physical and chemical laws. This harmony possessed a constancy in the short run but might alter gradually in the long run.

The bottom line in Weismann's argument was that the constitution of the organism itself produced the variability upon which natural selection worked. But here was his dilemma: how to explain this variability, "which always leaves open a door for smuggling in a teleological power"?[74] Whence these variations in mechanistic terms? Here, as in so many situations, Weismann built upon yet curiously altered Haeckel's viewpoint. He agreed with his colleague that nutrition and other external factors produced dissimilarities between developing embryos, but he made more explicit another factor: the historical past, which made the constitution of every embryo, every "germ," different. In considering sexual reproduction, Weismann discerned that where there "occurs a blending of the characters (or more precisely, developmental directions) of two *contemporaneous individuals* in *one* germ, so in every mode of reproduction there meet together in the same germ the characters of a whole *succession of individuals* (the ancestral series), of which the most remote certainly make themselves but seldom felt in a marked degree."[75] Where Haeckel's theory of perigenesis had emphasized external motions impinging upon the internal motions of dividing cells of the individual and of species, Weismann was groping for something in addition, namely, the accumulating inherited contributions of each ancestor. Variability was neither a special vital principle nor the sole result of the conditions of life of a developing individual but rather reflected a continuous chain from the past. There is an irony here, for Haeckel is known for the biogenetic law, the view that the phylogenetic past determines the ontogenetic present, and that this is mediated mechanistically through levels of biological individuals. He had not focused on how inheritance in itself might produce diversity. In retrospect, we can see Weismann and Haeckel approaching a point in their lives when their closely parallel research interests would diverge.

In this quasi-philosophical essay, Weismann was far more focused on defending Darwinian evolution from the corrupting influence of phyletic and ontogenetic vital forces, which seems the intention when he titled the third and last section of his essay "Mechanism and Teleology." This, too, was the title of the concluding section of von Hartmann's essay. Unlike von Hartmann, however, Weismann was not going to enter into neo-Kantian debates about the possible union of mechanism and teleology at some higher level. It was his predilection to keep to the empirical base of his studies and discuss what appeared to be discontinuities in the ontogenetic process. After all, his own observations of the case of axolotl turning into *Ambystoma,* Franz Eilhard Schmidt's discovery that eight-rayed medusae could give rise to twelve-rayed medusae was due to parasitism not species saltation, and the discovery that free-swimming, tailed cercaria were the larval form of trematodes, and many other apparent cases of heterogenesis had dissolved into routine cases of metamorphosis, alternation of generation, or parasitism.[76]

After assembling nearly 400 pages' worth of detailed biological discussions and another fifty or so pages of a quasi-philosophical summary about the relationship between mechanistic science and contemporary versions of a vital or phyletic force, Weismann arrived at a rather mild but logically safe compromise. He barely recognized and certainly did not follow the contemporary plunge into metaphysical materialism, nor did he support Haeckel's yet-to-be-drafted pantheistic union of matter and spirit. While resisting a dogmatic claim about the all sufficiency of a mechanistic worldview, Weismann also resisted pushing his science beyond what he could see, count, and measure at his bench and in the outside world. "The final and main result of this essay," Weismann concluded, "will thus be found in the attempted demonstration that the mechanical conception of Nature very well admits of being united with a teleological conception of the Universe."[77] There are no indications that he implied a formal spiritual or religious commitment in these words. The value of this position was that it suited his scientific style. He could concentrate on what concerned him most and what he did best. He moved from field to museum, to the microscope, and even to laboratory experiments when possible. Even when his poor eyesight hindered his ability to read, he tried to devote at least two hours a day to "reviewing the literature." While pursuing well-defined problems in developmental and evolutionary zoology, he articulated low-level, then somewhat higher-level generalizations from his findings that responded to contemporary biological concerns. Though he occasionally alluded to the metaphysical position expressed in the final essay of his *Studien,* he would not again delve into the philosophical maelstrom.

6

Daphnia Research and the Ecology of Lakes

It is hard, perhaps artificial, to disentangle Weismann's research that led to the publication of the five major monographs that collectively became the *Studien zur Descendenztheorie* (1875–1876) from his research and publications that eventually appeared as seven *Beiträge zur Naturgeschichte der Daphnoiden* (1876–1879). At first blush, the overlap of the two areas of study may seem self-evident from the dates of these publications and from the different objects examined: Lepidoptera and axolotl on the one hand, and daphnids or cladocerans on the other. A detailed examination of Weismann's works, some of his correspondence, and the common themes embedded in these two sets of monographs reveal a larger story. It is a story about Weismann securing his place as an outstanding naturalist, reestablishing himself as an embryologist concerned with frontline developmental issues, and as a modern "wissenschaftlichen" (scientific) zoologist dedicated to meticulous observations and modest, carefully directed experiments. The monographs overlapped in time more than their publication dates suggest, and in style they both reflected Weismann's exhaustive (and exhausting) propensity to anticipate objections and nailed down his own arguments with a broad display of facts drawn from personal studies, contemporary literature, and a sense of the history of the problems. The most important difference between the two sets of monographs appears to be a matter of orientation. The *Studien,* as their full title implies, entailed arguments about phylogeny and evolution. The *Beiträge* displayed a systematic examination of natural history and were ultimately directed at the confusing phenomena known collectively as "heredity." It would be a mistake, however, to overemphasize these differences, for Weismann was working in a space that by the end of the century would be identified by several contiguous disciplinary boundaries. Instead of "working on the boundaries," as is commonly and often correctly portrayed by historians and sociologists of science, Weismann, and many like him, were reevaluating and destroying old boundaries as well as creating new ones. His choice of organisms and his drive to solve specific problems allowed him to articulate well-defined subjects and pursue themes that would lead to boundary reformation, which provided a research space for developing his ideas about heredity and evolution. His work took him out of Freiburg to his and Mary's vacation retreat southeast to the boundaries of Germany, Switzerland, and Austria.

The Bodensee

Snug along the border between northwestern Switzerland and the southern extent of modern Baden-Würtenburg lies the Bodensee, or Lake Constance. The lake, including the inland port cities of Constance, Ludwigshafen, and Friedrichshafen, lies largely in Germany, but the eastern tip, culminating in the city of Bregenz, is Austrian territory. Switzerland controls the southern shoreline and the mouth of the upper Rhine. The lake is often spoken of as two lakes. The western, downriver end is referred to as the "Obersee," with arms into the Überlingersee and down the exit arm of the Rhine at Constance. The lake below Constance on to Bregenz is called the "Untersee." The length of the lake from Bregenz to Stein-am-Rhein beyond Constance is roughly 46 miles; at its widest, the lake is roughly 10.5 miles broad. The Obersee is much the deeper of the two areas, with a maximum depth of 252 meters; the Untersee reaches a maximum depth of 46 meters.

The lake itself is only an indirect consequence of the rise of the Alps. Its form and depth are more directly the result of the Alpine glacial caps. During the Quaternary period, when four successive ice ages serially covered much of western Europe, a topographical trough north of Switzerland became excavated by ice and runoff from the receding ice cap. In recent times, the area of the lake has been known for the mildest climate in Germany and for supporting luxurious Mediterranean vegetation. The climate of the western end encourages the growing of grapes, other fruits, and vegetables; the eastern end supports feed and livestock. We may easily understand how Adolf Gruber, the successful German businessman living in Genua, and his wife Julie, née Schönleber, would be tempted to purchase land on the Bodensee for a vacation retreat. The land would be easy to get to by traveling ever northward from

Genua via Milan, Lago di Como, and the upper Rhine valley to Bregenz. Continuing northward across the Swiss border and into Westphalia, they built an elegant house near the German town of Lindau. Their son, August Gruber, assistant at the Zoological Institute, would eventually inherit the house, but other members of the family were attracted to the same holding and built other houses. Mary Gruber and her husband August Weismann stayed off and on in a number of the houses. Tille, Mary's younger sister, and her husband, the Freiburg anatomist Robert Wiedersheim, also spent their vacations there. Both were interested in the natural history of the lake.

Daphnid Research

For the novice there is no taxon of animals that is more confusing yet intriguing than the large class of Crustacea. They play a pivotal role in the economy of nature both as predators and prey. Being primarily but not exclusively aquatic, they exist in the open oceans, the shores, lakes, streams, ponds, and puddles of the world. There are those that are littoral; others are bottom dwellers or permanently pelagic. Some are extremely mobile; others are planktonic. They may be parasitic or sessile, and there are some that stare up at us from the dinner plate through compound eyes. In this boiled condition, they confront the novice with their bewildering head of fused segments, attached to which are pairs of antennae, mandibles, and maxillae. When cooked, their red exoskeleton is dominated by a large carapace that conceals part of their highly segmented thorax and abdomen, and from which an array of attached appendages of varying sizes, shapes, and ramifications requires a specialized vocabulary and a keen sense of homology to fully appreciate.

What the novice does not see from his mealtime perspective are the stages of development that start within the egg and advance through several larval stages to miniature lobsters. Then it takes a half-dozen molts over several years before they attain dinner-plate size. What the novice cannot appreciate—as he wipes his mouth and probably his chin and hands after tossing aside the various chitinous segments, broken claws, squashed appendages, and the disarticulated carapace, tail, and rostrum—is how important these ravaged structures are for an understanding of the basic components and morphology of all crustaceans.

While working at his family's estate on the Bodensee, Weismann directed his attention to the Cladocera, an order of the crustacean subclass Branchiopoda, that is, the order of "gill feet."[1] The name Cladocera, derived from the Greek terms for "branch" and "feeler," is anatomically more descriptive but less poetic than the term "Daphnoidea," which was still in use in Weismann's day. Weismann employed Daphnoidea in the title of his monograph series, but in his text he more commonly used the newer term cladocerans. With some minor differences, both designations identified the taxon of water fleas or *Wasserflöhe*, which had first been seen by the Dutch microscopist Jan Swammerdam in the seventeenth century and picturesquely described as *"Pulex aquaticus arborescens."* In the post-Linnaean period, Swammerdam's *Pulex* was picked up in the binomial name *Daphnia pulex,* which in turn soon became the representative member of the filter feeders that form the majority of the daphnids, Weismann's cladocerans.[2] Most of the adults of these filter feeders are small for crustaceans and are far simpler than the dinner lobster. *D. pulex* measures 3.3 mm long, excluding its extended, ramified second antennae, which serve as oars and produce the water currents that are useful for the activity of feeding. The carapace that wraps about its rounded thorax and abdomen with its extensive brood chamber reflects the animal's clumsy, jerky locomotion. The adult, but not the embryo, possesses a reduced number of segments compared with other crustaceans. Unlike most crustaceans, cladocerans for the most part are transparent. This feature protects them from predators and in turn enhances their ability to prey on other invertebrates, including other cladocerans. Transparency, however, makes their behavior hard to see, so their collection is difficult for zoologists.

During his first summer of collecting in the Bodensee, Weismann thought he had discovered a new and unusual species. It was three times as large as the common *D. pulex,* and its streamlined shape, numerous segments, raptorial first thoracic limbs, and diminished carapace marked it as a predacious feeder. After a literature search, however, he learned that the Swedish zoologist Wilhelm Lilljeborg had already described and named this predaceous cladoceran as *Leptodora hyalina.* Weismann also ran across and used an account in Danish by the Danish naturalist P. E. Müller that confirmed many of Weismann's observations. This minor work, as we will note, turned out to be one of the pieces of evidence that would later bring Weismann's friendship with Claus to an abrupt end.

Weismann spent many hours in the summer of 1873 seining with muslin nets for plankton. Others before him had collected plankton in freshwater but with other objectives in mind. Viktor Hensen, a recently appointed Ordinarius at the University of Kiel, had begun sampling saltwater plankton in the North Sea in 1871. The connection between Weismann's work and Hensen's is not yet clear.[3] From the way Weismann described his initial activities, his was at first simply an exploratory quest "to get to know the inhabitants at the surface and in the depths," but in the process one of his earliest discoveries was the circadian vertical rise and fall of a concentrated layer of small organisms in coordination with the fall of darkness and the first appearance of morning light. Jürgen Schwoerbel has drawn attention to a charming description of this discovery from a free public lecture Weismann gave before an audience "of mostly women" in the university's auditorium during the winter semester of 1875–1876. One of Weismann's passages is so informative and picturesque that it is worth quoting in full:

> When a number of years ago I began to become confident in the animal world of our lake, I first began to search in the bright sunshine the surface with a fine net. But instead of the expected rich harvest my net contained next to nothing, and as often as I repeated the search, I always got the same result.
>
> Because I was now convinced that a large number of animals must be in the lake, I came to the thought that these animals perhaps were shy of the all too bright light, therefore they remained during the bright day at a certain depth and only rose to the surface during the night.
>
> I then fished during a quiet dark night. After every drag I spilled the unrecognized content of the net into a jar and observed these only after the return to the shore and to light. Instead of a few small animals, which I had expected, I found that the water was filled with thousands of small animals; it appeared milky thick merely from the quantity of small organisms, which it held. They hopped, climbed and flew so much about one another, that one could become dizzy while looking into the whirling crowd.
>
> Later I have often used this mode of seining and on many a still, star clear night, but also on many a pitch-dark and stormy one I spent hours in a boat in order to be able to get my hands on the interesting small denizens of the lake.[4]

We know that Robert Wiedersheim, who had become his brother-in-law the previous year, assisted Weismann during the first summer both in collecting in the field and by sketching specimens.[5] As the research gathered momentum, Weismann expanded his collection to ponds and lakes in the area of the Bodensee, in the vicinity of Freiburg, and across the Alps into Lago Maggiore. Further, he kept selected colonies at his institute on which he ran tests relating to their cyclical reproductive patterns. His collecting and associated experiments extended from 1873 well into 1879, at which time he

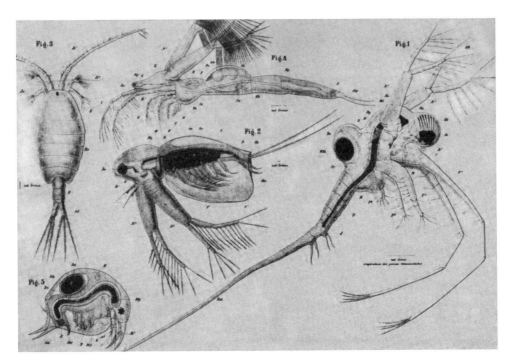

Figure 6.1. Illustration of dipterans Weismann collected in the Bodensee. Note particularly the "water flea" (*Leptodora hyalina*) in the center, which Weismann at first incorrectly thought was a new discovery and stimulated him into further collections.

Klaus Sander, ed. *August Weismann (1834–1914) und die theoretische Biologie des 19. Jahrhunderts. Urkunden, Berichte und Analysen.* Freiburger Universitätsblätter, vols. 87, 88. Freiburg: Rombach Verlag, 1985. Jürgen Schwoerbel, "Weismann und die Erforschung des limnischen Zooplanktons," p. 54.

completed and sent the seventh and last *Beitrag zur Naturgeschichte der Daphnoiden* to his publisher, Wilhelm Engelmann.

Leptodora hyalina

It was *Leptodora hyalina* in his collecting jars that first riveted his attention. Its size and structure clearly set it apart from the other cladocerans. As described previously, its discovery was attended with all the excitement of a possible newly found species. As also mentioned, when Weismann returned to Freiburg for the winter semester, he determined that the Swede Lilljeborg, the Dane P. E. Müller, and the Russian Nicolaus Wagner had all published in the 1860s accounts of this species in their respective lands and in their re-

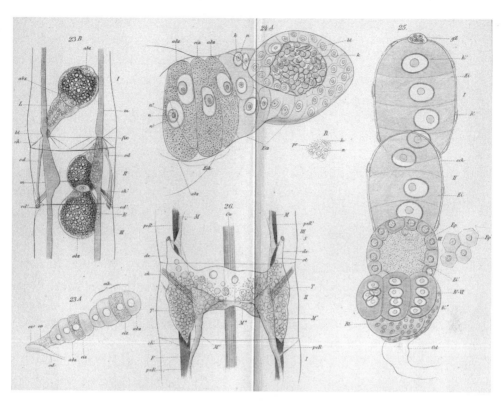

Figure 6.2. Side view of a young female, where "bl" designates the protoplasmic sheath (*blastemscheibe*) with free nuclei, and "k" denotes the forming egg cell. The egg cells (*Eiz*) are arranged laterally along the body; "abz" designates abortive egg cells. The small drawing labeled *"B"* beneath 24A is a segment of the protoplasmic sheath after the influence of the acetic acid.
Plate 38, figure 24a, from August Weismann, *Zeitschrift für wiss. Zoologie* 24 [1874], Bd XXIV, Taf XXXVIII.

spective languages. Furthermore, he realized that Gerstäcker had extended Lilljeborg's original description and illustrations in one of the most recent fascicules of Bronn's *Klassen und Ordnungen des Thierreichs*. Weismann duly reported all of this in his own account of *Leptodora* the following spring, which to be sure reflected the standard norm of scholarship for the period.[6] Ostensibly, Weismann was supplementing these earlier accounts, and he was doing so in his usual, thorough manner—in this case, in sixty technical pages. He presented a general description of the body, the external carapace and musculature, the nervous system, and digestive canal. He commented on the lack of gill structures, but he assumed the organism must breathe through the entire body. He wrote sections on the fat-containing bodies, the circulation apparatus, its intestinal track, and reproductive organs, both male and female, and finally turned to an examination of *Leptodora*'s production of eggs. At this point, he seemed to run into problems. An addendum indicated that in the following April he had focused, without much success, on the reproduction of *Leptodora* in Lago Maggiori.

It was accepted at the time that in the spring and summer female cladocerans multiplied parthenogenetically. (Cladocerans, water fleas, denominate Weismann's "Daphnoiden.") Today, it is recognized that the young are hatched from small, unfertilized, yolk-impoverished eggs (technically known as subitaneous eggs) and reared in an internal brood chamber on the dorsal side of the mother. When hatched, the young are expelled into the surrounding water. In the fall, by way of contrast, males begin to appear in the colony, and the females start producing morphologically larger and darker eggs that migrate under and then are enclosed by a protective portion of the mother's carapace known as the ephippium. This ephippium encasement and eggs are expelled together by the mother into the surrounding water. These protected or "winter" eggs (more accurately referred to as "resting" eggs) might drop to the lake's bottom and overwinter or end up in the desiccated mud of dried puddles and ponds and begin the cycle again when rains come and replenish the water.[7]

This cyclical process appeared parallel to the well-established seasonal cycle of aphids known to naturalists since the eighteenth century. In the 1870s, it was assumed that their parthenogenetic reproductive stage permitted the rapid increase of the population when the water was plentiful. It was also assumed that the resting eggs were fertilized and carried the species over until the cycle began anew. There is an obvious difference, however, between the aphid and cladoceran cycles, in that a species of the latter might have a double cycle during the year, with the conditions changing at the onset of each cycle. One of the confusing issues in Weismann's day concerned what physical conditions triggered the switch from parthenogenetic to sexual reproduction. Another involved the exact relationship between the appearance of males and the production of winter and resting eggs. Was copulation sufficient to induce the formation of the ephippium and winter eggs? Or must fertilization with spermatozoa, which in the case of cladocerans are without a flagellum and hence are not mobile, induce the transition in the oviduct from the formation of parthenogenetic to winter eggs? Further complicating the picture was another fact that there is a latency period between what we understand as fertilization and the onset of development of the resting eggs, and the fact that some females might produce both subitaneous and resting eggs at the same time.

"Daphnoidia" (*Beiträge* I–VI, 1876–1879)

Weismann could not answer all the questions about cladoceran cycles, but he approached them through painstaking examination of the formation of winter eggs.[8] The processes surrounding their formation and fertilizations, however, were not easy to disentangle from both a microscopic and conceptual point of view. Claus and others had earlier determined that the summer eggs arose from a foursome of germ cells. Such a quadrille, they recognized, traveled together in a single egg capsule along the oviduct, but only the third of the four germ cells actually developed into an egg and then embryo. The other three were reabsorbed—presumably to provide additional nourishment to the future embryo, which would eventually develop in the brood chamber. (See Plate 6.)

The formation of winter eggs was more complex. The egg chamber expanded into a brood chamber filled with obscuring yolk. Weismann, like P. E. Müller before him, determined that this yolk represented the dissolution product not only of three of the foursome of germ cells but of the entire quadrilles of further germs from neighboring egg chambers. These were all reabsorbed by amoeboid cells from the surrounding epithelium, which in turn delivered their contents as nutrition to the growing winter eggs. Because of this additional nutrition, the winter eggs could grow considerably larger and be far better provisioned than the summer eggs. Consequently, when laid by the mother into more hostile surroundings, they were better able to sustain themselves in a resting state. Weismann explained that he could follow the entire sequence in a single individual only after he had lightly stained the oviduct with acetic acid and learned how to keep the specimen alive under his microscope—in fact, for as long as fourteen days! Thus, he managed to follow the development in vivo of dozens of winter eggs of *Leptodora*. He identified three stages in the egg formation through measurements and comparisons of the growing egg and the degenerating nourishment cells in dozens of specimens.[9] The examination was tedious and technical. It was restricted to the methodological impossibility of conclusively explaining a developmental process without recourse to experimentation. In the long run, it was momentous only for those who cared about the life cycles of crustaceans. The work, however, demonstrated that by 1874 Weismann's eyes were well enough for him again to spend extended time at the microscope.

His collecting of *Leptodora* led him to realize that males also existed throughout the summer, though in far smaller numbers. This fact complicated the accepted beliefs that the summer eggs were parthenogenetic and that the males played a necessary role in the production of winter eggs. Focusing on the winter eggs, Weismann addressed two questions: (1) was the rise of winter eggs dependent upon copulation, and (2) could unfertilized winter eggs develop? Recognizing that it was essential to distinguish between the act of copulation and the event of fertilization, he proceeded to design simple experiments with *D. pulex,* and in this he followed in the footsteps of the English naturalist John Lubbock, who had recently carried out many isolation experiments with the same ends in mind. Because Lubbock's results had led to conflicting conclusions, Weismann felt that further experiments were justified. He performed over 200 trials in which he isolated females derived from resting eggs. His examination was not a straightforward matter; in order to get meaningful numbers, he needed to raise both isolated, single females and colonies of carefully selected females (the two sexes are easy to distinguish from one another). The strands of living algae used as food and oxygen replenishment needed to be sterilized for a few seconds to kill any unwanted males. Brood after brood was produced without a single viable resting winter egg being laid. Moreover, Weismann discovered that the females of *D. pulex* could on occasion produce both ephippia and nonviable winter eggs.

During the same period, Weismann always found a few males and viable resting eggs in his control groups, which he had kept under similar conditions. He extended and ratified his results with experiments of two further species of the genus *Moina.* To his astonishment, he noted the appearance of more winter eggs and ephippia but found that they, too, degenerated and disappeared without the presence of males. He concluded that the "anlagen" of winter eggs must already exist in females throughout the cycle. The matter of male presence and fertilization became critical, and more isolation experiments followed. He finally ended with four conclusions: (1) winter eggs had to be fertilized to develop; (2) the appearance of winter eggs, however, was not dependent upon the presence of males; (3) unfertilized eggs degenerated, but depending upon the species, they degenerated either in the oviduct or after migrating to the ephippium; and (4) the winter eggs stimulated ephippium development.[10]

The whole series of experiments and conclusions seems in retrospect somewhat archaic, but we must bear in mind that the experiments were performed and written up a year before Oscar Hertwig's momentous discovery that in free-floating sea urchin eggs fertilization consisted of a union of male and fe-

male gametes. When he learned of them, Weismann heralded Hertwig's results. In Weismann's cladoceran investigations, the actual process of fertilization, although pertinent and recognized occasionally with references to an
"internal constitution," remained unexplained. At the time, Weismann was
primarily concerned with understanding seasonal dimorphism in butterflies
as well as parthenogenetic and sexual reproduction cycles in Cladocera. One
important revelation of his investigations turned out to be a demonstration
of the irregularities in the parthenogenetic cycle. A few males were always
around; a few fertilized and unfertilized winter eggs were always produced.
There existed a sloppiness in the reality of the cyclical process that contrasted
to the neatness of the human generalizations concerning them.[11]

Two more of Weismann's *Beiträge zur Naturgeschichte der Daphnoiden* appeared in the same 1877 volume of the *Zeitschriften für wissenschaftliche Zoologie*. They provided supplementary studies to his findings on the complex
process of germ cell reabsorption in the winter eggs of *Leptodora*. The first of
these, *Beitrag* II, was comparative in nature and began with the examination
of winter and summer egg production among six different genera distributed
among four families of Cladocera. He found the germ cell reabsorption process in all of them more pronounced and complicated in the winter than
summer eggs. He speculated briefly about a physiological significance of this
consistent but indirect manner of increasing the yolk in cladoceran eggs, but
concluded that "I at least knew . . . [not] to ascribe [it] to any physiological
meaning worth mentioning and wanted to conceive of it only as a phyletic
reminiscence."[12] In *Beitrag* III, Weismann demonstrated that the summer embryos, developing in the yolk-impoverished eggs resting in the brood chamber,
were nourished from "amniotic" fluid. It was a study that explored the expansion in size of the brood chambers and the decrease of the yolk of the
summer eggs.[13] Both studies may be described as comparative functional
embryology.

It took another two years to complete and publish the last three *Beiträge*
of the series. They appeared during the years when Weismann was shifting
his research away from cladocerans to hydroids—we explore this transition in Chapter 7. But first, a brief description of the remaining *Beiträge,* an
interpretation of the rapport between Weismann and Claus, and some final
words about the importance of the cladoceran work in Weismann's career are
in order.

As previously indicated, the first of these final *Beiträge* appeared in a supplementary volume of the *Zeitschrift für wissenschaftliche Zoologie*. In retrospect,

it appears to have been further from Weismann's immediate subject but is the most interesting because it led most directly to his work on hydroid polyps.[14] Devoted to the ornamental colors of cladocerans, this paper also showed in detail the impact of Darwin's *Descent of Man,* which had appeared in German in 1871–1872. (Weismann used the third German edition of 1875.) Weismann was enormously impressed by Darwin's concept of sexual selection and exploited it for his own ends.

As mentioned earlier, daphnids are transparent organisms in the zooplankton, yet a few species sport external and internal flecks of brilliant colors. The pattern of these flecks is species specific; it may appear in one or both of the sexes, and it may be found in both the sexual and parthenogenetic producing females. Weismann saw this as an opportunity to examine Darwin's theory of sexual selection in the context of the annual cycle, which might contain one, two, or even more subitaneous egg-producing periods. As his argument went, the females producing parthenogenetic eggs would be under no selective pressure to sport decorative colors (*Schmuckfarben*) for attracting the males. Nevertheless, they sported the same colored flecks as their resting-egg-producing sisters.

His first conviction was that these flecks must be ornamental, for he could conceive of no advantage of their presence other than sexual selection, and he could imagine many disadvantages. Moreover, in the six different species he examined, the initial appearance of the flecks was different in the two sexes and occurred at different stages of development. In fact, in one species the flecks appeared in both sexes at an early stage of development throughout the yearly life cycle. In a closely related species, a different pattern appeared later in ontogeny. Generally, the flecks appeared first in the males, which suggested to Weismann that during the period of sexual reproduction they competed for the attention of the females. On the other hand, there were species where the flecks appeared first in females, and this suggested that during the onset of the sexual generation they outnumbered the males and must compete for their attention. That parthenogenetic-reproducing females and their parthenogenetic-reproducing embryos might also have flecks indicated to him that a gradual phylogenetic transfer of this hereditary trait to both sexes and all stages had taken place.

In the context of this period of biology, a comparative method appeared adequate and doable, particularly because it had earlier been supplemented by experiments on copulation and fertilization. Weismann relied, as he had with his study of the coloration of sphinx larvae, on recapitulation and backward displacement (*Zurückrücken*) of the flecks to provide him with a hypothetical phylogenetic lineage. Unlike in his recent experimental work, his conclusions of sexual selection were based on exclusion principles rather than physical manipulations and testing. The important point was that both the appearance of colored flecks in the sexually and parthenogenetic reproducing individuals and their contrasting appearances in the ontogeny of the two sexes called for a mechanistic explanation of inheritance. Haeckel might have explained the generational patterns of flecks by invoking his law of homochronic inheritance modified by a phylogenetic backward displacement, but this must have appeared to Weismann as a general description rather than a causal explanation.[15] Darwin seemed to offer two independent laws of heredity, but their lack of connection disturbed Weismann. Instead, he wanted "eine grosse Grundgesetz" ("a basic law").[16] In reference to the sexual reproductive breeding coloration in certain fish, Weismann vented his frustration in a long, involved passage:

> I believe, the answer for this simply shouts aloud: the transmission of the courtship clothes to the non-courting individuals has taken place because it is not prevented, because neither external or internal grounds or causes were present which resisted the transmission. This is to be sure only a guess, but a plausible one, the moment it may be firmly affirmed, that the fact of inheritance rests on internal causes, on laws, which lie in the nature of the organism itself. This can now—it seems to me—not be doubted.
>
> Nevertheless, if this be correct, it then follows that the tendency must be present for the transmission of the older characters everywhere and always and in respect to each character, which therefore in each single case, in which this does not happen, internal or external causes must have interfered with the transmission. There is no natural law to start with which contains its own exceptions; rather these occur only through the resistance of other forces, and it would be the task of a theory of heredity to identify these.[17]

Despite its ambiguities, Weismann's long passage reveals a good deal about his understanding of transmission. He operated under the same assumptions as did Haeckel and many others among his contemporaries. Inheritance was an automatic process; it signified the default condition. Thus, an established trait—let us say, a colored fleck along an appendage—would be transmitted to the next generation unless a force, let us say a physiological process or natural selection, prevented it from so doing. Furthermore, although not alluded to in this passage which dealt with sexual selection in fish, the colored flecks would in certain daphnids be transmitted to parthenogenetic generations because there was no reason for them not to be. Finally, given Haeckel's and Darwin's law of homochronic heredity and a general belief in the tendency of a backward displacement of traits over a number of generations, the flecks will appear increasingly early in ontogeny. All of this seems compatible with what I have called Haeckel's "overgrowth" conception of heredity. It reflected the conservative side of the commonly accepted balance between transmission and adaptation. It denoted the heredity aspect of Haeckel's palingenesis. There will be more on this "overgrowth" later.

Beitrag VI is a short contribution and deals primarily with the contrasting sizes of sperm, eggs, and the individual species. Weismann concluded that the morphological differences are unimportant compared with the numerical ratio of egg and sperm cells. "The brilliant discoveries of Hertwig and Fol now suffice to show that the physiological value of sperm cells and egg cells are the same, I mean, that they are in the proportion of 1:1." Weismann added that, in the one species that he had been able to examine sufficiently to support the conclusions of his colleagues, "not only does the nucleus of the sperm cell, but also its whole protoplasm unite with the egg cell." (Weismann was referring here to the fertilization of the resting eggs of a species of *Bythotrephes*.) The fertilization of summer eggs, he pointed out, is precluded because the gonopore ("Begattungscanals") in the female is obstructed. As a consequence, Weismann concluded, the fertilization process must be primary whereas the parthenogenetic condition of the subitaneous eggs must be a secondary acquisition.[18]

Cyclical Reproduction

The seventh and last of the *Beiträge* served not only as a general commentary on Weismann's investigations but placed them more systematically in a larger context. It is clear that Weismann had not ventured far from the main road in investigating cyclical reproduction in cladocerans, but his conclusions differed from those exploring the connection between external conditions and sexual cycles. Many, such as John Lubbock, the Swiss physician and naturalist Louis Jurine, and F. W. von Ramdohr, had examined the effects of the annual changes in the ambient temperature, had observed and experimented with the seasonal fluctuations in nutrition, and had supported its influence on the cyclical changes in reproduction. Herbert Spencer quickly picked up on Lubbock's results in his *Principles of Zoology*. Kurz had argued that the summer desiccation of puddles and small ponds brought changes in chemical composition and temperature of the disappearing water and consequently influenced the mode of reproduction. Before Weismann had done so, Lubbock had observed a low level of ever-present males and resting eggs throughout the spring, summer, and fall. Including but extending beyond the sexual cycle of Cladocera, Vladimir I. Schmankewitsch in a series of papers in German, Russian, English, and French in the 1870s had studied and experimented with the effect that a rise in salt concentration had on the form and sex of the brine shrimp *Artemia* in central Asia. As the acknowledged authority on Crustacea, Claus followed and commented upon recent studies in papers and in updating his *Grundzüge*.[19]

Weismann, however, would not subscribe to the general belief that the external conditions directly caused the contrast between parthenogenetic and sexual reproduction cycles. In addition to further experiments, which again challenged the influence of external conditions, Weismann remained skeptical of the results claimed by others. Springtime temperatures, when parthenogenetic reproduction dominated, were often lower than temperatures in November when sexual reproduction prevailed. Experiments performed by icing the water in midsummer did not change parthenogenetic to sexual reproduction. In marshes, there were often a summer and fall period of sexual reproduction although the external conditions were vastly different during these seasons. Furthermore, how did Cladocera detect the condition of "drying up" in summer ponds? Reporting on another experiment with *Daphnia*, Weismann documented his rejection of the widespread belief in the direct action of the environment via further experiments at his institute. "The external conditions of life in this case were as similar as one can possibly make them, similar glass vessels, same water and similar amounts of water and aeration of water (the latter by means of green Algae). Nonetheless in one of the glass containers winter eggs were formed and males produced; while at the same time in the other container neither appeared."[20]

If experimental morphology was too limited in scope to solve what turns out to be a complex chain of events, Weismann could fall back on natural history. It appeared to him that there were three kinds of cycles. First, puddle, pool, and shallow marsh dwellers tended to exhibit a "polycyclical" pattern, such as *Moina rectirostris* and *paradoxa,* which alternated between asexual and sexual generations throughout the year. Second, there were the "monocyclical" species that tended to live in ponds and lakes and possessed a single annual alternation from parthenogenetic to sexual reproduction. Here, Weismann identified eight species of daphnids; their annual cycle corresponded most closely to the situation with aphids. Finally, there were a few species that for all intents and purposes needed to be described as "acyclical," but Weismann felt certain they, too, would eventually reveal a sexual generation.[21]

With this schema in mind, he again experimented with the influence of temperature on laboratory colonies. As he had expected, the lake dwellers were less tolerant of rising temperatures than the pond and puddle dwellers. All he could conclude from such results, however, was that rising temperature was an indicator of bad times to come for the colony, which then turned to sexual reproduction and the production of resting eggs. Temperature was a factor, but so too, he thought, was desiccation, excessive vegetation growth, and an increasing presence of predators. Although he did not say so in modern terminology, it appears that he felt that the members of colonies of given species were adapted to respond to signals of impending cycles of environmental degradation.[22]

Further experiments followed in which he tried to determine the length of the latency period of resting eggs, but again the complexities of the situation placed the problem beyond him. The results of Weismann's many experiments, his careful examination of living conditions in natural settings, his awareness of the differences between species and between colonies of single species, and his efforts to classify these differences in terms of morphology and the clustering of functional types revealed a zoology that was not single-minded in its approach nor dogmatic about its conclusions. Above all, Weismann appeared tolerant of partial explanations and always left the door open for more to come—he was, as noted earlier, a patient positivist.

His conclusions about the phylogenetic rise of these cyclical processes continued a mixture of empirical claims and fundamental assumptions about evolution. As he saw it, this placed the collected phenomena in a different category from the alternation of generations found in hydroids and the seasonal dimorphism he had investigated in butterflies. Associated with the rise of the cyclical mode of generation in Cladocera were many anatomical changes necessary for the traditional fertilized eggs and their parthenogenetic counterparts.

Thus affected are the simplification of structure, the alteration of parts of the egg in subitaneous eggs, enlargement of the resting eggs, the introduction of parental care for subitaneous eggs, as well as parthenogenesis (the suppression of males), the change of monomial reproduction into a polynomial cycle all together and with one another, in order to call forth today's manner of reproduction of Daphnoiden.[23]

By examining the reproductive cycles of Cladocera, Weismann sorted out more clearly than he had previously his deliberations over the contrast between metagenesis and heterogenesis. As we saw in Chapter 4, he had discussed these concepts with respect to seasonal dimorphism in butterflies. With hydroids and trematodes, an alternation of sexual and asexual generation made sense, and to speak of the asexual generation(s) as "nurses" or *ammen,* as Steenstrup had done, immediately set up a hierarchy between primitive asexual and more advanced sexual forms of generation. The hydroids again offered an unambiguous case of metagenesis. In contrast, Weismann felt confident in identifying parthenogenesis as a form of sexual reproduction. The key was seeing subitaneous eggs as a sexual product that did not need fertilization. What followed from this physiological attribute, as he pointed out in the above quotation, were anatomical modifications, such as the differing structures of the egg, the brood chamber, and closure of the gonophore. Among the Cladocera, the "asexual" generations were not nursing generations; the eggs they produced were simply secondary adaptations of conventional eggs, which no longer needed union with a sperm cell to initiate development.[24] This secondary adaptation was simply an extension of heterogenesis.

The Weismann-Claus Affair

His conclusions emphasized the difference between Claus and himself. The former used development and natural history to articulate the functionally different forms of reproduction. Weismann used development and natural history to determine the contrasting phylogenetic emergence of metagenetic and heterogenetic patterns. It was a point Weismann had made in 1875; it was now a point that underscored the estrangement of the two German zoologists.

Perhaps it was inevitable that Weismann and Claus would eventually collide. Although we have only Weismann's side of their correspondence, it appears that during the period of the writing of the first five of his six extant letters the two former students of Rudolf Leuckart were on their way to developing a friendly and fruitful professional relationship.[25] They were, after all, only a year apart in age; the younger of the two, Claus had completed his dissertation at Giessen and habilitated at Würzburg two years before

Weismann had approached Leuckart for assistance in retraining from a physician to a zoologist. As Weismann somewhat cryptically reminded Claus in 1874, when the latter had evidently visited him in Freiburg, "Above all I see in your indefatigable work that you have come around to the opinion which you spoke to me about here years ago, when we walked together on the Schönberg, you may have no truer friend in zoology."[26] We can only guess at what the conversation was about on the steep mountainous paths above Freiburg. We may confidently surmise, however, that it included their mutual positive evaluation of Darwin's theory. Since that day, Claus's career path took him to Marburg and Göttingen before he was called to Vienna in 1873 as Ordinarius of zoology. That was the same year Weismann became Ordinarius on the philosophical faculty in Freiburg.

Thus, both had achieved institutional success by advancing to a Lehrstuhl at their respective universities, both were well-respected naturalists recognized for detailed investigations into the development and life cycles of arthropods, and both had become strong supporters of the Darwinian mechanism of natural selection. When Weismann published his general study on *Leptodora* in 1874, he placed it in the most prominent journal for such anatomical and developmental studies of the day.[27] Claus, who had already become an expert on crustaceans, was bound to see it as soon as it appeared, and he immediately wrote a friendly letter to Weismann in which he both applauded the latter's recovery and return to active research and offered a few minor anatomical corrections to his Freiburg colleague's detailed descriptions.

In turn, Weismann was clearly pleased to renew professional contact with Claus. He accepted the criticisms and shared with Claus some of his thoughts about the anatomical features of Cladocera. The exchange was open and professional, and even included a telling agreement about a recent "naturophilosophical swindle" ("naturphilosoph. Schwindel"). It is possible that Claus's words, endorsed by Weismann, were a reference to Haeckel. Nyhart has suggested that they were directed at Haeckel's *Anthropogenie*, but it is equally possible and just as pertinent that they were written in the context of Claus's very sharp criticism of the substance of the first part of Haeckel's gastraea theory. That Weismann appeared to agree with Claus's negative assessment throws new light on his warm but tangential support of his Jena friend's new theory.[28]

By the end of 1874, Weismann was thanking Claus for a copy of the second and much expanded edition of his *Grundzüge:* "Your book among similar ones to appear is the best," he applauded. By this time, Weismann felt comfortable enough in his correspondence with Claus to voice his political opinions about Germany and the rest of the world and about the inconvenient fact that important scientific papers such as P. E. Müller's on *Leptodora* were being published in a minor language such as Danish. Turning to cladocerans, Weismann also mentioned that he had confirmed certain features of egg formation found with *Leptodora* in other daphnids and asked Claus for the eggs of the two fairy shrimps, which at the time were also considered cladocerans.[29] (Fairy shrimps, like the saltwater shrimp *Artemia,* are today considered noncladoceran brachiopods.) Claus evidently complied, as we learn in Weismann's letter of two months later.[30] The samples helped convince Weismann of the generality of cladoceran winter egg formation. But he delayed publication of what would later become *Beitrag* II, for he felt compelled to visit Naples and check on some saltwater cladocerans and certain other crustaceans.

Eleven months passed before Weismann's next communication with Claus. In the interim, he had been unable to visit Naples in 1875 and 1876—in fact, it was not until March 1877 that he would finally make the pilgrimage to Dohrn's new marine station. His letter of January 8, 1876, to Claus disclosed both optimism and caution about the progress of his work. The following passage reveals much more: "I am now at the point," Weismann wrote his Vienna colleague, "of beginning the publication of my Daphnid-work, about which I already wrote to you earlier. First comes something new about *Leptodora,* on whose reproduction I have wracked my brains for two long years. It is, however, linked with really remarkable events. Unfortunately I cannot send you the work immediately after its appearance, but only first at the end of the year, if a Part 2 and 3 are likewise printed."[31]

These three sentences and their context are worth further comment. First, Weismann had informed Claus that he would be publishing on cladocerans, and that he had "wracked his brains" for two years on "something new" about reproduction with *Leptodora.* He had revealed nothing more about the contents of his monograph, but Claus would quickly learn more, for it soon appeared in print that spring as *Beitrag* I of Weismann's studies on the natural history of daphnoids and was devoted to the winter eggs of *Leptodora.* From previous letters, he had also learned that Weismann intended to investigate a few saltwater cladocerans before he would publish further on the subject and had been told of the importance of P. E. Müller's Danish paper, which anticipated some of Weismann's discoveries. Besides this information, Claus would learn nothing about the details of Weismann's research until 1877 when *Beiträge* II–IV appeared as a unit in the *Zeitschrift f. wiss. Zoologie.* The last of the series of five letters from Weismann was short, written in November 1876 just before the budding relationship fell apart. It was a cordial letter, complimenting Claus on an address and in a friendly way suggesting that Weismann might re-educate Claus about the goals of morphology. It closed with the promise of sending to Vienna the second part of his *Studien,* which had just appeared in print. (These included the monographs on the markings of caterpillars, on phyletic parallelism, and on axolotl as well as the essay on the mechanical conception of nature.)

Claus was an enormously prolific writer. Since initiating his correspondence with Weismann in the fall of 1874, he had thoroughly revised his *Grundzüge* with its details about every major group of animals and published nine papers on a variety of invertebrates, including infusoria, ostracodes, copepods, gastropods, and a polychaete. Earlier he had written on *Leptodora* and the two freshwater fairy shrimps that had interested Weismann. Understandably, he had considerable pride in his knowledge of development and basic functions of the invertebrate world and particularly of crustaceans. It is reasonable to expect him to be interested in what new findings Weismann would announce, but he did not wait. He had learned enough from their correspondence to know that Weismann was concentrating on the process of reproduction in Cladocera. He could read in Weismann's sixty-page monograph the details of the production of winter (resting) eggs of *Leptodora,* which included the discovery of the complex reabsorption process of germs into migration cells and then into yolk, of the contrast between the brooding chamber of the summer eggs and the nutritional chamber of the ovary and ephippium of the winter eggs, and of the independence of winter egg formation and fertilization. Furthermore, he realized that Weismann was in the throes of establishing similar processes with many other Cladocera and related polyphemids, including those Claus himself had worked on. Finally, he had learned from Weismann's letters and from his just published account of *Leptodora* that P. E. Müller had worked out many of these details in an obscure Danish publication. With this information in hand, according to Weismann's later complaint, Claus restudied his material and "hastened" to lay his new findings before the Vienna Academy in a preliminary form on October 26, 1876. This transaction was first distributed as a provisional or "vorläufigen Mittheilung," which Weismann received on November 11.[32]

This was the same November 11 on which Weismann finally packed off *Beiträge* II–IV to the publisher. As he explained, he felt obliged because of Claus's "vorläufigen Mittheilung" to add on November 23 an introductory account to the series, in which he detailed the completion and publishing sequence of his *Beiträge*.[33] It appeared to Weismann, at least through Claus's preliminary report, that Claus had documented Weismann's findings only on saltwater polyphemids and had only in part paralleled Weismann's own *Beitrag* III. Nevertheless, this was enough to upset him. He regretted that Claus's full report and his own *Beiträge* were to appear at the same time—Weismann's in the *Zeitschrift für wissenschaftliche Zoologie* and Claus's in the *Denkschriften* of the Vienna Academy of Science. The "coincidence" was something Weismann could not have anticipated, for he had not even an inkling that Claus was restudying and would rush into print material he had not touched for several years. This coincidence was even more suspect because the two had been exchanging "friendly letters" and Claus had known full well what Weismann was up to.[34]

Not to belabor the incident, but it is worth noting that both Claus and Weismann soon published a "Correction and Defense" and a "Vindication," respectively, of the "coincidence" from each one's perspective.[35] Claus accused Weismann of being possessive about a subject on which Claus had earlier published. He claimed theirs were independent studies, which was a healthy thing for science, and he ended up by printing large sections from Weismann's letters to him and pointing out that Weismann had been dilatory in publishing his results: "the publications of Herrn Weismann might have been delayed for years!" Weismann agreed with Claus's claims but gave them a different spin. He may have been slow to publish, but he had been careful and complete; whereas Claus in his rush to print had overlooked many things, including at one point confusing the brood chamber of summer eggs and the nourishment chamber of the oviduct of winter eggs. "Had Claus spent more time, he would have without doubt avoided this 'fundamental' error."[36] With the use of a dictionary, Weismann emphasized, he had worked his way through Müller's Danish paper and its rich contents whereas Claus seemed unaware of even its Latin abstract until Weismann had pointed it out.

Full of sarcasm, indignation, and self-righteousness, it was not a glorious exchange for either participant. The editor wisely decreed that he would not extend the altercation. It is tempting to attribute the entire fiasco to the reputation and competition of two full professors in academe, but such an explanation would miss some of the internal dynamics involved.

First, there certainly existed a muted tension between their contrasting personal styles. Claus was a dynamic lecturer, a rapid investigator, and a compiler of masses of information that were announced through papers and broadcast in the revisions of his very successful *Grundzüge*. He justified his return to the subject of egg production among Cladocera and polyphemids because Weismann's study of the formation of winter eggs compelled him to fill in a gap ("eine Lücke") in his own earlier studies. As it turned out, he was unable to examine many cases of winter egg development, but as we shall see, he leveled some strong criticisms of Weismann's morphological approach.[37] Unlike Claus, Weismann had not yet earned a national reputation as an outstanding university lecturer although he would soon do so. He was doggedly analytic and weighed every alternative he could think of, but like his friend Haeckel, he always kept a bigger picture in mind. His two very substantial monograph series, the *Studien* and the *Beiträge,* had identified him as a thoughtful investigator and an innovative generalizer.

Second, the two had different understandings about the role of morphology in unraveling ontogenetic processes. This is stated forcefully in Claus's "Kenntnis des Baues," where he took Weismann's conclusions about the nutritive cells of winter eggs to task. Weismann's descriptions of three of the quadrille of germ cells in a single egg chamber, as well as the quadrilles of cells in neighboring egg chambers, becoming nourishment cells depended on the assumption that the single germ cell destined to become the egg was restricted by an internal capacity of "maximal growth" and so required an outside augmentation to its nourishment, an addition of the cell to augment its nourishment by means of its own internal "Kraft." Claus's traditional account, according to Claus, would have relied on the internal capacity, but he asserted that Weismann must show that the traditional mechanism of cellular growth, which found its roots in Schwann's original cell theory, was not sufficient to augment the egg's size. Furthermore, Claus insisted that Weismann had yet to explain how it was that the crustacean nauplius larval stage had phylogenetically disappeared from the developmental cycle of cladocerans and polyphemids at the same time that the egg cell lost its capacity to attract its own nourishment and supplementary germ cells and became converted into nourishment cells. Claus demanded an explanation of the evolution of a complex and novel set of developmental mechanisms within a special group of crustaceans, which neither he nor Weismann could possibly provide through experimental means.[38]

If Claus's "vorläufigen Mittheilung," which was received in early November by Weismann, corresponded with his published account, Weismann's negative reaction is understandable. Nevertheless, his short letter to Claus on November 14, 1876, sent mixed messages. Because some of the growing animosity developing between Weismann and Claus must also hang on their contrasting understanding of morphology, I provide the final letter of the series here in full.

Freiburg i. Br.
14 Nov. 1876

Dear Sir!

Your speech "Darwin versus Galiani" gave me so much pleasure and I am so much in agreement with many points of its enunciated views, I nevertheless want to try to change your mind to a somewhat more favorable conception of organic morphology and morphologists. In any case I want to be able to show you that there are moreover those among the zoologists whom the brilliance, radiating from the theory of descent, has not so blinded that they cannot see where the frontier lies to which it directs us.

You will receive with this letter a book, the first three essays of which I am not so presumptuous to suggest you read; they are individually contributed investigations, which should provide the material for the conclusions of the last essay. If you, however, should find the inclination and leisure to read through this final one, I would be very pleased.

In esteemed respect, yours respectfully
Dr. Aug. Weismann, Prof. of Zoology

First, note that Weismann was both friendly and cautiously guarded. Unlike the collegial salutations and closings of the first four letters, those in this letter were distant and coldly formal. Second, the first paragraph deals with the important subject of the scientific value of morphology. Without access to Claus's previous letters, however, there is a problem. The title referred to, "Darwin versus Galiani," is the title of a recent, well-known address by Emil du Bois-Reymond, and therein lies an ambiguity, for Weismann clearly identifies it with Claus. Claus may have recommended it to Weismann, or Weismann may simply have used the title as an analogy for something Claus had mentioned in his correspondence, or it may be a reference to an unidentified talk Claus had sent him.[39] Whichever is the case, Weismann's point is clear. The famous Berlin physiologist had compared the fundamental grounding of physics in mathematical-physical laws to the science of morphology, which was reputedly grounded in Darwin's theory of evolution by natural selection and Haeckel's so-called biogenetic law. Although highly complimentary of Darwin's achievements, du Bois-Reymond considered natural selection as neither proven nor disproven—it was neither an empirical rule nor a mathematicophysical law. On the other hand, he argued that the "buildungsgesetz" of morphology was nothing more than an empirical rule. Claus must have associated himself with du Bois-Reymond's view of morphology, and Weismann equally clearly wanted to change Claus's opinion. His second paragraph contained a collegial gift to Claus of the second volume of his recently published *Studien*—whether this was or was not the lesson in morphology that Weismann promised Claus, with its exacting observations in natural history, its systematic experiments, and its general discussion of mechanism and teleology in the fourth essay, it was the best that Weismann had to offer.

With the appearance of Weismann's *Beiträge* II–IV at the end of January 1877, particularly with the added introduction to *Beitrag* II, permanent damage had been done on both sides of the relationship. Claus quickly penned his "Abwehr" in February, and Weismann replied with his "Rechtfertigung" in September. Science, however, was the real loser in the academic spat between these two gifted zoologists, for between them they might have sorted out the strengths and weaknesses of pursuing the natural history of development for the better understanding of phylogeny and the Darwinian mechanism of natural selection. Instead, each, in defending his turf, deferred to personal criticisms. Weismann's dilemma was that he had not found a way to respond to du Bois-Reymond or Claus. Was there more to morphology than simply empirical generalities about gross and fine anatomy and biological behavior?

7

From Germ Layers to the Germ-Plasm

The Study of Hydromedusae, 1877–1883

Upon receiving a copy of Haeckel's *Gastraea-Theorie,* Weismann praised its challenging contents but admitted to his "Friend and Colleague" that he had already heard disgruntled objections at the recent Versammlung in Wiesbaden. "It is so easy," he protested to Haeckel, "to present no theories under the pretense that the necessary foundation of facts is still missing. As though theory must not show what facts are to be looked for!"[1] These sentences reflected a recurring theme in Weismann's pursuit of scientific understanding, to which he would return again and again throughout his career. He never wavered from the belief that a productive theory requires painstaking thought and supporting evidence, which consumed an inordinate amount of time and energy. Casual criticism, on the other hand, was cheap and generally missed the mark. Weismann continued his letter with the prescient comment that "the homology of the germ-layers is or contains such a theory [the gastraea theory]. Whether it is correct or not, the facts to be discovered must teach us."[2] Over the next twenty-five months, the two friends were to exchange four additional letters about the newly appearing parts of Haeckel's *Gastraea-Theorie.* One unresolved question between them concerned the extent to which Weismann's earlier work on the development of Diptera enhanced Haeckel's grander vision of metazoan phylogeny. Neither of them saw how they might reconcile the superficial cleavage characteristic of eggs of insects and many other arthropods with the regular and irregular cleavage patterns found in the eggs of all other animals.

Weismann always remained cautious in the shadow of Haeckel's enthusiasm for his own evolutionary ideas. By temperament and because of his retinal handicap, he worked with a different style and at a different pace than did his Jena contemporary. Nevertheless, they shared the same goal of promoting a mechanical evolution theory through detailed zoological investigation. Although their research paths and objectives radically diverged, their public image grew in contrast, and their informal friendship waxed and waned. The historian must not underestimate their fraternal bond nor the influence Haeckel's work had on Weismann. This becomes quite evident as we follow Weismann into his next research project: the study of gamete formation in hydromedusae.

The Captivating World of Hydromedusae

We must envy the scientists of the nineteenth century. How satisfying, yet impossible, it would be to begin one's technical monograph today in the following manner:

> The creatures which I am about to describe and delineate in the following monograph are animals of very simple organization and beautiful form. They are members of the lowest section of the Animal Kingdom, and are intimately allied to the polyps, as we shall see when we come to consider their classification, which will be best understood after we have examined their structure. They are mostly minute, often microscopic, though many of their nearest relations, such as the great stinging Medusae, grow to a considerable bulk. They are active in their habits, graceful in their motions, gay in their colouring, delicate as the finest membrane, transparent as the purest crystal. They abound in the sea, but are not equally plentiful at all seasons. They have the power of emitting light, and when on a summer's evening the waves flash fire as they break upon the shore, or glow with myriads of sparks as they curl and froth around the prow of the moving ship or under the blade of the striking oar, it is to delicate and almost invisible Medusae that they chiefly owe their phosphorescence.[3]

These words, penned in 1848 by the English naturalist, geologist, and poet Edward Forbes, expressed sentiments of delight and respect for the organic form and the poetry found in natural form and processes, as he subjected the medusae of hydroids to exacting scrutiny. His was not the reverential awe commonly expressed by the natural theologians but a love for the beauty and diversity of nature, which was to succor Forbes, many of his contemporaries, and later generations of naturalists as well. These were the same feelings expressed by Haeckel in a letter to his fiancée, Anna Sethe, eleven years later when he described the luminescence of marine organisms in the Blue Grotto of Capri.[4] This combination of love for the neoplatonic mystery of nature, followed by collection, ordering, and synthesis, was common at midcentury at a time when the term "naturalist" was more frequently used than the late-eighteenth-century neologism "biologist." The science of both Forbes and Haeckel was influenced to some degree by the German tradition of romantic nature, but this did not blunt the sheer delight of experiencing the beauty of the natural world on its own terms.[5] The notion of a "love" of nature even penetrated into the professional monographs of the more analytical twentieth century. Alfred Goldsborough Mayer added to his twentieth-century world catalog of medusae an equally lyrical passage at the end, where he expressed the quintessential motivation of Forbes, Haeckel, and later biologists including himself: "Love, not magic, impels the naturalist to his work."[6]

Hydrozoa

There are few organisms that have elicited such responses more regularly than hydromedusae. When one surveys the great nineteenth-century monographs of these tiny jellyfish and their polyps, one is immediately struck by the love and care that is lavished upon them. Forbes added to his work thirteen hand-colored plates portraying the delicate, crystal-like organisms and regarded them as worth the proverbial thousand words for demonstrating the love of nature that informed his studies. In the same fashion, Sir John Graham Dalyell, one of the most active students of polyps and medusae in the 1830s and 1840s, embellished his volumes with exquisite hand-colored lithographs that captured both the delicacy and richness of these animals. In his case, moreover, they also reflected his delight in God's creation.[7] One of the most sumptuous works produced by the great Swiss naturalist Louis Agassiz consisted of a two-volume study of these "headless animals," the *Acalephs*. His collaborator, Henry James Clark, executed the detailed and beautiful plates, which continued to be of aesthetic and scientific value long after Agassiz's God-centered system of classification became obsolete.[8] With its two oversized volumes, this work exuded all the enthusiasm for nature that was characteristic of Agassiz's career as a teacher and scientist. George James Allman's monumental and scholarly volume on *Gymnoblastic or Tubularian Hydroids* of 1872 presented not only the most comprehensive guide to the known species and families of "athecate" hydroids of the North Atlantic but displayed the polyps and medusae together in twenty-three large and magnificently colored lithographs.

The ultimate of lavish and artistic presentations of medusae may be found in Haeckel's *System der Medusen*. The two volumes of this uncompleted work sported an atlas of forty colored lithographs, which might be imagined as the zoologist's counterpart to Wagner's contemporary Ring Cycle. Rich with color and detail, Haeckel's plates come alive with the exotic forms that had drifted into view from unknown places. Serpentine tentacles and voluptuous medusae bells filled up the space on the page with graceful and repetitive curves. So much were these illustrations an expression of Haeckel's emotional as well as analytic involvement that one of the most gorgeous of the plates he produced memorialized his recently deceased wife. If the gods of Valhalla, who had first burst forth on the stage of the Beyreuth Festspielhaus three years before Haeckel's publication, had chosen vessels for their earthly play, they might well have fashioned them in the mode of Haeckel's romanticized medusae. The scientific merits of the *System der Medusen* were less clear. Although complimentary in his evaluation, Weismann did not adopt Haeckel's elaborate nomenclature and system of classification. Mayer in the next century was somewhat more brisk: "With every respect for Haeckel's great work, it has appeared to me that its subdivisions are often too precise to be convenient, and too artificial to accord with nature."[9]

Aside from the sheer pleasure that they provided the nineteenth-century naturalist, these beautiful and intricate marine organisms also presented Weismann with a number of formidable scientific challenges: How should one classify the wondrous diversity? What were the limits and implications of their complex life cycles? And how should one understand their development in terms of the contemporary interest in primitive germ layers? We shall see that each of these questions helped shape Weismann's own studies of hydromedusae. Consequently it is worth reviewing the status of each question as all of them converged in the late 1870s.

Classification and Anatomy

Classification in itself might seem to the historian of science to be a dry and unrewarding enterprise. With the publication of Mary P. Winsor's *Starfish,*

Jellyfish, and the Order of Life, however, we have learned to appreciate that below a caramelized surface of Latinized names and diagnostic features boils a broth of biological and philosophical issues.[10] Because Weismann was not deeply interested in classification for its own sake, we will not pursue this issue in depth, but it serves a purpose to sketch out the taxonomy of these organisms as it appeared to zoologists by 1880, and how it became stabilized by mid-twentieth century.[11]

"Hydromedusae" was an antiquated and ambiguous term by the 1880s. It referred to small medusae, ranging between an eighth to two inches in diameter. The group stood in contrast to the "syphomedusae" or the larger and more familiar jellyfish that wash up on beaches around the world. In both cases, the terms referred to the free-swimming stage of their compound life cycles, which often but not always included a polyp stage, that is, a hydropolyp or scyphopolyp. It was not uncommon, however, to find the terms used in a formal taxonomic sense, as Lankester and G. Herbert Fowler did as late as 1900.[12] These taxonomic uses were being displaced by the terms *Hydrozoa* (Huxley) and *Scyphozoa* (Haeckel) by the 1870s, which emphasized more appropriately the "polyp-like" and "cup-like" shapes of the sessile stages in their development. (Haeckel coined the term Scyphozoa and included in this taxon the Anthozoa (i.e., sea anemones), the scyphopolyps, and scyphomedusae as subtaxa.)[13] Both hydrozoans and scyphozoans have a compound life cycle, which begins with a newly hatched planula larva. This swims in the zooplankton and crawls on the bottom until it attaches itself to a substratum of rocks, shells, driftwood, or other submerged or floating objects. We will follow this life cycle in a moment. Throughout I will use the terms "hydromedusae," "hydrozoans," and "Hydrozoa" interchangeably. These terms include the hydroids and siphonophores, both of which Weismann worked with, but not the scyphozoans, the traditional jellyfish.

To add further context and confusion to these terms, the freshwater *Hydra* is a solitary hydrozoan. It presents an exception to the common colonial lifestyle, yet as a common, generic designation this term represents a general type, which often serves as the pedagogical prototype of its more complex cousins. Such a traditional and reflexive use of *Hydra* unfortunately carries with it obvious taxonomic, evolutionary, and developmental assumptions, both then and now.

Nevertheless, like *Hydra,* the few other freshwater and the many marine hydromedusae are diploblastic or double layered with an epidermis and a gastrodermis, or to use the frequently employed embryological terms, an ectoderm and endoderm. These layers are separated by a gelatinous mesoglea, which is particularly pronounced in the medusae stage. As mentioned previously, hydromedusae are colonial and lead a sessile existence attached to the substratum by means of a common base or hydrorhiza. The individual polyps of the colony are attached to a central pinnate or arborescent stalk, referred to as a hydrocaulus or central hydranth. Within a single colony, the polyps are also dimorphic, often even polymorphic. The different forms take on different functions that benefit the entire colony. Thus, some polyps with tentacles surrounding a mouth assume the function of feeding. These are known as gastrozooids or hydranths, and with their tentacles and distal mouth surrounded by a hypostome, they superficially resemble the simple *Hydra*. Other zooids, known as gonophores, carry on the function of reproduction; still others among certain species are armed with stinging nematocysts, which serve protective functions.

Weismann's principal concern was with the gonophores or reproductive zooids, which exhibit a variety of forms and products. In some species, they contain the rudiments of free-swimming medusae that, when set free, disperse and develop into intricate medusae that sexually produce the eggs of the next generation. In other species, the gonophores retain medusoid bodies on a central blastostyle and produce eggs internally, or they may simply produce gametes and fertilized eggs from the blastostyle, or in certain species the gonophore is limited simply to a sporosac, which liberates planula larvae directly into the seawater. In short, the hydromedusae present an entire spectrum of sexual and asexual modes of reproduction, from budding through fission to varieties of sexual reproduction and alternation of generations. The problem confronting the nineteenth-century zoologist was how to disentangle the evolutionary sequence of these variations of the gonophores and how to understand the status of these structures with respect to the other zooids.

When liberated, the hydrozoan medusa floats in the ocean like an inverted bell or cup. These medusae, in contrast to the larger medusae of the Scyphozoa, possess a tissue-thin membrane or defining vellum, which extends inwardly from the periphery of the subumbrella toward the manubrium. Not only is this structure diagnostic for the class, but it provided the grounds for the older designation of Crespedota. It also played an important role in the arguments we will examine concerning the phylogenetic origin of the gonophore. The upper and lower surfaces of the medusae, designating the exumbrella and subumbrella, respectively, are ectodermal in origin. Internally there exists a gastric cavity with various modified radial and circumferential extensions or canals lined by a layer of cells of endodermal origin. Sandwiched between both layers is the same gelatinous mesoglea that gives the scyphozoan jellyfish their jellylike consistency. A mouth enters the gastric cavity from the surface of the subumbrella; the orifice itself consists of an extension called the manubrium.

Generation and Primitive Germ Layers: 1878–1883

Upon closer examination it is readily apparent that the medusae of hydrozoans possess a similar structure to the polyp. As Allman and the Hertwig brothers pointed out in the early 1870s, one can mentally invert the medusa and easily homologize its basic structures with those of the simple polyp.[14] Thus, the double germ-layer structure with its supporting mesoglea is readily identifiable in both. The tentacles that hang from the circumference of the medusa's subumbrella may be interpreted as being homologous to the hydra's tentacles; the manubrium of the medusa appears similar to the hydra's oral cone or peristome. Such successful comparisons elegantly reinforced the morphological tradition of the nineteenth century. The accepted homologies made it possible for Weismann and others to proceed with a more refined examination of the origin of germ cells in the hydromedusae. When he turned to his hydromedusae research, Weismann relied heavily upon Allman's descriptions of polyps.

When they are formed, the free-swimming medusae provide a conspicuous component in the life cycle of the hydromedusae. They are the animals celebrated by Forbes and his contemporaries. With their free-floating existence, bell-like structures, delicate tentacles, and pendulant gonophores, they give the cursory appearance of taxonomically different organisms in contrast to the inconspicuous polyp with its multiple zooids. Most of Haeckel's generation understood the cyclical relationship between the sessile polyp colonies and the medusae, and between hydrozoans and scyphozoans; however, even seventy years later, the great American systematist Lybbie Hyman summed up the difficulty that hydrozoans still presented:

> The hydroid and medusoid generations may evolve independently, so that often very similar medusae are budded from quite different hydroid types and vice versa. Because of these difficulties it is not yet possible to erect one single scheme of classification for both hydroids and medusae, and often medusae and the corresponding hydroid colony bear different generic names.[15]

Such a statement bears witness to the fact that even in the middle of the twentieth century hydrozoan taxonomy remained dependent upon traditional morphological guidelines.

A final word about the anatomy and taxonomy of the hydromedusae will be useful for what follows. In Hyman's classic text *The Invertebrates: Protozoa through Ctenophora,* the modern hydrozoa are divided into five orders. Allowing for some discrepancies in the categories used, the mid-twentieth-century classification is essentially the same as that used by the zoologists who organized the class in the 1870s and 1880s. For example, except for the fact that he included the Acalephae (Scyphozoa) with the class, Carl Claus in his *Grundzüge der Zoologie* of 1876 provided a convenient and acceptable classification of the hydromedusae at the time.

Today, the prevailing view of the origin of the coelenterates holds that the Hydrozoa, Scyphozoa, and Anthozoa all evolved along divergent paths from a primitive planktonic medusa that resembled the larva of the order Actinulida.[16] This interpretation results in the conclusion that the hydroid polyp is, as Hyman puts it, "a persistent larval state" that has become the dominant, even exclusive, stage. The more advanced hydrozoans, according to this view, would be those that emphasized the sessile polyp form at the expense of the medusa. It also suggests that the medusae of the Hydrozoa and Scyphozoa were independently derived and so are analogous rather than homologous stages. Böme first suggested this phylogenetic sequence in 1878, but it was the American zoologist William K. Brooks who raised it to the status of a well-documented theory in a masterful monograph in 1886.[17] Through embryological studies of the pelagic Narcomedusae and Trachomedusa (Trachylina), Brooks concluded that the ancestor of the hydromedusae was a "solitary swimming . . . Actinula" that first through metamorphosis became a highly organized free-swimming medusa and then, in its larval stage, adapted itself to a sessile mode of living and intercalated a polyp generation between larva and medusa. He viewed the subsequent additions and deletions as variations upon this pattern. In 1900, Lankester and Fowler added the Hydrocorallinae as a sixth order. From these descriptions of the anatomy and classification of the Hydrozoa, it should be clear that a mastery over the life cycles in their totality was essential for the naturalist's sense of order. It was equally essential for zoologists interested in development and reproduction.

When Weismann began his research on hydromedusae, there were practical reasons why these and other coelenterates were of great interest to contemporary zoologists. Vertebrate anatomists, such as Wilhelm Waldeyer,

Wihelm His, and Carl Semper, had made substantial progress in tracing the origins of the germ cells back to undifferentiated cells of the germinal epithelium associated with the rudiments of the urogenital tract of both higher and primitive vertebrates. They had argued, as Waldeyer emphasized, that the early embryo must be hermaphroditic before further development directed the embryo onto either a female or male track. They had further argued, as Semper insisted, that the presumptive germ cells of the germinal epithelial were sexually indifferent "Ureier" and that only after these migrated to the presumptive gonadal regions attached to the Müllerian and Wolffian ducts could one legitimately speak of specified sex cells. Theirs was in keeping with the reigning conviction that development of the individual rather than a sexual entity preestablished at conception determined the production of eggs or spermatozoa.[18]

Coelenterates, however, were different and provided an advantage to the investigator that the vertebrates did not. In addition to their general appeal as a diverse and beautiful group of organisms, they and the sponges were the simplest metazoans and the only ones with a bilaminar structure. This rendered them natural models for exploring the relationship between the primitive germ layers and adult structures and in particular for resolving questions about the place and the determination of the germ cells. In theory, it should have been simple to sort out these questions on the bilaminar level, but things are rarely straightforward in even the simplest of organisms. Beginning with Thomas Henry Huxley, who had pointed out that medusae, polyps, and colonial siphonophores all possessed similar bilaminar structures, which he in turn analogized with the two primitive germ layers of vertebrates, there were uncertainties and ambiguities. Huxley at first thought the gametes arose in the interstitial lamella, but ten years later, reasoning functionally that they needed to be expelled into the surrounding ocean, he argued that the gametes must be products of the ectoderm.[19] In contrast Kölliker, Franz Eilhard Schültze, and Haeckel were inclined to believe in the endodermal origin of the sex products whereas George James Allman, who had baptized Huxley's two layers the "endoderm" and "ectoderm," waffled on the matter. Again, Keferstein, Ehlers, and Claus considered the ectoderm a more likely candidate as the site of origin; while in his enormously influential monograph on *Hydra,* Dohrn's assistant Nicolaus Kleinenberg (1872) reasserted that the testicles and ovaries were formed in the interstitial layer, notwithstanding which he argued they were basically ectodermal.

There were a number of factors that led to these confusing and contradictory conclusions concerning the germ-layer origins of the gametes. The techniques of preservation and microscopy were only catching up during the decade to the demands of the questions posed. Second, the different species involved and the unexplained discrepancies presented by the different life cycles of even closely related species rendered it problematic when the investigator leaped from perfectly accurate observations on one species to a generalization about a whole family, class, or even phylum. Another difficulty was articulated by Eduard van Beneden, whose 1874 monograph on the hydroid *Hydractinia echinata* set a new technical standard for such investigations. There existed an intellectual barrier, or a "dogma" as he more provocatively called it, that he felt all investigators including himself had labored under. Up to the time when he became disillusioned, he had followed his peers in assuming that what must be true for the origins of the gonads and gametes of one sex of a given species must also be the case with the other. So if an investigator had focused on, let us say, the formation of the ovaries and the production of eggs, it was easy to transfer his conclusion to the testicles and production of spermatozoa, the development of which might have been harder or impossible to follow. In brief, van Beneden told of his experience: "When I returned to Ostende in order to try to follow a personal conviction about the question of the origin of the sexual products among the zoophytes, I was far from imagining that it would be different in the two sexes. At first I made all my observations on the female colony, because the eggs are much easier to distinguish in the midst of any tissue than the spermatic cells." After relating how surprised he had been to find that the male sporosacs did not follow the same pattern of development as the female, he added, "I had become so convinced in the belief of the commonality of the origin of the ovary and testicle that at the beginning I was more tempted to doubt my senses than the truth of scientific dogma."[20] The details of his subsequent report on his empirical findings served as a retelling of his new revelation.

As with most hydroids, the colonies of *Hydractinia* are dioecious, meaning they are either male or female. Furthermore, as van Beneden made clear, the two sexes are easily distinguishable by color, for the eggs and ovaries lend the female colonies a reddish hue whereas the testicles render the male colonies a milky white. Furthermore van Beneden perfected a technique of first soaking his specimens in a weak solution of osmic acid for preservation and then following this with a stain of platinum chloride and chromic acid. The combination allowed him to trace the germ layers both in the hydranths and sporosacs, and to distinguish the egg and sperm cells. It was not surprising that he found only the former in the female colonies and the latter confined to the male colonies. What seemed clear from his account was that, at all stages in the development of the sporosacs, the eggs could be easily associated exclusively with the epithelium of the endoderm; the spermatozoa were identified as cells that had invaginated from the ectoderm to form the endocodon of the male sporosac.

His lengthy report was rigorous and stunning; his general conclusions were more daring yet. Because the female sporosacs possessed the rudiments of the male endocodon (the invaginated ectodermal cells) and because the male sporosacs similarly possessed rudiments of the female endodermal epithelium, van Beneden concluded that "the sporosacs are therefore morphologically hermaphrodites." He further noted that the ectoderm and endoderm possessed opposing sexual significance and that each germ layer was charged with what he considered were sexually related physiological and social functions. Thus, the ectoderm was neural and muscular, and the endoderm was vegetative in both an embryological and gender-related sense. Furthermore, van Beneden, as had his predecessors, leaped tentatively from conclusions drawn from one species to the entire animal kingdom. If the germ layers were associated with opposed sexual functions, he suggested, fertilization was a mechanism ensuring the renewed union of those two capacities. "The new individual comes into reality at the moment when the union between the elements of opposed polarity function together completely as the molecule of water is formed by the union of the atoms of hydrogen and the atoms of oxygen."[21]

Despite the excellence of van Beneden's observations, they and his associated theory failed to disentangle convincingly the relationship between germ layers and germ cells of *Hydrozoa*. Gegenbaur accepted the conclusions as far as hydroids were concerned but added the afterthought that a migration of germ cells from the ectoderm to the endoderm might explain the apparent contrasting origins of the male and female gametes (reported in the second edition [1877] of his *Grundriss der Vergleichenden Anatomie*). A young student of Claus's in Vienna, however, challenged its application even within the hydroids. Reporting on work done in Trieste, he concluded that "with Tubularia the eggs as well as sperm develop from the ectoderm, but with Eudendrium the eggs develop from the ectoderm and the zoosperms from the entoderm."[22] Kleinenberg took up Gegenbaur's suggestion of germ cell migration by asserting that female germ cells in *Eudendrium* migrated into the endoderm and later returned to the ectoderm at maturation. His conclusions were based on "optical cuts" through living specimens. At the same time, he confessed that he could not recognize the developing eggs until they had increased in size in the endoderm, which left their origin still in doubt. He also pointed out that he never actually saw the germ cells passing from endoderm to ectoderm. Nevertheless, Kleinenberg emphasized the amoeboid mobility of the germ cells and their appearance in both germ layers and the lamella. He further hypothesized that the migration he proposed, from ectoderm to endoderm then back to endoderm, might be a process to bring the developing germ cells into closer proximity to the nutrition provided by the gastric cavity. His paper was a challenge to both the morphological notion of stable germ layers and to van Beneden's theory of their hermaphroditic nature.[23]

Hydroids were demonstrating the difficulty of forcing germ cell production to conform to the dictates of germ-layer specificity. Balfour summed up the frustration that zoologists must have felt in trying to arrive at a unified picture of the production of gametes in hydroids, let alone the coelenterates in general:

> In view of the somewhat surprising results to which the researches on the origin of the genital products amongst the Coelenterata have led, it would seem to be necessary either to hold that there is no definite homology between the germinal layers in the different forms of Coelenterata, or to offer some satisfactory explanation of the behaviour of the genital products, which would not involve the acceptance of the first alternative.[24]

Weismann on Hydromedusae

Early Work

In the spring of 1868, Anton Dohrn made a point of stopping at Freiburg on his way to Switzerland. The visit made good sense. He had just completed his Habilitationsschrift in Jena and had been impressed by Weismann's embryological work on Diptera. When he began publishing the results of his own research on the freshwater isopod *Asellus aquaticus,* Dohrn also received a highly complimentary letter from Weismann. As he traveled west from Jena, Dohrn would have realized that in the past year Weismann had been appointed a salaried außerordentlicher Professor charged with the responsibility for teaching zoology at the university. A year later as he traveled to Italy with his family and father, Weismann returned the visit to Dohrn in Naples. This gave him the chance, as he reported to Haeckel, to hear firsthand Dohrn's plans to build a city museum and aquarium. The visits were the beginning of a life-long, though circumspect, friendship between the two zoologists. They not only evinced an enthusiasm for embryology and a strong belief in evolution, but they practiced their profession with the same methodical and focused style. Dohrn evidently found Weismann nervous and distrustful of others, but as time would tell, they both shared a love of music and the fine arts that ce-mented cordial relations and a mutual professional respect.[25]

Quickly the museum became a public aquarium, and the research institu-tion became Dohrn's driving concern. When he was at the stage of establishing the system of table subscriptions to help support the research side of the equa-tion, Weismann, as did many other zoologists around Europe and the United States, responded enthusiastically. He cautioned Dohrn, however, that the Uni-versity of Freiburg was not wealthy and therefore could provide only limited support.[26] Nevertheless, the association served Weismann well as he brought his daphnid research to a close in January 1877. By this time, he was already receiving medusae specimens from a Dr. Calberla in Messina and sharing sam-ples and information with Haeckel.[27] He became fully committed to his next research project, the study of hydromedusae, when he successfully applied to occupy the Baden table at Naples.

Hydroids

Hydroids and daphnids lie at opposite ends of the invertebrate spectrum. To turn from one to the other, however, was an understandable move for a keen naturalist and important for a young "außerordentlich" professor interested in demonstrating his command of life's diversity. More to the point, Weis-mann's use of both research organisms revealed a common concern in the bi-ology of reproduction. In retrospect, it seems clear that there was a natural continuity in the questions he posed: Whence the germ cells? What were the differences between fertilized and parthenogenetic eggs? What were the bonds between sexual and asexual generations? And what was the function of mul-tiple modes of reproduction? But it is less clear that these were the questions he started with as he arrived in Naples in the middle of March 1877. Unfor-tunately, he collected and examined hydroids for only three weeks when he was suddenly called back to Freiburg because of the fatal illness of his youngest daughter Meta who, as it would turn out, died three days after his return.

Despite the personal setback, Weismann continued over the next four years to collect and examine dozens of species of hydroids and siphonophores, or to use the common collective term, hydromedusae. In the spring of 1878, he journeyed to a laboratory at Marseilles for further specimens. (Not to be confused with the marine station of 1887.) Its young director Antoine Fortuné Marion and Weismann established a friendly working relationship, and by using Marseilles as a base during the Easter holiday, the visitor from Freiburg collected specimens along the Riviera coast. He appears to have both preserved specimens for future examination and studied many species in vitro while they developed before his inspection. Alas, some of his living specimens perished on the return to Freiburg when their glass container was inadvertently kicked over and most of its water spilled. Weismann commented to his French colleague that an emergency shipment of seawater from Berlin had failed to arrive in time to save his quarry.[28] Marion evidently offered to supply Weismann with additional specimens, both fresh and preserved, as Weismann began requesting species of *Gonothyraea, Campanulaea,* and *Plumularaea* and included instructions as to where to find them and how to preserve them. At that time, absolute alcohol and picric acid were his preservatives of choice. Furthermore, Weismann's correspondence to Marion makes clear that the two had discussed Weismann's further research plans in a general way, for he asked Marion to remind him about collecting points along the coasts of Brittany and Normandy.

Their correspondence, however, was not exclusively concerned with hydromedusae. Weismann requested the eggs of the fairy or brine shrimp *Artemia* so that he might attempt to reproduce the research recently reported by Vladimir I. Schmankewitsch, who had experimentally transformed saltwater forms into freshwater forms by changing the ambient salt concentration. Weismann's attempts to duplicate Schmankewitsch's results were aborted, but he was to return to the subject on a number of occasions in the future. Weismann's correspondence often switched to lighter subjects. He expressed to Marion a desire to see a French impressionist exhibit in Paris, and he groused about the time-consuming need to present popular lectures before a general audience in the auditorium of his university. More distracting yet was a recurring illness of his wife and his election as decan of the philosophical faculty in January 1879. That honor unfortunately entailed functionary duties, which forced him to curtail a return trip to Marseilles that spring when the hydromedusae would again pass through their reproduction stages. Finally, he still had to finish writing up his work with *Daphnia,* "so up to now my work of the recent year lies untouched," as he complained to his friend.[29]

Focus on Hydromedusae

By the spring of 1880, Weismann was free to focus exclusively on his hydro-medusae. He appealed to both Marion and Dohrn for new specimens, and this time he recommended different killing and dehydrating solutions that were superior to the picric acid used in the past.[30] The change may have indicated he was becoming more interested in his microscopic examination of the fine details of germ cell movements. Perhaps the most pressing concern, however, was a realization that other investigators seemed to be converging on the same set of phenomena that Weismann must have identified, so he drafted an overview of his hydroid work to date and sent it on April 10 to the *Zoologischer Anzeiger,* a new journal founded by J. Viktor Carus and designed for the rapid dissemination of new zoological literature, recent research reports, and news of the profession. Weismann promised Dohrn that this statement would contain "the quintessence" of his hydroid efforts and assured Marion, "I have finally come to a point to bring my observations on hydroid polyps together." He, however, intimated concerns about competition in the first paragraph of his report. Nevertheless, the *Zoologischer Anzeiger* was good to its promise, and the first of a series of short publications on his hydromedusae studies appeared on May 8.[31]

The first brief report was more of a survey of what Weismann had accomplished in the past two years than a presentation of new material. It seems to have been a preemptive strike to avoid a recurrence of the misunderstanding he had with Claus rather than a research prospective. With a subject that called for illustrations, it is significant that Weismann chose the *Zoologischer Anzeiger,* which emphasized speed of publication and accomplished that goal by accepting no illustrating plates. Weismann explicitly recognized Julien Fraipont's recent papers, one of which had been published three weeks earlier in the *Zoologischer Anzeiger* as well, as the stimulus for his writing.[32]

That he had been searching for a larger picture than simply recording where and in what germ layers the germ cells were first identified was made clear in Weismann's subsequent report published two months later.[33] Here he described the germ cell origins in three additional species (*Plumularia echinulata, Eudendrium ramosum,* and *Cordylophora*) that he had found in specimens lent to him by his "good friend" Franz Eilhard Schultze. His conclusions, however, began to sort out the problem in a different way. He noted that one could divide the hydroids into two types. Those in which the germ cell origin was found or assumed to be in the coenosarc, he designated "Coenogone." Those in which the germ cell origin was found in the blastostyle, he designated "Blastogone."

The distinction was to be short lived, but it reflected the Hertwigs' division of the coelenterates into "Ectocarpen" and "Entocarpen." The distinction also forced Weismann to think more carefully about the significance of the alternation of generations in the hydromedusae.

By 1880, Weismann would later say, his eyes had continued to get better. Nevertheless, he vividly described a complex of problems that would plague him for the rest of his life: "A high sensitivity a[nd] sudden exhaustion still remains, a[nd] along with work at the microscope reading and writing can only take place to a limited degree, a[nd] it still remains for me a really difficult task to follow the scientific production of the time."[34] Despite such a handicap, he remained hopeful for improvement and plunged into a period of intense research with more hydroid species, more requests for specimens from Dohrn and Marion, and further urgent appeals for the expeditious publications of his short reports in the *Zoologischer Anzeiger*. In addition, Weismann devoted the spring of 1881 to proofreading Raphael Meldola's English translation of his lengthy, two-volume *Studien zur Descendenztheorie*. Weismann returned the completed manuscript to Meldola on August 27, 1881, and signed the English edition's preface in November before leaving for Genoa on August 3.

The year was also punctuated by Weismann's submission of a lengthy monograph on the early development of insect eggs for a festschrift honoring his Göttingen mentor Jacob Henle. This would appear the following April. By mid-September 1881, the same entomological material became the subject of a short presentation to the zoological section of the Naturforscher-Versammlung in Salzburg. Two days later, Weismann had the honor to deliver his first plenary address at the Versammlung.[35] By early December, he sent a finished version, revised to accommodate a more general audience, to his new publisher Gustav Fischer. It had been a busy and productive year.

The scope of the plenary session talk and ensuing monograph, as we will see in Chapter 8, reached far beyond his hydromedusae and insect work and required a different kind of data collecting. To give the Salzburg talk was an honor, but it also suggests an irony. It was the first of three such plenary session talks Weismann was to give before the Versammlung during his lifetime and preceded by one year the last of Haeckel's three plenary talks before the same professional audience. Weismann's career trajectory, stymied at first, was beginning to overtake that of his Jena colleague in the view of Germany's most distinguished body of scientists. Above all, 1881–1882 marked a watershed in Weismann's conceptualization of germ cells and propelled his career into a profoundly new direction.

With specific attention to his hydromedusae, Weismann began 1881 by sending Marion illustrations and a three-page "supplement." He instructed that they should be appended to an article he had written in French that covered more extensively the work described in his two notes to the *Zoologischer Anzeiger* of the previous year. It appears that he was now feeling the pressure from Kleinenberg, who had described in the same venue the amoeboid movements of germ cells in the coenosarc, along the hydranth, and toward the gonophores. Kleinenberg had even recounted observing the germ cells of living specimens passing from one germ layer to the other. In a revealing passage, Weismann explained to Marion that it was only an "accident" that he himself had not seen a similar passage ("Durchbohren") of the eggs through the lamella from one germ layer to the other the previous summer. Explaining his failure, he mentioned that "at that moment I did not have the time," then adding almost as an afterthought, "moreover I could not expect to see something else other than what I had already seen with Eud. Capillare, i.e., E. ramosum."[36] In short, Weismann had not expected that the *Eudendrium* specimens in Naples would be different from those he had examined on the Riviera, but he needed to be precise about the trail of his discovery. It is possible that he wished to avoid another altercation similar to the one he had just weathered with Claus.[37]

His cryptic sentence about "Eud. Capillare" makes sense only in the context of the footnote and the supplement he had sent Marion earlier in the month. Weismann appears by then to have realized that there must be an orderly movement of germ cells from the coenosarc to the blastostyle and gonophores and that this must entail a passage through the lamella from one germ layer to the other. As indicated previously, he had confounded the species of *Eudendrium* collected from Naples in 1877 and Marseilles in 1878. With better results obtained from the latter, he had failed to double-check his older Naples samples; now he was presented with a set of necessary readjustments. Were the samples from Naples and Marseilles both *E. ramosum* as he initially believed? Might the new specimens from Naples be instead *E. racemosum* Cav.? By the early months of 1881, Weismann had come to realize that even *E. ramosum* from the north coast of the Mediterranean was, in fact, a variant of another species, *E. capillare,* described by Allman as inhabiting the coast of England. So upon reexamination of the gross morphology, Weismann concluded that he was actually dealing with two closely related but different species—*E. racemosum* and *E. capillare*—that were often hard to distinguish. For the question of germ cell origins, however, it was essential to do so. The correction allowed him to explain how his account of germ cell origins differed from those of Alexander Goette.[38] By then, there could be no question that female germ cells of *E. racemosum* originated in the ectoderm, and the female germ cells of *E. capillare* appeared to originate in the endoderm.[39]

There was more, as he suggested in his French supplement. "But this difference in the mode of the origin of the same sexual elements in two close species," he asserted, "shows again that it is impossible to grant a fundamental importance in the formation of the reproductive organs to a determined germ layer."[40] Within the hydranth or blastostyle there appeared no specified germ-producing germ layer. Once again, this suggested a verbal confusion between the primitive germ layers of egg cleavage and the ectoderm and endoderm of the polyp development. In other words, there was no need to assume the germ layers of the embryo and those of the asexual generation were functionally similar or even homologous.

There remains the historical question of whether Weismann's failure to distinguish between *E. ramosum* and *E. racemosum,* as suggested in his letter to Marion, led him to overlook a trans-germ-layer migration between 1878 and 1881. We may never be sure, but we know for certain that by early 1881 germ cell migration became a central theme in his research. In 1883, Weismann implied that he became aware of trans-germ-layer migration independently of Kleinenberg.[41] The supplement he sent to Marion not only described the migration of the germ cells from the ectoderm to the endoderm, but his additional illustrations and explications both clearly referred to an incident in which he captured on a single slide the passage of female germ cells through the lamella: "One sees half of this egg between the endodermal cells, while the other half finds itself still in the ectoderm."[42]

Revisit Naples

As indicated previously, Weismann was intensely occupied with his research and its implications in 1881–1882. By February and March 1881, he had sent three further brief "Observations" to Carus at the *Zoologischer Anzeiger*. Two of the three dealt indirectly with the thrust of his ongoing research. The third was a published statement about his taxonomic confusion with the two *Eudendrium* species.[43] He sent to Naples a slightly longer note on hitherto unrecognized urticating capsules, baptized "cnidophores," on the hydranths of *E. racemosum,* in part to placate Dohrn for the lack of publication of his work done in association with Dohrn's Zoological Station.[44]

It was clear, however, that he needed to revisit Naples and its rich sea life. After having to cancel a planned trip to the Zoological Station in the summer semester, when *Eudendrium* would be at the peak of its breeding season, Weismann secured permission from his colleagues and the Baden administration to occupy the Baden table during the winter semester of 1881–1882. Groeben has written a wonderful account of Weismann's four-month visit.[45] Drawing upon his entries in his Tischkalender and Notizbücher as well as institutional documents, she has been able to identify people he met, from the supportive staff at the Stazione, the German ambassador, and the African explorer Gustav Nachtigal to Prince Heinrich of Prussia. She mentions his many outings, several of which were more than a day in length, to areas around the Bay of Naples from Ischia and Cape Miseno west of Naples to the Island of Capri, Sorrento, and Positano southward across the bay. It is not surprising that the former alpinist would enjoy the delightful views of the city, the rural villages, surrounding hills, and Mt. Vesuvius, as well as the many spectacular views of the bay and its many islands. The station provided other opportunities for relaxation, which included evenings of enjoyment at a local restaurant. For twelve days in February, Weismann also left the Bay of Naples altogether for a vacation with Mary in Sicily.

Groeben details Weismann's adventure on the Monday after his arrival of descending into a grotto wearing the Zoological Station's new diving helmet. He dove twice that day, the first time for just 10 minutes and the second time for an hour. Not given to romantic superlatives, Weismann nonetheless expressed enthusiasm for the "marvelous view of the rock walls of the grotto overgrown with sponges, astroids, [and] hydroids." Frustrating, however, is the lack of specificity in Weismann's additional daily entries after his arrival. His brief sentences only give a sense of daily routine—at least through December. He expressed satisfaction with the arrangement of his work area and the available equipment. He appears to have spent the morning until the midday meal at the Zoological Station, probably both at his table and in the library. He also recorded the many letters received from and written to Mary and his father-in-law. He prepared manuscripts and illustrations to be mailed to Gustav Fischer and Wilhelm Waldeyer.[46] Weismann appears to have spent many afternoons walking in the neighborhood, perhaps visiting a fortress, often in company with Paul Mayer or other guests at the station. Groeben has documented that Weismann had the opportunity to meet the American microscopist Charles O. Whitman; the British comparative embryologist Francis M. Balfour; perhaps with the plant cytologist Eduard Strasburger;

Figure 7.1. August Weismann in 1882.

Klaus Sander, ed. *August Weismann (1834–1914) und die theoretische Biologie des 19. Jahrhunderts. Urkunden, Berichte und Analysen.* Freiburger Universitätsblätter, vols. 87, 88. Freiburg: Rombach Verlag, 1985. Helmut Risler, "August Weismanns Leben und Wirken nach Dokumenten aus seinem Nachlass," p. 29.

the contentious Haeckel student Otto Hamann; and the Italian marine lieutenant Gaetano Chiechia, who was learning the craft of specimen collecting and preparation. Given all that can be gleaned from his correspondence, his *Tischkalender,* his *Taschenbücher,* and the zoological station's archives, we get the sense of a very fruitful four months. When Weismann returned to Freiburg in early May for the start of the summer semester, he appeared relaxed and ebullient about his experience. Both August Gruber and Robert Wiedersheim reported back to Dohrn what a successful trip it had been for their brother-in-law.[47]

Hydromedusen of 1883

In 1883, two substantial monographs on medusae appeared. One was written by Carl Claus and concerned the development of and phylogenetic relationship within the scyphozoans.[48] The other was Weismann's full-blown study of the sex cells in hydromedusae. Despite their embryological approach and common taxonomic concerns, there could not have been a greater difference between their two works.

As mentioned in Chapter 6, Claus had studied and published on polyps and medusae for half a dozen years. In this monograph of just under a hundred pages, he focused on a detailed morphological study of three species of scyphozoans while briefly comparing the class with the other coelenterate classes. This is particularly evident in his examination of the origin of the germinal epithelium and germ cells in the "Schirmquallen," as he called the scyphozoans. In the three species he studied in fine detail, he determined that the germinal epithelium was endodermal in origin and to this extent he confirmed the Hertwig brothers' claim that scyphozoans and anthozoans were "Entocarpen." In fact, Claus tended to agree that the scyphozoans and anthozoans were more closely related to each other than to hydrozoans. He, however, did not endorse the Hertwigs' major taxonomic divide between the Entocarpen and Ektocarpen, which implied a diphyletic origin of the coelenterates and the assertion that the two forms of medusae were the result of convergent evolution.

Overall, it was a monograph with a limited objective. Claus felt that by focusing on the very earliest stages of development in a few species he could test and, in fact, refute the Hertwigs' major taxonomic claim based on the contrasting origin of sex cells. In style and content, Claus's monograph was a traditional study in comparative and descriptive embryology. It followed the questions first broached by Haeckel, the two Hertwigs, and the recent wave of studies of hydromedusae and scyphozoans. Claus, however, was clearly more sensitive than most to the limitations of the new morphology in interpreting phylogeny.

Weismann's *Entstehung der Sexualzellen bei den Hydromedusen* was dramatically different not only from Claus's monograph but from everything Weismann had previously published. His text was three times longer than his contemporary's, and it included an atlas of plates as well. It was handsomely printed and meticulously illustrated by the author. Its format included the traditional overview of predecessors that imparted a historical perspective to that learned century. The body of the text presented a detailed description of Weismann's microscopical examinations. By the time he had completed his study, Weismann had examined in detail forty species of marine hydroids. These were distributed among three major groups: nineteen tubularians (the modern Anthomedusae), fifteen campanularians (the modern Leptomedusae), and six siphonophores. He also incorporated into his discussions the freshwater *Hydra,* a species of which had been thoroughly studied by Kleinenberg in the early years of the decade.[49] Weismann's collecting had been confined to the Bay of Naples, the Riviera, and along the coast of Brittany from the mouth of the Loire, and to Rostock; so he did not examine any pelagic Trachomedusae (the modern Trachylina) that were to be so effectively employed three years later by Brooks for a complete phylogenetic revision of the entire coelenterate phylum, as mentioned earlier.[50] It is doubtful, however, that Brooks's discoveries would have altered significantly the structure and goal of Weismann's own investigation.

Weismann's *Hydromedusen* began as a taxonomic work. In fact, even a cursory study of the 200-page "Specieller Theil" (the monograph's presentation of its research data) tells much about Weismann as a thorough and systematic comparative morphologist. Nevertheless, the focus of his microscopical work was a detailed comparison of the origin and development of germ cells in forty-odd species of hydroids. Most of the information was drawn from Weismann's own studies over a five-year period on both live and preserved specimens, but nearly always he commented on and included relevant information gleaned from his predecessors. As might be expected, the information varied in depth from one species to the next. Some sections, such as his twenty-two-page examination of the two species of *Eudendrium,* upon which he had exerted so much effort, provided the reader through its text and figures a vivid and complete image of the germ cells, their development, and their movement along a specified germ-layer track of the main hydranth to the side hydranths and then into the side hydranths to the point of blastostyle formation and further into the gonophores. With other species, such as *Dendroclava Dohrnii* (now known as *Turitopsis dohrnii* or TAXID 308579 of Max Planck Institute for Molecular Biology), the germ cells may have originated in the gonophore itself and migrated into what would become the manubrium of a free-swimming medusa. This new species was brought to the Zoological Station by coral fishermen in the summer of 1881 and was recognized by Weismann as a new species, which he named to honor the director of the Zoological Station.[51]

With other species, Weismann's account was briefer but could also contain very different observations, such as a three-page description of the genus *Obelia,* in which he examined "many colonies" ["viele Stöckchen"] of *O. dichotoma* where he found the germ cells arising first in the manubrium of its free-swimming medusae. He often spoke of the necessity of making physical slices, both longitudinal and transverse, through preserved specimens at a time when many of his contemporaries relied solely on optical "cuts" on living specimens. All his accounts, long and short, detailed and sketchy, were the result of painstaking preparations and countless observations; collectively, they testified to the time and care that Weismann must have spent at the microscope. This point emphasizes again that the business of microscopy was an exacting and time-consuming affair, but that the required labor led to its own rewards.

Weismann reserved his theoretical message for a seventy-page concluding analysis of the data. It was here that he led his readers through a series of complex arguments that wound their way to conclusions far beyond the modest lives of the hydromedusae. It is here we must follow him in order to understand how different this work was from those of Claus and other contemporaries. As with his *Studien* and *Beiträge,* it was rigorously analytic and pushed to novel conclusions.

Dislocation of the Germ Site

Weismann's original intent had been to produce a comprehensive comparative study, including not only gross morphological features but the histology as well.[52] The reproduction of the tubularian genus *Eudendrium,* however, focused his interest on the complex processes of reproduction and allowed him the opportunity to follow the question of germ cell development, a subject that he had intently pursued with daphnians.[53] *Eudendrium* crystalized the possibility that the male and female germ cell arose in different related species in different germ layers. Furthermore, in the cases of *E. racemosum* and *E. capillare* he determined that in the first species the egg cells arose in the ectoderm of the coenosarc, but in contrast the egg cells of the second species appeared to arise from the endoderm of the root hydranth. He found, too, that as the polyps developed and the primary hydranth budded, secondary and tertiary orders of hydranths, the germ cells in both instances, migrated from their point of origin— that is, from their germ sites—crossing into the endoderm, if they were not already there, to the endoderm along the branch hydranths to where the sporosac (the reproductive body) emerges. Once there,

they matured and crossed over to the ectoderm for final formation, liberation if needed, and fertilization.

Weismann tells us that his initial intent was to render a complete histological study of the entire class of Hydrozoa. This was at first an ambitious
plan without a real focus, and at best would have provided simply a comparative histology—if that had not proved too impractical. During the first year
of his investigations, he noted certain anomalies having to do with the origin
of the sex cells, which brought him back to one of the principal interests he
had developed with daphnids: the formation of germ cells. The most important of his specific findings concerned two species of the tubularian genus *Eudendrium,* first studied on the Riviera, later examined from the Bay of Naples and on the Atlantic coast from the Loire estuary to the northern coast of
Brittany.[54]

As discussed previously, Weismann discovered that the egg cells of *E. racemosum* arose in the ectoderm of the coenosarc whereas the origin of the egg
cells of *E. capillare* remained uncertain but he could find them at the earliest
stages of polyp development in the endoderm of the root of the hydroid. In
both species, the germ site was far removed from the gonads or germ-maturation
site in the ectoderm of the sporosac. As the hydroid colony grew, the distal
location of the germ site in both species extended outward when the primary
hydranth produced branch and second-order branch hydranths. The germ cells
migrated from the germ site (*Keimstätte*), crossed into the endoderm in the
case of *E. capillare,* and became situated along the branch hydranths at the
points where the blastostyle budded and the sporosac took shape. Weismann
was fully aware of how the dynamics of colony growth changed the location
of both the germ sites and the germ cell maturation sites in the ectoderm of the
gonophores.

It was a magnificent display of organizing data which at first had seemed
arbitrary. Weismann had been able to fashion his sequence of site dislocations
by combining a detailed understanding of the natural history of hydromedusae with an exacting microscopical investigation of the germ cells themselves. Yet what had he accomplished? What problems had he solved through
these correlations? And what new problems had he introduced? Above all, had
Weismann simply placed the facts he had harvested into the most orderly sequence he could imagine, or were there organizing principles guiding his
correlations? To answer these questions we must recognize that Weismann
necessarily touched upon a controversial contemporary issue in hydroid
morphology: the hierarchical and developmental relationship between the
polyp, the gonophore, and the medusa.

Evolution of the Sporosac

In his massive, two-volume study of Acalephs, his class designation for all coelenterates and ctenophores, Louis Agassiz, the Swiss/American anti-evolutionist, approached the relationship between polyp and medusa in accordance with the principle of complexity—the medusa was an increasingly more dominant and complex structure in both the Discophorae (our Scyphozoa) and Ctenophorae than of the hydroids and Siphonophorae.[55] Increasing complexity was the fundamental principle, which helped him organize this class and indeed the entire animal world.[56] One did not have to rely on cosmological principles, however, to support a hierarchical view of the medusa. George Allman, recently retired Regius Professor of Natural History at the University of Edinburgh and lifelong student of marine invertebrates, was cautious in assigning anything but a developmental sequence to the reproductive structures of hydroids. At one point, however, he seems to have tipped his hand by asserting that the mode of reproduction of the "simplest" gonophores is to be found in *Hydra,* and that there was an "advance in complexity" until the gonophore of *Corymorpha* and *Syncoryne* "can become free and lead an independent life in the open sea."[57] A generation younger than Agassiz and Allman, Weismann's nemesis Carl Claus endorsed an evolutionary sequence from hydra to medusa in his exhaustive *Grundzüge:*

> The sexual bodies appear at many different levels of morphological development, for they at first form simple swellings of the body wall filled with sexual materials (*Hydra*), at a more advanced stage they take over as a prominent extension of the body cavity or radial canal, in whose surroundings the sexual materials accumulate (*Clava squamata*), in an even more advanced stage there exists at the periphery of the bud a mantle-like covering with more or less radial vessels (*Tubularia, Coronata, Eudendrium ramosum Van Ben.*), and finally it comes to the formation of a small releasable medusa with mouth, floating bladder, tentacles, and peripheral bodies (*Campanularia gelatinosa Van Ben., Sarsia Tubulosa*).[58]

In the 1870s, whether an anti-evolutionist such as Agassiz or comparative morphologists such as Allman and Claus, a progressive sequence from sporosac to gonophore to free-swimming medusae appeared to measure the natural hierarchy in siphonophores, hydroids, and scyphozoans.

There was, however, another picture that stressed that medusae had regressed to become the gonophores. This had first been suggested by Haeckel's doctoral

student and assistant Gottlieb von Koch in 1873, and was later endorsed by
another of Haeckel's assistants (and later antagonist) Otto Hamann and by
the brothers Hertwig.[59] Gegenbaur presented a more balanced picture in his
Grundriss, where he accepted that an embryological division of labor produced the free-swimming medusae in the first place but that a later phylogenetic degeneration of the medusa resulted in the sessile medusoid bud.[60] All
of these scenarios, less direct than the progressive view, were turned on their
heads. Weismann had predicated the ordering of his four groups of maturation site locations on just this indirect regression, and by the time he had completed his analysis, there could be little doubt that a phylogenetic degeneration of the medusa was the true origin of the sessile gonophore.[61] His analysis
was intricate and involved a detailed comparison of gonophores and medusa
ontogeny.

Weismann was arguing that evolution was not a symmetrical process. Forms
did not come down the scale of complexity exactly as they had climbed it.
Moreover, as he reconstructed the evolutionary stages that the degeneration
of the medusae must have entailed, he found himself describing a sequence
that was unlike the observed ontogeny of the medusae. "One sees," he asserted
in a key sentence, "that this reconstructed phylogenetic development of the
entocodon is not merely a simple reversal of today's ontogenetic stages." In
short, the ontogeny of degenerate structures did not recapitulate the phylogenetic past. His commitment to the belief that the medusoid gonophores and
sporosac represented the evolutionary degeneration of the free-swimming medusa, however, did not eliminate his conviction that the medusa evolved from
polyps. It was likely he felt that it developed out of the hydranth head, instead.[62] Furthermore, Weismann insisted that each evolutionary stage bore
some functional relationship to the life of the organism. This became clearer
as he offered a neo-Darwinian scenario of the evolution and later degeneration of the medusa. Weismann's phylogenetic claims and neo-Darwinian
scenarios have important implications for our perception of his use of the
biogenetic law.

Weismann's reevaluation of the evolution of the sporosac forced him to conclude that the medusa underwent a dislocation of its maturation site from the
ectoderm of the manubrium to the entocodon, that is, to the ontogenetic precursor of the ectoderm of the manubrium and subumbrella. A parallel dislocation, due to the requirements of space, he argued, must have occurred with
the germ site, which in the first stage coincided with the maturation site. Weismann's carefully constructed comparative tables of the location of the two

sites of the forty-odd species told the rest of the story. As the medusa phylogenetically degenerated into a medusoid structure of the gonophore and then
into the gonad of the sporosac, the germ site was forced to move backward:
first into the entocodon, then into the endoderm of the gonophore bud, further into the endoderm of the blastostyle, until in the most advanced evolutionary species the germ sites appeared in the endoderm and ectoderm of the
coenosarc. (Weismann also noted that the female germ cells advanced in this
centripetal dislocation before the male germ cells.)[63]

Migration of the Germ Cells and Specificity of the Germ Layers

This phylogenetic separation of the germ site from the maturation site necessitated an ontogenetic migration of the germ cells. Because the germ cells were
generally, but not at all times, distinctly different from the other cells, it was
a migration that the trained histologist could easily follow for most of its course.
Moreover, as he never found germ cells wandering into gastrozooids and as
there always seemed to be the correct number of germ cells congregated at the
point where a blastostyle with a set number of gonophores would develop,
Weismann argued that there must exist a preestablished migration route, or as
he called it a "Marschroute," from the germ site to the gonads. This "migration
route of the single cells," he wrote, "must be imprinted in a rather special way."[64]
What determined the destiny of the cells and the germ layers? In 1883, there
could be only one nonteleological answer for him. It was heredity or "Vererbung," a word that had multiple meanings in this period. We will take this
issue up in Chapter 8.

   With most of his species, Weismann found the germ site in the endoderm.
This location resulted in the migration route, whether short or long, existing
for the most part along the endodermal layer until the germ cells reached the
ectodermal entocodon. Did the endodermal germ site indicate that the germ
cells were themselves endodermal in origin? The question forced Weismann
to confront two of the most topical issues of his day. Did the germ layers possess integrity and specificity? And did the germ layers of hydroids correspond
to the primitive germ layers? Suffice it to say for the moment that Weismann
had been very much impressed by the recent studies of Oscar and Richard
Hertwig on the germ-layer theory. Although they had vacillated in their commitment toward the doctrine of specificity, the Hertwigs had spoken reso-
lutely of the ectodermal origin of the germ cells of hydromedusae.[65] Weismann's
biology demanded another explanation. On the basis of its simple and

noncompound nature, Weismann reasonably, although in retrospect we realize incorrectly, maintained that *Hydra* was the most primitive extant hydroid. On the basis of the principle of the division of labor and the contemporary understanding of the alternation of generation as a mode of combining
asexual and sexual reproduction, Weismann held that the medusae of the
Hydrozoa must be a phylogenetic stage and ontogenetic generation beyond
the colonial hydra-like polyps. Many of these also possessed germ sites associated with the gonads in the ectoderm of the manubrium. Only when evolution reached the point when the medusa retrogressed into gonophores did
this ectodermal association cease. Weismann argued that for the want of space
and for the need of accelerating the reproductive process, the germ site was
forced out of the entocodon into the adjacent endoderm.[66] Similar pressures
brought a further dislocation to the endoderm of the blastostyle, and then to
the side and main hydranths.

Did Weismann's observation of a migration of the germ sites imply that
the germ cells had an endodermal origin? Such a conclusion would have been
contrary to Weismann's biology. "How and by what means can . . . the endoderm cells differentiate into sex cells just as the ectoderm cells have done
up to then? It is no exaggeration," Weismann emphatically claimed, "for us
to regard this sort of thing as impossible. Thus if the cells in the pertinent
[that is, the endodermal portion of the] gonophore bud demonstrate the property of differentiating into germ cells, the conclusion is unavoidable that they
must have [first] migrated from the ectoderm, WHETHER THIS CAN BE
CONFIRMED BY OBSERVATION OR NOT."[67]

Weismann was adamant that the germ cells of hydromedusae must originate in the ectoderm and migrate into the endoderm whether this could "be
confirmed by observation or not." To justify the lack of ocular proof in some
species, he spoke of "Urkeimzellen" that were indistinguishable from the embryonic cells of the ectoderm and that migrated into the coenosarcal endoderm. Only somewhere along their "Marsh-route" ("Marschrouten") to the
ectodermal entocodon did they become microscopically evident. The term Urkeimzellen itself veiled the same ambiguity that Carl Semper's term "Ureier"
had when he sought the germ-layer origins of male and female germ cells in
elasmobranches.[68] With Weismann, the uncertainty concerned not the hidden
distinction between embryonic male and female germ cells but his inability
to distinguish the female germ cells from other undifferentiated cells at their
site of origin. As for observational proof of such a trans-germ-layer migration,
only once did Weismann secure tangible evidence with his microscope of such
a tortuous migration.[69]

Weismann observed germ cells in both the ectoderm and endoderm of the coenosarc of a female *E. ramosum*. Might this simply mean a random wandering on the part of the cells? Only once did Weismann catch, immobilized by his chromic acid and absolute alcohol, two germ cells, which he interpreted to be in the act of boring through the interstitial lamella on their way from the ectoderm to the endoderm, whereupon they presumably migrated to the gonophores. Weismann illustrated and described this slide not only in the "Atlas of the Hydromedusen" but in the supplement he had sent Marion from Naples in 1881—hence his eagerness to have the French paper published as soon as possible.[70]

An argument by analogy from other species where the egg cell could be seen first in the ectoderm then in the endoderm reinforced his conviction and supported the prevailing doctrine of specificity of the germ layers, Weismann drew up a comparative table in which he systematically presented the pattern of the dislocation of germ cell origin sites and the stage of the medusa degeneration in related species.[71] Species by species, stage by stage, page by page, Weismann documented the various Marschrouten he had followed in part under his microscope. His argument was well supplemented by schematic figures in the text and an atlas of lithographic drawings of his microscopic observations.

Alternation of Generations

After addressing the nature of the dislocations and resolving the apparent germ-layer inconsistencies of germ cell origins and maturation sites, which had so exercised the students of hydroids, Weismann turned to one last question: Did hydromedusae violate the standard definition of the alternation of generations? If the classic definition still held, if the alternation of generations were strictly an alternation of asexual and sexual beings, a metagenesis, Weismann's work appeared to have created more confusion. Thus, among the species of the genus *Clava*, for example, the male germ cell site is in the gonophores while the comparable female site lies in the coenosarc. Did this imply that the male gonophore was the sexual generation and the nearly identical female gonophore was merely a bud off the stem hydranth? Weismann's answer was emphatic: "That would be not only unpleasant but useless and senseless."[72] It added to his argument that Brooks had found a similar inconsistency due to germ cell dislocations within the ascidian genus *Salpa*.

Weismann's answer reflected the spirit of his maturing zoological approach. Words and expressions were human conveniences and should not constrain the way we think about new phenomena. Perhaps "alternation of generations" could still apply in both the cases of hydromedusae and ascidians, in spite of their germ cells dislocations, but the zoologist must recognize that evolution has brought changes in structures and processes over time. Dislocation was a product of evolution; thus, a rigid definition of the "alternation of generations" must be relaxed enough to accommodate new zoological structures and processes. Understanding the mechanics of those changes was what counted, "and it will depend completely" Weismann insisted, "on the living conditions, the general growth and a thousand other conditions, whether and in which direction a displacement of the anlage for the gonads takes place."[73]

Immediate Aftermath

Weismann wrote *Hydromedusen* for the specialist. In size and scope, it followed in the tradition of the monographs on medusae and hydroids by Agassiz, Allman, and Haeckel. Instead of being a natural history as the others were, however, it was one long argument about the phylogenetic dislocation of the germ and maturation sites and the migration of germ cells. The details and subsidiary arguments were elaborate, and unless the reader was familiar with the material and literature, he probably skipped much of the "Special Part," wherein lay the data of his argument, and skimmed the "General Part," if only to get the structure of Weismann's basic claims. By May 1883, when his monograph appeared, Weismann was clearly moving ahead with the ramifications of its conclusions. In June, he delivered an important address, "Heredity," at the inaugural of his two-year term as prorector of the university, and soon thereafter in his capacity as prorector delivered an expansion of that talk for the birthday celebrations of Friedrich, the grand duke of Baden.[74] Both of these addresses were natural outcomes of the hydromedusae work. We will turn to their content in Chapter 8. In the meantime, Weismann must have felt the problem of communicating with zoologists in general about his hydroid findings. He did not take the opportunity to make a formal presentation at the following September's Versammlung, but he probably had other things on his mind, for he and Freiburg were the host for these annual meetings.

Autoreferat; Clemens Hartlaub and Johannes Thallwitz

By the end of the year, Weismann had put together an "Autoreferat" on his *Hydromedusen,* which appeared on the first of March in *Das Biologische*

Centralblatt, a widely distributed weekly journal of biological reviews and position papers.[75] Rather than a critical review of the monograph, Weismann's report offered a convenient summary of his basic argument and expanded beyond this frame to some contemporary studies of germ cells of insects, daphnids, and even a frog. In addition, Weismann enlisted two of his students to write their dissertations on certain unresolved questions. Clemens Hartlaub from Bremen spent the fall of 1883 in Helgoland examining two new species of *Obelia* and their free-swimming medusae. He found a centrifugal dislocation of the gonads between the radial canals and the manubrium, thus tying up a loose end in Weismann's argument, which Weismann was happy to acknowledge in his review and earlier in his university recommendation.[76] Johannes Thallwitz from Saxony pursued a microscopical study of the male germinal track and maturation of spermatozoa in hydroids at the institute in Freiburg. The male germ cell was the sexual side of the germinal track that Weismann felt less secure about, but time and again Thallwitz reconfirmed his professor's observations on a dozen selected species. Moreover, using magnifications of over 1,000 power and an array of nucleophylic stains, Thallwitz examined the nuclear changes of spermatozoa in the gonads and gonophores. Although neither he nor Weismann could have realized it at the time, some of his descriptions and figures came close to depicting the quartet formation of meiosis.[77]

Neither dissertation marked an outstanding contribution to the literature, but they signaled how Weismann enrolled many of his students in a supportive role.[78] After the completion of his dissertation, Hartlaub went on to become an expert of marine invertebrates and eventually served as curator of the zoological division at the Prussian biological station in Helgoland. Thallwitz, whose microscopic skill Weismann admired, disappears from historical view.

It was a daring, perhaps outrageous extension of the carefully documented conclusions of his *Hydromedusen,* to argue that a continuous germinal track must be a general feature of all animals and perhaps even plants. The crux of such a sweeping generalization was that by 1884 Weismann understood the early stages of this germinal track to be a cluster of molecules of the germplasm. This remained microscopically undetectable in undifferentiated cells (Urkeimzellen) far from their eventual destination in the maturation site, the gonads. "The Hydromedusen thus provide a proof of the continuity also for those cases in which the germ cells are not yet divided from the somatic cells during embryological development."[79]

Henry Nottidge Moseley; Case of Hydromedusen *Used throughout His Career*

Having sailed with the HMS *Challenger* Expedition of 1872–1876 and having edited an associated report that included the hydroids, Henry Nottidge Moseley was an appropriate person to review *Hydromedusen* for the journal *Nature*.[80] At the time (1881–1891) he was the Linacre chair of human and comparative anatomy at Oxford, and in a few years he would become closely involved in inviting Weismann to the annual meetings of the British Association for the Advancement of Science (BAAS). The opening sentence of his review immediately identified Moseley as a strong admirer of the natural history of the Freiburg professor: "all of the memoirs which he publishes are of extreme scientific importance, and abound in original views." For four and a half double-column pages, Moseley summarized the details of Weismann's monograph. Alas, so enthralled was he by the intricate particulars and demonstrations of the Specieller Theil, that by his own admission it was impossible to do justice to the "most able summary chapter." He did manage to point out that Weismann ended up supporting the Hertwigs' belief in the ectodermal origin of germ cells in all hydroids. He recognized the dilemma Weismann posed for the conception of alternation of generations, and he pointed out the more general importance of Weismann's arguments about the phylogenetic shifting of both the germ and maturation sites.

It was a well-informed, positive review overall, but he did not see beyond the delightful migration of hydroid germ cells to the larger consequences. Weismann's "Autoreferat," however, was still four months in gestation, and the larger implication of the hydroid's germ cells for the greater problem of heredity would not yet have been easy to foresee.[81]

8

From Egg to Heredity

1876–1885

The decade between the publication of the second part of his *Studien* and his dramatic declaration that there must exist a cytological element that served both as a hereditary link between generations and a developmental structure that operated throughout the life of the individual was perhaps the most rewarding period for Weismann. He had successfully demonstrated that he was back on the academic track. He had completed and published the second part (with four very different sections) of his *Studien zur Deszenztheorie* in 1875–1876 (*Studies of the Theory of Descent,* 1882); he had embarked on and completed an in-depth seven-part series on both freshwater and saltwater daphnoids (1876–1879); he had done the same in-depth study for hydromedusae, which led him to proclaim the doctrine of the continuity of the germ cells; and he had presented to the public two revolutionary statements about the "Duration of Life" (1881) and "Heredity" (1883)—the second serving as his Prorector address in Freiburg in the presence of the duke of Baden. In addition to his successful reentry into solid and provocative biological research, Weismann traveled to Naples via his in-laws' home in Genua.

A closer look at his activities during this decade shows that Weismann not only had recovered his ability to pursue research but also had undertaken a range of pursuits both in Freiburg and on the national level. He felt a true personal loss in 1877, when his youngest daughter Meta (his fifth child, in fact), who was just over a year old, died. The loss was later mitigated by the birth of his sixth child and only son, Julius, at the end of 1879. Julius later became a concert pianist and composer and lived in Freiburg as a close family member with his wife and two children.[1] Locally, Weismann belonged to a band of hikers, self-dubbed the "Philabulatorische Gang." They often climbed to the hinterlands of the Schlossberg with its ruins of earlier fortifications. In 1879, Weismann was elected Decan of the university, which required him to make public presentations. Shortly thereafter, in October 1880, Weismann's father passed away in Frankfurt at the age of seventy-six.

Weismann enjoyed the summer trips with colleagues, visiting Paris, Orleans, the Loire valley, and with Delages in Roskoff. In October and November 1880, he traveled to Leipzig, Dresden, and Berlin. The trip seems to have coincided with a professional exploration of a possible position in Breslau and where he visited Carl Hasse's anatomical institute staffed by young professionals such as Wilhelm Roux, Gustav Born, and Hans Strasser. Weismann attended the Versammlung in Salzburg in 1881. The same year we find him traveling to Rome and Naples via Genua, presumably visiting his wife's family on the way. Dohrn's zoological station provided Weismann with a place to stay over Christmas. The following year, in 1882, he made another trip south to Naples and Sicily, clearly this time with Mary. In June 1883, he delivered the prorector's address, had the opportunity to deliver a talk on Lorenz Oken at the Okendenkmal in Offenburg, and helped host the Versammlung in September which convened in Freiburg that year. By the following year, he had assumed the chair of the academic Krankenkasse—the time when he was reading proofs for his *Hydromedusen*. It was the same year during which he thought seriously about moving on to Munich. His decision to remain in Freiburg came with the conviction that he would have more time to work on his scientific problems in its less-demanding university town. His decision to remain in Freiburg was celebrated by a student Fackelzug, or torchlight procession. By 1885, he began studying English in earnest with the wife of his mathematical colleague Jacob Lüroth. A trip to Strassburg, Metz, Luxemburg, and Louvain and later to Cologne and Bonn with Strasburger enriched his love of art. Back in Freiburg, he received the invitation to become an honorable member of the Swiss Natural History Society. It was particularly a year to celebrate, for Weismann opened his new institute building in November as required in his negotiations when he decided to turn down the offer of Munich. "Saal eingerichtet Offen brennt endlich!!!" ("The furnace of the hall finally is on!!!") he pronounced, somewhat relieved.

It was a short-lived celebration: Mary died the following October, and his eye ailment returned. The historian today may take the liberty to conclude, rightly or not, that this was no coincidence.

Trajectory of Research, 1876–1885

When one looks at Weismann's publications from the time when he had completed his *Studien* of 1875–1876 to when in 1885 he made his first statement of a germ-plasm theory, one notices a pattern that began with introductory presentations and informal talks. These were soon rewritten and submitted for publication to professional journals. Later a substantial monograph followed that filled out his thoughts. It was a natural sequence, undoubtedly followed by many academics. Thus, we find a sequence of publications on daphnids, which started off with a popular lecture given at the university's auditorium in the winter semester of 1875–1876; a paper on the winter eggs of what he thought might be a new species in the *Zeitschrift für wissenschaftliche Zoologie;* a paragraph comment on a paper by August Forel on freshwater fauna; and a single-page description in the fiftieth report of the Versammlung Deutscher Naturforsher und Aerzte. These were followed by a series of substantial essays on daphnids focusing on egg formation, embryonic development, the production of winter eggs, coloration, spermatozoa and copulation, and cyclical reproduction. All seven of these lengthy contributions, including the first one on the winter eggs of *Leptodora hyalina,* all published in the *Zeitschrift für wissenschaftliche Zoologie,* were then assembled and published together as *Beiträge zur Naturgeschichte der Daphnoiden* by Wilhelm Engelmann of Leipzig in 1879. Between his first lecture in the auditorium and the final book form of his *Beiträge,* Weismann published a combined paper on new species of daphnids with his brother-in-law, a reply to Carl Claus on certain species, a report with Robert Wiedersheim on the zoological and anatomical institute in Freiburg, and three incidental reports on bird migration, butterfly "Duftschuppen" (i.e., sent scales), and the significance of zoology for medicine. A similar pattern unrolled in his selected studies of hydrozoa and the eventual appearance of his monograph on hydromedusae.

Later in the 1880s the same pattern is detectable as he studied the fine nuclear details of maturation division. In this case, however, the pattern is concealed somewhat by his responses to multiple critics about his support of natural selection and the development of his germ-plasm theory, which terminated in his major monograph on *Amphimixis* in 1891. It seems clear that Weismann had thought through this pattern early. Not only did he inform his university and professional peers of his ongoing research, but he staked down his research terrain in a way that notified potential interlopers. One cannot say that the end result was in view when he began. His research style remained that of an explorer. One can, however, detect from his studies of daphnids and hydromedusae and from his claims about the duration of life and heredity a trajectory that drew him into the microscopic details of reproduction. This trajectory will be the subject of chapters 9, 10, and 11.

On Life and Death

Weismann read the proofs for his monograph on hydromedusae between February 7 and July 23, 1883. As suggested in Chapter 7, the work was the capstone of two decades of intensive research on selected organisms: butterflies, daphnids, and hydromedusae. It was a capstone because it highlighted for the most part a style of research that joined a deep pursuit of natural history with a focused examination of specific problems, the results of which could be generalized to evolution and development. It also served as a bridge directing Weismann into the new problem area of heredity. To be sure, heredity had always been in the back of his mind. As had Haeckel, he saw the process as one of two sides to the comprehension of the shape of the organic world: heredity and adaptation, stasis and change, palingenesis and cenogenesis. As comparative embryologists, however, both zoologists had failed to envision how to examine empirically the heredity side of the evolutionary equation. Embryologists, after all, are used to studying histological change rather than the stasis characteristic of heredity.

In his autobiographical sketch of 1896, Weismann confirmed that the hydromedusae had completed a phase in his scientific development for other than simply intellectual reasons. His exhaustive and exhausting examination of germ cell origin and maturation sites exacted its toll. "These, in part, really eye-straining researches would exert an unpleasant influence on my working ability." Clarifying his comment, he added, "To be sure it seemed at first that my eyes had not been hurt, but a few weeks after the completion of the editing of the aforementioned work, symptoms in the eye appeared, with which I had exclusively used with the microscope, symptoms which frightened me and made it necessary in the near future to give up entirely my own researches."[2] As the optic malady threatened to reappear, however, he struck the more optimistic note that he had already begun new investigations that simply focused "on the right connection and thinking through of already known facts; at least for the moment."[3]

Weismann wrote his assessment of the transition in his research interests retrospectively in the early 1880s, some thirteen years after the fact, but a glance at his activities and publications around these critical years supports his claim. As noted earlier, before Weismann signed off on his *Hydromedusen* of 1883, he spent the winter semester of 1881–1882 in Naples. Between 1878 and 1883, we see the appearance of less specialized presentations designed for a general scientific and public audience. A paper on the migration of birds appeared in a series of popular scientific papers edited by Rudolf Virchow and Fr. V. Holtzendorff.[4] His address before the second plenary session of the Versammlung in Salzburg in 1881 consisted of a general discussion on the duration of life, and Gustav Fischer published an expanded version with a more biologically oriented title of the address the following year.[5]

As he assumed the prorectorship of the university in 1883, Weismann was obligated to present a general inaugural address. He seized the opportunity to talk specifically about heredity. The talk has become a benchmark for the ensuing decade of technical research into the subject. Life and death were important enough themes for Weismann to frame another public talk on the subject at the fiftieth birthday celebration of Grand Duke Fredrich I in the summer of 1883.[6] (I will have more to say about all of these related presentations in a moment.) A celebration in the neighboring Baden town of Offenburg in July 1883 provided him with the opportunity, perhaps even obligation as Freiburg University's prorector and professor of zoology, to give a brief address at the commemoration of a statue of Offenburg's famous zoology son and a former Freiburg student, Lorenz Oken.[7] Finally, in September of the same year, Freiburg University served as host to the annual meeting of the Versammlung, for which a few words of welcome from Weismann would have been expected.[8] During these same years, Weismann also read through the English translation of the entire two volumes of *Studien* ably executed by the English chemist and lepidopterist Raphael Meldola. The edition included a new introduction, a few additional footnotes, and a short preface by Charles Darwin written shortly before the latter's death.[9]

So the years of 1881–1883, in addition to Weismann's determination to complete his monograph on hydromedusae, were full of professional obligations and personal necessities. They entailed a shift in emphasis, a different style of research, and a new opportunity to deal "with the right connection and thinking through of already known facts; at least for the moment."[10] His thoughts embraced simultaneously two themes: the origins of natural death and the nature of heredity.

E. R. Lankester on Natural Death

There is no hint in either his published monographs or in the brief papers that lead up to his treatise on the origin of germ cells in hydromedusae, that Weismann would devote his plenary address at the 1881 meetings in Salzburg to the subject of life expectancy and natural death. The subject, however, was not an unfamiliar one in biological and statistical circles. This was amply demonstrated in 1869 when a twenty-two-year-old Oxford student who had already earned a first in natural sciences and published some technical zoology papers won an essay competition on the subject of "The Comparative Longevity of Different Species of Lower Animals and the Longevity of Man in Different States of Civilization." Edwin Ray Lankester's winning essay appeared the following year in book form with essentially the same title as the competition statement.[11] There is no reason that the book would be referred to by Weismann eleven years later, and it is doubtful that he even knew of its existence until somewhat later still.[12] Nevertheless, it is an essay worth reviewing for a number of reasons. First, it reveals how little was known at the time about the subject, despite the fact that, in Lankester's somewhat disarming prose, "the field is a well-trodden one."[13] In fact, he observed, one can trace information and speculations about longevity back through Herbert Spencer, the Prussian physician C. F. Hufeland, and Francis Bacon to Aristotle. Nevertheless, Lankester recognized that much of the information amassed through the ages was unsystematic, factually suspect, and often simply anecdotal. Second, Lankester's work, dependent as it was on Spencer's *Principles of Biology,* summed up the "well-trodden" field in a way that provides the historian with a useful contrast to the direction in which Weismann would later take the subject.

Lankester, his contemporaries, and predecessors conceived the problem as an environmental, behavioral, and physiological one. It was a problem that was more concerned with the comparative potential longevity of individuals of the same species than with the comparative average longevity in different species or other categories of classification. What was known about the longevity of individuals, races, and species? The subject included the importance of the expenditure of life's vital powers on growth, differentiation, reproduction, and other physiological processes. Theirs was a concern about recorded life spans in individuals and species, and an attempt to establish the conditions determining these results. Lankester was adept at dealing with the known record and at dissecting out its various components. He made distinctions between individual longevity and the potential longevity of species, between absolute longevity, average longevity, and mortality and senility among individuals. He gathered data from a host of classical, modern, and contemporary authorities on major groups of invertebrates and the five vertebrate classes, and he reproduced forty-nine numbered paragraphs of information from Bacon's *History of Life and Death* (1638). He combed statistical records—particularly those on races of human beings found in the journal of the British Statistical Society. He borrowed from Spencer a crude technique of generating algorithms for comparative calculations of average and absolute longevities, a procedure Lankester did not pretend produced a complete picture. For example, parthenogenesis, various stages of life cycles, and the suspension of life in particular forms of encystment and hibernations needed to be included in any general calculus. At one point, Lankester even deferred to the Creator's "purpose" as a guide to "the greater probability of truth."[14] It is likely that the appeal was more designed to assuage the prize committee than an expression of personal conviction. At the same time, Lankester deferred to Spencer's understanding of individual differentiation and increased fertility as costs in the overall calculation of the well-being of the organism and recognized that they might be offset by more beneficial behaviors, nutrition, and environments. Only in passing did he refer to "inherent longevity," and once, in a footnote, he indicated that evolution and the Darwinian mechanism of natural selection would add an important dimension to the subject:

> There is a vastly important and most interesting aspect of longevity which has been but little written on or considered as yet, and which it is to be regretted we cannot now enter upon. The longevity of races and species (as such) is the subject to which allusion is made: in connection with the Darwinian theory and struggle for existence, the longevity of species will prove a most fertile field of research.[15]

It was a prize essay, to be sure, but it was largely a library research paper focused on summarizing obvious classical and contemporary sources. It was particularly accepting of Spencer's *Principles of Biology,* which at the time was often used as an advanced textbook for students of biology. Whether reflecting Spencer, papers on human mortality and longevity statistics, or Bacon's catalog of information, Lankester's essay captured a common approach to the problem of comparative longevity. There was little in it, other than a single passing footnote, that suggested there might be another dimension to the subject.

Weismann on Natural Death

A decade later in Salzburg, Weismann alluded to much the same information as Lankester concerning the life span of organisms from birds and mammals to many orders of insects and some mollusks. Only a small fraction of
such material was immediately supportive of the text as he read it. He included
the bulk of his data in the appendix of the expanded version, which appeared
the following year.[16] It must have become quickly apparent to his listeners and
certainly to his readers that Weismann was presenting a different kind of analysis. Size, metabolic activity, and the complexity of the organism, according
to him, could only partially account for the differences in life spans of various species. The critical factor for Weismann was in understanding the reproductive success of an individual, and this meant understanding reproductive strategies. It was the perpetuation of the species not the life span of an
individual that mattered. Of course, some species live long after they reproduce, but then one needed to understand the efficacy of tending to one's young
or having another reproductive period the following year. Among birds and
mammals in particular, the individual's life span needed to be lengthened because of the relatively small clutch or litter sizes and the advantages of attending
to the young. *Ceteris paribus* natural selection in a paradoxical fashion must
favor the reduction in the life span of the individual, as we find well illustrated with insects.

The hierarchical pyramid with more complex organisms at the apex, as recognized by Lankester and Spencer, was reinterpreted by Weismann to be multiple pyramids of the utilities of lifestyles weighed in a balance of multiple
conditions by natural selection. The same argument applied when one examined the great differential in the life span of the two sexes in, let us say, the
colonial insects (e.g., termites, bees, and ants) where the males possessed very
short life spans compared with the egg-laying females. Again Weismann felt
this must be interpreted as an adaptation with respect to the conditions of
life. "Duration of life, like every other characteristic of an organism, is subject to individual fluctuations. . . . As soon as the long-lived individuals in a
species obtain some advantage in the struggle of existence, they will gradually become dominant, and those with the shortest lives will be exterminated."[17]
Durations of life, both long and short, were evolutionary adaptations.

Weismann went further. He deepened his analysis to the cellular level. Cells
multiply and regenerate in both lower and higher organisms, but there appeared to be a limited number of divisions before such activities slowed and

then ceased. Such phenomena, he argued, were physiological, but there seemed to be no single condition that universally brought about such a process as natural death. Instead, variations in longevity of individuals and the utility of their death to the species as a whole must be the accepted alternative explanation. "The above-mentioned hypothesis on the origin and necessity of death," he asserted, "leads me to believe that the organism did not finally cease to renew the worn-out cell material because the nature of the cells did not permit them to multiply indefinitely, but because the power of multiplying indefinitely was lost when it ceased to be of use."[18]

This was a bold claim that contradicted current speculations, and Weismann reluctantly admitted he could not prove it. "I consider that this view, if not exactly proved, can at any rate be rendered extremely probable."[19] He could, however, offer a phylogenetic scenario that made his suggestion more plausible. Single-celled organisms, such as an amoeba, did not die a natural death but divided into two daughter cells. The parent, as an individual, simply ceased to exist by becoming two individuals. In one sense this might be construed as the death of the parent individual; on the other hand, fission might be viewed as the perpetuation and immortality of life. The awkwardness in such an argument could not be avoided, for the perpetuation of life is not the same as the immortality of an individual. One life might end at the same time another life begins, but a natural death does not, as normally conceived, occur with fission. According to Weismann, amoebas were too simple to experience aging; only violent death and accidents eliminated individuals. Weismann went further in his rationalizations. He considered colonial organisms, such as *Volvox,* an intermediate step between single-celled organisms and metazoans, but he failed to discuss this transition with any precision. Simply put, they were multicellular organisms that phylogenetically "soon lost their homogeneity."[20]

The metazoans had been the exclusive object of Weismann's research and now became the focus of his discussion. With a division of labor among the cells, he reasoned, cellular differences and then specialized tissues and somatic structures could emerge. Only a few cells remained unspecialized to carry on the function of germ cells; with such a developmental and evolutionary step, "normal death is possible, and as we see, has made its appearance, among multicellular organisms in which the somatic and reproductive cells are distinct." Embedded in the same discussion was the hint that a law of heredity was to be found in the fact that cells "could only give rise to cells which resembled themselves."[21] So what allowed for differentiation of tissues? Simply the division of labor? Weismann had reached the limits of his knowledge. He was prepared to accept Rudolf Virchow's dictum that cells had the power to modify their own growth through nutrition and Wilhelm Roux's just published *Kampf der Theile,* which explicitly depicted a selection at the cellular and molecular levels. He considered that differentiation must be told not on the cellular but on the molecular and chemical levels within the cell. He suggested that the surface-mass ratio of the developing embryo limited the size of all metazoans and that the specific size for each "is in reality pre-determined, and that it is potentially contained in the egg from which the individual developes [*sic*]."[22]

Weismann's arguments were in reality variations on what he had earlier maintained or what one might find in Haeckel's *Generelle Mophologie:* heredity versus adaptation, coenogenesis versus palingenesis, the routine invocation of a process designated "the division of labor," and the overgrowth assumption embedded in cell reproduction. They were not informed by the meticulous and extensive research that characterized his *Studien* and *Beiträge* and would soon again be displayed in his *Hydromedusen.* As a scientific mechanist, without resorting to Haeckel's materialistic and antireligious embellishments, Weismann was convinced of the logical necessity that organisms initially arose through spontaneous generation. He insisted that by "immortality" he really meant simply "an immensely long time." Unlike his Jena colleague, who wanted full answers, Weismann continued to insist that scientific research "is the quest after perfected truth, not its possession, that falls to our lot, that gladdens us, fills up the measure of our life, nay! hallows it."[23] By venturing into a more speculative mode, Weismann inadvertently invited equally speculative rebuttals.

Alexander Goette and Haeckel on the Egg

A more sustained challenge came from the newly inaugurated professor of zoology at the University of Rostock. Six years younger than Weismann and Haeckel, Alexander Goette had in 1872 made the transition from being a private scholar investigating development in vertebrates in Tübingen to a privatdozent in Strassburg, where he taught and assisted in Oscar Schmidt's Institute for Zoology.[24] He established his reputation as a productive embryologist three years later with the publication of a monumental study of 964 pages with an atlas on the development in the fire frog "Unke."[25] This monograph, however, was not simply a conventional descriptive embryology. As its subtitle implied, it provided a self-declared "Foundation of Comparative Morphology of Vertebrates." At the time of its publication, the recognized dean of comparative morphology, Carl Gegenbaur, took Goette's pretensions seriously enough to write a forty-six-page critical review in the first volume of his *Morphologisches Jahrbuch.*[26] He had nothing but praise for Goette's empiricism and illustrations, but he took detailed issue with the claim that a frog, which is a highly modified amphibian, could provide the basis for determining vertebrate homologies and phylogenetic relationships. Part of Gegenbaur's protest revolved around the doubtful identification of significant structures throughout the entire vertebrate phylum. For example, the notochord is derived from different groups of cells in different classes; rib and muscle formation is hard to generalize even within the fish themselves; and Goette's presentation of the formation of the sternum in higher vertebrates was based on "imperfect observations" and "arbitrary comparisons." As for his interpretation of limb formation, one of the reviewer's specialties, Gegenbaur claimed he could not follow the author's generalizations at all.[27] Finally, Gegenbaur pointed out that Goette failed to understand the importance of the functioning, adult structure for judging homologies, and he questioned Goette's recourse to mechanical-causal factors when he simply described rather than determined the causes.

It was a brutal review and implied in particular that Goette was a talented but maverick investigator. It centered in general on the overconfidence of contemporary embryologists that they should be the final arbiters of constructed phylogenies. The critique, however, did not hinder Goette's career. The latter soon became außerordentlicher Professor in 1877 and director of the city zoological museum of Strassburg. He accepted the offer in 1882 to become professor of zoology in Rostock, from which position he wrote a review of Weismann's *Über die Dauer des Lebens.* Goette returned to Strassburg in 1886 after Weismann had reluctantly indicated to his friend Anton de Bary that he would not entertain a job offer to succeed Oscar Schmidt.[28] From this Lehrstuhl on the west bank of the Rhine, Goette continued to labor in the field of descriptive embryology and to write from unconventional perspectives until his formal career ended when Alsace and Lorraine were returned to France at the Treaty of Versailles. None of his subsequent publications gained the attention that his early monograph on the fire frog had attracted.

Goette's essay *On the Origin of Death* was a direct response to the expanded publication of Weismann's Salzburg address.[29] As Hans Querner has pointed out, the two essays and Weismann's later rejoinder are not well known in the historical literature.[30] This is indeed a shame, for the exchange between these two came at a juncture when both embryology and evolution theory were undergoing a rapid reorientation owing to the advent of nuclear cytology. I will have more to say about this exchange later in this chapter and in Chapter 9. Needless to say, the implications of this historical conjunction were hardly clear at the time, and each of the arguments of the two zoologists simply fore-shadowed what later became more obvious. The immediate issue, of course, was the establishment of natural death. Weismann, as we have seen, had suggested that this was an adapted trait that had evolved at the time of the evolution of cell colonies and the differentiation of somatic structures, which in due time would outlive their usefulness and die. In contrast, the germ cells continued the old "immortal" ways of single-celled organisms by multiplying through direct division, forming gametes, and creating new individuals generally through sexual reproduction.

Goette built his counterarguments on concepts he had developed earlier in his monograph on the fire frog. One must make a distinction, he asserted, between metazoans, where the physiological division of labor has led to an organism with separate parts with different functions ("Kräfte")—that is, cells and germ layers—and the whole organism with its own integrated efficiencies ("Arbeitsleistungen"). Natural death of the organism, Goette maintained, reflected the disintegration of the whole into its parts. On the other hand, he regarded the death of the parts as a death of individual cells, as a piecemeal process that might occur slowly over time after the whole, integrated organism itself had ceased to exist. By way of elucidation, Goette maintained that with single-celled organisms, such as with the amoeba, in which the cellular part and the whole being were identical, no distinction between the two kinds of deaths might be made (i.e., the death of the organism versus the death of cells) as they both occurred at the same time.

If one looks at natural death, as Goette did, as a prelude or concomitant to the process of reproduction, one will recognize that in both single-celled organisms and metazoans natural death was a physiological attribute. It was inherent to life itself. Such a view challenged directly Weismann's insistence that protozoa were metaphorically immortal, that their fission was simply a continuation of a life flowing into two lives. Instead, Goette claimed that there existed a three-step evolutionary continuum: (1) starting with single-celled organisms, in which natural death comprised the disintegrating process of the initiating cellular fission into the component molecules of the protoplasm; (2) then undifferentiated colonial organisms, "homoplastids," which disaggregated and whose component parts (cells) produced by means of fission the onset of new colonies; and (3) last, differentiated "polyplastids" (metazoans), in which organic death preceded a rejuvenation through some sort of homogeneous encystment and from whose structure new germ cells emerged to reproduce the next generation. ("Homoplastid" and "polyplastid" were Goette's terms and continued to be used into the twentieth century.)[31] Goette reinforced his argument by calling upon the example of Haeckel's *Magosphaera planula,* a colonial flagellate that could alternate between asexual and sexual reproduction.[32] In both single-celled organisms and metazoans, natural death of the whole organism coincided with a "rejuvenation" of the protoplasmic chemistry to form a new beginning. With higher metazoans, the association between death and reproduction might chronologically be a distant one, but the reproduction of a second generation appeared to be always functionally bound to a natural death and a protoplasmic rejuvenation. Examples, however, of a temporally close connection of the two were abundant, particularly among the insects, such as mayflies, moths, and bees, where the death of the male coincided with or quickly followed copulation and where the death of the female followed the completion of egg laying. In these cases, a physiological state rather than an evolved adaptation seemed evident.[33]

Less clear but critical to his account was Goette's understanding of "rejuvenation" and of the egg. On both counts he separated himself from Weismann and others. The first of these terms, "Verjungerung" or rejuvenation, was a concept illustrated by Carl Brandt, who at the time was a young assistant in du Bois-Reymond's Physiological Institute in Berlin.[34] His dissertation of 1877 had examined in detail an encystment process in the heliozoan protozoa *Actinosphaerium eichornii.* As Goette understood it, encystment consisted of a disappearance of the nuclei and other internal structures and spines and the formation of an encapsulated cyst followed by a resting stage. After new nuclei developed, a silicate capsule formed around nuclei clusters, and new actinosphaeria emerged.

The cycle became a model for Goette in describing death followed by a rebirth. Goette associated death at the polyplastid level with the dissolution of the germinal vesicle of the germ cell and rejuvenation with the appearance of a new nucleus at fertilization. In the early 1870s, it was not an uncommon way of understanding gamete maturation and fertilization. Reinhard Mocek has pointed out that Wilhelm His and Eduard van Beneden, among others, described fertilization as a rejuvenation process.[35]

More in keeping with the themes presented here, Haeckel illustrated what he called the non-nucleated, undifferentiated "Monerula" stage of the rabbit in his *Anthropogenie* and commented, "At present, therefore, the majority of observers assume that between the original nucleated egg-cell and the known nucleated parent cell there is a stage in which there is no real cell-kernel or nucleus, and in which, therefore, the form-value of the whole organic individual is no longer that of a true nucleated cell, but that of a non-nucleated cytod." Although by 1879 Haeckel was well aware that the investigations of Eduard Strasburger, Oscar Hertwig, Leopold Auerbach, and Otto Bütschli had radically transformed the picture of fertilization, his emphasis on the monerula as the beginning point of each generation had not changed. Consider two further, somewhat inconsistent comments on Haeckel's part: "Although morphologically we can see no defined constituent parts in the Monerula, yet chemically we must regard the latter as a complex product of at least four different constituents [i.e., two protoplasms and two nuclei]," and "Either before or immediately after fertilization, the egg cell loses its original kernel [i.e, nucleus] and appears for a time in the form of a kernel-less cytod, as a monerula." (In Haeckel's terminology the monerula was the non-nucleated cytod and represented the beginning point of ontogeny and phylogeny.)[36]

At the time there seemed nothing radically wrong with Goette's or Haeckel's conceptions of gamete formation and fertilization. The identity between single or multiple follicle cells and the emerging germ cell remained unclear. The nuclear membrane of the egg cell does disintegrate; the nuclear sap appears to amalgamate with the surrounding protoplasm. Any chromatin material of the nucleus disappears. For a time the germ cell does appear simpler and more homogeneous given a certain degree of magnification, the uncertainty and complexity of the preparation and staining processes, and the variations among different species—all three factors helped obscure a general pattern in egg formation. The perspectives of both zoologists were understandably focused around development from the germinal epithelium of the parent to the initial appearance of the offspring, from a mass of follicle cells to a single protoplasmic mass in or expelled from the oviduct—with or without a vitelline membrane. A long tradition, stretching back through Waldeyer, to Remak and Schwann to Baer, had emphasized the homogeneous nature of the newly emerging egg. In the 1870s, this, of course, coincided with a description of ontogeny recapitulating phylogeny—let it be remembered, however, that investigators were dealing with the very first ill-defined stage in both the ovum and phylogeny.[37]

In his monograph, Goette ended his lengthy chapter on the formation of the egg in the ovary of the fire frog with a declaration that the initial egg possessed life but was not an individual. It was a structure clearly originating from

cells but was not itself a cell. His position was noted by Weismann: "The fer-
tilizable egg of a *Bombinator igneus,* according to Goette, cannot be considered to be a cell either wholly or in part; and this is equally true of it at its
origin and after its complete development; it is only an essentially homogeneous organic mass enclosed by a membrane which has been deposited externally. This mass is 'unorganized and not living,' and 'during the first phenomena
of its development all vital powers must be excluded.' "[38] Eight years later, this
vision of an in-between stage of life without form or individuality provided
the core of Goette's criticism of Weismann's adaptationist theory of death.
Ultimately, Goette claimed, the continuity of life could only be understood
in the chemistry of the continuing protoplasmic substrate of cells rather than
the cells themselves.

Weismann and "The Duration of Life"

In fulfillment of his duties as prorector of the university, Weismann addressed
his assembled Freiburg colleagues and guests in a program dedicated to the
seventy-fifth birthday of his highness the Grand Duke Friedrich I of Baden.
It was a special occasion. Not only was it a birthday celebration for his royal
highness, but it came at a time when the Baden government was making a
concerted effort to support the sciences and medicine at its "second" university in a manner that would be commensurate with its support of its older
and premier university in Heidelberg—and indeed with the other second-tier
universities in the Reich. It also came a few months after Weismann had received in the name of the grand duke a written commitment that his annual
stipend, not including the statutory living compensation, would increase to
4,500 marks, or an increase of 40 percent over his stipend set in 1880, and
that his institute would receive an additional 1,500-mark subvention.[39]

Whether by design or accident, Weismann had already been selected as principal spokesman for the senate of the university. It may be fairly questioned,
however, whether he actually read the entire seventy-nine pages of his address
to the assembled coterie of royalty, civic notables, and academics. The title of
his talk nevertheless had an appropriately elevated, even celestial ring to it:
"Ueber die Ewigkeit des Lebens" ("On the Eternity of Life").[40] No matter what
transpired during the ducal birthday celebration, Weismann's address served
as a necessary sequel to his talk on "The Duration of Life" delivered two years
earlier in Salzburg. It turned out to be a scientific polemic, structured as a
counter-critique of Goette's analysis.

Although Weismann strongly disagreed with Goette's position, he was more professional in handling his younger colleague's opposition than he had been in dealing with Claus's alleged encroachment on his discoveries with daphnids. After reviewing and "refuting" Goette's peculiar views about encystment and rejuvenation, he felt challenged enough about the subject to pursue the issues further: "for we shall thus reach many ideas worthy of consideration."[41] First, he reflected on the whole question of how Goette explained the phylogenetic transfer of the attribute of death from monoplastids to the polyplastid Magosphaera. The problem appears to be one of definition. By interpreting the dissolution of the colony into multiple cells, Goette had identified a form of death comparable to encystment. Extending Goette's argument to its logical conclusion, Weismann pointed out that death in heteroplastids must be equated with the formation of the germ cells. This led to the highly problematic conclusion that death of the soma must be a totally different phenomenon. Goette, in fact, had concluded as much when he claimed that cellular death and organismal death were different in kind. For Weismann, however, "nothing concrete dies in the dissolution of Magosphaera; there is not death of a cell-colony, but only of a conception."[42]

Still on the transition between mono- and polyplastids, Goette had made a big issue out of the death in the "worm-like" organisms known as Orthonectides. These had been carefully investigated by Charles Julin, a Belgian student of van Beneden, who in 1882 had described in detail the reproduction process of these minute organisms. As marine parasites, Orthonectides consist of a simple outer layer of ectodermal cells, muscle cells, and a body cavity filled with germ cells. Although nearly without structure, the Orthonectides have both sexual and asexual life cycles and were, and still are, generally assigned to the "enigmatic" phylum of Mesozoa, a taxon with a highly problematic phylogenetic position.[43] Of relevance for Goette is the fact that the body, consisting of only a few dozen cells, breaks apart when the germ cells are liberated to reproduce either through fission or a sexual union. Goette had interpreted the cycle as the obvious conjunction of the death of one generation and the birth of the next. Death and rejuvenation appeared in this example to be part of the same process and necessary before the germ cell for the next generation came into being. Weismann countered that the coincidence of death and birth was the outcome of the parasitic lifestyle and that the leftover fragments of the parent were simply too small to be viable. For him, Orthonectides were prime examples not of a phylogenetic transition from encystment to sexual reproduction but of a degenerated metazoan, in which the soma died while the germ cells continued.

Second, Weismann examined in more detail his own conviction that natural death first appeared phylogenetically among the polyplastids. It was an adaptation, he claimed, brought on by natural selection because of the advantages that would accrue to the species as a whole. Goette had resorted to a common accusation against such a utility argument. How could natural selection explain the origin of death? "The operation and significance of the principle of utility consists in selecting the fittest from among the structures and processes which are at hand, and not in directly creating new ones."[44] Weismann's response was typically Darwinian: gradualism, a range of variations, and the belief that without changing surroundings, the species would remain at a constitutional equilibrium. There was a twist, however, to the selection of the particular trait in question. How could natural death, which came into operation after the reproduction period and was by definition lethal to the bearer, become established—let alone arise? Weismann had to walk a tightrope between an embryological and a population explanation. The appearance of a variation that brought death to the soma had to be balanced against the positive advantages of life of the remaining members of the species. Only by invoking an unequal division of labor among the colony's cells and finding compensating advantages in specialization during development could he postulate the possibility of a restriction in the ability of somatic cells to live indefinitely. But it was not a satisfactory answer and left confusion. By the time of the preparation of the English translation of his essay, Weismann recognized that his concept of panmixia would partially solve the dilemma. "As soon as natural selection ceases to operate upon any character, structural or functional, it begins to disappear." Was he, however, now thinking about the attribute of death or of life?[45]

Nevertheless, his argument brought him to consider the nature of the germ cells themselves, an issue that he had dealt with in daphnids and hydroids and that allowed him to go on the initiative from several angles. As we have just seen, Goette considered the egg of the fire frog to be without structure, to be a homogeneous mass between life and death and without individuality. It was the counterpart of the encystment of Magosphaera. According to Weismann the germ (*Keim*) included, "every cell, cytode, or group of cells which, while not possessing the structure of the mature individual of the species, possesses the power of developing into it under certain circumstances."[46] His emphasis here was on development. A germ was not simply a fission product followed by growth, as commonly found among many protozoa or monoplastids, but rather was characterized by developmental stages. So unlike Goette or Haeckel, Weismann rejected a germ, or more appropriately an egg, as the biogenetic starting point of phylogeny: "the fundamental law of biogenesis does not apply to the Monoplastides."[47] A new wave of protozoologists, including Otto Bütschli and his brother-in-law August Gruber, backed him up on the microscopic details leading to such conclusions. Goette's identification of encystment with a newly formed egg must fall by the wayside. In keeping with his original claim, Weismann remained satisfied that natural death was the product of the evolution of polyplastids and the evolution of sexual reproduction.[48]

The interchange between the two men reveals a lot about ways of understanding the ovum and its maturation. Neither one had yet become involved either directly or theoretically in the nuclear cytology set into motion by the microscopy of Bütschli, Oscar Hertwig, Fol, and others. One participant appears to have remained unconcerned about the possibility of investigating a level of biological organization below that of the cell; the other was about to change his research orientation.

Weismann "On Heredity"

By all rights, Weismann's paper "On Heredity" deserves to be considered a classic paper not only in regards to his personal development but also with respect to changes in the field of heredity theory. It is rarely so considered, for such a judgment is generally made looking backward from a modern roost, perched on the enormous success of classical and population genetics and evolution theory. When he wrote his paper, Weismann also looked backward but from the vantage point of 1883. He drew lessons from his work on seasonal dimorphism, his examination of colored flecks on daphnid legs, and his discussion about the mechanical conception of nature. He was fervent about his contemporary conception of natural death, and he was likely planning his public talk on the subject for the grand duke and his colleagues, when he delivered his inaugural address as prorector. He certainly was casting about for the appropriate expression for a fundamental idea that appeared in both; thus the migration of germ cells of hydroids became the "continuity of the protoplasm of the germ cells," which in turn at the end of his essay became "the continuity of the germ protoplasm." On the one hand, Weismann rejected with confidence, though not with complete certainty, the prevailing assumption that acquired characters could be inherited. On the other hand, he considered the "value" of experiments, which explored the relationship between the inheritance of acquired characters and the germ protoplasm, to be problematic. He mentioned having begun a new series of butterfly experiments to pursue that very relationship, but he considered the experiments that had been done previously on that relationship to be unclear, "careful collections and arrangements of facts" in regard to the question.

At the outset, Weismann explored the meaning of "heredity" but got little farther than simply reaffirming his commitment to Haeckel's overgrowth conception of germ cells. It denoted "a growth over and above the mass of the individual." ("Ein Wachsthum über das Maass des Individuums hinaus," as he emphasized it in the German original—"a growth above and beyond the extension of the individual.") Thus, reproduction, "Fortpflanzung" and "Vererbung," remained for him closely overlapping concepts. The only difference now was that Weismann saw this overgrowth to be limited to special cells—the germ cells.[49] This conclusion nevertheless forced him to question the prevailing assumption that acquired characters could be inherited. Never an enthusiast for the possibility, Weismann now recognized that the isolated germ cells of metazoans directly confronted this age-old belief. On a personal level, he was challenging the convictions of two of his zoological heroes, Lamarck and Darwin. Each, after all, had made a concerted effort to explain the inheritance of acquired characters in mechanical terms.[50] How now to devise a mechanism that transferred acquired characters of the soma to the germ cells? Neither a special force nor material gemmules, as articulated in Darwin's theory of pangenesis, could pass muster because each required an organizing or teleological factor. For Weismann, the only alternative appeared to be a rejection of the belief all together. It was, of course, a deductive argument dependent upon the validity of his claim that a lineage of germ cells was significantly distinct from the soma. As we shall see in later chapters, Weismann would spend the rest of his career defending this new concept against those who insisted that the inheritance of acquired characters was necessary for evolution to take place and who believed it had been amply documented by casual observations and occasional experiments.

Weismann would have none of it. He gladly recognized that others before him, such as Viktor Hensen, Emil du Bois-Reymond, and Eduard Pflüger, had questioned specific cases of inheritance of acquired characters. There were other doubting biologists he might have named, including Francis Galton, Wilhelm His, and Georg Seidlitz. Nevertheless, Weismann realistically added, "but no one has hitherto attempted to cast doubts upon the very existence of such a form of heredity."[51] To emphasize the tenuous nature of a belief in "such a form of heredity," Weismann listed and easily dismissed some of the often

cited anecdotes purporting to document the inheritance of acquired pathological conditions. The transmission of mutilations, scars, lost horns, and syph-
ilis were commonly referenced but had not been rigorously demonstrated. For
example, there were the many experiments on guinea pigs by the American-
French physiologist Charles Edouard Brown-Séquard, who, after producing
the symptoms of epilepsy by severing key nerves, claimed to have induced
the same symptoms in later generations without using a scalpel. Two clini-
cians, C. Westphal (1871) and Heinrich Obersteiner (1875), appeared to confirm
these experiments, but the results always remained complex and problematic.
At the time, Weismann was not alone in finding them so; others would later
become increasingly adamant about their deficiencies in both methods and
claims.[52]

The stock-in-trade example of the day was of the inheritance of features
acquired through use or disuse of anatomical parts or mental exercise. Lamarck was well known for promoting such cases. The stretching of the gi-
raffe's neck and the webbing of the feet of water fowl were his most celebrated
examples. Du Bois-Reymond had already critiqued the influence of practice
on the fingers and possibly the nervous system of seasoned pianists, yet the
issue remained unresolved. Weismann, however, proceeded further: "The increase of an organ in the course of generations does not depend upon the
summation of the exercise taken during single lives, but upon the summation of more favourable predispositions in the germs."[53] The other side of the
coin was the issue of disuse. Weismann devoted much of his essay critiquing
the notion of the inheritance of degenerated organs, instincts, and talents.
He also challenged convictions that the induced effects of the external environment on the soma could be transmitted to the germ cells. At best one
might consider the possibility of a parallel induction that affected the undeveloped germ cells as well.[54] Repeatedly he would respond that the effects
of pathological states, of use and disuse, and of the environment could be
explained through coincidence or the action of natural selection. It was not
a proof against use or disuse inheritance but the promotion of the efficacy of
natural selection.

Above all, a general principle emerged by the end of his essay, which more
profoundly separated him from his many detractors. This principle had always been present in his discussions of evolution. It was first pushed to the
foreground with Weismann's discussion of atrophied structures. At this point,
he coined the term "panmixia" for a process he equated with the "suspension
of the preserving influence of natural selection" and explained it in terms of

"the greater number of those variations which are usually attributed to the direct influence of the external conditions of life."[55] Variations of all structures at all levels were the key to understanding Weismann's perspective. He again emphasized the presence of variations at the end of his address when he turned to the small variations in microscopic yet visible parts, variations such as the ones he had earlier discovered made up the colored skin stripes in caterpillars and the colorful flecks on the legs of daphnids. "All these quantitative relations are exposed to individual fluctuation in every species; and natural selection can strengthen the fluctuations of any part, and thus cause it to develop further in any given direction."

The stochastic nature of Weismann's perspective became clearer as he continued: "From this point of view, it becomes less astonishing and less inconceivable that organisms adapt themselves—as we see that they obviously do— in all their parts to any condition of existence, and that they behave like a plastic mass which can be moulded into almost any imaginable form in the course of time." It, of course, could only have been natural selection that "moulded" this "plastic mass . . . into any imaginable form." Pursuing his thought, he added, "If we ask in what lies the cause of this variability, the answer must undoubtedly be that it lies in the germ-cells." Weismann's conception of the changing germinal constitution was fundamental and the change was a gradual as well as quantitative process. "Just as any character in the mature organism vibrates with a certain amplitude around a fixed central point, so the predisposition of the germ itself fluctuates, and it is on this that the possibility of an increase of the predisposition in question, and its average results, depend." Finally he concluded, "We must however be clear on this point—that the understanding of the phenomena of heredity is only possible on the fundamental supposition of the continuity of the germ protoplasm."[56]

Historian and philosopher of science Rasmus Winther has written a valuable in-depth analysis of the development of Weismann's conception of variation on both the somatic and germinal levels. The challenge he set was to understand how Weismann perceived the mechanism of variation production in the germ cells at the same time he envisioned a barrier between the line of germ cells and the many acquired somatic traits that he had just examined. At this point in Weismann's career, Winther concludes, he "still only provided a vague externalist outline of the source of variation in the hereditary material."[57] I concur. Although Weismann recognized part of the problem—"how is it possible," he asked of evolving protozoa, "that one kind of cell in such a colony [the germ cells in Magosphaera] can produce the other kind by division?"[58]—he was far more concerned with challenging the prevailing belief in the inheritance of acquired somatic characters than considering different mechanisms for germinal change. His "vague outline" remained that which it had been throughout the 1870s: one of externally induced hereditable change but restricted now to the unspecialized germ cells.

I believe it would be wrong to maintain that Weismann arrived at a rejection of the inheritance of acquired characters in the logical fashion he presented in his essay. The surviving documents do not show precisely what led him to recognize why such a mechanism of inheritance must be abandoned. Fifteen years of consciously integrating Darwin's mechanism of natural selection into the natural landscape as he so carefully studied it; fifteen years of embryological and cellular research in which he investigated the origin, development, and nature of the germ cells in particular; fifteen years of insisting that an "internal constitution" or "innate tendencies" must somehow be embedded in the organisms that he so expertly examined; a decade of pioneering cytology of germ cells, including his own on dipterans, daphnids, and hydroids; a critical mass of zoological discourse on development and evolution generated by university colleagues and investigators at the marine zoological laboratory in Naples; and Weismann's own penchant for examining most questions with a methodical, thorough, often tortuous warping of arguments—all contributed to Weismann's mental framework that compelled him to challenge the prevailing belief about heredity.

No matter how the pieces fell together in Weismann's mind, by the time of his address "On Heredity," he recognized that the separation of the germ cells from the soma in the Metazoa solved two fundamental problems, one phylogenetic, the other developmental. "Only in this way," he seemed to remind himself as well as his audience, "can we render to some extent intelligible the transmission of those changes which have arisen in the phylogeny of the species; only thus can we imagine the manner in which the first somatic cells gradually developed in numbers and in complexity." What these sequestered germ cells consisted of he could not say in 1883 other than referring to "variations in the molecular structure of the reproductive cells."[59]

9

A Perspective on Heredity

The historians of biology Müller-Wille, Rheinberger, and colleagues, who are participating in a long-term, multivolume work on the rise of the modern notion of heredity, have already come to the valuable conclusion that "heredity" was not a biological concern until after the mid-nineteenth century. As they point out, other historians have for a long time recognized but not systematically explained biologists' lack of attention to heredity before the middle of the nineteenth century.[1] Instead of accounting for this phenomenon in the traditional way in the history of science by invoking either a gradual accretion of new information or identifying a "revolution," they have adapted Michel Foucault's metaphor of an "epistemic space," while wisely stripping this metaphor of the implication of sharp discontinuities that simultaneously reach across different domains. These authors have redefined "revolution" to refer to a gradual shift and expansion of a conception or belief into hitherto not recognized problem areas. Thus, in line with this new metaphor, over time heredity became a legitimate problem in many areas as one domain of study gradually inoculated another. Their collective enterprise draws upon many case studies that have indicated the transference of the concept of heredity/*hérédité* from seventeenth-century legal and property rights literature forward to eighteenth-century French medicine, late eighteenth-century and early nineteenth-century natural history, to anthropology—particularly race studies, and to agriculture.[2]

Accepting this historical picture, what needs to be explained at greater depth is how the transference of "heredity" to biology came about and how biology became the dominant domain in this concept's change over the next 150 years. Müller-Wille and Rheinberger intend to pursue this goal, but we can address how the scientific career of Weismann, stretching as it did from 1857 to 1914—from just before Darwin to the advent of classical genetics—provides an important window onto the first half of this history. The historian, however, must tread with care: "heredity" in the modern sense of a transmission from one individual to an offspring existed in other senses and metaphors at the time.

The Biological Nature of Heredity: The Overgrowth Principle

Karl Ernst von Baer and the "Overgrowth Principle"

In discussing the nature of biological heredity in the nineteenth century, we are brought back to 1828, seven years before Weismann's birth, five years after Oken had first assembled the Versammlung in Leipzig, and the year following Baer's momentous discovery of the mammalian egg. As we have already seen, the year was notable in the history of biology for it was the year Baer published the first volume of his *Entwicklungsgeschichte der Thiere.* This and its six Scholia, or added generalizations, contained therein, were presented in Chapter 2 as one of the premier examples affecting the study of embryonic development. Left out in the previous discussion, however, was Baer's understanding of reproduction, which appeared in Scholium II. This scholium was dedicated to the self-promotion of the embryonic chick. (The title of Scholium II was "Die Ausbildung des Individuums im Verhältniss zu seiner Umgebung." Although the word "Scholion" has the connotation of a law, as shown in the example of Newton's *Principia,* "generalizations" or "conclusions" seem more suitable in Baer's case.) Baer brought up the phenomenon of heredity when in the second scholium he discussed reproduction in both lower and higher organisms:

> In the lower animals, where there is no contrast between the sexes and each individual thus completely contains the idea of this animal form, it requires only maturity in order to generate. Here generation is simply the extension of growth over and beyond the borders of the individual and reproduction is nothing more than a further growth beyond itself. In such animals in contrast, which possess either a double sex, or consist of separate sexes, the growth generates in one of the sexual organs the Anlage for the new germ as a part of itself, and the influence of the opposite sex exercises control over the first.[3]

Baer's language, not surprisingly, is that of a morphologist—or, more specifically, an embryologist, the most distinguished one of the century. The passage speaks of the idea of animal form ("die Idee dieser Thierform") in both
lower and higher animals, of asexual and sexual generation ("Zeugung"), of
this generation being an extension of growth over and beyond the borders of
the individual ("Verlängerung des Wachsthums über die Grenzen des Individuums hinaus"), and of reproduction ("Fortpflanzung") all being an extension of growth. It goes on to identify sexual reproduction in a somewhat different fashion, for in this case the "overgrowth" in one sex generated the
"Anlage" for the new germ as part of itself ("neuen Keime als einen Theil von
sich"), while the other sex completed the fertilization by an "overgrowth" that
exercised control over the first. In the corollary to Scholium II, Baer further
developed the difference of asexual and sexual reproduction and added cases
of what would later be known as parthenogenesis. Not only is the vocabulary
part of the embryologist's language of the day, but the thought behind it emphasized growth and form, which were central principles for contemporary
embryologists. All that was needed to round out the continuum between parent
and offspring was a principle of differentiation, but as we saw in Chapter 2,
Baer's four laws of development articulated in Scholion V and his and Pander's focus on germ layers provided an important step in that direction.

Baer's passage and his embryology of growth, germ layers, and generation,
however, are a long way from that which by the 1870s and 1880s biologists
would describe as "heredity." Nevertheless, it is easy to discern a model for
intergenerational transmission of species form and individual variation. Such
a model closely identified growth and reproduction, the latter simply being a
product of the former.

Theodor Schwann and Cell Reproduction

Between the time of Baer's discovery of the mammalian egg in 1827 and Darwin's *Origin of Species* there was no more important contribution to the midcentury model of heredity than Matthias Schleiden and Theodor Schwann's
cell theory.[4] As Schwann indicated in his title and as historians have pointed
out many times since, the primary achievement of these two investigators was
to unify the microscopic morphology of the plant and animal worlds by identifying cells as their common unit of structure and function. Schwann particularly stressed the similarity in the growth and multiplication of their cells.
He buttressed his comparison by extensive reference to the early development

of animals and to what Schleiden and others had found with plants. Thus, he determined that the cells of the notochord and cartilage of larval frogs, fish, and mammals, despite differences in their surroundings, revealed marked similarities in structure and development with like structures in plants.[5] A lengthy review of the cells associated with animal ova, germinal membranes, and "subsequent" tissues revealed the same cellular formations and development.[6] His study, in short, has often been described as an embryology of cells, and this embryology told the same story in vascular and nonvascular tissues, in different species, and in general in all animals and plants.

The model Schwann then constructed highlighted the similarity in this cell embryology. It began with the precipitation of one or more nucleoli (*Kernkörperchen*) from the surrounding, nutrient-bearing fluid he called the "cytoblastema." Concentrically about this "spot" formed the walls of a nucleus containing a clear, fluid-filled cavity. Next, according to him, the walls of the cell gradually developed, looking similar to a watch glass in contact with the nucleus. As the wall extended farther out, it developed into a vesicle filled with its own fluid; at the same time, the nucleolus and nucleus, their formative role for coming into being having been fulfilled, simply vanished. The cell itself expanded further into a mature cell with its own unique form in accord with the adult tissue it soon became a part of and with which it functioned. Later, the cell, like any organism, aged, degenerated, and eventually died.

Differing in one major respect from Schleiden, Schwann argued there were not two but three ways in which this cell formation occurred: (1) endogenously, within mother cells; (2) exogenously, outside of any cell (Schwann's addition); and (3) by a direct cell division. Key to the first two modes of formation was the cytoblastema. Schwann conceived of this having different consistencies, solid, gelatinous, or fluid, depending upon the tissue involved.[7] In short, the cytoblastema served as the organic mother lode for the generation of nucleoli, nuclei, and cells and thus indirectly for the organism itself.

As is well known, the concluding section made clear that Schwann was serious about his analogy between cell formation and crystal precipitation. Such an analogy reflected the belief held by his cohort of fellow students at Müller's institute that life could be understood in physical and chemical terms. In addition to crystallization by apposition at the surface of the cell's formative structures, Schwann recognized that cell formation and growth also included an intussusception of nutrients and cytoblastema into the interior of the nested vesicles. This made possible both growth and endogenous and exogenous cell formation. In addition to a developmental life cycle, assimilation and growth were the fundamental processes linking conceptually all cells and all life together.

This leaves the impression that the ovum was a growing and developing set of vesicles first within the oviduct then externally at incubation. These vesicles were derivatives of the cytoblastema, that is, of chemical fluids. This cytoblastema might be different from vesicle to vesicle; nonetheless, it had to be derived from the fluids of the ovaries and determined by the assimilative capacities of the various orders of cells. The cytoblastema might, if Schwann chose, be considered a simple overgrowth of the parental cytoblastema. To be sure, the ovum became structured as it developed, but minimally so. The stress Schwann placed was that of fluid and a precipitation therein of concentric vesicles. Because ova, too, were dependent upon precipitation in a cytoblastemic matrix and were literally an "overgrowth" of the parent, no cellular parts of the body were assumed to be derivative of cellular activity. The picture became complete in the 1850s when the picture was simplified by Robert Remak and Rudolf Virchow, both of whom insisted that cell multiplication was almost always the result of direct cell division. Again, growth and overgrowth were the primary mechanisms producing cells.

Johannes P. Müller's Handbuch

Johannes P. Müller had been bookish in his youth with a predilection for botanizing and entomology. He had excelled at the gymnasium in Coblenz, and upon maturation he had studied medicine at Friedrich-Wilhelm University in Bonn, where the faculty was inclined toward romantic "Naturphilosophie." After writing a romantically inspired dissertation, he had won support to study under the tutelage of the Berlin anatomist Carl Rudolphi. While directing him toward a career of careful observation and introducing him to many new laboratory instruments, Rudolphi had also weaned him from the excesses of romantic science and directed Müller toward a career of careful observations in physiology and comparative anatomy.[8] After a successful return to Bonn as a privatdozent, Müller had developed his own research style combining observation, experimentation, mathematization of experimental results, and an integration of scientific method and philosophy. It was in this Rhineland city that Müller had published some of his most important work in the physiology of vision, motor and sensory nerves, and endocrinology and within the decade had advanced to a full professorship. At this time Müller developed the law of specific nerve energies and clarified the Bell-Magendie law distinguishing the function of the posterior and anterior nerves of the spinal column. So when Müller returned to Berlin in 1833 as Rudolphi's successor in anatomy and physiology, he had already become a celebrated physiologist.

To be sure, Müller is known today primarily as a neurophysiologist, pathologist, and later as a marine zoologist. His research and teachings in these areas inspired generations of German biologists. The fields of biological research and teaching in the first half of the nineteenth century, however, were not demarcated and institutionalized in the same compartments as they were by the last third of the century. As we will see in a moment, he immediately began publishing his influential *Handbuch der Physiologie.*[9] The *Handbuch,* in combination with his teaching style and his range of functional interests, encouraged the next generation to move beyond him in pursuit of a mechanistically oriented experimental physiology.

Müller, however, had broader interests. Ten years Baer's junior, he followed in the latter's footsteps in a number of important ways. He conceived of comparative morphology in a sweeping manner; he encouraged the study of the natural history of marine invertebrates and the examination of protozoa. He was persuaded that the study of development was the only way to understand fully the goals of organic form and function. He emphasized the importance of observing the fine structure of organisms and human pathology by means of the microscope and incorporated the recently articulated "cell theory" fashioned by his student Theodor Schwann and by Matthias Schleiden, an advanced student in Johannes Horkel's neighboring botanical institute. Despite his deep involvement in the development of the study of biology, Müller continued to think of the organism in both teleological and mechanistic terms—Müller has been identified as an important member of a "teleomechanist tradition" in Lenoir's analysis.[10]

The point here is that Müller and his immediate contemporaries were scientifically steeped in developmental processes at a refined level, but none of their extraordinary work at the beginning of the decade anticipated the advent of the cell theory at its end. None of the embryologists working in Baer's descriptive tradition was more affected by it, when it emerged on the scene, than Schwann's mentor Müller himself.[11] Even before the publication of the complete version of Schwann's *Microscopical Researches,* Müller had adopted Schwann's cell theory for an overhaul of his own concept of cancer and other morbid growths.[12] Where prior to 1838 he had tried to describe cancer in terms of chemical and microscopical discriminations and to organize their distinctions into thirteen tumor types, Müller now saw all growth and differentiation in terms of the normal and abnormal development of cells. The "Entwicklungsgeschichte" of cells, to emphasize Schwann's expression and theory, also became Müller's guide for understanding generation. He quickly integrated the new perspective into the second volume of his *Handbuch,* which appeared in 1840.

Starting with the facts of asexual reproduction both in plants and many invertebrates, Müller envisioned the basic processes of growth in the terms of all three forms of cell multiplication: exogenous, endogenous, and cell division. Müller saw, too, that division, budding, and excrescences of cells led to multiplication of organisms. He was to use the same scenarios as a model for sexual reproduction. All three of these processes, as we have seen, had been sanctioned by Schwann. "Asexual reproduction," Müller insisted,

> accomplishes just the same thing that sexual reproduction does. [Only] here one can disregard all the mysteries of sexual reproduction and start with the fact that an organic body forms a multitude through division and budding, yes through growth, that further the cells themselves, the basic parts of the organic body, partly through forming new cells internally to themselves, partly externally to themselves, partly through division of cells and forming fragmenting outgrowths of the cells the same as are inside the organism, and that there finally are organisms where every cell is a germ, which through the outgrowths of a cell again reproduces all the germs of the species.[13]

Cells were clearly the fundamental unit or "Urteilchen" for understanding these many modes of reproduction. Cells provided a mechanism to bind growth with the multiplication of further cells and to bind these in turn with the multiplication of organisms of the same species. Adding to the images conjured up by this passage and the inference of growth through metabolism so deftly described by Schwann, Müller's work on normal and abnormal differentiation of cells completed a unified picture. The reader was provided with a model for understanding growth, embryogenesis, neoplasms, regeneration, and asexual reproduction. His notion of reproduction combined Baer's "overgrowth" with Schwann's cell theory. It was not hard for Müller to insert an understanding of heredity into this model.

As already intimated, sexual reproduction for Müller possessed some mysteries both from his and our perspective. The first was bound up in Müller's concept of the organism as a unique union of material and "Kraft." Here was a Kantian concept related to the purposefulness of the functioning whole of a living organism rather than the separate parts, but Müller's discussion was not always consistent and interchanged with abandon the singular (*Kraft*) and the plural (*Kräfte*). In his mind, matter and Kraft were notably united in the germ that produced the next generation and ensured the realization of the same family, generic, and species predispositions or "Anlagen" that were previously expressed in the parents. It was clear, however, that the organism was not just a mechanical synchronization found with a human-constructed clock. Instead, the purposefulness was to be conceived of in the Kraft of life, of function, and of reproduction. It was ratified by the whole not the parts. It was "die Kraft des Keimes," that is, "the force in the germ," that ensured the reproduction of the form of family, genus, and species in the offspring to come. In contrast to mechanical clocks, Müller insisted that

> organisms come into being not merely through a chance combination of their elements, but generate also the organs necessary for the whole through forces [*Kräfte*] of the organic material. This rational creative force [*vernünftige Schöpfungskraft*] expresses itself in every animal according to a rigid law, as the nature of every animal requires; it is already present in the germ, before these have separated and the later parts of the whole are present, and it is the one that truly generates the limbs, which belong to the concept of the whole.[14]

Müller added that the Kraft was not the analog of consciousness ("Geistebewusstseyn") but must be viewed as a creative force ("Schöpfungskraft"). Above all, he insisted that the force and organic material were irrevocably bound together in such a way that they should not be construed in the terms of a dualistic union. The fusion of Kraft, cells, and organic material was central to Müller's famous prolegomena to volume 1 of his *Handbuch*.[15]

There is more to Müller's understanding of sexual reproduction. The paternal and maternal germs, he felt, were materially and functionally significantly different. "Sexual propagation has this as its essence," he asserted, "that with it, to be sure, related germs propagate the characters of the genus, species and even the individual, but the organization of the germ cannot be fully fulfilled without the influence of a related germ, the semen, which is different in material from the first germ."[16] Thus, he viewed egg and semen as germs but of different kinds. The egg was either a single cell or multiple cells; it grew and multiplied, and after fertilization it developed into the embryo of the next generation. The semen was largely fluid but also included granules ("Samen-körnchen") and most of the time spermatozoa. In 1840, prior to Kölliker's demonstration that the spermatozoa were derived from cells, Müller presented two general interpretations of these morphologically diverse and curious objects. The Berlin protozoologist and Müller's colleague Christian Gottfried Ehrenberg and his followers considered them as small, parasitic animals, as had Baer, Cuvier, and Leeuwenhoek before him. Treviranus, on the other hand, considered them the animal analogs of pollen grains. Müller appears to have preferred the latter interpretation but not with complete confidence. More important for him was that the fluid, granules, and spermatozoa collectively comprised the active male germ, which bore the family, generic, specific, and individual attributes.

After dropping the subject of propagation and devoting the final part to development and birth of the new embryo, Müller completed the *Handbuch* with closing remarks, almost afterthoughts, on animal and human varieties. This was a natural transition, for generation in its broad physiological sense included both developmental and transmission processes of all forms, from reproduction to alternation of generations, from fertilization to regeneration, and from innate to environmentally induced attributes.

As one surveys the structure and content of the *Handbuch,* it is not immediately evident why this was such an important tool for students and educators of human physiology. Emil du Bois-Reymond, an assistant then privatdozent in Müller's institute between 1843 and 1848 and Müller's successor in the physiological half of his Chair of Anatomy and Physiology in Berlin, posed exactly this question in his widely read memorial.[17] He pointed out that externally Müller's two volumes were poorly printed in a small, cramped font, which may be explained, but hardly excused, by the fact that the Coblenz publisher was his father-in-law. Between the covers it contained no index or illustrations except for the final book on development. In many places, it was repetitive and digressive. It lacked extensive citations, and it failed to display the breadth and concern for scholarly learning found in multivolume physiology texts from Albrecht von Haller's *Elementa physiologiae corporis humani* (1757–1763) to Karl Burdach's *Physiologie als Erfahrungswissenschaft* (1826–1840). Despite these shortcomings, according to du Bois-Reymond, the *Handbuch* was avidly read because it drew together Müller's highly regarded experimental research on sight, sympathetic and motor neural systems, muscle contraction, vocalization, and development, work that had been scattered throughout many sources. It was informed, as Baer's *Entwicklungsgeschichte* had been before, by a comparative method that brought together vertebrate and some invertebrate physiology. Moreover, it combined physiology with a philosophical understanding of the contrast between the material inorganic and functionally purposeful organic.

Both du Bois-Reymond and Köller, a century later, wrote only a perfunctory line or two on Müller's books on generation and development.[18] It is clear, however, that Müller had well-articulated, up-to-date opinions on both subjects, which according to du Bois-Reymond incorporated the soon-to-be-published research of his student Karl Bogislaus Reichert.[19] Telling in both Müller's *Handbuch* and in Reichert's dissertation is an enthusiasm for Schwann's cell theory. As Lenoir has detailed, theirs was a continuation of the teleomechanistic tradition begun by Kant and Blumenbach and furthered by Baer and Schwann.[20]

What one finds in Müller's volume on *Zeugung* is a presentation of a generalizable morphological structure, the cell, to explain growth, development, and reproduction. Collectively the cells offered what the germ layers had left in abeyance. They provided a model for asexual reproduction, which was easily extended to contrasting genders, and a mixture or "Vermischung" of two germs. They suggested that individual variations were to be discussed within the context of this Vermischung. Because of its involved "Schöpfungskraft," the model included more than a mechanistic insight. Müller's *Handbuch* would be unacceptable to a younger generation intent on turning biology into a "Wissenschaft."

Müller's closing remarks added, if only in passing, that the differences among the traits of generic, specific, and racial categories were part of the passage of Kraft and matter from parent to offspring in humans and domesticated animals. That which grew and developed into a given individual could also be passed on by the germs for the next generation. Variations at all levels were transmitted through propagation. "Overgrowth" explained heredity, which resolved simply into "like producing like." Heredity, or "Vererbung" as it would increasingly become to be known, was not a part of Müller's vocabulary, but his *Handbuch* furthered a physiological context for discussing what soon became its formal study.

As mentioned earlier, Müller turned from experimental physiology to marine biology after completing the fourth and final revision of the first volume of the *Handbuch* in 1844. He was still an active middle-aged scientist, aged forty-three. He guided a renowned program of anatomy and physiology within the medical faculty in Berlin, but he looked forward to performing less vivisection and pursuing more natural history in the little-explored world of the ocean. He contributed significantly to the natural history and embryology of marine invertebrates before he died prematurely in 1858. Much had happened in biology since his *Handbuch* first appeared. As we have seen, Schwann's cell theory had forced a new discourse onto anatomy, physiology, and pathology.

A cohort of Müller's students, perhaps the most famous, declared their emancipation from the teleological implications of Müller's prolegomend and the general employment of a vitalistic Kraft throughout the text. Among them, Hermann von Helmholtz enunciated the famous first law of thermodynamics in his *Erhaltung der Kraft* of 1847, which had lasting implications not only for physics and chemistry but for biology as well. Another member of the cohort, Karl Ludwig, wrote a new physiology textbook on mechanistic principles that in 1852 replaced Müller's *Handbuch* as the preferred text for students and researchers of physiology.[21] The cell theory also underwent significant changes in the 1850s associated with Robert Remak, a former Müller student and an independent investigator in Berlin, and with the newly appointed professor of cellular pathology in Berlin, Rudolf Virchow. The upshot of their parallel investigations was that Schwann's three routes to cell multiplication, described as endogenous, exogenous, and direct division, were whittled down to one. Direct cell division, excepting a few pathological situations, was soon recognized as the sole mechanism of cell multiplication. Remak also modestly claimed in a comparative study of vertebrate development that "das Eie eine Zelle sei" ("the Egg is a Cell") with the value of other single cells.[22] Finally, it almost goes without saying that the year Müller died was the year that Charles Lyell and Joseph Hooker presented to the Linnaean Society Charles Darwin and Alfred Russel Wallace's paper on evolution by natural selection. When Müller's last student picked up the torch of the Schwann-Müller model of reproduction and heredity, he did so in a very different scientific climate.

Ernst Haeckel and Vererbung

This brings us once again to Ernst Haeckel who assimilated all of these events, and much more, in his comprehensive, quasi-philosophical and highly schematic *Generelle Morphologie*. It might have been merely an act pro forma when Haeckel thanked his "unforgettable teacher Johannes Müller" for exposing him to the empirical foundations of his science and the reigning perspective of his dualistic morphology, a position Haeckel quickly disavowed in the pages of his *Generelle Morphologie*. He also recognized his debt to the critical influence of his "honored teacher and friend" ("hochvererten Lehrers und Freundes") Rudolf Virchow and to the organic view provided by his cellular pathology—this praise despite the public run-in with Virchow three years earlier over the teaching of evolution in schools. There could be no doubt, however, that it was Darwin who caused "the scales to fall from his eyes."[23] Whatever the motive, these were the mentors Haeckel chose to recognize in the foreword as he promised to save morphology from "sweeping unclarity and a Babylonian babble of tongues" ("weitgehende Unklarheit und eine so babylonische Sprachverwirrung") now rendered more incomprehensible by an unphilosophical empiricism. Fundamental ideas were needed to advance morphology: "The thought of the unity of all organic and inorganic nature, the thought of the general truth of mechanical causes in all recognized appearances, the thought that the arising and developing forms of organisms are nothing other than the necessary product of the unavoidable and eternal laws of nature."[24]

Whatever might be professed in the foreword, the thousand-plus pages of his two-volume masterpiece suggests to us how his understanding of heredity was directly derived from Schwann, Müller, and Virchow. Adding to what has already been pointed out earlier about his fundamental conceptions, it is clear that Haeckel considered heredity as one of two countervailing processes in life. "Vererbung," as he called it, identified the conservative aspects of phylogeny, which guaranteed a continuity and repetition of attributes from generation to generation through life's lineages. Congruent with the formalizations of his text, he designated "Hereditas" as the Greek-derived equivalent of Vererbung. The capacity for Vererbung was "Erblichkeit" or heritability, and as he explained it, this reflected a virtual force or "virtuelle Kraft," which existed bound to the organic material. Now it is a moot question whether Haeckel had completely abandoned Müller's physiological forces, including the "Schöpfungskräfte," and had updated his ontology to accommodate the revolution in physics signaled by Helmholtz's classic paper of 1847.[25] He was clearly cognizant of the latter, and later in his *Anthropogenie* he relied on Helmholtz's popular paper on "Wechselwirkung der Naturkräfte" (1871).[26] Nevertheless, there remains an awkwardness in his use of "Kraft," which harkens back to the usage if not the ontology of Müller's *Handbuch*. Again, let us look at a statement from his *Anthropogenie:*

Just as the motive force of our flesh is involved in the muscular form-element, so is the thinking force of our spirit involved in the form-element of the brain. Our spiritual forces are as much functions of this part of the body, as every force is a function of a material body. We know of no matter which does not possess force, and, conversely, of no forces that are not connected with matter.[27]

Haeckel may have left behind the Schöpfungskraft employed by Müller to explain the generation (literally creation) of specific forms, but he failed to understand fully Helmholtz's use of "Kraft," which soon became identified with the term "Energy." For Haeckel, "Kraft" appeared to be a motion specifying the functions of material. Heredity, that is, Vererbung, was simply a basic function or property of propagating organisms.

His examination of Vererbung did not stop at a simple description of its role in phylogeny, for Haeckel fashioned it into a mechanical model based on the cell theory and an explicit use of Baer's description of the bond between growth and generation. Haeckel was equally direct: "Reproduction is a maintenance and a growth of the organism over and beyond the individual mass, one part of which is elevated to the whole." For convenience, I have referred to this concept elsewhere as Haeckel's "overgrowth principle" and will continue to do so here.[28] Vererbung as envisioned by this overgrowth principle became the explanandum in Haeckel's mechanical model. This began with the cell theory, as amended by Remak and Virchow, and with the phylogenetic claim that the simplest and most primitive mode of reproduction must be the direct cell division of single-celled organisms. Thus, heteroplastic life arose through cell growth and cell division whereas reproduction consisted of overgrowth beyond the physical limits of one, or in sexual reproduction two, individuals. Thus, Haeckel's model was really a model of transmission—better yet, transmission by default. Because offspring were simply the overgrowth of the parental growth, differentiation and cellular divisions and heredity of parental traits automatically followed a transmission of parental material and Kräfte through the necessary sequences of cell division. Because the parents changed throughout their lifetime, the potential offspring shadowed those changes, whether the overgrowth was in the form of single-cell releases (e.g., spores), groups of budding cells (e.g., in hydroid colonies), or germ cells (e.g., ova and spermatozoa). In short, "like produced like."

Later in his career, Haeckel made one brief conjectural excursion to suggest what might be transmitted in those overextending cells. In the spring of 1876, four years after he had first appropriated for his own purposes the germ-layer doctrine, Haeckel published his *Perigenesis der Plastidule,* a short treatise written in a semipopular style. Ostensibly he presented a "provisional hypothesis of Perigenesis" as an explicit alternative to Darwin's own provisional hypothesis of pangenesis. In actuality, he took the opportunity to explore a monistic cosmology. In more concrete terms, Haeckel filled out the ontological suggestions about the contrast between the inorganic and living worlds first voiced in the *Generelle Morphologie.* He presented his objections to any form of dualism and made a stab at bringing material and consciousness together in one entity by arguing that material atoms have souls.[29] The inspiration for Haeckel's discussion turned out to be the notion in Goethe's 1809 novel *Die Wahlverwandtschaften* and an address given in Vienna by the Prague physiologist Ewald Hering. The first had discussed selective affinities in human affairs; the second had likened the process of heredity to memory.[30] Haeckel exploited both in order to clarify the conservative nature of heredity. Mother cells divided to produce daughter cells, and their likeness was ensured by repetition inherent in the motions of the basic protoplasmic molecules, the plastidules. "Inheritance," Haeckel explained, "is the transmission of the motions of plastidules, [the] reproduction of the individual molecular motions of the plastidules of the mother plastid to the daughter plastids."[31] And so it went for indefinite generations of division. Variations entered the system as slight disturbances in the molecular motions caused by external influences. It was fitting that the subtitle of Haeckel's treatise was *The Propagation of Waves of Living Particles.*[32]

From the perspective of post-Darwinian biology, Haeckel's "provisional theory" provided an explanation for heredity in its broad mid-nineteenth-century context. *Perigenesis der Plastidule* repeated, sometimes verbatim, the major assumptions embedded in the *Generelle Morphologie.* Because cell division was the primary mechanism of multiplication, the transmission of molecular motions from one cell generation to the next provided a tangible explanation for transmission between morphologically and physiologically different individuals. Divergence among the plastidules during differentiation, among individual somatic characters, among polymorphic individuals in a complex society, and among species during evolution all could be attributed to the general principle of the division of labor. To speak of external influences causing an alteration in the motion of the plastidules was Haeckel's way of rendering this universal process into familiar mechanical terms.[33]

If Haeckel remained ambiguous on the matter of whether the principle of the division of labor was a basic law that required or did not require further reduction, he repeated his familiar insistence that reproduction must be understood in terms of a growth over and beyond the individual. He reminded the reader that recent discoveries of reproductive modes broadened our appreciation of the asexual end of the reproductive spectrum. As for sexual reproduction, biology "now allowed one to trace the mystery of sexual reproduction again back to a form of growth and division of labor of the plastids."[34] Finally, Haeckel envisioned phylogenetic branching as simply an extension of the diverging pattern expected by multiple lines of transmitted molecular motions under different circumstances. The biogenetic law simply reflected the conservative as well as the progressive nature of these compounding motions. Haeckel's theory of perigenesis was not a theory of heredity in any modern sense; it was, as its second subtitle implied, "an attempt at a mechanistic explanation of the elementary developmental-processes." At the time when Weismann had wrapped up his monumental *Studien* on descent and had embarked on a series of microscopical examinations of reproduction in daphnids, Haeckel was plunging deeper into the ramifications of his *Generelle Morphologie.*

Overall his discussions of heredity on both the cellular and phenomenal level were hypothetical, qualitative statements about transmission. They were deductive and anecdotal. Nevertheless, Haeckel's model, featuring the cellular overgrowth from parent(s) to offspring, was well known in German biological circles. Georg Seidlitz, entomologist at Dorpat and one of the earliest and strongest proponents of natural selection, voiced cautious reservations while accepting the concept of overgrowth:

> Now to be sure the special processes, by which the transmission of form and character is always connected to a growth over and beyond the individual, may still be unknown to us,—it is never denied that the heredity of characters remains a solid fact; it will, however, have been misunderstood if one has asserted that it would cause absolute similarity between parents and offspring, which as we have already seen, is an impossibility.[35]

A former student of Müller in the early 1850s, Haeckel's one-time colleague and confidant in Jena, and after 1873 a professor of anatomy in Heidelberg, Carl Gegenbaur was the foremost comparative anatomist of the day. In commenting on the provisional state of understanding of heredity, Gegenbaur endorsed an "overgrowth principle" at the lowest levels of life:

> If we take any one of the numerous lowest forms of life (below the Protista) which multiply exclusively by division, there the products of division represent new individuals, they gradually grow, multiply again in

the same way, and so on. Is it not necessary here that in propagation by division the existence of material from the maternal body brings with it a similitude of behavior? If in both cases the same material substrate is present why should it not produce the same phenomena?[36]

Weismann had no hesitation in explicitly endorsing the general conception of overgrowth in his lecture on heredity in 1883. In ascribing the principle to Haeckel, he clearly had overlooked how general and physiologically logical the "overgrowth" principle was. "Haeckel was probably the first to describe reproduction as 'an overgrowth of the individual,' and he attempted to explain heredity as a simple continuity of growth. This definition might be considered as a play upon words, but it is more than this; and such an interpretation rightly applied, points to the only path which, in my opinion, can lead to the comprehension of heredity."[37]

Continuity of Substance as an Explanation for Transmission

If Haeckel and his predecessors provided an important context for Weismann's own discussions of heredity, the subject was broadened by Darwin's discussion of heredity found in his *Variation of Animals and Plants under Domestication*. This appeared two years after Haeckel's *Generelle Morphologie*.

Charles Darwin and The Variation of Animals and Plants under Domestication

Even as early as his first "Transmutation Notebook," Darwin had wrestled with many of the issues that had to do with heredity. By 1839, somewhat frustrated by the confusion of information about what the process entailed, he drew up a short pamphlet for distribution on "Questions about the Breeding of Animals," but as one recent historian has rightly maintained, this effort to grapple with heredity was "very limited indeed."[38] The whole subject appears to have been suppressed in Darwin's publications until after he had completed *The Origin of Species,* but this low profile certainly did not imply that heredity was a secondary concern to him. We now know that when Alfred Russel Wallace's famous letter from the Malay Archipelago arrived at Downe House, Darwin was compelled to leave aside what he had assembled on the subject of "inheritance" while he turned to writing the "abstract" that became *The Origin of Species.* His extended ideas appeared nine years later in *The Varia-*

tion of Animals and Plants under Domestication.[39] I will limit my examination of Darwin's ideas to this work, for this, after all, is where most biologists learned of his thoughts on the subject.[40]

Darwin's theory of pangenesis is too well known to require describing in any detail.[41] Briefly put, he argued that each cell of the organism, embryonic as well as adult, constantly sloughed off molecules into the fluids of the body. These "gemmules," as he called them, were formative elements bearing the imprint of the cell whence they came. Eventually they collected in the gonads or other generative organs where they became incorporated into reproductive cells and so became the formative basis of a new generation, or while still circulating the gemmules might play a formative role in the regeneration of parts.

The theory was ingenious. It allowed Darwin to account for the lasting features of heredity, for he could invoke "latent" and "ancestral" gemmules that became active on specific and perhaps rare occasions. It also allowed him to explain the labile features of heredity, for the population of gemmules reflected the tissues and cells as they underwent changes because of the changing conditions of life. Darwin also argued that the gemmules might fuse, thus explaining blending phenomena and continuous traits. As organic molecules, the gemmules allowed Darwin to associate closely the processes of growth and reproduction. Asexual forms of reproduction simply entailed one set of gemmules. In sexual reproduction, the gemmules from both parents commingled, and in keeping with the principle of elective affinities—like attracting like—the gemmules derived from similar parts combined and together became the basis for that part in the future offspring.

No one can read either the 1865 manuscript on pangenesis or the expanded version in *Variations* without being struck by the marked imprint of mid-nineteenth-century concerns.[42] In both works, Darwin demonstrated his awareness of the broad spectrum of modes of reproduction that had the consequence of deemphasizing the difference between sexual and asexual forms of reproduction. By the second edition of *Variation,* he strongly identified sexual, asexual, regenerational, and developmental processes with growth: "Sexual and asexual reproduction are thus seen not to differ essentially; and we have already shown that asexual reproduction, the power of re-growth[,] and development are all parts of one and the same great law."[43] He wrote of the gemmules as excess material that could produce new growth, fission, or collect in the gonads. They were the surplus production of cells. Darwin also wrote of "the powers" of inheritance and distinguished these from the phenomena of variations known to occur in both sexual and asexual modes of reproduction.[44] They helped him explain forms of blending inheritance and the transmission of acquired traits. By designing a theory of gemmules to explain the many generative phenomena, Darwin held true to his positivistic belief to produce a mechanical explanation. With all of these features in an articulated theory, Darwin, despite all his originality, joined many contemporaries in identifying the mechanism of heredity with growth.

Herbert Spencer and a Philosophy for Heredity

When in a moment of exhilaration his colleagues at the highly exclusive X-Club chose pseudonyms for one another—such as "Xalted Huxley," "Xquisite Lubbock," and "Xcentric Tyndall"—there surely must have been a moment of felicitous cheer when they hit upon "Xaustive Spencer."[45] Of all the nineteenth-century followers and commentators on science, Spencer certainly deserved this cognomen. His interests spanned the sciences from astronomy to sociology, and he was a major commentator on ethics, political science, and education. When involved in a subject, he plumbed its depths, and his works now occupy about a yard of shelf space in most standard research libraries.

But only recently have historians of science found Spencer's works valuable as statements of contemporary scientific aspirations. There should have been no doubt about this as far as the social sciences are concerned. Spencer had explored and contributed to this area from the outset of his literary career, and his *Principles of Sociology* (1874–1896) played an important role in staking out the boundaries of this new discipline. When we realize, however, that William James used Spencer's *Principles of Psychology* (1870–1880) as a textbook for his course in psychology at Harvard and when we learn that *The Principles of Biology* (1864–1898) became a required text at Oxford, we have more than enough reason for taking seriously Spencer's collected works as a measure of nineteenth-century life sciences.

In addition, there are other compelling reasons for examining Spencer's biological ideas. Perhaps more than any other writer of the century, Spencer was a passionate promoter of a universal law of evolution; in fact, he was the first to use this term in its modern sense. Seventeen years before the appearance of *The Origin of Species,* Spencer had envisioned an advance in human society. The same progressive process became the metaphysical last upon which he shaped his psychological and biological theories. It was after he had expanded and more thoroughly developed an evolution theory that Spencer confidently charged into battle against Weismann in the 1890s to defend the notion of the inheritance of acquired characters. We will examine that episode when dealing more fully with Weismann's own mature evolutionary theory. Here it is valuable to sketch out how Spencer developed his ideas of heredity. Not only do they uncover a response to the same set of biological assumptions that we have found with many of his contemporaries, but they reveal Spencer, the philosopher, striving for the broadest application of the assumptions they all used. Moreover, he presented his ideas in an accessible, albeit tedious, fashion. All his volumes in his systematic philosophy were quickly translated into German.[46]

The first of the mentioned considerations, the vital phenomena of the individual organism, consisted of a 200-page analysis of the processes of genesis, heredity, and variation. Indeed, if Spencer had chosen a more descriptive title for the section than "The Inductions of Biology," he might have called it "Generation of the Organism." He returned to the same subject at the end of *The Principles of Biology* with a concluding hundred-page section on the "Laws of Multiplication." Thus, in terms of pages allotted, Spencer devoted more attention to the complex of phenomena associated with generation and heredity than to any other biological process, including the evolution of species. The reason for this emphasis becomes clear when we seek the very core of Spencer's biological system. A balance between growth, development, and reproduction explained all else in the organic world. Spencer linked growth to the surplus accumulation of nutrition, defined development as increasing heterogeneity or "the increase of structure," and argued that function determined structure. Spencer insisted that any and all of life's processes—including exercise, mental activities, correlated changes, and reactions to and interactions with the environment—brought about adaptive modifications in the labile organic form. This belief was also in keeping with Spencer's concept of life as a "moving equilibrium" and that a cessation of such activities found the organism returning to its original form. Spencer concluded his "Inductions of Biology" with chapters on the various modes of reproduction (he chose the more general term "genesis"), on heredity, on fixed variations, and on the organic linkage among all three. It was in these final chapters of this central section that Spencer developed his theory of "physiological units" that bore a minor resemblance to Darwin's theory of pangenesis.

What becomes readily apparent to the patient reader—and Spencer requires considerable forbearance—is the extent to which the principles already laid down more than a dozen years previously remained operative in and central to Spencer's entire biological system. If we were to put Spencer's theory of heredity together in a logical sequence beginning with his most basic concept of the organism, the constellation of processes would reveal the following features. (1) There exists a direct connection between parental growth, development, and life's regulative activities and the reproductive activities. This relationship necessitated a balance between the expenditure of material and energy (not well defined) consumed during the life of the parent(s) and those materials and energy devoted to the production of offspring. (2) A corollary of this automatic, accountant-like reckoning maintained that the mode of production appeared in an inverse relationship to the physiological and structural complexity of the organism. Thus, the asexual modes dominated at the lower end of the biological spectrum, but only the sexual mode appeared possible at the upper end. (3) Both claims 1 and 2 flowed from Spencer's general understanding of the offspring as simply an overgrowth derived from the surplus of the parents' own development and mature activities. (4) Spencer did not consider sexual reproduction a unique process but claimed it constituted merely a "coalescence of a detached portion of one organism, with a more or less detached portion of another." (5) It followed from this generalized view of sex that the gametes "have not been made by some elaboration, fundamentally different from other cells," and that the spermatozoa and ova were simply "cells that have departed but little from the original and most general type."[47]

Indeed, the laws of reproduction led Spencer to the happy conclusion that "changes numerical, social, organic, must, by their mutual influences, work unceasingly towards a state of harmony—a state in which each of the factors is just equal to its work."[48] It has rightly been pointed out that Spencer's evolutionary ideas, although initially inspired by Lamarck via Lyell, were basically derived from his progressive and optimistic social beliefs. The biological imprint on synthetic philosophy, however, may be judged by Spencer's deep involvement with the problems of generation and heredity. We might conclude that the study of contemporary biology with its focus on growth, development, and reproduction was a necessary exercise that allowed Spencer to buttress with empirical force the optimistic social traditions of his Derby upbringing. We will also see in Chapter 20 how it was that Spencer was emboldened to challenge Weismann, who in the 1890s offered a contrasting view about heredity and evolution.

William. K. Brooks, a Darwinian Update

It may seem odd to turn from Spencer to the American William K. Brooks, whose small, largely deductively argued book on heredity appeared twenty years after the first installment of Spencer's *Principles of Biology*.[49] Neverthe-less it seems fitting for a number of reasons. First, *The Law of Heredity*, coming as it did in 1883, actually reflected the end of the midcentury era of heredity theories, where so much stress was put on heredity being explained by nutri-tion, accretion, and growth (i.e., overgrowth). Second, appearing in the same year as Weismann's monograph on hydromedusae, it contrasted in style and content with Weismann and many of his German contemporaries who were struggling to work out a new way of understanding transmission. Third, al-though it reads like an armchair treatise rather than a research monograph, Brooks's small volume is thoughtful and well informed within the narrow range of what he considered to be the problems that needed to be surmounted in order to find a satisfactory general explanation. That Brooks makes it abun-dantly clear that his theory is a refinement of Darwin's theory of pangenesis, that he is imbued with a sense that searching for biological mechanisms is the only legitimate direction for biology to proceed, and that he appears as an uncompromising proponent of natural selection provide his volume with a contemporary flavor. Finally, Brooks's book contained a statement about con-tinuity of germinal material, which elicited a response on Weismann's part. We will examine this and similar coincidences in Chapter 10.

As an expression of the waning years of the midcentury concept of he-redity, Brooks's book on heredity appeared almost out of nowhere. It clearly showed the proliferation of ideas on transmission at a time when speculative suggestions were common and evidence rarely was gathered into a systematic and coherent argument. Brooks had taken up the subject ten years earlier at the very outset of his scientific career when in the summer of 1873 he was introduced to marine invertebrates at Penikese Island by Louis Agassiz. Fol-lowing Agassiz to Harvard in the fall, Brooks enrolled in a graduate program in biology. (Agassiz passed away at the end of the fall semester, so Brooks com-pleted his dissertation under the son, Alexander Agassiz.) It was an exacting and important study on the development of the pelagic tunicate *Salpa,* and it earned Brooks a fellowship at the newly founded Johns Hopkins University. His status was soon upgraded to an "Associate in Biology," which was Brooks's title at the time his book on heredity came out.[50]

His study of *Salpa* and of mollusks soon earned Brooks a reputation as an evolutionary morphologist who was capable of executing first-rate observa-tions on the development and metamorphosis of invertebrates. His volume on *The Law of Heredity,* however, emphasized the other side of his career. It presented a deductive explanation for a range of phenomena that were to be encompassed under an envisioned common law of heredity. The list, designated by Brooks as "requisites of a theory of heredity," included explanations for (1) various forms of asexual and sexual generation, (2) parthenogenesis, (3) fertilization, (4) the taxonomic limitations to fertilization, (5) the obvious similarity between ancestors and descendants, (6) the contrast in complexity, in reversions, and new traits in sexual and asexual reproduction, (7) the similarity and differences between male and female gametes, and between them and other cells of the organism, (8) the influence of changed conditions on subsequent generations, (9) the differences between the two sexes, particularly with regard to higher organisms, (10) the differences between crosses and their reciprocals, and (11) the differing effects of the two sexes in trans-specific hybrids.[51] For all of the armchair overtones of this list, Brooks was expressing a common outlook, shared by Darwin, Spencer, and Haeckel, that heredity embraced a cluster of what seemed at the time to be interwoven phenomena. To comprehend one was to weave rather than unravel its connections with the rest.

To the modern eye, Brooks's list offers a disconnected ramble through a patchwork of biological phenomena: genetic, developmental, morphological, physiological, and phylogenetic. The informal presentation unquestionably enhanced the impression of a disorganized understanding of the problems at hand. And yet it is just this widespread combination of issues that we need to keep in mind when dealing with the innovations and reactions to these phenomena, which were often subsumed under the term "heredity" during the second half of the nineteenth century. Brooks left no doubt about the source of his inspiration. The theory was heralded as a modified theory of pangenesis, and without qualms Brooks drew much of his factual information from Darwin's *Variations*. Furthermore, he gratefully dedicated his volume to the memory of Darwin, who had died the previous year.

Parthenogenesis, discussed in detail, was a key phenomenon for understanding reproduction; heredity and the production of variations were antithetical processes; and the list of phenomena, which he intended to explain, showed that Brooks was more concerned with the traditional problems of generation than with the mechanism of transmission. He found Darwin's theory of gemmules appealing but refashioned it to explain exclusively the appearance of novelty. Because the testicles and seminal vesicles were the only place, he argued, where these gemmules congregated, it seemed to Brooks that characters of the type were transmitted by growth and cleavage of the conservative female ovum. On the other hand, new variations were introduced into the lineage by the production of gemmules in the male. Brooks was not dogmatic in excluding the females from the production of progressive gemmules, "but the female differs from the male in having no specialized organ for the aggregation and transmission of gemmules."[52]

The juxtaposition of his meticulous descriptive morphology and his sweeping assumptions about heredity is worth noting. Brooks, perhaps in the wholesale manner from Darwin, had completely internalized the midcentury perspective.

10

Carl Nägeli and Inter- and Intragenerational Continuity

Carl Nägeli's career and work needs to be examined in considerable detail. It represents a very German response to the increasing need to explain hereditary transmission, quite different from that of Darwin, Spencer, and Brooks. Nägeli was more concerned with scientific methodology than were his English-speaking peers, and his work was thorough and based on a lifetime of highly respected academic botanical research. Nevertheless, there were elements they all held in common. They wanted to explain evolution in a mechanistic fashion, and they saw that the mechanics of hereditary transmission provided the key to success. More than this, Nägeli's major publication on this subject set down elements of dealing with the problem that clearly influenced the shape of Weismann's own ideas. To know Nägeli's idioplasm theory in some detail is to gain considerable insight into Weismann's germ-plasm theory.

Early Career

Although it was not published until 1884, Nägeli's monumental *Mechanical Theory of Descent* appeared to reflect the same understanding of heredity that Darwin's and Spencer's works had earlier.[1] In a strange way it nevertheless influenced many morphologists following in the wake of the discoveries of Flemming, Strasburger, and other nuclear cytologists. Belonging to the same generation as Rudolf Leuckart, Nägeli had been a productive botanist and influential personage on the German academic scene for over forty years. Born in 1817, educated at the university in Zurich, and inspired by Oken's lectures in natural history, he forsook medicine for a career in botany. With three years of studies in Zurich under his belt, Nägeli traveled to Geneva to work under the noted Swiss plant morphologist and physiologist Pyrame de Candolle and graduated in early 1840 by completing a study on the Swiss exemplars of the evening primrose genus *Circia*. With the doctoral degree in hand, Nägeli made a pilgrimage to Berlin to drink the waters of Hegelian philosophy at their source. The lack of empirical content quickly disabused him of both Naturphilosophie and its Hegelian offshoot, but there always remained in Nägeli's science, which for the most part was meticulously executed and rigorously tested, the quest for a unified picture of the natural world.[2]

From Berlin, Nägeli traveled in 1842 to Jena to spend a year and a half in close association with Schleiden. In later years, the famous cofounder of the cell theory was to accuse Nägeli of Hegelian leanings, but during this visit the two collaborated in publishing the short-lived *Zeitschrift für wissenschaftliche Botanik*. This journal conveniently served as a vehicle for a string of Nägeli's important microscopic studies.[3] It was here that Nägeli announced his discovery of "spermatozoids" in ferns and rhizocarps, presented his earliest examination of apical growth, and reviewed for the first time the taxonomy of the genus *Hieracium*, the line of research that bore directly on his later interaction with Gregor Mendel. More important, it was here that Nägeli began his exacting studies on the nucleus and on the formation and growth of plant cells. According to Robert Olby, "These studies of cell formation illustrate Nägeli's striving for general laws, the strong influence of Schleiden on him, and his eye for detail."[4] This modern judgment, noting both Nägeli's theoretical and empirical accomplishments, signals the danger on the part of others making a facile characterization of this multifaceted and methodologically astute botanist. Before he left for Jena, Nägeli had habilitated at the University of Zurich. When he returned to Zurich in 1845, he was certified to teach and in three years advanced to the rank of außerordentlicher Professor. He then spent three additional creative years as an Ordinarius at the University of Freiburg and another two back in Zurich at the same rank. Nägeli's reputation as a systematist, as a microscopist, and as a plant physiologist with a marked leaning toward chemical and physical problems of development was by that time firmly established. Justus Liebig, who had wooed Leuckart to Giessen in 1850 for similar sympathies, performed a parallel service for the University of Munich when he persuaded Nägeli to succeed Carl Martius as the Ordinarius in botany. Nägeli spent the next forty years teaching in the Bavarian capital. It was in this cultured city he built and directed the botanical institute and herbarium, and here he supervised the establishment of a strong physiologically oriented program for botany.[5]

Throughout his career, Nägeli did not hesitate to examine the methods and goals of his science. An exemplary statement of his mature thought may be found in his address to the Versammlung of German Scientists and Physicians, which had gathered in Munich in the late summer of 1877. His talk was in essence a response to the famous essay of Emil du Bois-Reymond, in which the Berlin physiologist had laid down certain epistemological limits to the purview of science and had concluded his strictures with the oft-repeated aphorism "Ignoramus und Ignorabimus." ("We do not know, and will not know.") Like du Bois-Reymond, Nägeli was intent to steer a course between the idealistic conclusions fostered by Naturphilosophie and a crude materialism that held that all of nature could and would be described as the direct effects of matter in motion. Like du Bois-Reymond, Nägeli also believed that science progressed when phenomena became dissected into their chemical and physical components. Where the former, however, drew the bounds of a complete knowledge at our inability to understand causality and consciousness, Nägeli argued that the indeterminate spatial and temporal beginnings and end of the universe and the uncertainties concerning the classical atom or smallest unit of matter were what restricted our ability to describe a complete Laplacean world system. Nägeli's argument might at first appear as an ontological one in which the limitations to scientific knowledge were determined by the structure of the universe rather than by the human intellect, but this would miss the Kantian thrust of his message. Nägeli was emphatic in insisting that "the limited capacity of the 'I' therefore permits us only an utterly fragmentary knowledge of the universe."[6]

Nägeli's point and the lecture's relevance to his biology were soon evident. Nägeli followed Kant. "We are hence able to perceive [erkennen] the real things with certainty only insofar as we realize in them [the] mathematical notions, number and size, with all that mathematics deduces from them. Natural knowledge thus rests on the application of mathematical methods to natural phenomena."[7] In short, limits to quantification alone, not qualitative material, allowed for precise knowledge. Nägeli denied that the biological domain need be treated any differently than the inanimate. The difference between the two spheres was only one of degrees. Special effects, such as life, sensitivity, and consciousness, were no more and no less beyond the bounds of science than the inorganic. Chemical and physical phenomena were understandable insofar as they were analyzable into quantifiable attractions and repulsions between molecules. The methodological reduction may not guarantee us a complete world picture, but it did make science possible. The same method held for the more complex organic responses—even "intellectual force is the faculty of the particles of material to influence one another."[8] Despite the limitations to knowledge that Nägeli accepted, his message was an optimistic one for the advance of science. It was grounded in the ability to describe inorganic and organic phenomena in terms of molecular actions. In response to du Bois-Reymond's resigned submission to eternal ignorance, Nägeli concluded on a more positive note: "Wir wissen und wir werden wissen." ("We must know. We will know.")[9]

The phrase and the entire lecture aptly described Nägeli's own drive to dissect the phenomena of life—including heredity. Nägeli's thirteen years of study of apical growth, during which he developed equations to describe cell divisions and lineages, culminated in "the realization of Schleiden's [and his] hope that the development of plants would someday be expressed in mathematical laws."[10] Equally as relevant for his understanding of heredity was Nägeli's investigation into the fine structure of organic matter. Starting as early as 1850–1851 and collaborating the following year with his student and later biographer, Carl Cramer, Nägeli painstakingly worked out the molecular structure of starch granules.[11] He undertook this ambitious project long before modern tools such as x-ray diffraction methods and the advanced physical chemical understanding of bonding were available, but his was at least a happy choice of subject matter. Through various simple chemical and physical tests and through microscopic investigations that exploited recently developed techniques of using polarized light, Nägeli was able to draw surprisingly sophisticated inferences about the molecular structure of starch, which existed well below the resolving power of the light microscope. The late historian and microscopist J. S. Wilkie, who reexamined Nägeli's research in detail, concluded that many of his generalizations about fine organic structure were validated in general terms fifty years later with the onset of more modern techniques.[12] We need not cover Nägeli's work (or for that matter Wilkie's excellent historical recapitulation) in any depth, but it is important to note that Nägeli's elaborate theory of micellae, emerging from this solid empirical groundwork, reflected the physical and mathematical orientation of its creator. A number of the specific details in Nägeli's model of micellae, however, are of value for our understanding of his heredity theory.

The term "micellae" was invented by Nägeli in 1877 to capture his notion of an organic crystal-like state.[13] The German expression was "Micell" in the singular and "Micelle" or "Micellen" in the plural. In English, the singular and plural became "micelle" and "micellae." Etymologically Nägeli derived the term from the diminutive Latin word for crumb, "mica." "Micella" consequently had the German connotation of "Krümchen" or "Teilchen."[14] It was an unusual crystal, for it grew by intussusceptions rather than apposition. Water molecules were absorbed into the interstices of the starch and intercalated themselves between the molecules of starch. Two sets of forces held the micellae together: one that bound the starch molecules to one another, the other that bound the water molecules to the starch. Nägeli thus envisioned each starch molecule as surrounded by an "envelope" of water, the two held together into a network by a balance of the two competing forces. Nägeli's model of micellae was dynamic as well as structural. The amount of water in a given micelle was inversely proportional to the size of the micelle. Finally, he envisioned groups of micellae adhering together into micellar associations that superficially resembled the protein polymers described in 1861 by the great German organic chemist Friedrich A. Kekulé. Most important, Nägeli freely applied his micellar model of starch grains and the notion of micellar associations to the cell wall and other "crystalloids." As we shall see, they also became the basis of his hypothesized hereditary substance.

In his meticulous fashion, Wilkie has also provided us with a review of Nägeli's most important contemporary critics. The major German botanists of the day, Wilhelm Hofmeister, Julius von Sachs, and Eduard Strasburger, all took issue with various aspects of Nägeli's thought, but what emerges from Wilkie's account is that admirers and critics alike had an enormous respect for the rigorous manner in which Nägeli developed his theories. In more than one way we find biologists building upon Nägeli's innovative approach and elaborate model.[15]

Toward a Theory of Descent

None of the discussion of Nägeli would be of particular relevance to the development of the notion of heredity were it not for the fact that Nägeli published in 1884 his substantial tome on heredity and evolution theory. He tells us that his *Mechanische-physiologische Theorie der Abstammungslehre* was an outgrowth of his 1877 lecture on the limitations of scientific knowledge. His objective was to put into practice the positive message of his talk—to take two timely scientific topics, the nature of physical forces and of biological descent, and to pursue them within the boundaries set by his empirical and mathematical approach. The book itself was written in fits and snatches. Nägeli's health deteriorated; the same year that he completed the volume, he began to

withdraw from his teaching obligations.[16] The work was long and often repetitive: 550 pages swollen by another 300 pages of appendices and reprints
of two earlier papers, confirming Nägeli's own assertions that the text was
neither well organized nor tightly structured. The sheer bulk of the volume
forces the historian to wonder whether it was read in detail and its arguments
suitably assimilated at the time. Despite its weakness, there can be little doubt
that Nägeli's work was a major effort to come to grips with the important
issues of evolution theory. In so doing, he drew extensively upon the theory
of micellae so painstakingly developed in his earlier studies of starch grains.

"The origin of the organic world," Nägeli boldly announced at the onset of
the *Mechanische-physiologische Theorie,* "belongs to the most inner sanctuary
of physiology."[17] So, too, did speculations about the contrast between the organic and the inorganic, the nature of life, nourishment, growth, reproduction, heredity, and a host of other basic biological matters. Nägeli set out to
show how through his physiology these various phenomena were all interre-
lated, and he ended by producing a full-blown picture of the origin of life and
a phylogeny of the plant kingdom. Given this picture, we might consider it a
contradiction that Nägeli also invoked a perfecting principle or "Vervollkommnungsprinzip" to help explain the progressive pattern of phylogeny itself.
But this was no vitalistic maneuver on Nägeli's part. Just as Herbert Spencer
had based the development of life on a physical law concerning the transformation of the homogeneous to the heterogeneous, Nägeli sought through the
physiology of a hypothesized "Idioplasm" an explanation of heredity and a
progressive phylogeny. Drawing upon the text of his pre-Darwinian lecture
on "The Individuality in Nature," written during the same years that Spencer
first proposed his own evolutionary theory, Nägeli presented the evolutionary
process in a monetary and hence mathematical metaphor:

> The individuals transmit to their offspring the inclination to become sim-
> ilar to them; the offspring, however, are not completely similar to the
> parents. The inclination to change must also be transmitted. When all
> conditions are favorable, an Anlage must always be able to improve it-
> self further through a row of generations, just as capital increases with
> its annual addition of interest.[18]

Apt as this metaphor was for the 1850s, it suited even better Nägeli's ambitions of the 1880s. We find intimated therein the years of research Nägeli
had spent on starch grains. Embedded in the idioplasm (hereditary substance),

the *anlage* (plural, *anlagen*) represented the molecular bearer of potential traits and inclinations. These were the smallest segments of the idioplasm that possessed their own motions and forces. Their number and order reflected the idioplasm's phylogenetic history, and like invested capital the anlagen possessed the capacity to increase or progress under the inertia of organic growth and reproduction. The anlagen of 1856 were easily transformed into the unique configurations of micellae of 1884.

From the onset of his *Mechanische-physiologische Theorie* Nägeli assumed that the micellae and their collective embodiment the idioplasm were distinct from the rest of the organism. To heighten the contrast, he designated the rest of the organism the "Ernährungsplasm" or nutritive protoplasm. The distinction was an assumed rather than a demonstrated claim, and it appears derived from considerations that the germ must contain the characters of all the ancestors. As will soon be seen, Nägeli envisioned nonfunctional micellae over time deteriorating and disappearing. Nägeli also invoked a rather feeble morphological justification for the distinction based on the difference in size between the egg and the spermatozoa. Because the parental inheritance seemed nearly equal, his argument went, the idioplasm must be equal in quantity and hence could only comprise a small portion of the total substance of the egg. It is also possible that, steeped in Kantian philosophy, Nägeli had simply taken Kant's and Blumenbach's notions of a Stamm and anlagen and reified them into the idioplasm and micellae. In this respect, Nägeli lashed together by means of his idioplasm theory both ends of the century. Whatever its origins, this basic distinction between idioplasm and Ernährungsplasm provided the source from which flowed the rest of Nägeli's treatise on heredity, evolution, and development.

Nägeli devoted a considerable portion of his text to describing the function and structure of the idioplasm.[19] His task was to account in molecular terms for both the constancy and change that exists throughout the biological world: germ returns to germ, like produces like but only imperfectly, and change accumulates progressively as phylogeny. Nägeli relied on his earlier starch grain studies to calculate the idioplasm's size, function, and structure. He wrote of its threadlike form that brought it into contact with and allowed it to penetrate all the cells of the organism. He reasoned that the idioplasm grew in length during ontogeny and that its cross section reflected the intercalation of different micellae or anlagen. Environmental and physiological circumstances determined which anlagen became active and which remained latent. As anlagen became reshuffled during reproduction, or altered and added as a result of molecular dynamics, the cross section of the idioplasm changed and guaranteed variations among the offspring as well as increasing complexity over many generations. Using calculations based on the ideal gas law and assuming a minimal molecular chain of seventy-two carbon atoms for the micellae and 80 percent water content of the idioplasm, Nägeli made some rough calculations about the minimum size of each anlage. The resolving power of the light microscope provided the upper limit for the size of the cross section of the idioplasm. These and other calculations, however, appear more useful as guidelines than as a prescription. No matter their accuracy, the results revealed that "it is evident . . . that there does not exist, as one commonly though falsely assumes about molecular relationships, an unending quantity of material particles, especially with respect to the substance, in which all characters are transmitted from an individual to its offspring, the number of those particles has a rather narrow range."[20]

Given these constraints in dimensions, Nägeli made an innovative suggestion. Instead of assuming that there existed a direct correspondence between the multitude of expressed traits and an equal number of anlagen, he argued, though provisionally, that if the organization of the idioplasm was a significant feature, then "a limited number [in its parts] would be sufficient, in the same way as language is composed of a limited number of words, music of a limited number of notes." It was the permutations and combinations of the micellae, Nägeli ventured, that established the anlagen.[21] It would take, however, a new century and more certain knowledge about the "idioplasm" to exploit the suggestion in the way that it deserved.

It was clear from Nägeli's discussion that the phenomena of heredity must be grounded in a material substrate. He had carefully thought through the various problems associated with any such molecular structure that might serve multiple purposes: to assimilate new material, to multiply, to function subsequently as the substrate for transmission, and to form the foundation for differentiation. Not one to insist that he had found the answer to these very complex problems, Nägeli simply claimed that the eventual mechanism could only be found through chemistry. He insisted that a theory of heredity must make physiological sense, that it had to unravel the processes down to the molecular level. Nägeli had felt all along that the shortcoming of Darwin's theory of pangenesis was its neglect of the physiological requirements of the gemmules. As for Haeckel's theory of perigenesis, Nägeli simply relegated it to a world of a poetic fantasy more akin to the Naturphilosophie he had rejected as a student.[22]

Nägeli and Heredity

These discussions of the idioplasm, of spontaneous generation, of the causes and locations of change, and of the contrast between ontogenetic and phylogenetic development were a necessary introduction for Nägeli to the subject that today we would consider the traditional matter of inheritance. As a botanist and plant breeder, he was all too aware of the complex and perplexing behavior of visible characters. They were often present in invariable form, sometimes latent for multiple generations, at times associated with given external conditions, and an indicator of sexual determination, essential for demarcating races, varieties, and species from one another, and collectively recording the progression of types in the evolutionary process. Visible characters ("merkmale") necessarily occupied the center of any concentrated discussion of transmission. That Nägeli found the subject only one of many to present in his *Mechanische-physiologische Theorie,* however, suggests how far removed his theory was from being a forerunner to classical genetics. Yet his discussion reveals a keen realization that his idioplasm theory must accommodate the elementary behavior of simple traits.

The key word in Nägeli's discussion was *anlage.* He introduced this word as a designation for the potential visible characters within the organism, but he immediately made clear that the anlagen should be construed as structural variations in the idioplasm. "Each visible character is present as an Anlage in the Idioplasm; there are therefore just as many forms of Idioplasm as there are combinations of characters."[23] It is important to note that Nägeli was not committing himself to a unifactorial theory of heredity by this conception, for changes in the arrangement and interaction between the micellae of the idioplasm were as much a device for creating characters as for the intercalation of new micellae. As Nägeli speculated upon the relationship between the anlagen and the visible traits, he constantly bore in mind the manner of idioplasmic change, whether it was through adaptation ("Anpassungsanlagen") or through progressive development ("Vervollkommungsanlagen"), which alone called for additional micellae.

Activation of anlagen was the key to appearances. The quantity of the micellae corresponding to a given anlage, the length of time the anlagen had been in existence in the idioplasm, the frequency of its previous expressions, its relative strength, and the propinquity in relationship between the anlagen of different parents were among the subsidiary principles Nägeli invoked to explain which anlagen became active and which remained dormant in the idioplasm. Thus, where the environment acted as a contributing factor in the evolutionary process of adding or subtracting anlagen, the makeup of the idioplasm itself became the proximate cause in determining trait expression. "It was as if the Idioplasm knew," Nägeli wrote in reference to regeneration, "exactly what happened in the remaining parts of the plant, and what it must do in order to restore the integrity and the life capacity of the individual."[24] This was only a figure of speech, to be sure, but it symbolized the burden that Nägeli was willing to place on the content and structure of the idioplasm.

Activation constituted the "mechanische-physiologische" explanation for trait appearance. At its core, therefore, Nägeli's explanation was a physiological and developmental rather than a morphological and unifactorial theory.[25] When he explicated specific hybridization phenomena, Nägeli always did so in terms of the interaction and balance of the various anlagen involved. For example, with the color distribution of the inflorescence in *Hepatica* in the first and later hybrid generations, a situation in which Nägeli even used the expressions "dominating" and "latent"—as in the case of the angora cat, which when mated with a normal cat produces normal cats in the first hybrid generation only, though the angora trait may reappear later—or the ubiquitous examples of the 50/50 sex ratio and the apparent equality in paternal and maternal contributions to the offspring.[26] As we will later see, Weismann was inclined to resort to similar tactics as he developed his own explanation of hybridization.

Nägeli's most extended hybridization research was done with the hawkweed genus *Hieracium.* This was the genus that he persuaded Mendel to take up after the latter had sent him his famous paper on *Pisum.* Since the rediscovery of Mendel's laws at the turn of the century, much has been written about the relationship between these two researchers: the one a university professor, the other an Augustinian monk; one a searcher for the foundations of evolution, the other an unappreciated herald of classical genetics. Nägeli advocated that the future lay in amassing varieties of *Hieracium* and exhaustively detailing their breeding capacities with one another; Mendel unwisely took up the same search and allowed the clear results of his *Pisum* experiments to become submerged in a morass of uncertain generalizations about *Hieracium.*[27] Let us, however, forget modern genetics, as it biases comparisons between the two men. Rather, let us look briefly at Nägeli's *Hieracium* research in the context of his mechanico-physiological aspirations and his scientific accomplishments.

Nägeli's book provided a specific theory about the cause of descent. His conception of the idioplasm and the elaborate discussions about its structure and behavior in reproduction possessed little purpose if they could not be lined up with the problems of adaptation, change, and ultimately evolution. In this respect, Nägeli followed in the footsteps of Darwin, Spencer, and Haeckel. Nägeli's earliest studies on *Hieracium* flowed directly from the doctoral work on the contrast between species and hybrids in *Circia*.[28] With the advent of Darwinism, Nägeli tells us, he renewed his interest in the genus, this time with the intent of exploring the origin of varieties and ultimately of species. For thirteen summers, between 1864 and 1876, he and his son collected *Hieracium* in the mountains; after 1876, his assistant, Dr. A. Ibert Peter, extended the search. Collectively they had sampled the genus not only in the central Alps but in the Apennines, the maritime Alps, and the mountains in Monrovia, Silesia, and Galicia. Many of the plants or their seeds were transferred to Munich so that by the time of his writing the *Mechanische-physiologische Theorie,* Nägeli could report that he had transplanted about 4,450 specimens and still possessed nearly 2,500 of them in his botanical garden. It was a heroic as well as systematic effort to sort out the difference between inconstant and constant characters, or as he had described it earlier in his book, the differences between the modifications of the Ernährungsplasm and the idioplasm. His garden, his herbarium, the raising of offspring from seeds, and the standardization of soils in which the various breeds were raised became the instruments and techniques of his quest.

One particular trait that might be considered variable and site dependent was the season of blooming. The same variety that existed in Bavaria bred a month earlier in the south and a month later in the north. Yet was this a feature determined by external influences or was it an idioplasmic factor? Nägeli made it a standard procedure to record the blooming date of his cultivated breeds and their descendants raised from seeds. Thirteen years and 16,000 comparisons later, he asserted that when the statistical variations had been averaged out, the blooming time was found to be constant within a given variety. It therefore must be associated with an anlage in the idioplasm. By 1884, he and Peter were convinced that where earlier botanists had counted the total number of species and varieties at four dozen, they had established no less than 2,800 constant varieties. No wonder that Nägeli had urged Mendel to turn his attention to the breeding of *Hieracium,* "about which we will soon know more than any other species."[29]

The second half of the *Mechanische-physiologische Theorie* concerned a general theory of evolution. Nägeli began with a long critique of the Darwinian mechanism of natural selection. When he compared it with an internal perfecting process that worked within the idioplasm, he found the former wanting. In a ninety-page sequel, Nägeli presented an analysis of the phylogeny of plants, in which he found eight lawlike patterns of cell production and tissue extension. These he insisted reflected the more basic perfecting processes of the idioplasm. A much shorter chapter followed on the phylogenetic significance of the alternation of generations, a subject that concerned botanists as much as zoologists during the nineteenth century. In the final full chapter, Nägeli offered some remarks about the significance of the science of morphology for unraveling phylogeny. As the uncertain lineage of the phanerogams suggested, these morphological determinations were at best problematic. Nägeli closed with an extended thirty-page summary of his work; it was the only portion of his work that found its way into an English translation.[30]

Between the first and second halves of Nägeli's massive monograph lies the most telling section for our purposes. This insert was a brief addendum to the four-chapter exegesis of his concept of the idioplasm. Here Nägeli examined in the abstract his contributions to heredity theory. Heredity, or "Vererbung" according to the "Darwinian school," as Nägeli labeled his immediate predecessors, was considered a conservative principle. Change or "Veränderung" was construed as its opposite, a principle permitting progress. For Nägeli, this was an oft-repeated but false mantra based on external appearances. The theory of the idioplasm in contrast focused attention on the internal events, the real inertial processes in the biological world. Seen in this way, heredity and change constituted a unified process. "Heredity," Nägeli somewhat cryptically explained, "when taken as a general concept, is actually nothing more than appearances that are necessarily bound with the transition of one state to the next, and the entire ontogenetic and phylogenetic motion consists of one continuous series of such transitions."

Both "ontogenetic and phylogenetic motion"? Yes, indeed, and this passage gives us an important insight into the confluence of biological ideas at mid-nineteenth century. Heredity was not only the transmission from one generation to the next, "but it is equally heredity when in the individual a cell, which divides itself, leaves all its properties to the two new cells, or when the plant stem annually produces twigs, leaves and flowers, or when an adult and an old man rise out of the child."[31] Cell division, development, metamorphosis, and the alternation of generations as well as reproduction were all exemplars of the hereditary process so conceived. It is easy to suppose how collectively these different dimensions of heredity led to a continuity concept. The next logical step, which because of his background Nägeli was preeminently prepared to take, was to ask in a materialistic vein, "What bound these dimensions together?" "Was wird vererbt?" Of course, "blos Idioplasma." ("What is inherited?" Of course, "merely idioplasma.")

At the risk of saturating the reader with too many excerpts from the *Mechanische-physiologische Theorie,* we reproduce Nägeli's vivid description of this "mere idioplasm" as it occupied the central position in the chain of life:

> For the sexual organisms the continuity from parents to offspring consists merely through the Idioplasm of the spermatozoa and egg cells, and the new individual only brings forth what the inherited idioplasmic Anlagen and the external influences, which it itself experiences, allow. The history of an ancestral line from the simplest up to the most complicated plant, from the lowest up to the highest animal is really nothing more than the history of the idioplasmic system, which in the course of time will always be more richly divided [*gegliedert*] individuals. The entire ancestral line is fundamentally a single continuing individual which exists of Idioplasm and which grows, multiplies, and thereby changes, and which puts on a new dress with every generation, that is; it forms a new individual body. In continually new and more complex ways it periodically fashioned [*gestaltet*] this dress, according to its own change, and every time it exchanges the dress it relinquishes the largest portion of its own substance.[32]

The image was perfect. Weismann had fashioned a similar one from his earlier study of hydromedusae. Phylogeny could be viewed as a series of stages in the life of a single individual; evolution could be seen as that individual's developmental process. In his youth, Nägeli rejected the idealistic overtones of Hegelian cosmic philosophy, but at the height of his career the mechanistic physiology required, on a less grand scale to be sure, the same grand perspective of a single developing entity to explain the multiplicity of life's processes. The perspective gave a whole new thrust to heredity theory. Nägeli himself sensed the revolutionary implications of the continuity concept: "the image basically stands reality on its head."[33]

Nägeli and Weismann

Nägeli had already explained that the biologist had to distinguish among modifications, which were the effects of the environment of the developing "Ernährungsplasma," and facial variations, which were simply the reshuffling of the idioplasm and which did not introduce new components into the idioplasmic system, and finally individual and varietal variations, which reflected the slow continuous transformation of the idioplasm.[34]

It is difficult to assess what impact the *Mechanische-physiologische Theorie* really had. Ernst Mayr has argued that "because Nägeli had speculated about every conceivable aspect of the process of inheritance and development, he exerted an enormous influence."[35] The problem is to identify both the extent and nature of that impact.

There is no question that as Nägeli envisioned an independent idioplasm from the rest of the organism, he fulfilled the fundamental requirements of a mechanical theory of both intergenerational and intragenerational continuity. He explicitly distinguished the phenomenon of transmission from that of development, and in the process he identified each with unique activities of the molecular substrate. Moreover, he described this substance as an independent entity that existed from zygote to gamete and from generation to generation, and he recognized that its architecture is essential to the various forms of the change that every organism undergoes. Nägeli took issue with the midcentury view that heredity constituted an antithetical process to variation and adaptation; in so doing, he began to wrench apart those categories of thought. He also recognized that sexual reproduction and molecular realignments introduced variations to the idioplasm and that over generations both affected the change of species, which became increasingly complex. To the extent that these ideas could be discovered and clearly grasped by any bold reader of the *Mechanische-Physiologische Theorie,* Nägeli must have had an important impact.

In that he used his experience as a plant breeder and as an explorer of the fine structure of the organic, Nägeli anchored his theoretical discussions to a supported empirical base in a way that most contemporary creators of a notion of continuity had not done. However, the idioplasm, coursing an intracellular network through the entire organism, contravened recent histological experience. The problem was that for that period Nägeli had secured his ship to the wrong shore of facts, to chemistry rather than to embryology and cytology.

As far as Weismann, overgrowth, and continuity were concerned, we have seen that he had arrived at a concept of continuity based on the morphology of hydromedusae before Nägeli's tome appeared. We also find that Weismann read extensively in Nägeli's text when it arrived; and even though he disagreed with most of its details, Weismann more than once paid homage to its suggestive insights. Nevertheless, Weismann was alert to a different movement reshaping biology, which drew him away from the overgrowth principle of heredity. This new direction is the subject of Chapter 11.

11

The Emergence of Nuclear Cytology

Casual historians of the life sciences of the nineteenth century put most of their emphasis on the rise and repercussions of Darwin's and Wallace's works on evolution and the concomitant rise of modern physiology. With some reservations, this is fully understandable and appropriate. What gets lost in this picture, however, is the impact of cell microscopy on the understanding of organisms, their grosser structures, the details of their development, reproduction, and ultimately their evolution. This chapter focuses on the extraordinary accomplishments of cytologists and how their microscopic examinations of the processes of growth, development, gamete production, and fertilization further shaped our understanding of animals and plants. Not only was this dimension of biological research necessary, but it played an essential role in the advance of biology on all levels. It encouraged, even forced, investigators to examine the relationship between the organismal whole, the cellular, and the microstructures of cells, their nuclei and other subcellular elements. By the 1870s, that is thirty-plus years after the rise of the unified cell theory, the development of microscopic techniques permitted and even forced biologists to take the microscopic discoveries on the subcellular level seriously. Weismann had speculated since his early studies under Henle at Göttingen about the fine structure of the organism, but casual speculations they mostly remained until the end of the 1870s.

Weismann's "Nachtrag" on Insect Development

Weismann's contribution to a festschrift commemorating the fiftieth anniversary of Jacob Henle's doctorate is easily overlooked in a discussion of Weismann's career. Gaupp mentioned it in passing but considered it merely a "Nachtrag" (an addendum) to Weismann's earlier embryological studies of twenty years earlier.[1] Nevertheless, insect embryology between 1862, the year of Weismann's Habilitationsshrift in Freiburg, and this celebration of his former histology mentor reflected enormous technological and factual advances. Embryologists of the accomplishments of Elie Metchnikoff, Alexander Kowalevsky, Otto Bütschli, and Alexander Brandt had dramatically altered the understanding of early insect egg development. For example, it was long since clear that one of many complications in early insect development concerned the formation of what Weismann had initially called a "Faltenblatt." The appearance of this folded layer, the first differentiated structure on the surface of the insect egg, prompted Weismann to conclude that there was no germ layer correspondence between insect and vertebrate development. It was soon recognized, however, that the faltenblatt represented a fusion of two extra embryonic membranes, the serous layer and the amnion, and that the embryo-to-be lay internally to these superficial layers. By the time he published his comprehensive *Treatise on Comparative Embryology,* Balfour could generalize about segmentation patterns across much of the animal kingdom and identify the meroblastic (incomplete) pattern of segmentation and the early superficial embryonic membranes with the centrolecithal structure of most insect eggs.[2] What had appeared to the young Weismann as a lack of homology between early insect and vertebrate development had been refuted by a wealth of detailed information about the embryology of the insect egg in a comparative and functional setting.

Now twenty years older, Weismann took Henle's festschrift as an opportunity to respond to some of his critics and to correct his earlier misconceptions. He gladly agreed with others that germ-layer formation in insects, beginning with the anterior and posterior pole nuclei and the formation of the blastoderm, could be brought into harmony with vertebrate development.[3] He also disavowed his former belief in the free formation of nuclei and consequently of cells during early insect cleavage. To have done this openly and to have documented the necessary revisions on four additional species of insects was indeed professional, but for this reason alone his essay deserved being labeled simply as a nachtrag.[4]

On closer inspection, we discover that this essay was not simply a casual contribution to honor his former Göttingen friend and mentor. Weismann worked on insect development during the summer mornings of 1876–1877 and during the winter of 1878–1879. This research could not have been initiated as a short-range undertaking but might have signaled the beginning of another meticulous and exhaustive study in development. The project, however, never advanced beyond this contribution. Weismann explained at the time that he recently learned that V. Graber had announced plans to produce a large work on insect development; nevertheless, that challenge in itself seems unlikely to have deterred Weismann from a new extended study on his own.[5] It is possible he was again beginning to feel the strains of microscopy. These insect eggs were small and required a magnification of up to ×400 power to observe nuclear migrations and segmentation. Moreover, working with live specimens necessitated sustained observations. Weismann was now using the "Schnittmethode," though both in these studies and with his hydroids, he preferred to immerse the freshly fertilized insect eggs in olive oil (which allowed him to observe early development in vivo, as he found it gave him a better feel for the sequence of developmental events).

Klaus Sander has provided a more subtle interpretation of Weismann's insect embryology.[6] Pointing out Weismann's earlier misconceptions as well as the inadequacies in his 1882 contribution, Sander comes to the reasonable conclusion that Weismann's findings and corrections were driven more by theoretical commitments than new factual material. For example, in reference to Weismann's failure in the early 1860s to generalize the vertebrate germ layers to insects, Sander cites several later passages in which Weismann recognized that the theoretical background for making such a judgment had changed in the course of twenty years. One such passage is particularly revealing, for it indicates not only Weismann's full understanding of the changing context but also how fundamental the germ-layer theory had become for determining the facts. "Today to be sure we observe everything with totally different eyes," Weismann insisted and then continued, "today we know that the germ layers hold true for all the remaining types of metazoans and this would therefore also hold for all arthropods, even though research into them had not taken place."[7] For Sander, this was putting theory ahead of facts, and I do not believe Weismann would have disagreed.

Throughout his life Weismann insisted that theory was necessary for the progress of science. Writing from Freiburg to Haeckel at a time when he was just beginning to feel confident again about pursuing microscopy and at the same time when his friend's gastraea theory was being critically examined, he shared his feelings about the status of Haeckel's theory in science:

Do not, however, doubt that a mighty cry will be let loose over the theory "constructed out of the blue" etc! Since I have heard my honorable colleagues talk this past fall in [the Versammlung in] Wiesbaden (e.g., Schneider and Greef), I have been convinced above all in this regards! Because they cannot themselves make one, others should not do so either. It is so easy to present no theory under the pretense that it still lacks the necessary basis of facts. As though the theory must not first indicate the way in which facts are to be sought!

After commenting to Haeckel that facts will eventually decide the issue, Weismann continued with his own situation:

I would have the desire once again to get back to the embryology of insects—if the eyes would endure it, which, however, still need to be protected.[8]

The lesson preached in his 1874 letter also applied to Weismann's acceptance of a homology between insect and vertebrate germ layers in 1882. Weismann was a mid-nineteenth-century positivist. Theories and facts were separate categories; the former framed the problem area in which one looked in nature for the latter. Their mutual interaction and correction ensured the progress of science. He and his colleagues remained blissfully unaware of how in the late twentieth century the two categories merged when facts would become theory laden and when theory appeared to vacillate between the realm of intuition and raw desire.

Finally, his contribution to Henle's festschrift provides the only document Weismann wrote that suggests how his research interests shifted from the examination of polymorphism, heterogenesis, and metagenesis, from the embryology of germ layers and production of germ cells to nuclear cytology. By the late 1870s, the study of insect egg cleavage required an exacting analysis of how apparently isolated nuclei in the yolk of the centrolecithal ovum multiplied, organized, and eventually migrated "amoeboid-like" to the internal periphery of the egg to form the cellular layer of the blastoderm. It was generally assumed that the nuclei were surrounded by cytoplasm that was distinct from the yolk, but cell walls were not evident; during the migration and multiplication processes, it was easy for the viewer to confound the two. Both processes are complicated because the nuclear walls disappear, and the fragments of the nuclei dissolve and then reappear. Weismann singled out a recent work by Alexander Brandt to drive home the problem.[9] Although Brandt was reporting at the time on development in the nematode, *Ascaris,* and the gastropod, *Limnaeus,* Weismann found an instructive parallel in Brandt's descriptions and those in insect cleavage cells.[10]

He [Brandt] attempted to put the amoeboid motion in the place of Karyo-kinese, as has been maintained by so many investigators in the area of the animal and plant cell. The "nuclear spindle" would therefore be nothing other than the falling together of the "folded nuclear" membrane which has emerged amoeboid-like from the nuclear substance, the radiating arrangement of the yolk particles around the pole of the nuclear spindle demonstrates itself to be determined by pseudopodia of the cleavage nuclei . . . although now the "protoplasmic and as such admittedly viscous nuclear substance" serves to explain the karyokinetic processes and to extend itself during division into threads, exactly like syrup or glutinous resin.

His paraphrase of Brandt's confusion warranted an immediate, sharp criticism as Weismann continued:

It is hard to understand, how one with such rather crude comparisons can believe to be promoting an understanding after all the recent researches, especially those by Strasburger and Flemming, have demonstrated what uncommonly complicated and standardized processes one must be dealing with here.[11]

Weismann insisted that the two processes, the "amoeboid" motions of cleavage cells and the indirect cell division manifested in segmentation, must not be confounded with one another. As they approach the periphery of the protoplasm, the nuclei of insect eggs become independently and internally activated into division and attract protoplasm to each new nuclear center. To recognize this division in all its complications and exactitude bespoke a much more important relationship between nucleus and protoplasm. Weismann's focus had clearly turned to the details and significance of indirect cell division.

The investigators mentioned in Weismann's criticism were the botanist Eduard Strasburger and the zoologist Walther Flemming. They were the two microscopists who more than any others worked out the details of karyokinesis, which was renamed in 1882 "mitosis." There is no way that a historian of cellular morphology of the period can avoid dealing with the investigation of these "uncommonly complicated and standardized processes" of the period.

The Emergence of Nuclear Cytology

When Balfour wrote the section on the internal segmentation of the ovum in one of the early chapters of his *Treatise,* he clearly indicated a familiarity with the recent cytological literature coming from the continent.[12] Works by Büt-schli, van Beneden, Fol, Flemming, and Strasburger adorned the bibliography and are mentioned in his text. Published in 1880, however, the first volume of the *Treatise* appeared months too early to reflect the overpowering demonstrations of indirect nuclear division that came with the books, though not articles, of Strasburger and Flemming. In fact, Balfour's account of first cleavage revealed the factual and theoretical chasm between classical descriptive embryology at its very best and the new, emerging perspective of nuclear cytology.

The two fields, of course, shared common ground. Traditional descriptive embryology dealt with egg formation, fertilization, and patterns of segmentation, all manifestations of cell multiplication. Nuclear cytology of the 1870s and beyond dealt with the internal cellular dynamics of the same basic phenomena. It was easy and natural for biologists who had been trained in the microscopy of the former to extend their art and quests to the latter. The first, to emphasize with a human metaphor, served as the godparent of the latter. A ratification of indirect cell division and its implications for heredity turned out to be the future of the godchild.

Balfour's description of first segmentation reveals a bridge across the traditional and the emerging field. He presented seven stages of the cleavage process, which included general comments on the appearance of "radial striae," the elongation of the fusion nucleus and the clear protoplasmic substance that surrounded it, the formation of clear, starlike rays bursting from both poles of the nucleus, "a clear mass surrounded by a star-shaped figure," the "well-known spindle" form extending from pole to pole with a thickening at the center where the nuclear reticulum from the original fertilization nucleus metamorphosed into a plate of granules. Furthermore, he observed that the center of this thickened "spindle separates into two sets," each of which moved to the opposite pole while remaining bound together by "filaments." The ovum, Balfour added, begins to constrict perpendicularly to the long axis of the nucleus, and in the final stage "the spherical vesicles" at either pole "begin to unite amongst themselves, and to coalesce with the neighbouring granules."

Balfour interpreted these stages as showing that "the new nucleus is therefore partly derived from the division of the old one and partly from the plasma of the cell." So it is not surprising that he added a cautionary caveat: "The view that the nucleus is a single centre of attraction, and that by its division the centre of attraction becomes double and thereby causes division, appears to be quite untenable." The protoplasm of the whole cell, not the nucleus, was for him the motive force of cell multiplication. He recognized Flemming's discovery of longitudinal splitting of the "threads" in the "so-called nuclear plate," but he was not struck by any special significance of indirect nuclear division. Balfour appears to have concentrated on the dynamics of attractions rather than the splitting of the spireme.[13]

What seems obvious in retrospect is Balfour's dependence upon unstained material for his own observations, despite the fact he knew that others had depended on reagents to make the nuclear material clearer. He lacked a clear distinction between what Flemming called the "chromatin" apparatus and the clear striae, spindle, and other purported derivatives of the protoplasm. Balfour's serial account of segmentation, drawn particularly from a recent account by Hermann Fol, was outdated before it appeared, yet it is eye-opening to the historian, for it suggests how rapidly the new field was changing and how radical the change would be in the immediate future. Balfour had been in touch with German advances in particular; he had worked at the Naples Station, and his own *Treatise* was immediately translated into German because of its broad comparative perspective—the Berlin anatomist Wilhelm Waldeyer considered it the first successful attempt at a complete comparative anatomy.[14] I have no doubt that Balfour would soon and quickly have adjusted to the new understanding of cell division, but this was not to be as he died while climbing in the French Alps in the summer of 1882.

A gap between general comparative morphology and nuclear cytology had emerged, in part, because of the rapid development of microtechniques and, in part, because of the significance of what Flemming in the same year would designate mitosis. The story of the rise of nuclear cytology has been told many times, but it is a valuable subject to review because it became so central in the development of heredity theory and to Weismann's own understanding of biology.[15]

The ten years from 1873 and 1882 were extraordinary in the annals of biology. With respect to the development of nuclear cytology, the decade possessed all the excitement and multiple revelations of the decade preceding the discovery of the structure of DNA some seventy years later. In its own way, the achievements by 1882 were as significant as those leading up to and including the two epoch papers by Francis Crick and James Watson in 1953.

Nuclear cytology was forged out of a convergence of new microscopic techniques, including the use of preservatives and stains.[16] It entailed a general exploration of the nuclein in a variety of organisms—lower and higher plants, protozoa, invertebrates, and vertebrates all—that helped resolve issues of the importance of details; it included a recognition that newly discovered structures and their transformations were universal and had to be organized in some temporal sequence, and it was promoted by a critical mass of microscopists eager to spend the time and energy for hours on end in front of the laboratory bench patiently preparing and studying both living organisms and sectioned specimens. Microscopists throughout the Western world participated in this undertaking; nevertheless, the overwhelming number came from the German Empire. They were not driven forward by fantasies of a Nobel Prize, but in Wilhelmine Germany, at least, there were pressures on both the hopeful and established academics to gain status in an important new field. Above all, there was the simple excitement of exploration and discovery at an unexplored level of organic structure.

These social and technical forces are reflected by the number of investigators who took up the study of the nucleus of the cell. These forces are also reflected in the number of statements and restatements of what had been found and in the exacting details of the lengthy monographs being published. They are echoed in the careful reviews of the recent literature appended to many of the monographs published. It is not that the pace of research slowed after 1882, nor did it after 1953, but in each case a fundamental biological process had been established beyond a reasonable doubt and upon which new fields could be constructed. In fact, the realization in the nineteenth century that there was a complex and precise mechanism for indirect cellular division, which included a longitudinal splitting of each "chromosome" and the segregation of its two halves to the two daughter cells foreshadowed the discovery of the double helix and its replication and distribution in the twentieth. Where the twentieth-century episode directly resulted in the mechanical explanation of heredity, the nineteenth-century episode resulted, in a manner of speaking, in the realization that heredity might be mechanically explained.

The history of this decade of the nineteenth century, however, can be distorted by the outcome. The identification and establishment of indirect cell division is too easily seen only as a necessary step toward understanding heredity, but this would be reading the decade backward. Even if many investigators had held that possibility in the recesses of their mind, it was not until the very end of the period that they mentioned that the impact indirect cell division might have on the notion of heredity. Other issues, many demanding searching study in their own rights, appear to be more important at the time: the contrast between free cell formation and direct cell division, the contrast between free nuclear formation and direct nuclear division, the dynamic relationship between the nucleus, nucleolus, yolk, and protoplasm, and the derivation and relationship between what would be called achromatic and chromatic structures were all issues of intense investigation and discussion during these ten years. The nature of fertilization, which called forth studies of the egg and spermatozoa, necessarily became part of the mix in understanding cell division. Then there were the polar bodies, those curious products of ova maturation. Were they universal? Were they necessary ejections or rejections by the egg? Were they abortive cells? How were they to be incorporated into any general scheme of cell multiplication?

One of the challenges then is how to organize for today's audience, what appears to be an abundance of studies, teeming with details and replete with multiple objectives. Three contemporary texts, Eduard Strasburger's *Zellbildung und Zelltheilung* (1880), Edward L. Mark's "Maturation, Fecundation, and Segmentation of *Limax campestris,* Binney," (1881), and Walther Flemming's *Zellsubstanz, Kern und Zelltheilung* (1882), may serve as guides to the erratic course of cytology during the decade. All three texts, written at the end of the decade in question, presented résumés of earlier literature, but in forms that are difficult for the historian who may desire a lineal synopsis of the history leading up to the recognition of indirect cell division and its importance for theories of heredity and evolution. All three authors participated in the research of the day. All three gave the impression of wanting to validate their own understanding of cell structures and division through historical reflections as much as presenting their own research to their colleagues.[17] They parsed contemporary research into categories that made sense to them, not to today's historian. Furthermore, they signified their many concerns by discussing a given work in two or three different places. Finally, all three of them paid scrupulous attention not only to the years of publication but to the dates of submission, a practice that suggests that the issues of priorities and influences lay close to the heart of the investigator.

The decade may be more easily understood by subdividing it into three, albeit unequal, parts. The first year, 1873, appears to present a break from the past.[18] Previous to that time, direct "nuclear" division had implied one of two alternatives. On the one hand, in the tradition of direct cell division perceived by Remak on embryological material in 1852 and Virchow on pathological material in 1855, one might mean a simple fission of what they took to be the "nucleolus," half of which normally relocated to each daughter cell.[19] Or one might dismiss a "nuclear" division altogether by considering the nucleus as a transitory structure that came newly into being at each cell division. There was persuasive evidence for both options. Four zoologists, working independently of one another in 1873, "rediscovered" the intracellular rays and realized that the nucleus underwent a complex "metamorphosis" during cell division.[20] Whether this metamorphosis was simply a variant of the second alternative or not, it set the following research agenda for the rest of the decade.

First, in a paper that was not well known until later, Anton Schneider, at the time in Geneva, described his examination of maturing eggs of the flatworm *Mesostomum*. (By the time Waldeyer wrote his overview in 1888, Schneider had become professor of zoology in Bresslau.) He resorted to acetic acid to make the disappearing "outline" of the nucleus evident, and he thereby gained a sense of a nuclear transformation into "cords" arranged in line with two astral poles. His paper described a dynamic relationship between nucleus and the refractive material of the cytoplasm.[21]

Second, at twenty-eight years old, Hermann Fol was a native of Geneva, a one-time student of Haeckel, and a doctor of medicine from Heidelberg. While examining the hydroid *Geryonia* and confirming his findings on mollusks and worms, Fol insisted there was no direct connection between the germinal vesicle of the unfertilized egg and the first segmentation nucleus. Nevertheless, he found that the stellate figures that grew out from the poles of the disappearing nucleus at the onset of first cleavage became centers of attraction for the formation of two daughter nuclei. He, too, used acetic acid to elucidate the segmentation nucleus and the appearance of the new nuclei.[22]

Third, Otto Bütschli, just twenty-five at the time and not yet a privatdozent, examined fertilization and cleavage of the nematode *Rhabditis dolichura*.[23] He used live material exclusively and became convinced that the spermatozoa disintegrated upon contact with the vitelline membrane of the egg. Nevertheless, he was able to observe at fecundation two structures, which he and others would later designate as "pronuclei," approach each other in the center of the egg and "melt" together. "Knobs" associated with threads then appeared at the side of the fusion nucleus and separated to opposite poles as the nucleus vanished from view. The intimate relationship between the knobs and nucleus convinced Bütschli that there was a direct connection between the two, so he maintained that the nucleus or its derivatives were persistent through segmentation. According to his view, the knobs, probably centrosomes in retrospect, became the centers of attraction for the two daughter nuclei at the time of segmentation.

Fourth, Walther Flemming, the oldest of the group at thirty and at the time an außerordentlicher Professor at Prague, applied slight pressure to the cleaving eggs of a freshwater mussel. This produced a clear area in the center and rays with granules radiating toward two poles. As did Fol, he found that these soon established the location of the two new daughter nuclei prior to first segmentation.[24]

An effort to understand fertilization and early segmentation was the common starting point for three of these four contributions; the germinal vesicle was not understood as simply a cell nucleus, but as a changing structure whose fate appeared quite different. All four contributors used different organisms for their studies, so any generalizations were problematic. Only two of these four used reagents to enhance their view of the nucleus. All of them appeared to recognize a dynamic sequence at first segmentation, but there was no consensus as to what belonged to or was derived from the nucleus and what was protoplasmic in origin.

Very quickly others became involved in investigating the nucleus and its relationship with egg maturation, fertilization, and cell division. Between 1875 and 1877, carmine and hematoxylin came into common use on preserved and sectioned material. Nevertheless, living organisms, sometimes lightly tinctured, continued to be valuable objects of study. Three noteworthy investigators clarified the process of fertilization, but in a curious way they rendered the process of "karyokinesis," a term coined in Schleicher, more complicated and therefore less comprehensible. A fourth and fifth investigator redirected attention to the nuclear elements and their many forms. A synopsis of their work during these years is again useful.

First, Leopold Auerbach was twenty years older than the other authors. (Auerbach was born in 1828, Bütschli in 1848, O. Hertwig in 1849, R. Hertwig in 1850.) He was a practicing physician in Breslau and had teaching rights at the university as an unpaid außerordentlicher Professor. Despite the fact that he divided his time between patients, devoted students, and an active family life, he managed to assemble a private laboratory—much in the fashion and for the same social reason as his Berlin mentor Robert Remak. With a minimum of equipment, he had already accomplished important microscopic work demonstrating the cellular nature of protozoa, describing the sympathetic ganglia of the vertebrate limb, and documenting the nucleated nature of the cells of blood capillaries.[25] He joined the enterprise of examining fertilization and cell division with a lengthy monograph on the eggs of the nematodes *Ascaris* and *Strongylus*.[26] Using a "compressorium" on the living eggs, which slightly squashed the specimen, he observed the fusion of pronuclei and their combined dissolution into granules, which lay along rays in a dumbbell shape (*Hantelförmig*) of clear yolk. Auerbach not only confirmed Bütschli's union of pronuclei but supported their transitory nature.[27] He furthermore argued that one could find the nucleus in tissue cells when treated with the appropriate reagents and that within the nucleus one could find many small particles of varied sizes or "Nucleolen," as he called them all. Writing in 1882, Flemming considered Auerbach's publication the "first general monographic treatment of the nucleus." Nevertheless, he pointed out that which Auerbach had interpreted as "Nucleolen" and which had appeared to him to be identical to the cell substance of young cells with the original nucleus itself serving as their brood chamber turned out to be from Flemming's later perspective nothing other than the nuclear apparatus for cell division.[28] In the same vein, Oscar Hertwig, who admired much of Auerbach's account, criticized his explanation of the reconstruction of the segmentation nucleus out of the surrounding protoplasm. Auerbach had dubbed the process "karyolysis" in contrast to "karyokinesis" and justified it as a palingenetic stage in cell evolution.[29]

Second, Bütschli presented seven more contributions through 1877, but his 250-page monograph detailing the cleavage process in snails and nematode eggs, cell division in assorted tissues, and conjugation in infusoria stood above all as an important surviving classic. Submitted in November 1885, the work served as his Habilitationsschrift.[30] Bütschli demonstrated beyond doubt that infusoria were single-celled organisms and that their conjugation, though not identical with, could be analogized with the fertilization process in higher metazoans.[31] His analysis anticipated Hertwig's more elaborate demonstration of fertilization in the sea urchins. Although Bütschli did not use stains, he was able to distinguish that which Flemming would later call the chromatic and achromatic elements of nuclear division. His illustrations presented in retrospect what might be construed as sequential stages of cell division. At the same time Bütschli contradicted his earlier assertion by maintaining that the cleavage nucleus arose epigenetically from the egg yolk and consequently could not be derived from the pronuclei.[32]

Third, Oscar Hertwig provided an important statement of the problems during these middle years of the decade. His major contribution is worth analyzing in detail.[33] Hertwig's inspiration had been Auerbach's *Organologische*

Studien of the previous year. A young man of twenty-six in 1875, Hertwig submitted his study of the fertilization of the egg of the sea urchin, *Toxopneustes lividus,* as his Habilitationsschrift. This was recognized in its day as the incisive demonstration that under the artificial conditions of flooding the unfertilized egg with spermatozoa, male and female pronuclei could be found, radiating like suns as they approached each other at the egg's center and "copulated" in the final act of fertilization. The use of *Toxopneustes* was a calculated choice based on its availability, the ease with which one could induce fertilization, and the transparency of the relatively small egg, which made it possible to follow the internal transformations and emergence of key structures.

The process of fertilization, as Hertwig described it, consisted of a circle of clear material appearing at the internal periphery of the egg close to the external location of the tail of a spermatozoon. This "sun" grew in size and quickly migrated to the center of the egg where it joined the similar sun of the pronucleus of the egg. Although the actual penetration of the spermatozoon head through the vitelline membrane of the egg had not been observed until Fol did so in a series of papers between 1877 and 1879, Hertwig felt convinced by the timing and circumstances that what he had seen was some element of the sperm. The pronuclei merged and together disappeared from view. Shortly thereafter, rays extended to the periphery of the egg itself and a newly formed cleavage nucleus emerged, the apparent product of the two pronuclei. Hertwig confirmed his observations and conclusions on sectioned material preserved in osmic acid and stained with Beale's carmine—a formula that emphasized in rosy pink the nucleic material.

As its title indicates, Hertwig's Habilitationsschrift encompassed much more than a description of the fertilization event, which has made it famous.[34] In fact, the account of fertilization comprised only the middle, and shortest, section of three. In all, Hertwig was addressing his colleagues about the phenomena of maturation, fertilization, and cell division, from the germinal vesicle of the unfertilized and unreleased egg to first cleavage in the new embryo. He asserted that only the nucleolus or "Keimfleck" survived the transformations and became, he argued, the cleavage nucleus of the fertilized egg. Justifiably considered a "classic" of modern genetics because it invalidated stock theories about the fertilizing motions and energies of the spermatozoa, it nevertheless conformed with contemporary efforts to understand the details of egg maturation, fertilization, and nuclear division as physical events leading to blastomere segmentation.[35] His account did not describe a continuity of the complete male and female nuclei after fusion, and nowhere in the 1870s did Hertwig address the related but very different question of the transmission of individual traits between generations, nor more importantly, within the individual through the maelstrom of developmental events. With respect to an understanding of the process of heredity, Paul Weindling has correctly pointed out that in the Darwinian context of genetically linked phylogenies, a theory of heredity necessarily formed a backdrop not only for Haeckel, but Hertwig himself. As we shall argue in Chapter 20, Hertwig was not yet in a position to present a revised theory of heredity.[36]

Using solutions of acetic and chromic acid on preserved material, Hertwig devoted the final section of his monograph to the process of first cleavage. The basic dilemma, as Hertwig saw it, was that which had dogged the matter of cell division since the 1840s: did the nucleus dissipate before cell division or did it persist by means of a direct division? The first position, he argued, was held by most botanists and a few zoologists, particularly Reichert (1846); the latter view was held by many zoologists, including Baer (1846), Müller (1852), Haeckel (1869, 1874), and van Beneden (1870). Of his immediate predecessors, Hertwig argued that Fol (1873), Flemming (1875), and Auerbach (1874) all appeared to have adopted the position that the nucleus decomposed (*sich aufgelöst*) whereas Bütschli (1873, 1874, 1875) and Strasburger (1875) supported a nuclear persistence through division.[37]

Hoping his work would further the understanding of cell division, Hertwig summed up his lengthy paper with two basic conclusions. (a) A metamorphosis of the nucleus explained its apparent dissolution. In reality the various parts of the nucleus during division were derived from the original "Keimfleck" of the mother nucleus. (b) The nucleus demonstrated a physiological value as it became the center of force—which after segmentation became two centers of force.[38]

Fourth, in a series of short articles, both in German and Polish, from November 1875 and through 1876, Waclaw Mayzel described cell production in the regeneration of epithelial cells of amphibians and mammals. He generally used living specimens but on occasion stained his material with a weak (0.01% chromic acid). He was able to distinguish three forms of the nucleus during division: a tangled skein-like form ("Knäuelform") with occasional corpuscles, a starlike formation ("Sternform") made up of achromatic structures and a central nuclear plate consisting of stained corpuscles, and a daughter nucleus form ("Tochterkernform"). Flemming later noted that Mayzel was not always certain about the sequential order of these forms and apparently did not generalize beyond the case of regenerating tissues. Nevertheless, his observations demonstrated the similarity between the early segmentation process and adult tissue regeneration.[39]

Fifth, drawing upon the riches of his microscopical experience in botany, Eduard Strasburger brought plant reproduction and cell division into the general discussion of cell multiplication. There were important differences in the two fields. Botanists supported free cell and nuclear formation more readily than zoologists. Strasburger's own research, particularly on preserved specimens of pine embryos and on living specimens of the alga *Spirogyra*, convinced him that the nucleus underwent a complex change. He described bundles of threads, the placement of small rods into a plate at the center of the bundles, their separation into two poles, and their eventual formation into two daughter nuclei. His terminology was somewhat different from that used by zoologists. He spoke of both nuclear and extranuclear filaments and was persuaded that, in plants anyway, the substance of the old nucleus was replaced by, rather than metamorphosed into, the substance of the new nuclei. This conclusion led him to follow a sequence of variations to the simple division he had first described: cases of multiple cell formations, of multiple nuclear formations, and of the retention of the old nucleus during new cell formation. All of these variants reinforced Strasburger's conviction that plants were capable of free cell and nucleus formation.[40]

Lest we conclude from these foregoing accounts taken from the middle years, 1875–1877, that the history of nuclear cytology of the decade was simply a gradual unfolding of the steps to the event that would become known as mitosis, we need to correct our view by looking at Richard Hertwig's concept of the nucleus (*Kern*).[41] A year younger than his brother, R. Hertwig had already become a privatdozent in zoology in Haeckel's institute and had worked closely with Oscar on the fertilization of the sea urchin and other microscopical investigations. What becomes immediately clear is that there was no generally accepted understanding of the nucleus. Hertwig emphasized the importance of using stains to establish the boundary between the stuff of the nucleus and the protoplasm of the cell, and he sought to assemble the common features the nucleus shared throughout animals, plants, and single-celled organisms, their tissues, and their unfertilized and fertilized eggs. At one point, he likened his endeavor to the activity of systematists, who separated that which was dissimilar from that which was held in common.[42] Beyond that point, however, his effort at a generalization veers, at least to the modern eyes, in an unfamiliar but not an unfounded or illogical direction.

R. Hertwig's goal was to establish the nature of the primitive nucleus, which according to evolution theory should be common to all life. He concluded that this primitive object consisted of albuminous material, the "Kernsubstanz," which was not unlike but distinct from the protoplasm of the cell body. This substance was saturated in various manners by a fluid, the "Kernsaft." The primitive nucleus, typified in the mature and fertilized egg, was reminiscent of Haeckel's concept. One found both substance and fluids bound by a transitory "Kernmembran," and that the fluid contained internal vacuoles or "Kernkörperchen," corpuscles. Finally, R. Hertwig considered the nuclear apparatus or "Kerngerüst" to be a nutritive protoplasmic network. This penetrated the nuclear membrane through pores and infused the nuclear fluid with nutrients. By far the most important element of Hertwig's nucleus was the nucleolus. Following his brother's claim, he argued this alone possessed a continuity between the unfertilized and fertilized egg. It was the nucleolus, according to him, that controlled the functions of the nucleus—just as the latter controlled the functions of the cell. So Hertwig condensed the elaborate metamorphoses of the nucleus, described by so many of his contemporaries, into a simplified taxonomy of transitory elements and hypothesized functions. Events, however, rapidly overtook such a simple and straightforward scheme.

The third period of this decade, 1877–1882, is marked by the clarification of the basic stages of nuclear metamorphosis. This included a recognition that the resting nucleus contained a network of stainable material that metamorphosed into the stained cords or rods of that which Flemming was to call "chromatin" and the determination that these cords became elongated threads, which split lengthwise before separating to the poles of attraction at nuclear division. The advance was signaled by W. Schleicher, a doctoral student in Geneva, who in 1878–1879 coined the term "Karyokinesis" to emphasize the movement of nuclear elements during cell division.[43]

By the time R. Hertwig had completed his review, Walther Flemming had left Prague for Königsberg and within a year had again moved, this time to Kiel, where he was installed at the university as Ordinary Professor of Anatomy and director of its anatomical institute. He remained in this Baltic city through his retirement in 1901 and died there four years later. It was here that he carried out some of the most exquisite cytological research of the period, complete with commentaries on contemporary nuclear studies, microscopical techniques, and the logic of hypothesizing about structures that might simply be interpreted as artifacts of the employed preservatives and stains. Flemming responded to R. Hertwig's assessment of the primitive nucleus with gentle praise of the talented microscopist seven years his junior and with a demonstration of his own much broader experience in nuclear cytology.[44] After all, he had been focused on the subject of nuclear metamorphosis for four years and had come to know and further develop the use of various stains. He had established optimal concentrations, the conditions under which to employ certain preservatives, and the method of differential staining pioneered by F. Hermann, which consisted of an overstaining of specimens followed by a differential washing out with alcohol of the stains from neighboring tissues and cellular structures.[45] Above all, he was familiar with many organisms and the reactions of their various tissues to given procedures. Persuaded by some reports that a structured network or reticulum might be lurking unseen within the resting nucleus, Flemming set out to find ways of investigating its existence. He found particularly useful in this quest the epithelial lining of the bladder of a common German salamander (*Salamandra maculata*), the large cells of which possessed nuclei that could be easily observed both in living and preserved conditions. He found that potassium bichromate preparations stained with hematoxylin brought the network of the nucleus and its wall into a surprisingly sharp, deep blue relief. Other reagents combined with the same differential staining technique also produced positive results.

There was, however, a challenge. Was what he saw simply the results of coagulated albumin and therefore an artifact of his techniques? Flemming argued empirically against such a charge by drawing upon his vast reservoir of experience. The network must really be preexisting, he argued, because it appeared in a consistent form and order, which spoke against an incidental chemical agglutination. There were also limits to the variations in the network's form, and when it broke up into rods at later stages, the rods always appeared associated in the same central location relative to the nuclear wall. He easily dismissed contrary interpretations by Bütschli and Auerbach.[46] In short, the meticulous employment of multiple techniques and a variety of specimens mitigated in his favor.

Flemming's repeated confirmation of a stainable network in the nucleus negated the idea of a primitive nucleus supported by R. Hertwig and championed by Haeckel, Goette, and perhaps even Balfour. Of R. Hertwig's four concluding points, Flemming insisted that one could no longer argue that the primitive nucleus was a "naked clump" of nuclear sap that differentiated into a nuclear membrane and nuclear substance by the in-pouring of a protoplasmic network through pores in the membrane.[47]

Sometime between the winter semester of 1877–1878 and August 1878, Flemming turned to *Salamandra*.[48] Their large cells made it possible not only for him to study earlier stages of nuclear metamorphosis but to observe a longitudinal splitting of the threads that had risen out of the skein of stained substance at what we would now call the late prophase of cell division. Much has been made of this discovery in twentieth-century histories of this period, generally because identifying the separate "chromatids" was essential for a post-Mendelian understanding of chromosomes in mitosis and meiosis, but none of these technical terms had been fashioned in 1878, and their role in heredity had hardly been fathomed.

For his part, Flemming was cautious about his generalizations. More important for him at the moment was the completion of a schema for nuclear metamorphosis. He could now illustrate by way of his microscope a complete cycle of nuclear transformation: (1) a resting stage, which when properly prepared revealed a stained reticulum; (2) a skein of "fine thread," the "Knäuelform"; (3) a thickening of the thread and loosening of the skein; (4) a separation of the skein into a wreath of central and peripheral loops, the "Kranzform"; (5) the formation of the loops to make a star in the mother nucleus, the "Sternform"; (6) a longitudinal splitting of the star's loops; which led to (7) a fine-rayed star formation.[49] The termination of this metamorphosis, for Flemming, was represented by the formation of the equatorial plate with its doubled loops at the center of the nucleus. He then envisioned the next seven phases reappearing in the reverse order to complete the division into two daughter nuclei. Thus appearing in sequence came (7) "reverse": a fine-threaded half-barrel form; (6) reverse: the possible fusion or "Verschmelzung" of every other double strand;[50] (5) reverse: coarse threads of the half-barrel of the star form of the daughter nuclei; (4) reverse: the possible end-to-end union or "Vereinigung" of the rays to form a wreath of central and peripheral loops of the daughter nuclei; (3) reverse: a thinning and tightening of the wreath; (2) reverse: the skein of fine thread, again the "Knäuelform," in the daughter nuclei; and finally (1) reverse: a return to the resting stage of the two daughter nuclei, each with its reticulum. Mapping out a regular pattern for the metamorphosis of the nucleus forward and back during cell division was an enormous achievement. The full cycle emphasized a systematic unrolling of the nucleus and return to its resting state.

Flemming diplomatically found ways to both agree and disagree with his contemporaries about the process. Among other things, he could not accept the implications of Auerbach's theory of karyolyse, which assumed that the protoplasm was the formative force and substance of the nucleus and entire process. He rejected Strasburger's claims that the resting nucleus consisted of a homogenous mix of granules and that nuclear division was merely a stretching out and direct separation of two antagonistic materials between the poles and equatorial plate. He pointed out that many of his contemporaries falsely conflated the radiating structures of the asters and spindle with the star formations. They were not homologous structures, for the one was "monocentric" and appeared to be protoplasmic in origin (as revealed in egg fertilization studies) whereas the other was "dicentric" and, as he had rigorously shown, must be nucleic in origin.[51] Flemming remained cautious about the fate of the split threads, but he identified them as possibly fusing at "reverse" 6, rather than seeing them as a means to ensure the equal distribution of the loops to the two daughter cells. They were thus instrumental in a doubling of the bulk of stained material, but they did not function as a mechanism for assortment or for differentiation. Such distinctions became part of later discussions. Not surprisingly, his concept of a symmetrical pattern of stages and reverse stages was quickly criticized.

By 1878, the evidence Flemming presented overwhelmingly argued for the persistence of a nuclear structure during the resting stage of the cell. The direct cell and nucleus division championed by Remak and Virchow over a quarter of a century earlier could no longer be generally accepted. Its more recent alternative, a repeated new formation of the nucleus promoted by many of Flemming's contemporaries, such as Auerbach, Strasburger, O. Hertwig, and Bütschli, must also be rejected.[52] The cycle of nuclear transformation and the longitudinal splitting of the threads and rods affirmed a new alternative, which Flemming dubbed "indirect cell division."

There could be no more forceful testimony to Flemming's dogged persistence and ultimate achievement than the importance he placed on how comparative nuclear cytology had documented indirect cell division in other organisms and in fertilization.[53] Only after preparing and observing hundreds of samples of nuclear division and studying the detailed descriptions by others had Flemming recognized that the prevailing assumption that the nucleus divided simply through constriction must be false. Even purported cases of direct cell division, such as in blood-cell multiplication and regeneration, had not be proven. In response to a recent critical, even sarcastic statement by the elderly physiologist and microscopist Theodor Bischoff that questioned the advantages of nuclear cytology, Flemming defended his empirical approach. Only through the morphological and chemical details of the nucleus could one hope to gain a picture of general processes. "Without this idea," Flemming submitted, "I would see no reason to use my microscope any further." As could be anticipated, he proceeded to justify his labors: "In order even to make assertions over the mechanics of these processes, one must first know exactly what their observable appearances in form are and which of these is regular, which variable."[54]

The principal results of Flemming's research were largely morphological. He continued to refine his picture of indirect cell division during the next three years. He sparred with Strasburger, Fol, Klein, and others about how far to generalize the phases he had identified in *Salamandra* and other amphibians. Strasburger significantly began modifying his own views. In 1879 he softened his stand about repelling and attractive forces within the spindle during the phases of daughter cell separation. The third and fully altered version of his *Zellbildung und Zelltheilung* appeared in 1880. He had thoroughly revised the work, and it remains a tour de force detailing contemporary work on cellular and nuclear division in both plants and animals—much of which was supported by his own observations. He catalogued in greater detail the known cases of "free cell formation" and was now persuaded that the phenomenon was far less common in plants than in animals, among which the eggs of insects and crustaceans loomed as the prime examples.[55]

He accepted Flemming's distinction between chromatic and achromatic nuclear materials, and equally important he agreed that direct nuclear division had not been convincingly demonstrated.[56] Furthermore, he appeared to adopt Flemming's description of various phases in the indirect division process.[57] At the end of his lengthy and disjointed descriptions and summaries of contemporary papers, he systematically set down his own views: cell division and nuclear division were different and sometimes independent processes; the protoplasm plays the active role in cell division; it also incited the nucleus to divide; it collected at the two poles of the nucleus and penetrated to the nuclear plate (*Kernplatte*) in order to form the spindle; he felt it induced a tension within the nuclear material, which led to division. Division itself, he argued, was promoted by the two halves of the nuclear plate (*Kernplatte*), which glide in opposite directions along the spindle fibers. He assumed the presence of these spindle fibers whether he could observe them or not.

He felt the nuclear plate separated through division (*Spaltung*), the spindle fibers formed the primary connective fibers, and that in plants additional connective fibers developed from the protoplasm. He urged that the daughter nuclei originate from the chromatin halves of the nuclear plate, and that they underwent a variety of changes, which in most cases did not simply reverse the processes of the transformation of the mother nucleus as Flemming had envisioned. Strasburger further maintained that the two halves increased in size through nourishment from the protoplasm. He argued that the cell division, following nuclear division, took place through the mediation of the cellular plate or through constriction.[58] He figured that the cellular plate in higher plants might arise only from the connective fibers but elsewhere might arise directly from the cytoplasm.[59]

This catalog of positions reveals, on the one hand, an agreement in the morphological details of the nuclear division in plants and animals. On the other hand, it underscored the gulf that still lay between the orthodox view of direct nuclear and cell division as represented by Strasburger and Flemming. Strasburger still believed that the contents of the resting nucleus comprised a mix of chromatin granules complemented in some instances by a chromatin network.[60] He considered the separation of the chromatin material at the equatorial plate to be either a constriction among the collection of granules or small collections thereof; if the equatorial plate was composed of fused threads, then a longitudinal splitting seemed possible.[61] Much of the disagreement between the two investigators centered on interpreting what happened just before and after the formation of the equatorial plate. Of plants, which at a later period would offer prime classroom examples of cell division, such as the alga *Spirogyra* and the anther cells of the spiderwort *Tradescantia*, Strasburger described and illustrated strands of chromatin dividing at the equatorial plate by pulling apart in a transverse constriction.[62] Here lay the factual and mental chasm between the two premier cytologists of the day. Finally, such examples led Strasburger to question the homology between plant and animal cell divisions at a time when Flemming urged its morphological reality.[63]

Flemming's *Zellsubstanz, Kern und Zelltheilung* appeared two years later. Its very title emphasized the difference between him and Strasburger. The words of Strasburger's title, *Zellbildung und Zelltheilung,* depicted in a general way the variations in cellular forms and modes of multiplication, yet the title had little to do with the organization of the book or whatever general message Strasburger intended. Flemming's title emphasized the variety of stuff in the cell and nucleus, as it was relevant to cell functions and division. The three nouns of his title corresponded with the book's structure, with place of pride coming in the third and by far the most elaborate and lengthy section. Flemming's book as a whole was a "morphology and biology of the cell," made possible through the new microtechniques in preserving, embedding, and staining and the employment of water and oil immersion lenses and better light condensers.

Flemming was candid with his readers that he considered the "true morphology" was essential to provide a secure foundation for a physiology and pathology of the cell. He would have been aware that he reiterated an argument that had played out earlier on the macro level between morphologists and physiologists.[64] In fact, contrary to his intentions, his book might have served as a textbook of contemporary cytology, but its strength lay in its central message: cells normally divide only after an indirect division of the nucleus. This could be demonstrated through the ubiquitous pattern of nuclear metamorphosis starting with the nuclear reticulum becoming visible at the end of the resting stage.

He reproduced the schema of the phases as he had presented them in *Beitrag* I but with some important differences: the segmentation of the skein of chromatin came earlier at the "Knäuel" phase, and the segments bent into loops with the angle first directed toward the center and later toward the poles, with a pulsing in the loops at the "Stern" (i.e., the star phase) that preceded their longitudinal division. Flemming gave up the longitudinal "Verschmelzung" of each pair of threads at the reverse phase, but now left their fates to speculation.[65] Pointing out that each daughter nucleus contained the same number of loops as the mother nucleus, he suggested, "One might consider that of the two sister threads one would be designated for one, the other for the other sister nucleus. For the moment this has not been proven, nor has it been refuted."[66]

During the next two years his suggestion about the equal distribution of threads was to be verified by Emil Heuser, Guignard, and Strasburger. At the daughter poles, the threads thickened into loops and again fused, most likely at their ends, to fashion the new skein of chromatin. This in turn mixed with achromatic material to form the network of the resting stage of the daughter nuclei. Strasburger finally signaled his endorsement of Flemming's latest schema of what he now called mitosis by suggesting functional rather than morphological designations of the phases: thus, prophase, metaphase, and anaphase.[67]

For Flemming, this indirect nuclear division became a fundamental maxim. It was an essential morphological and biological discovery that, he felt, would help reorient physiology and pathology. He saw it contributing to a better understanding of growth, embryology, and phylogeny, and from a higher perspective he was convinced that it would elucidate the understanding of life.[68] In a section presenting the morphological details of the egg, he concluded with an expectation that would have certainly caught Weismann's attention:

If, to take only one important example, the protoplasm of the egg were nothing but a morphologically homogeneous mass with dispersed yolk granules, or even more a fluid, as one has in all seriousness recently called it, then we must leave any answer to the question over the conditions of development, which the egg brings with it, to chemistry. Should the substance of the egg, however, have a structure, then this and the creation of threads in particular areas of the cell body can be different, so also therein can a basis for the predestination of development be looked for, in which one egg distinguishes itself from the others; and this search will be possible with the microscope—*how far, no one can say, but its goal is nothing less than a true morphology of heredity.*[69]

From Mitosis to Structural Continuity

Before Weismann or anyone else took up Flemming's challenge, two additional publications helped frame the issue for all of them. They appeared between August 1883 and April 1884. Their authors arose from quite different research traditions within morphology, and they barely tapped the same wells of information. Conceivably they might have met in Jena in 1871 when both were associated with Haeckel's institute—one as a medical student, the other as a visiting, newly appointed professor from Liège. Where Flemming had only suggested in passing that the details of mitosis pointed toward a mechanistic understanding of heredity, these two appeared to shout that possibility from the rooftops. Because of their different approaches to heredity, it makes sense to discuss these publications in reverse order of appearance.

Eduard van Beneden: Fertilization as Rejuvenation

We introduced van Beneden in Chapter 7, where we saw that his 1874 work on the origin of the sexual cells of the hydroid *Hydractinia echinata* provided the background for Weismann's own project to study the origin of the germ cells of hydromedusae on a much grander scale.[70] Beneden had argued that the male germ cells were ectodermal in origin while the female germ cells were endodermal. Upon this claim he constructed the general hypothesis that the germ cells were polar opposites and that all organisms were physiologically hermaphroditic. In preparation for fertilization, he asserted, both the female and male germ cells eliminated the component of the opposite sex. Fertilization, the union with the other sex, consisted in the reestablishment of the original hermaphroditic state. In the next two years, van Beneden supported his hermaphroditic hypothesis in substantial monographs on the egg and early development in the rabbit, bony fish, and bats. His work corresponded in subject matter with Oscar Hertwig's Habilitationsschrift on a sea urchin, but with a different conclusion. Beneden felt he had demonstrated that the female pronucleus had not been derived from the disappearing germinal vesicle but was a structure *de novo*. As for the male sperm, neither Hertwig nor van Beneden had observed the actual penetration into the egg. Hertwig had assumed such a single penetration had taken place; van Beneden sanctioned the possibility that multiple sperms entered to form the male pronucleus. As one of van Beneden's biographers, Carl Rabl, explained it, "where Hertwig had assumed a morphological continuation from one generation to the next, van Beneden had supposed a chemical continuity, that is, the chemical matter of maleness and femaleness joined at fertilization."[71]

A serious alpine accident forced van Beneden to postpone further publications for three years. Nonetheless, he was able to return to research, and in the 1880s once again revealed his skill in tracing the subtle changes in the protoplasm and nuclei during fertilization and egg maturation—first with mammals, then, in those days before the advent of automobiles, with the omnipresent horse parasite, the nematode *Ascaris megalocephala* (today designated *Parascaris equorum*). His large monograph of 1884 was a continuation of the excellent cytological work being pursued in Liège by himself and his colleague Charles Julin.

It was already known that *Ascaris* offered tremendous advantages for the cytologist pursuing questions of reproduction. Beneden took pains in his first monograph on this organism to explain that during the 1850s parasitic nematodes played an important role in controversies between leading naturalists and microscopists debating the nature of fertilization.[72] In the early 1870s, both Bütschli and Auerbach had turned to free-living and parasitic nematodes, including *Ascaris*, in their efforts to understand fertilization and early development. When he reviewed its potentialities as a research organism, van Beneden found the anatomy of *Ascaris* was simple, and its genital organs, comprising two coiled, threadlike ovaries, two uteruses, and two oviducts, was easy to dissect. Although *Ascaris* eggs, when laid, are considerably smaller than those of the sea urchin, its spermatozoa are longer and have at least ten times the volume.[73] More significant was van Beneden's discovery that, when appropriately treated, *Ascaris* could be induced to produce thousands of eggs at the same stage of development. Preserved intact, these allowed for carefully

considered and replicable observations. The eggs were easy to stain and section. An advantage that *Ascaris* had over the popular sea urchin at this point
in time was that egg maturation with the expulsion of polar bodies came after
the penetration of the sperm and formation of the male pronucleus. It thus
became easier for van Beneden to disentangle the relationship between maturation and fertilization. In retrospect, we can see that another advantage lay
in the fact that the number of chromatin elements found in the nucleus was
only four in what was soon to be recognized as the bivalent form. Although
van Beneden at the time had not always found consistency with the numbers, the small number of his "anses chromatiques" certainly allowed him to
observe polar body formation with a new understanding.[74]

Beneden's major publication on *Ascaris* appeared in early 1884 as a 375-
page monograph entitled "Researches on the Maturation of the Egg and Fecundation: *Ascaris megalocephala*."[75] Hamoir has explained how this monograph introduced bibliographical confusion. By October 1883, van Beneden
had completed 230 pages and had sent them off to his printer. With a stroke
of luck, however, he turned to female specimens that he had preserved in a
solution of weak alcohol several months earlier. "The alcohol [had] made its
way so slowly through the perivitelline layers that its penetration required
several weeks. During this period of time, the embryo developed slowly;
new stages occurred along the genital tract. These new data were hastily
included . . . [and] . . . Printing was finished in March 1884." In brief, van
Beneden quickly added 125 new pages and three new plates, which filled
in the missing developmental steps. Because he was one of the editors of the
Archives, the delay in publication was easily arranged.[76] The new monograph
again revealed the great care with which van Beneden traced the slightest
change in development. When he had stopped the presses in October 1883, he
had realized that he had failed to describe or illustrate the copulation of the
two pronuclei or to follow with precision the *anses chromatiques* of polar body
expulsion. Now with his recovered specimens he could do both. As Hamoir
cleverly remarked, "Quel paradoxe de voir se poursuivre une évolution normale dans des parasites morts!" ("What a paradox to see a course of normal
evolution in dead parasites!")[77]

The sequences in van Beneden's plates are so elegant and his descriptions
so rich in detail that it is easy to be persuaded as to the accuracy of his account. In reality, they reveal a problem with the entire enterprise of nuclear
cytology at this stage in its development with its multiple new tools and reagents. What was important to examine with the utmost patience and what

could be passed over as artificial, accidental, or simply trivial? Cytology could easily become overburdened by the nonrelevant as every change had to be noted. Besides, the nonrelevant for one researcher became the relevant for the next. As we have seen, van Beneden was not alone in producing large monographs of detailed observations and lavish illustrations. More than most he seemed to want to record it all in the glory of its finest details. It turns out that many of his observations, however, were heavily theory laden—but then so were those of Flemming, who had repeatedly castigated his contemporaries for developing explanations of processes that were not to be read from the microscopic slides. This was certainly the case with van Beneden in his elaborate description of an "Ypsilon" formation connecting the disintegrating female pronucleus and the spindle of the first polar body formation. This structure strongly suggested to him that there was no continuity between the chromatic elements of each and so implied that the *anses chromatiques* of the polar body were a composite of achromatic and chromatic material drawn from the unfertilized egg.[78]

Beneden's examination of the first maturation division itself was brilliant, but as we will see, it was also strongly influenced by a more basic assumption. He described and illustrated in exhausting detail the transformation in the achromatic bundle and noted that during division two chromatic disks, each with four "agglutinated corpuscles," doubled the number of corpuscles found in the original germinal vesicle. Nevertheless, he insisted that what he had repeatedly observed was not an indirect cell division. "One must not compare the genesis of the first polar body to an indirect cell division," and so he called it instead a "pseudokaryokinesis."[79] As for the second maturation division, van Beneden was convinced that it was of a totally different kind, for the polar body ended up with two *anses chromatiques* rather than four. For this reason, he decided that the second polar body, as a whole, must have the value of the female pronucleus. It could not therefore be considered a whole cell at all.[80] There is no reason to believe, however, that the observations van Beneden carefully made were inaccurate, and the reduction in chromatin elements at the second maturation division appears in retrospect to be remarkably prescient. His interpretations, however, as we shall soon see, came from a different perspective from that which we might expect in retrospect and toward which many of his German and Swiss contemporaries seemed to be headed.

What could not be denied about van Beneden's long-term achievements was his recognition that *Ascaris* confirmed something that he had already noted with mammals. Hertwig had described a fusion or "Vermischung" of the two pronuclei with the sea urchin. *Ascaris* now persuaded van Beneden otherwise. Although the male and female pronuclei made contact and might conform in shape to one another, their contents did not mix. Furthermore, the chromatic nuclear elements from each pronucleus—the two paternal and two maternal *anses chromatiques*—proceeded to undergo the mitotic division of first cleavage in a coordinated yet independent fashion. They remained distinct within that which had been their respective pronuclear membranes. Beneden even suggested that this condition must remain in all tissues as the embryo developed so that the male and female elements could separate again at the forming of the gametes for the next generation.[81]

Moreover, *Ascaris* convinced van Beneden that fertilization implied more than simply two sets of *anses chromatiques,* one from each parent, coming together at fertilization and maintaining this "diploid" condition throughout the tissues of the developing organism. Fertilization, for him, was analogous to conjugation among single-celled organisms, by means of which rejuvenation rather than multiplication seemed to be the immediate consequence:

> The study of the maturation of the egg, of fecundation and of cellular division has convinced me of the idea that cell nuclei are hermaphroditic and that that is much more than the male chromatin merging with the female chromatin. If the male pronucleus and the female pronucleus merit these denominations, which imply their sexuality, the cellular nuclei are manifestly hermaphroditic. The cells of the tissues partake of this character with the protozoa and the protophytes.[82]

In the production of gametes, according to van Beneden, there was an expulsion of the paternal entity with the second polar body. Fertilization then constituted a provisional joining of the two pronuclei in order to augment the lost paternal material. The reverse was assumed to happen with the production of spermatozoa. Such an augmentation guaranteed a newly rejuvenated hermaphroditic organism. The crux of van Beneden's explanation of fertilization, cell division, and polar body formation was really far more an explanation of life itself than of heredity. "La fecundation est la condition de la continuité de la vie. Par elle le générateur échappe à la mort." ("Fertilization is the condition for the continuity of life. By this process the engenderer escapes death.")[83] His was an additional expression of physiological continuity between generations as well as throughout development.

Wilhelm Roux: The Struggle of Parts

Although only a young privatdozent when he published his short contribution in August 1883, Wilhelm Roux was not an unknown figure in embryology.[84] Prior to writing his study on the struggle of parts, Roux had studied medicine in Jena where he sat in the lectures of Haeckel, the anatomist Gustav Schwalbe, and the physiologist Wilhelm Preyer and participated in Rudolf Eucken's seminar on Kant. He enjoyed the philosophy seminar very much, and it was evident that as he pursued a career in embryology, he carried with him a keen interest in scientific method and a mechanistic ontology. After training further in Berlin and Strassburg with the pathologists Rudolf Virchow and Friedrich von Recklinghausen, respectively, he returned to Jena where he pursued a doctorate under Schwalbe. The latter directed him toward a project on the mechanics of growth and its impact on the developing form of the organism. Focused on this assignment Roux studied the influence of blood flow on the form and branching of blood vessels. His work encouraged analogies between hydrodynamics and hemodynamics and promoted Roux's lifelong commitment to understanding functional adaptations.[85] After a brief stint in Leipzig at a hygiene institute, Roux submitted a Habilitationsschrift on the relationship between Darwin's theory of descent and the formative, adaptive effects in the developing embryo.[86] The work was enough to secure him a position at Carl Hasse's anatomy institute in Breslau, where he spent ten years rising from a privatdozent to an außerordentlicher Professor and director of his own institute for embryology. It is here where he first became identified with experimental embryology or, as he nominated and proselytized for it, "Entwicklungsmechanik."

Before this would happen, however, Roux published a curious, 244-page book entitled *Der Kampf der Theile im Organismus* ("The Struggle of Parts in Organisms").[87] It was based on his previous research, but Roux complemented that with a flight of speculative fancy.[88] He argued, as had many before him, that Darwin's mechanism of natural selection among individuals did not explain the appearance of hereditary variations that were necessary for Darwin's mill of natural selection. The problem was simply left in the lap of scientists either to consider variations the products of random chance or to assign them to a teleological corner of Darwin's mechanistic edifice for evolution. The problem was really twofold: how to rationalize the appearance of variations, and how to explain their transmission to successive generations. In his *Kampf der Theile,* Roux attempted to explain the first in terms of an equally mechanistic selection, which he reasoned must occur on different levels within the developing and mature body of each individual. He assumed the second, transmission, after discussing at length and accepting the inheritance of acquired adaptations.

The work offered a provocative challenge to biologists. Haeckel praised the work as an extension of many of his own ideas, which was certainly the case. Although he found the German difficult to read, Darwin wrote to Georges John Romanes that "as far as I can imperfectly judge, it is the most important book on Evolution that has appeared for some time"—incidentally, a comment Roux was fond of citing. Darwin suggested that Romanes review the book, yet he concluded with the qualification "but I may be *wholly* mistaken about its value"—a remark that Roux did not cite.[89] In his essay on "Heredity," Weismann found Roux's struggle of internal parts a useful mechanism, in addition to natural selection, to explain the atrophy and eventual disappearance of organs, such as in the case of atrophied eyes among cave-dwelling animals.[90] Roux's own mentor, Schwalbe, however, warned the young author never again to publish such a " 'philosophical' book, or else you will never become an Ordinarius of anatomy."[91] Even though its contents had little to do with Roux's paper of 1883, it is worth looking at the *Kampf der Theile* in somewhat greater detail because it demonstrates Roux's dynamic embryology at the center of both his evolutionary and hereditary discussions. Moreover, Weismann was to have recourse to Roux's intracorporeal competition a decade later.

As a young embryologist, Roux was very conscious of pushing his discipline beyond the limits of pure description. He now put his examination of functional adaptations discussed in his first line of research into service for his speculations on evolution and heredity. The embryo appeared to him as the important stage of the individual when the chemical assimilations and deletions, which effected changes in morphological attributes, could be transmitted to the next generation. It was the period when the effects of internal chemistry, rather than of external factors, could collect in the developing gonads. Thus, the chemistry of one generation constructed the variations for the next. Such a conclusion, however, violated the ideal version of the biogenetic law, which dictated that all eggs, while recapitulating the past, began with the same primitive form, and that while developing into the adult form, each embryo traced its ancestral lineages. This recapitulation would be impossible if the chemical changes responsible for old and new variations alike were first manifested as embryonic changes and incorporated into the gametes as such.[92] Consequently, the chemical precursors of newly arisen adult attributes had to be condensed backward over time (over many generations) to the earliest embryonic stages.

> By the reduction of acquired changes of form to chemical changes and by their easier transmission to the sperm and to the egg as a chemical exchange, which takes place between them and the father with respect to the mother, the problem of inheritance as such will be raised and led back to the appearance of a more general problem, where it is reduced to the chemical processes of the structure, which is the foundation of the whole of biology.[93]

There could be no doubt that Roux unconditionally endorsed a mechanistic biology with its reduction to chemistry. He would devote his career to the proclaimed science of *Entwicklungsmechanik*.

The next chapter lent its conception to the title of the book as a whole: "Kampf der Theile im Organismus." It was here that Roux drew upon Virchow's concept of the organism as a society of cells. In so doing, he emphasized Virchow's proviso that such a society was not a single interdependent unity ("Einheit") but rather an association ("Genossenschaft") of semiautonomous units that *inter se* were slightly different and had the leeway to assimilate and grow in competition with one another. The functional and purposeful unity of the organism, so heralded by generations of physiologists and anatomists enthralled by the "Ganzheitsbegriff" (the concept of the whole), was reduced to an internal Darwinian struggle for space and nutrition. Roux envisioned this struggle occurring on four levels of the organism: between molecules, cells, tissues, and organs. The smallest advantage one unit had over another on the same level became magnified during embryonic development and ensured both differential growth and embryonic differentiation. On the levels of molecules and cells, growth and differentiation were solely the outcome of this competition for space and nutrition. On the two higher levels, functional adaptation and stimulation by the neural system reinforced the struggle. Thus, the harmonious whole of the organism reflected a mechanistic competition and self-regulation. The structure of the organism appeared no less purposeful for it.

Where he devoted the first half of his book to a priori speculations about the struggle between parts, in the second half Roux drew upon his experience of development as studied in the laboratory for the empirical support of functional adaptations. Collectively, the science had already established the importance of functional stimuli on assimilation and growth in muscles, nerves, and glands, on the pathological growths of tumors, on supporting tissues and organs, and on metabolic structures. Roux's own study of the shape and branching of blood vessels and connective tissues provided further examples of the effects of functional adaptation, of function producing form. Roux furthermore examined in detail the effects of regulating mechanisms, particularly of the neural system. He recognized that the consequences of such functional adaptations had been predicted by Virchow. "We have at hand," his Berlin mentor had claimed, "the ability to form not only the entire individual but in particular organs and systems and thus to shape the individual characters in this or that direction. . . . Among the ways to produce men with more flesh, blood and neural mass are certainly above all stimuli, the means to excite. Without a stimulus there is no organic work, no reception of new formative stuff, no development."[94] In contrast to Virchow, however, Roux believed he had anchored his ideas to the Darwinian mechanism of a struggle between competing units.

It is easy for us to argue that Roux failed to establish the identity of the internal competition with the reproductive dynamics embedded in Darwin's mechanism of natural selection, but this would miss two central points. First, Roux strove to find an internal mechanism to explain developing form. Second, he believed that the competition of molecules and cells and the functional adaptations of tissues and organs were based on trophic changes.

Through it all, Roux assumed that heredity was simply the result of assimilation and growth and that gametes were simply vessels for the transmission of stable and altered chemical molecules. Moreover, he repeatedly emphasized that molecular and cellular selections and functional adaptations must regress to the embryonic level before they could be impressed onto the gametes and passed to offspring. This roundabout way of transmission of juvenile and adult traits meant that the inheritance of an acquired character might take many generations—perhaps "thousands."[95] Nevertheless, when in detailing the essence of the organic and inorganic realms, he unabashedly echoed a long tradition. Reproduction in either realms proceeded through "the so-called growth over and beyond the individual mass. It is at the same time dependent on the character of assimilation." What distinguished, for him, the organic from the inorganic was the former's unbroken continuity of reproductive forms and the property of self-regulation.[96]

A few months after publishing *Der Kampf der Theile,* Roux resumed the task of reviewing recent literature on embryology and reproduction. He had done

so in the past, perhaps because it offered a young scholar not only the chance
to survey the literature but to promote his own work. Roux at least used this
opportunity to critique his *Kampf* as well as survey over 200 additional titles.
While doing so, he clearly indicated his dissatisfaction with his earlier treatment of reproduction. The dilemma was clear: how could acquired changes
in a highly differentiated organism be passed on through undifferentiated eggs
and sperm to the offspring?

> As a result of this simplicity of the direct propagative bodies every ac-
> quired characteristic at the time of being carried over to the egg or to the
> sperm must be transformed into an undifferentiated quality whether it be
> acquired at the stage at the beginning or at the completion of develop-
> ment. . . . This return transformation of the manifold, the developed,
> [and] the explicit into a simple entity, an undeveloped entity, and an im-
> plicit entity must be seen as the essence and therefore the peculiar problem
> of development, as far as there is a transmission of "acquired" changes. . . .
> Up till now this has been overlooked in the pertinent works.[97]

Given the current advances in nuclear cytology, the holistic approach found
in the overgrowth explanation of transmission no longer satisfied him. Considering Roux's background, we may better understand how radical the change
was when Roux wrote in the summer of 1883 his well-known paper on the
significance of indirect nuclear division. It appears that he himself had little
firsthand experience exploring nuclear division, but it is hard to believe that
he had not followed it in the laboratory as he investigated the first, second,
and third cleavage in several species of frogs. He does not speak of nuclear
stains, which would have given him a better grasp of the chromatin segments,
nor does he seem to have followed the metamorphosis of the nucleus closely
at fertilization or egg maturation. The truth of the matter is that he was
interested in the fate of the blastomeres not the structures of the dividing
nucleus—so specialized had nuclear cytology become compared with the
mainstream of descriptive and experimental embryology. Nevertheless, he had
a firm grasp of what Strasburger and Flemming had described in their recent
articles and books, and he was struck by the elaborate sequence and mechanisms involved in indirect cell division. (At this point Roux had not adopted
either Schneider's expression "karyokinesis" or Flemming's "mitosis.")

Roux got right to the point. He saw that the Remak-Virchow belief in a
direct division of the nuclear mass no longer reflected reality. Instead, the

Flemming-Strasburger details of the metamorphic stages of the nucleus and of the longitudinal division of the chromatin threads, despite the differences in detail between species and between animals and plants, appeared to be universal. Because there had to be a biological cost in time, energy, and structure for such a roundabout process, Roux reasoned that "there must be more to indirect division than simply the halving of the nuclear mass. Our judgment will be otherwise, if the goal of nuclear division is not merely an arbitrary halving of the nuclear mass but is also some sort of designated separation of 'qualities,' which comprise this mass."[98]

Most of Roux's discussion centered on the mechanisms for sorting and distributing multiple "qualities" to two daughter nuclei. He compared the cell and its nucleus to a physical-chemical factory. He argued that supportive achromatic threads must secure the chromatin during nuclear division, commented upon the size of the chromatin corpuscles at the equatorial plate, and calculated the value of a single plane of attraction from pole to pole for holding rigid the entire arrangement. He figured that the "Schleifen" held the chromatin and "qualities" together and eliminated the need for an unwieldy number of threads in the spindle. He speculated on the need to split those "qualities" and their arrangement to ensure that half of each went to the opposite daughter nuclei, and he foresaw the function of the spireme, wreath, and star phases as guarantors of an equational distribution.

It was completely a post hoc set of arguments derived from Flemming's and Strasburger's factual presentations. Nonetheless, it emphasized the mechanics of indirect division and its hypothetical importance in dividing and transmitting "qualities" to ensuing cell lineages. Almost as an afterthought, Roux noted that the same mechanism might also ensure a qualitatively unequal ("qualtitativ ungleich") sorting of the nuclear material, a process in accord with his current studies on the fate of second- and third-cleavage blastomeres in different species of frogs. Finally, Roux seemed emphatic in his hypothetical claim that the "qualities" were chemical and far too small to observe through the microscope; consequently, the chromatin, which he argued contained the hypothesized metastructure, may always appear homogeneous to the cytologist.

Postscript to Part One

We have now followed the career of Weismann up through the completion of his monograph on the sexual cells of hydromedusae and his two general lectures on life and death and heredity. All his work from his Habilitations-schrift to *Hydromedusae* drew upon his study of development, natural history, and evolution. His studies came at a time when biologists were wrestling with the meaning of diverse reproductive modes, alternations of generation, and sexual and seasonal polymorphisms. Collectively these were embryological in expression but reproductive in consequence. It seemed very natural, given his broad interests, for Weismann to move from the fact of reproduction to spec-ulations of how traits were ensured intergenerational transmission. The avail-able model to explain such transmission appeared to piggyback on one of the most obvious physiological functions of life: on growth. "Overgrowth" served as a description of reproduction, development, and the phenomena of heredity. As a consequence, heredity was patently obvious but rarely argued. We have found this with Baer, J. P. Müller, and Schwann, and it was made explicit by Haeckel. This overgrowth applied equally to metazoans as to unicellular organisms; it applied to both sexes, to the alternation of generations, partheno-genesis, regeneration, and from one cell generation to the next. Even Darwin, Spencer, Brooks, and Nägeli, in different ways, saw heredity from one generation to the next as the automatic outcome of growth of the organism "over and beyond itself."

Concomitant with this emphasis on growth was the assumption that he-redity designated the traditional or conservative side of reproduction and was identified by truisms, such as "like begets like" ("Aehnliches erzeugt aehn-lisches") and "chip off the old block" ("Der Apfel fällt nicht weit vom Stamm)," and "the child is flesh and bone of the parents" ("Das Kind ist Fleisch und Bein der Eltern").[99] These affirmations, however, were a matter of perspective, whether one viewed the similarity between members of a species or focused on the fine-tuned differences of racial or family lineages. Where generational differences were evident, or where adaptations were undeniable, heredity and development appeared antithetical. The cell theory itself incorporated from the start an overgrowth perspective, in which nucleoli, nuclei, and cells assimilated and grew, expanded beyond parental boundaries, and simply divided by fission. When Spencer, Darwin, and Nägeli made efforts to un-derstand the transmission of individual traits, they relied on the growth of molecules, gemmules, or the idioplasm to expand beyond the traits of the individual to the next generation. When Haeckel toyed with his admittedly highly speculative theory of perigenesis, he had little concrete to say.

During the first three quarters of the century, the common-sense way of viewing heredity appears to have been holistic. Its scientific refinement focused on chemical or molecular processes, on the circulation of molecules and even cells, or on the spread of a molecular idioplasm. It focused on what biologists in the twentieth century would call phenotypic characters. These models rarely emphasized anatomical details, except those in gonads, spores, and gametes, and these in turn were viewed simply as bridges to the next generation. There was nothing special about sexual reproduction. It was merely a combination of two individuals, each of which, at least at lower levels, might reproduce asexually. Further diversity in reproductive modes, be they unicellular or multicellular, heterogenesis or metagenesis, regenerative or pathological, reinforced the feeling that heredity was the conservative side of reproduction. The continuity of features stressed an intergenerational connection, be it between cells, between life cycles, or simply between parent and child.

A second tradition took shape gradually after the late 1870s. First came the demonstrations that indirect rather than direct cell division was the near universal mode of cell division. This was quickly followed by the recognition that a longitudinal division of the "chromosomes" guaranteed equal distribution at mitosis of "chromatids" to both daughter cells. Such a distribution of multiple elements could not have happened any other way. Flemming's presentations were followed in the next two decades by a gradual clarification of the nature of the chromosomes and of both male and female maturation divisions. After the turn of the century, with the recognition of a concordance between maturation division and Mendelian ratios and with the ratification of the innovative side of heredity through classic chromosomal genetics, a newer tradition was secured—and overplayed.[100]

This newer tradition provided a structural understanding of heredity in contrast to the physiology of overgrowth. Individual, racial, and species differences were to be explained by changing structures within the nucleus. Born of the nuclear cytology that we have reviewed, this second tradition not only focused on the commonplace continuity between generations but forced the related problem of a continuity within the developing individual. The nucleus, and more particularly for Weismann, the nucleus of the germ cells, appeared to be the sole refuge where relative constancy could be secured against the maelstrom of activity of the rest of the developing organism. But as we shall see, this was only a relative constancy. Instead of being holistic, this newer tradition at its purest was "architectonic," or in the language of the day it was "morphological" rather than "physiological."

I have argued that Spencer, Darwin, Brooks, and Nägeli were transitional figures. They saw growth as an integral part of reproduction and consequently of heredity. The first relied on physiological units, the second and third on gemmules. None could be explicit about structure other than that an element grew, somewhere circulated, and somehow relocated into the next generation by mutual affinities. Nägeli, at least, located his idioplasm in time and space, but he did not associate it with the new cytological reality. Beneden and Roux in 1883–1884 appear to me to have been far more committed to an architectonic view. Roux, in fact, appears to have made the transition before our eyes between his *Kampf der Theile* and his paper on the meaning of nuclear division, "Kerntheilungsfiguren." Each in his own way, however, seemed to remain with one foot solidly embedded in the older tradition by locating the essence of heredity at an amorphous chemical level far below the level of nuclear morphology. It was at this point that Weismann began playing an important role in pushing forward the architectonic or morphological view, but as we will see in the second part of this book, even he held on to overgrowth and holistic elements in certain situations.

II

AN ARCHITECTONIC VIEW OF HEREDITY

12

A New Perspective on Heredity

Continuity and Heredity

Weismann spent almost his entire research career at his institute in Freiburg. He successfully applied through the Baden government and the Zoological Station in Naples for research time on the Mediterranean, and during his stays in Naples he had specific projects of collection and research, which would further his ideas about germ cell production and to a lesser degree strengthen his conviction concerning the correctness of Darwin and Wallace's theory of natural selection. A university professor (according to one account there were forty-one in all fields in Freiburg in 1876) had to do more than research. He had to run an institute for research and for advanced students, and give university courses for advanced students and oversee their careers not only in zoology but invertebrate anatomy as well. Faculty colleagues desired and expected to be informed firsthand about university affairs and to learn the details about each other's personal opinions on academic and local events as they logically emerged for personal consideration. In this context, the local professor was expected to give public talks to the university and educated community. We have already alluded to such obligations earlier. Besides special lectures, Weismann also took collecting trips to Marseilles, to the Loire estuary, and to Roskoff in France. A zoological colleague, van Rees, appears to have accompanied him in 1879 and 1880. A trip to Rome and Naples in 1881 and a trip in the company of Mary to Sicily in 1882 also occupied both aesthetic and zoological interests. Some of these travels were clearly connected with collecting and professional meetings. The Swiss landscape increasingly attracted him, and this was perhaps connected with his becoming an *Ehrenmitglied* of the

Figure 12.1. August Weismann on a collecting trip, 1880. It is believed that van Rees is on Weismann's left and an unnamed assistant is on Weismann's right.

Klaus Sander, ed. *August Weismann (1834–1914) und die theoretische Biologie des 19. Jahrhunderts. Urkunden, Berichte und Analysen.* Freiburger Universitätsblätter, vols. 87, 88. Freiburg: Rombach Verlag, 1985. Helmut Risler, "August Weismanns Leben und Wirken nach Dokumenten aus seinem Nachlass," p. 29.

Swiss Naturforschen Gesellschaft in 1885. The trip was made a year before the death of Mary in Lindenhof in October 1886.

For all his earnest and successful research, Weismann could not escape the public duties of a university professor. For Weismann, this entailed becoming chair of the Bibliotheca comity, and being elected as decan of the university in 1880 and chair of the academic Krankenkasse in 1883–1884. The popularity of his lectures, signified by the construction of and additions to a new lecture hall, the yearly talks to the community, and in the 1890s his specialized seminars and semester courses on evolution theory, which captured the attention not only of medical students but the university community as a whole, represented honors but also included time-consuming duties. At times, Weismann appeared weary of such attention. In writing to Haeckel in 1877 about his winter semester lecture course, he seemed both proud of his success but weary of the obligations of teaching masses of students, some of whom were also women and required undue respect concerning the biological material:

> Not that they [these courses] would be unpleasant to and for me unpleasant! To the contrary, they gave me pleasure, but their preparation costs—particularly for the zoologist—much more time and effort than the attained success is worth. As with beautiful themes, one must lay them aside unused, as once we had to put the rectum and reproductive organs to the side unnamed because of the women [in the course]. What then concerns the *Kosmos,* I am not actually disinclined to publish an article for it, however, I would not want to make a promise because I know by experience how disruptive one can be. At the moment, for example I would not be in the position to find the time for it. I don't know whether it is the same for you, but it is scarcely possible for me, if I am seriously involved in a work, at the same time to become involved in an entirely different line of work.[1]

In Search of the Ureier: Ultra-Organismic Continuity as an Issue

Chapter 11 emphasized the rise of nuclear cytology as an essential development for a new conceptualization of heredity. One aspect of this entailed a shift from a holistic to an architectonic way of conceiving the phenomenon. The advantage of such a shift was quickly recognized by Flemming and rapidly exploited by others, and it led to the gradual demise of overgrowth of the whole in favor of a specified chromatin structure. At the same time, there was a parallel route to this change that wrestled with a more traditional embryological problem. Both routes were important for Weismann, and both drew attention to the need to understand an intracorporeal as well as intercorporeal continuity.

In the past, I have referred to this second route as a "pursuit of the Ureier." When he sought to establish a continuity of germ cells through several alternating generations of hydromedusae, Weismann was traveling this path.[2] He was hardly alone in following such a research project; among his contemporaries were Richard Owen, Francis Galton, August Rauber, Julius Sachs, and Gustav Jäger.

Minor Statements of Intra-Organismic Continuity

Modern scientific innovations are rarely the inventions of a single mind. After he had published his major manifesto on continuity, Weismann became both the major spokesman for the concept of intra-organismic continuity and the target of others who claimed to have previously developed the same idea.[3] Consequently, during the ensuing years Weismann found himself involved in several priority disputes, and in response he wrote over the next decade four historical sketches of the development of continuity. He never denied that others had earlier put forth similar notions, but he was adamant about claiming recognition for exploring the full significance of intragenerational continuity. "The utility in an idea," he pointed out in one of his exchanges, "lies not merely in the fact that someone once had it, but that one has followed its implication as far as possible and tested its feasibility."[4] Weismann and his contemporaries, as they vied with one another, and modern historians in their examination of heredity theory have not sufficiently followed the contrast between the various enunciations of continuity and their implications. It is worthwhile beginning the exploration with sketches of some of the minor statements of intragenerational continuity.

Richard Owen

Geddes and Thomson in 1889 and Weismann following their account mentioned that Richard Owen was the first to have conceived of an intra-organismic continuity. Although they were not specific in their references, they must have been referring to Owen's provocative piece on *Parthenogenesis*. There is some merit to the claim. Owen described and illustrated development as a process of successive cell divisions, and at one point he wrote of "the retention of certain of the progeny of the primary impregnated germ-cell . . . unchanged in the body."[5] An isolation of a germinal line seems to be implied, but a continuity of the germinal material during development certainly was not. In the first place, Owen was primarily concerned with the production of successive asexual generations as earlier described by Steenstrup. The passage cited came in the context of explaining how alternating generations remained faithful to type. Second, Owen was very clear in his description of the first cells of the organism. They were not in his mind derived from a primary cleavage of the fertilized egg but from the consolidation of yolk material within that egg. In other words, these cells arose *de novo,* and a continual line of germinal cells was broken at the point of the initial development of each egg. Far more important to Owen than any possible continuous mass of germinal material was his concern for a "spermatic virtue" that caused the development of form. During the sequence of cell divisions and in alternating generations, Owen understood that this virtue dissipated its force. For Owen, as for many others later, the act of fertilization entailed an essential rejuvenating event following upon the heels of a determined number of enervating asexual cell reproductions. Included in Owen's concept was indeed a notion of continuity, but this consisted of the unbroken thread of an ill-defined spermatic force. We could return to Aristotle and identify a similar idea in the formative cause of the continuing male influence.[6]

Francis Galton

If Owen fails the test, there can be no doubt that Darwin's cousin Francis Galton passes. Through a long and distinguished career of collecting and comparing family lineages, Galton became an important contributor to heredity theory. Unlike the other minor commentators we examine, he focused on developing the techniques for the statistical analysis of the similarities and differences between parents and offspring. He was almost exclusively concerned with human heredity and not deeply immersed in the biological literature that established the view of heredity of the 1880s and 1890s. Nevertheless, Galton was well acquainted with his cousin's provisional theory of pangenesis and, as is well known, experimentally tested and refuted Darwin's idea of gemmules circulating in the bloodstream.

It is often assumed that that is where the matter rested, but Galton did adopt some of Darwin's perspective. Following his cousin's lead, he became a champion of a "particulate" theory of heredity (the expression was Galton's), and although generally skeptical of the inheritance of acquired characters, he remained willing to postulate a minimal circulation of "gemmules" to explain its rare occurrence.

Even before Darwin published his provisional hypothesis, Galton used similes that implied an intra-organismic continuity. Considering the origin of human life in a popular essay in 1865, Galton compared all of life as one continuous organic system:

if we consider our own embryos to have sprung immediately from those embryos whence our parents were developed, and these from the embryos of their parents, and so on for ever [sic]. We should in this way look on the nature of mankind, and perhaps on that of the whole animated creation, as one continuous system, ever pushing out new branches in all directions, that variously interlace, and that bud into separate lives at every point of interlacement.[7]

The simile might be construed as simply an endorsement of a phylogenetic continuity, but speaking seven years later, now within the framework of a particulate theory, Galton distinguished between the elements of the developing individual and latent elements that contributed to the initial "structureless" state of the offspring.[8] By the time he presented a specific "theory of heredity" in 1875, Galton spoke of the "sum-total of the germs, gemmules, or whatever" as forming a root, or "stirp," and that "as fertility resides somewhere, it must have been vested in the non-developed residue of the stirp."[9] There is no question that implicit in the last two of these statements lies a notion of a germinal material isolated from the rest of the developing body, but Galton never probed deeper than these general allusions. When in 1889 he presented his most detailed account of a theory of heredity, Galton simply had not progressed beyond such vague intimations; in fact, he even failed to refer to his earlier hypothesis of stirps by name. By that time, when continuity of germ cells and the germ-plasm had become an issue of professional debate, Galton ignored the question and its associated literature. He referred to Weismann only once—and then misspelled his name, at that! Ruth Cowan, who examined the development and orientation of his ideas, has rightly insisted that Galton was primarily interested not with biological problems per se but with the nature-nurture dichotomy as it concerned human society. With specific reference to Galton's development of the concept of continuity, she concludes that "in an important sense, Galton did not discover the continuity of germ-plasm [sic] at all."[10] Robinson, who has carefully collected Galton's comments on a particulate theory and on intra-organismic continuity, remarked that Galton "did not concern himself further with the cell or the mechanisms of heredity; his studies did not take that direction."[11] In a less critical vein, his biographer Karl Pearson believed that Galton had put forth "a doctrine probably for the first time in the history of science, which amounts to the theory of the continuity of the germ plasm."[12]

Galton would seem to present the example of an unfulfilled promise with respect to intra-organismic continuity. Nevertheless, when he heard of Galton's theory of stirps, Weismann accorded him all due recognition.[13]

Despite these limitations, Galton is worth bearing in mind. Given his different orientation and given his statistical examination of variations in sequential populations, he did bring about an important change in the concept of heredity. He ceased to consider "inheritance" and variations as competing forces, as Lucas, Darwin, Spencer, and Haeckel had done; rather, he thought of heredity as a description of ancestral and parental contributions to a given population of offspring.[14] This latter achievement, not the similes of continuity, propelled Galton into the revolution of heredity theory that was to come. He accomplished at the statistical and phenotypic level what Weismann and others were to fashion at the germinal and microscopic domain.

Julius Sachs

Considering the severe criticism that Weismann was to receive from botanists for his idea of intra-organismic continuity, it comes as a paradox to find Julius Sachs claiming priority for the notion. Two years Weismann's senior, Sachs was a distinguished professor of botany at Würzburg and had led his profession into the era of physiological experimentation with his work on various types of plant tropisms. He was also the author of several important textbooks and compendia on botany. The last of these, his *Vorlesungen über Pflanzenphysiologie* of 1882, contained a suggestion of continuity. When we examine the relevant passage, it turns out to be a digression in a lecture on fertilization and consists of a single lengthy paragraph in which Sachs described a "continuity of the embryonic substance."[15] What he had in mind is not altogether clear. Sachs likened the substance of the fertilized egg to the vegetative points, that is, the meristematic tissues, which continue to produce the differentiated structures in growing plants. Sachs claimed that the embryonic substance, wherever it might be, was derived directly from the similar substance of the fertilized egg and that this in turn came from the gametes of the previous generation.

At first blush, the notion appears to have features in common with the idioplasm described by Nägeli in 1884 and the germ-plasm pictured by Weismann in 1885. There was a lack of specifics to be sure, but at minimum Sachs had postulated a sequestration and persistence of an undifferentiated substance throughout the life of the plant. When he first read the passage, Weismann also saw the similarity, but upon reflection he argued against Sach's claim, and he later omitted Sachs from the historical sketch of the subject he wrote for *Das Keimplasma*. His objection considered three points, two of them closely related.[16] First, Sachs might have described a continuity between the fertilized ovum and the "embryonic substance" of various vegetative points of the plant, but this was a trivial sense of continuity. One could claim, after all, that a similar continuity existed between egg and all adult structures. Such would be, indeed, a consequence of Remak and Virchow's cell theory. Continuity in Weismann's sense implied more than a material descent of cells. Second, because botanists believed at the same time that the fate of most meristems was already predetermined to produce specific structures, such as leaves, thorns, or roots, they must consist of altered or differentiated material. The embryonic substance of the vegetative points could not be identical with the embryonic substance of the fertilized egg. Such an identity comprised for Weismann an essential feature of his own germ-plasm theory. Finally, Sachs believed that the embryonic substance of the male and female gametes must be different in kind. As argued later, this was contrary to the essentials of Weismann's vision. As far as can be told, Sachs did not defend his claim of priority after Weismann's critique. The fact that he made the claim at all suggests that confusion and difficulties surrounded the intra-organismic continuity concept.

August Rauber

Perhaps the most interesting minor statement of the intra-organismic continuity concept was put forth by the anatomist August Rauber. Rauber was a maverick on the German academic scene. Raised in the conservative and Catholic Bavarian Pfalz, he studied in Munich and in Vienna where he habilitated in 1870. After serving as a battalion field surgeon during the Franco-Prussian War, Rauber became an assistant to the anatomist and embryologist Wilhelm His in Basel and followed his chief when the latter became professor of anatomy at Leipzig in 1873. Rauber's research interests paralleled His's. He pursued gross anatomical and histological studies and performed detailed examinations on the embryology of the chick. As with His, he also had a deep concern for the mechanics of development, but unlike His, Rauber insisted that his anatomical and embryological research be placed in an evolutionary framework. His's work was known for its hostility to Haeckel's brand of morphology and evolutionary speculations. By contrast, Rauber "lebt in Ernst Haeckels und Gegenbaurs Gedankenwelt" ("lived in Ernst Haeckel's and Gegenbaur's world of ideas").[17]

By 1875, the methodological and philosophical tensions between the professor and his demonstrator were so great that Rauber resigned his position to become a private instructor with rank of Außerordinarius. For the next eleven years, Rauber struggled on with his investigations in a laboratory he outfitted in his home. Under the circumstances, we must consider his accomplishments astonishing. During the period, Rauber published over fifty papers and monographs, some of major scientific significance, and a two-volume study, largely of a derivative nature, on the prehistory of mankind from an evolutionary perspective. Despite his undeniably valuable anatomical and embryological work, Rauber was passed over for more conventional candidates as openings appeared at anatomical institutes. Only in 1886 did he finally secure a university position and his own institute, but it was at the cost of leaving Germany and living in Dorpat, Estonia, which under Tzars Alexander III and Nicholas II was being subjected to an increasingly intense Russification campaign. Despite the prevailing political winds, Rauber maintained a vigorous research program; he was a popular university teacher even toward the end of his career, when the auditorium of Russian students could barely understand his German lectures, and he reedited and later rewrote a major textbook on human anatomy.[18] Rauber, living through most of World War I in Estonia, died a month before the outbreak of the Russian revolution.

Rauber's first pronouncement of a concept of intra-organismic continuity appeared in a lengthy embryological monograph concerned in a general way with the construction and alteration of form.[19] In its nearly 150 pages, Rauber presented new observations on developmental deformities in fish, but his concern focused primarily on understanding the movement of cell masses and the commitment of tissues during organ formation. The work represented his major discussion of the impact of growth patterns on form and what he called "cellular mechanics." Within this context, Rauber portrayed the organism as a composite of two contemporaneous "generations":

The personal and germinal parts [*Personaltheilen* and *Germinaltheilen*] proceed from a fertilized egg. . . . Only a part of the entire material of the egg achieves by means of an extraordinary development the embodiment of a person in the narrow sense: the personal part of the egg. The germinal part of the egg in contrast does not achieve such a broad development; its individual components remain more in a primitive condition; all the individual components of the germinal parts also remain among themselves equivalent, while the components of the personal part experience as a group the highest differentiation.[20]

Rauber clearly intended both a separation of a germinal line and its inactivity during the development of the organism. His distinction of personal and germinal parts paralleled Weismann's distinction between the soma and germplasm. Despite this separation of his two "generations," however, Rauber did not balk at accepting the inheritance of acquired characters.[21] Only after Weismann and Nussbaum had left their mark on the public consciousness did Rauber assert his priority. His was far from a tactful reminder. He sarcastically suggested that any scholar interested in heredity should not have missed Rauber's comments of 1880.[22]

Where Weismann treated Sachs with kid gloves and would later go out of his way to commend Galton, he had no such patience with a long-unaffiliated researcher who had just a few months earlier secured a foothold at the margins of German academia. Weismann conceded that Rauber had envisioned a continuity in the same sense as he had, but then in an uncharacteristic taunting reproach, six pages in length, he skillfully destroyed Rauber's assertion that the latter's 1880 monograph had made intra-organismic continuity obvious to all who cared. Even Rauber, he noted, had failed to mention his innovation to heredity theory when he himself reviewed the critical section of his own monograph for a well-known Jahresbericht (annual report)![23] Weismann's sarcastic defense, however, served a deeper purpose than ridicule. It again gave him the opportunity to insist that scientific ideas had little value if they were not developed. It allowed him to point out that "new views over the foundations of heredity" could no longer be buried in a section devoted to "the creation of form and cellular mechanics." The implications were clear. By the 1880s, an innovative hereditary theory necessitated rigorous demonstration and extensive discussion.[24]

Intragenerational Continuity of Cellular Material: Moritz Nussbaum and the Continuity of Germ Cells

Of all of Weismann's contemporaries who independently developed a theory of a segregated germinal line, Moritz Nussbaum deserves to be nominated at least as cofounder of the concept of intragenerational continuity. Like Weismann, Nussbaum was a trained embryologist and microscopist. Unlike most other inventors of the concept, Nussbaum resented the later identification of continuity with Weismann's name, and he repeatedly reminded his colleagues and readers that his own version of continuity had appeared in print prior to and possessed certain advantages over Weismann's.

Born in Hoerde in Westphalia on November 18, 1850, Nussbaum belonged to the first generation of German Jews who were entering the universities in large numbers. Nothing is known about his childhood or his decision to matriculate in the medical faculty. With the exception of two semesters as a beginning medical student at Marburg, Nussbaum's entire academic career was associated with the Anatomical Institute at the University of Bonn. His migration from Hesse-Nassau to the Rhineland came at a time when the fortunes of the Anatomical Institute in Bonn had climbed to an apogee. Max Schultze had been its director since 1859 and along with the physiologist Eduard Pflüger had become pillars of the medical faculty. In order to meet calls from the University of Leipzig and the newly founded German University in Strassburg, the Prussian Ministry of Culture had given Schultze an elaborate new building, which upon its completion in 1872 represented the grandest anatomical structure in the empire. Done in the opulent style of an Italian palace, the building boasted a large, though impractical, vestibule, corridors of 3.5 meters in breadth, a large lecture hall on the second floor under the central dome, workroom ceilings five meters above the floor, and space for the weekly comings and goings of 1,200 students. If the building was not efficient by modern standards, it expressed the confidence of the university in its professor of anatomy and attracted national attention. It was to this modern monument to the advancement of the life sciences that younger scholars, among them the Hertwig brothers, made pilgrimages of a semester or two to work under Schultze's tutelage. Nussbaum, too, coming to Bonn in 1871, could anticipate exciting research opportunities in association with one of Germany's leading microscopists.

But the winds of fortune vacillate. In 1874, Schultze, who stood at the height of his powers, died suddenly of a perforated duodenal ulcer. The issue of finding a successor stretched out beyond a twelve-month period; finally, with the urgings of Pflüger, the cultural minister chose as Schultze's prosector Adolph Johannes Hubert Freiherr von la Valette St. George. The Lehrstuhl, which had been associated with Johannes Müller and occupied briefly by Hermann von Helmholtz and for fifteen years by Schultze, now fell to an easygoing Rhinelander with a flair for entertaining lectures.[25] La Valette had performed some promising anatomical work on spermatogenesis in the mid-1860s, on the basis of which he had become an außerordentlicher Professor. Since that time, however, he had not been a highly productive scientist. Johannes Sobotta, who in 1918 became director of the same institute, wrote critically of his predecessor's advancement. Nussbaum, who succeeded La Valette in 1906, was somewhat kinder in the official university obituary of his former chief, but even then he found it difficult to explain away the paucity of La Valette's scientific accomplishments.[26]

The cultural ministry must have had some misgivings from the outset, for they had taken the unprecedented step of appointing Franz Leydig as second Ordinarius for comparative anatomy and as codirector of the Anatomical Institute. Leydig was a first-rate microscopist and the author of an important textbook in histology. He had, in fact, been one of La Valette's teachers in Würzburg. The arrangement in Bonn, however, was not a successful one. Leydig voluntarily retired to his native Rothenburg ob der Taube at the age of sixty-five because of poor health, leaving the Anatomical Institute completely in La Valette's hands. Schultze's magnificent new building metamorphosed in accord with La Valette's personal interests into a fishery station.

Under these circumstances, a career tied to the Anatomical Institute in Bonn could not have been propitious. It would be nice to know whether Nussbaum made any serious efforts to extricate himself from the Rhineland university. That he was Jewish did not enhance his mobility, but there are also some suggestions that he was too easygoing to push vigorously for his own betterment. The famous American embryologist Ross G. Harrison came to know and admire Nussbaum when in 1892–1893 he concentrated his German Wanderjahr in Bonn. It was Nussbaum who taught him some of the most advanced microscopical and experimental techniques, and it was at the Anatomical Institute that he availed himself of La Valette's fish-breeding program to focus on the embryology of teleosts. After Nussbaum's death in 1916, Harrison began a short sketch of his friend's life. Perhaps inadvertently, he captured the predicament that a German academic Jew must have found himself during the Wilhelmine period: to be condemned as aggressive for pushing his career, then consigned to obscurity for not doing so.[27]

Nussbaum's career slowly advanced despite the circumstances. He received his doctorate in 1874 and habilitated the following year. He was promoted to außerordentlicher Professor in 1881 and within months became La Valette's prosector. As beholden as he may have been to the director, there are some indications that Nussbaum considered himself to be more a disciple of Leydig, an intellectual association that accords better with Nussbaum's scientific style and accomplishments. In 1888, he became curator of the anatomical collection. Upon La Valette's retirement in 1906, the disenchanted Berlin ministry invited the Greiswald anatomist Robert Bonnet to revitalize the institute. A feeling prevailed, nevertheless, that Nussbaum deserved recognition for his anatomical achievements. He was elevated to a personal Ordinarius and made the leader of the biological laboratory within the institute. In 1911, he was given the honorary title of "Geheimer Medizinalrat." Nussbaum died while still an active teacher and researcher two days short of his sixty-fifth birthday, on November 16, 1915.

In 1880, while he was still a privatdozent and assistant nominally under La Valette, Nussbaum published a major monograph on the subject of sexual differentiation in animals.[28] Despite his later involvement with experimental morphology, this maiden voyage into the development of the reproductive system was confined to a careful histological examination of the embryogenesis of the sex glands in two groups of vertebrates: anurans (frogs) and teleosts (bony fish). It was a natural sequel to his long-standing interest in the development and structure of excretory glands, among which would naturally be numbered the ovaries and testes. The study also paralleled in part the much earlier work on spermatogenesis done by La Valette. Even though the director of the Anatomical Institute had for all intents and purposes withdrawn from anatomical research, as the heir and chief editor of Schultze's prestigious *Archiv für Mikroskopische Anatomie,* La Valette could provide a vehicle for the ready publication of Nussbaum's 120-page discussion. The work represented a substantial monograph, which unquestionably contributed to Nussbaum's promotion to außerordentlicher Professor the following year and established his claim as an independent inventor of the notion of intragenerational continuity of germinal material.

The observations that Nussbaum presented in the main body or descriptive portion of "Differenzirung" joined the pool of information about the embryology of the germ cells that had been accumulating since Waldeyer's and Semper's earlier studies. Microscopically derived, histological in focus, and comparative in its quest with more than one major animal type forming the substantive base, Nussbaum's report was an accomplished though hardly stunning exercise in descriptive embryology. The proof of its value lay in the fact that its observations continued to be cited for the next two decades of discussions on the genesis of vertebrate germ cells.

The innovative dimension to Nussbaum's monograph lay in the general interpretations presented by its author in the sixth and closing section. They were to push Nussbaum further along the path that had been opened by Semper and others. In the first place, Nussbaum took issue with earlier anatomists and physiologists, such as Gegenbaur, Pflüger, La Valette, and we might add van Beneden, who in one way or another had deviated from the original teachings of Schwann and had considered the egg as a fusion of cells rather than as a single cell. In contrast, Goette's depiction of the egg of salamanders provided Nussbaum support for maintaining that the unfertilized egg was, as Schwann had originally claimed, a cell. He considered the spermatozoon, as had Kölliker forty years earlier and La Valette had recently confirmed, to be the direct modification of a cell as well. Second, Nussbaum insisted that he went beyond Semper in viewing the egg and the spermatozoon as originating from homologous and indifferent epithelial cells. Like a weather vane pointing toward the changing winds, Nussbaum directed attention to the significance of this second conclusion. The sexually differentiated state must now be considered a secondary feature. "One will therefore not conceive of the sexes as something different, not [conceive] of their rise as the expression of a previously given but latent opposition that does not appear in the phenomena."[29] Whereas Semper had only hinted at the consequence of the sexually indifferent Ureier, Nussbaum explicitly joined ranks with those who since the 1840s had focused on the primacy of asexual reproduction in a growing spectrum of reproductive modes.

Finally, if male and female germ cells were derived from indifferent homologous cells, Nussbaum reasoned that sexual fertilization must be simply an evolutionary step along the road of specialization beyond protozoan conjugation. The ontogenetic consequence of this phylogenetic transition, he figured, dictated that the indifferent primordial germ cells—he might have said "Urkeimzellen"—must segregate from the rest of the differentiation process. "Is it developmentally probable," he queried, "to expect that the [germ] cells . . . would be separated out of the fertilized egg for the purpose of preserving the species before any histological differentiation occurs and would remain uninvolved during the marked division of labor during germ-layer formation?"[30] His examination of sexual differentiation in frogs and bony fish provided Nussbaum with the expected affirmative answer to his question.

From this point on, Nussbaum's demonstration became both comparative and highly speculative. He used the conventional economic and embryological principle of the division of labor to argue that throughout the animal kingdom the germ cells, which he reasoned preserved the species, segregated from the cells of the differentiating embryo during early cleavage. "Both groups of cells and their offspring multiply although completely independently of one another, so that the sex cells play no role in the construction of the tissues of the individual, and not a single sperm or egg cell arises out of the cellular material of the individual."[31]

Nussbaum had arrived at a continuity theory, complete with the notion of a segregation of a line of unique cells that were both necessary and sufficient for the production of the gametes. He was able to do so because he carried with him much of the conventional baggage of the holistic view of heredity, including a belief in the primacy of the asexual model and in the division of labor metaphor. His pursuit of the development of sexual structures back to the point where he could speak of indifferent reproductive cells, that is, Semper's Ureier, gave Nussbaum the structure for his continuity model. The second component of this formula followed very naturally from the first. Nussbaum took the additional step by insisting upon the "absolute independence of the sex cells from any one of the three germ layers." In a phrase that he invoked on many an occasion, he explained that "after the segregation of the sex cells the domains [*Conti*] of the individual and the species are fully separated."[32]

Although he was not successful at demonstrating continuity of a germinal line back to early cleavage with either his frogs or fish, Nussbaum's discussion and embryological demonstrations gave momentum to the general belief that sex determination was the product of an embryological process rather that an act of fertilization. The last two decades of the century were to provide many embryological theories of sex determination.[33] One subject that the reader, however, will not find addressed in Nussbaum's monograph is the subject of what was transmitted in the germinal line. Only in passing did he indicate that factors influencing characters during development might equally influence the germ cells so as to account for the inheritance of acquired characters.[34]

In 1884, Nussbaum returned to the subject of continuity of germ cells.[35] By this time, Weismann also had made public his beliefs founded on his examination of hydromedusae. Nussbaum's topic was again on the early development of the sex cells, and he included an examination of *Ascaris megalocephala,* the organism so skillfully being exploited by van Beneden. This roundworm provided another example of the appearance of primordial germ cells prior to the sexual differentiation of the gonads. The microscopical study, however, lay a long way from providing empirical evidence that there existed a continuity from zygote to germ cells. Nevertheless, Nussbaum believed that *Ascaris* and his earlier studies on fish and frogs provided satisfactory, though circumstantial, evidence for the continuity he so badly wished to document.[36]

The difficulty of producing conclusive embryological evidence did not cause Nussbaum's ardor for the continuity of germ cells to wane. In 1884, he repeated in a clearer fashion the same deductive arguments that he had employed in 1880. Because differentiation could only be construed as an instance of the division of labor, following which time cells became committed to specialized tasks, and because the primordial germ cells appeared unspecialized and sexually indifferent and could be identified prior to the appearance of the presumptive gonads, it seemed to follow as a matter of course that a continuous germ cell track must exist prior to the point of germ-layer differentiation. As though to underscore his point, Nussbaum resorted to a metaphor for describing segregation that in a somewhat different form had already been used by Galton and was to be employed by Weismann. "The sex cells in higher animals," Nussbaum imagined, "represent the continuing foundation [*Grundstock*] of the species, from which the single individuals after a brief existence, fall off like withered leaves from a tree."[37]

One new argument bolstered Nussbaum's mixture of deductive claims. It rested on the combined demonstrations by Hertwig and Fol that fertilization consisted of the penetration of the spermatozoa into the egg and of the union of the "pronuclei" of each gamete. Here it is important to recognize that in turning his microscope on the same events in *Ascaris,* Nussbaum found in fertilization a confirmation for his belief in the equivalency of the gametes. "Fertilization is a copulation of two homologous cells, whose equivalent parts . . . fuse with one another," he insisted.[38] We might choose to select modern implications in this comment, but for Nussbaum the fusion of pronuclei represented an additional proof against a preferred status for either sex and a rejection of the uniqueness of sexual reproduction altogether. We have already seen his argument that homologous primordial germ cells implied that fertilization was simply a more specialized version of conjugation and that that, in turn, was simply a step in the specialization process above asexual division. Despite this unfamiliar emphasis, there can be no doubt that by 1884 Nussbaum had become actively engaged in one aspect of the new cytology that was revolutionizing biology.

The title and contents of his second paper explicitly revealed an interest in heredity as a transmission process. "Vererbung" by 1884 had become the new buzzword among leading German biologists. It remained an open question how far Nussbaum would run with the new events. For the moment, he considered the separation of the domains of the individual and the species as primarily a mechanism for preserving the constancy of the latter during the development of the former.[39] His commitment to continuity wavered closer to the level of a holistic concern than an architectonic description, and its rationalization remained the enduring feature of Nussbaum's heredity theory.[40]

"New Heredity"

There can be no doubt that the next three biologists combined both a conviction that a continuity of hereditary elements was a necessity during ontogeny and a belief that nuclear cytology provided a key to understanding the structure of those elements. That their solutions to both the intergenerational and intragenerational aspects of heredity varied from one another and from Weismann's contemporary solution suggests all of their discussions between 1884 and 1885 took place at a point in time when theory building and empirical research made fruitful but not definitive contact.

Eduard Strasburger

When we reviewed the contributions of Eduard Strasburger to nuclear cytology in Chapter 11, we followed his work past the third edition of his *Zellbildung und Zelltheilung* to 1882. By that time, he had accepted Flemming's basic claims about indirect nuclear division in both plants and animals. Strasburger serves as a bellwether indicating the rapid acceptance of many of Flemming's assertions, yet as he then made clear in his next publication, *Die Controversen der indirecten Kerntheilung,* there existed many unresolved details, such as the plausible cytoplasmic origin of the spindle rays, the relationship between the nucleolus and the segmented network of chromatic particles at early prophase, and the point in division when the nuclear spindle and the longitudinal splitting first occurred (Strasburger called these "Mikrosomen").[41] Where Flemming had at first maintained that the split threads had fused again, he later imagined that the result of longitudinal splitting was to guarantee an equal distribution of each thread to the daughter cells. Ever cautious, however, he resisted calling it a fact: "In the meantime, this has neither been proved, nor refuted."[42] Two years later, Strasburger had the benefit of preparations, a subsequent publication by his friend Emil Heuser, and new studies by the Parisian botanist and cytologist Léon Guignard, all of which revealed beyond a doubt that the pair of longitudinal strands were distributed in division to opposite poles of the incipient daughter cells. This demonstration, Strasburger claimed, was the reason for writing his first paper of 1884.[43] There he renewed examinations of many of his favorite botanical species but more importantly closed with a two-page endorsement of Wilhelm Roux's recent suggestion that a distribution of chromatin threads was the only logical way to achieve equal distribution of qualities to the next generation of cells.[44]

Strasburger's introduction to the realm of heredity theory followed later that year.[45] The title and foreword to his extensive monograph promised to apply the author's broad experience in the cytology of fertilization in higher plants to a theory of reproduction. New staining techniques in general and the new facts concerning the pollen tube of angiosperms prompted him to follow in more detail the progress of the vegetative and generative nuclei of pollen grains to the latter's union with the nucleus of the ovule. This waited for the event in the ovary at the far end of the micropyle. The technical anatomy of this journey alone is worth emphasizing if only to give a vivid sense of the intergenerational continuity involved.[46] At this point, Strasburger achieved for higher plants what Hertwig and Fol had achieved for animals. He went beyond his predecessors, however, and claimed the intragenerational continuity of the nuclear filaments. "Should, as I believe it must, the nuclear filament of every cell nucleus remain preserved in the reticulum of the nucleus, then prior to every cell division it will condense in a similar fashion, so that the nuclear filaments of all the following nuclear generations will contain approximately similar pieces of the nuclear filaments from the father and mother."[47]

Thus, with empirical reasons to believe in both the intergenerational and intragenerational continuity of nuclear structure, Strasburger offered a theory of heredity—he called it variously a theory of reproduction ("Theorie der Zeugung") and a theory of fertilization ("Theorie der Befruchtung"), terms that in the newer context somewhat missed the mark. Nevertheless, Strasburger examined many of the issues that had or soon would become associated with heredity or "Vererbung."

Starting off with three demonstrable claims about phanerogams, Strasburger insisted that (1) fertilization is the union of sperm nucleus with egg nucleus, (2) the cytoplasm is not involved in the fertilization process, and (3) both sperm and egg nuclei possess the status of true cell nuclei.[48] He drew upon recent publications by Nussbaum, van Beneden, and Weismann as foils for his own concept of the hereditary material involved. He reified Nägeli's notion of the idioplasm (a term Strasburger did not hesitate to employ) by describing what he actually could see through the microscope. Thus, the nuclear threads, he argued, consisted of a nearly transparent nucleo-hyaloplasm, embedded in which lay the stained nucleo-mikrosomen, other darkly stained particles, and a number of stained nucleolen.[49] All of this was surrounded by the nuclear sap, which was directly confined within the nuclear wall itself. Because this wall broke down and later reformed during nuclear division, the sap must be the product of the cytoplasm of the cell.

Such a conjectured microanatomy allowed Strasburger to explain processes as well as describe the associated structures. Ultimately, he envisioned a dynamic interaction between cytoplasm and the nuclear hyaloplasm that was mediated by the nuclear sap. This explained the processes of nuclear division, transmission, differentiation, and compensatory growth: "The cell nucleus determines the development of the cytoplasm, in that this now changes, and it delivers to the cell nucleus changed nutrition and which in turn causes its change. The advance of development depends on this reciprocal effect."[50]

There was nothing unusual in such an epigenetic concept of development. If the hereditary material, however, was specified by the reciprocal development of the cytoplasm and the nuclear chromatin, how did given cells acquire the status of original germ cells? In other words, recent nuclear cytology had forced him to address the problem of intragenerational continuity. Strasburger solved the matter by focusing on the commonplace reappearance of germ cells in the meristems of plants. In following the lead of Nägeli, he argued for a recycling of the development of certain cells back to the organism's starting point. The ultimate justification for his explanation was to be found in Haeckel's biogenetic law—at least insofar as the starting point of development of the offspring must recapitulate at least the starting point of the parents and near ancestors.[51]

Strasburger explicitly rejected an alternative solution offered by Nussbaum and Weismann that called for the isolation and continuity of a germ cell lineage. Alluding to the developmental events, Strasburger insisted that "these changes lead henceforth so that the substance of the generative cells and cell nuclei returns to that state which it possessed at the beginning of the ontogenetic development. It is to be accepted in any case that this return transformation should happen also with the cytoplasm of the generative cells, so that the necessary conditions for the corresponding nutrition for the sperm and egg nuclei must be reestablished."[52] Botanists, it appears, had a different perspective of the problem of intragenerational continuity than the two zoologists.

Within the framework of interacting cytoplasm and nuclear idioplasm and of changing nutrition and multiple segments accumulated at the equatorial plate, Strasburger found his explanations for the repeatedly often conflicting phenomena of heredity: that is, the constancy of species characters, the presumed inheritance of acquired characters, atavism, latency, the gradual diminution of ancestral traits, sexual determination, parthenogenesis, and hybridization.[53] One matter that Strasburger rejected, after some discussion, was the view held by van Beneden, Minot, and Balfour, that the maturation of the egg, terminating with the formation of polar bodies, entailed an expulsion of male material. He offered in its stead a lame explanation about the organism's need to prepare the cytoplasm of the egg for fertilization.[54]

Oscar Hertwig

Oscar Hertwig was another major nuclear cytologist who in 1884–1885 presented the world with an explicit theory of heredity.[55] I have noted in Chapter 11 that in his famous paper on fertilization in sea urchins and other papers on fertilization (all written in the mid-1870s) Hertwig did not deal with heredity, the transmission of traits. Once one reads his 1885 "Theorie der Vererbung," it is not difficult to see the logic in this historical claim. Hertwig himself recognized the disjunction between his earlier fertilization studies and a transmission theory. "According to my conception," Hertwig reflected, "the fertilization theory *when it is further pursued encompasses in addition a theory of Vererbung.*"[56] It is the passage from a fertilization theory to a theory of heredity that highlights the importance of the events described in this chapter.

Hertwig depended on very recent work by contemporaries to drive home three points. First, the substance of the nucleus was, in his opinion, unequivocally the material of fertilization. Nussbaum's and van Beneden's work on nematodes emphasized that penetration alone by the sperm nucleus was not sufficient to initiate cleavage. Only after the full maturation of the egg, including polar body expulsion, did the two pronuclei of these organisms approach each other and fuse. Material fusion of the nuclei not the physiological action of penetration was key.[57] Second, relying on Nägeli's discussion of the equality in the contribution of both parents to the next generation and on van Beneden's observation that the chromosomal number of the two pronuclei in *Ascaris* was the same, Hertwig insisted there must exist an equivalence in the contribution of the male and female gametes.[58] He also relied heavily on Nägeli's elaborate theoretical discussion and on recent cytological work by Flemming, Strasburger, van Beneden, and others in accepting a structural or architectonic concept of the hereditary material.[59] Third, Hertwig was deeply concerned about the relationship between the nucleus and the protoplasm. Again, he disagreed with Nägeli with the conjecture that the idioplasm was an extracellular structure. On the other hand, he found Roux's hypothesis that the mechanism of indirect cell division ensured the equal distribution of nuclear qualities to the two daughter cells to be particularly noteworthy.[60]

By placing his own work in the context of the recent half-decade of research, Hertwig came solidly down with the growing consensus that there was an "indissoluble" hereditary substance in the nucleus. The following excerpt, interrupted by too many dependent clauses and irritatingly tentative with its visions of processes, nevertheless hammers home the point, in part by virtue of being set throughout in large type:

The maternal and paternal organization will be transmitted through the reproductive act to the offspring by means of substances, which are themselves organized, that means, which possess a very complicated molecular structure in the sense of Nägeli. In the development of an organic chain there takes place no spontaneous generation [*Urzeugungen*], nor will this chain be broken by means of disorganized conditions, from which organization must arise again as in the case of spontaneous generation. In the sequence of individuals there take place only changes in organization, in their most inner nature and for us, to be sure, in an incomprehensible way, whereby forces in a law-like rhythm unfold and new forces of tension accumulate. We may consider the nuclei as the sites [*Anlagen*] of complicated molecular structure, which transmit the maternal and paternal characters, which appear in the sexual products as the single parts equivalent with one another, with which we observe in the fertilization act the only extraordinarily meaningful processes, and from which we alone may obtain the proof that the impulse for development emanates from them. During the development and maturation of the sexual products, as well as with the copulation of the same, *the substance of the male and female nuclei, as searching observations teach us, never undergo a dissolution but only transformations in their form to become the nucleus of the egg and sperm,* the one stemming from the germinal vesicle, the other from the nucleus of the sperm mother cell.[61]

It is clear that Hertwig envisioned a continuous intracellular and intergenerational "organic chain" of organized material. It was a chain that could be seen but hardly detailed in the nuclein of the nucleus. Furthermore, Hertwig felt that this chain underwent transformations between fertilization and gamete production. In other words, the substance had structure that changed during the organism's development but did not lose its fundamental form. He adopted a mode of continuity, inspired by but unlike Nägeli's idioplasm. Hertwig's continuous substance instead lay in the nucleus. He denied dissolution of its organization and a formation *de novo* of the nuclei of the egg and sperm cells. How he envisioned this transformation, he did not say. With his insistence on a change in the architectonics of the material, it is not surprising that he rejected Nussbaum's and Weismann's notion of a continuous lineage of germ cells isolated from the changing soma of the individual.

Hertwig closed with a chapter on the relationship between protoplasm and the nucleus.[62] If the nucleus still held mysteries, the interaction between the nucleus and protoplasm, and between nuclear division and cell division held even more. Hertwig, like Strasburger and Nägeli, recognized that the protoplasm must be the source of nutrients for the nucleus, but how did the protoplasm participate in the process of cell division? Hertwig was no stranger to this problem. In May 1884, he had discussed his evaluation of a contemporary work on the physical effects of gravity on the primary axis of the embryo. It was a research project stimulated by the claim of the physiologist Eduard Pflüger that gravity determined the first axis of the egg.[63] It is interesting, during the ensuing exchange of papers, how easily Hertwig could go from a strictly embryological problem to a nascent heredity issue. Hertwig was not the first nor last in German zoology to pursue the two together.[64]

Albert von Kölliker

If Strasburger, Hertwig, and Weismann confessed the influence of Nägeli on their own thinking about heredity, Albert von Kölliker's debt appears even greater. His association with Nägeli extended at least back to 1836 when the two began their university studies. Both were native Swiss, born within a month of each other in the canton of Zurich. Both attended the same gymnasium and then matriculated in the medical faculty at the University of Zurich, where they became fast friends as they studied and collected plants together under the tutelage of Oswald Heer.[65] When in 1839 Nägeli traveled to Geneva to further his botanical studies, Kölliker pursued anatomical studies first in Bonn under Müller and then with Jacob Henle in Berlin. During the fall vacation of 1840, the two students undertook, along with two other Swiss comrades, a collecting trip to the islands of Föhr and Helgoland. Their paths joined again in early 1841 when they made a short pilgrimage to Jena to establish connections with the new außerordentlicher Professor of botany, Matthias Schleiden. By this time Kölliker was close to completing his doctoral dissertation, but because the medical faculty in Zurich would not accept his work on the cellular origin of spermatozoa, he received his degree from the philosophical faculty. Accepting the priorities of the different faculties, Kölliker also submitted a medical dissertation on early development of two dipterans.[66] At the close of the 1841–1842 winter semester he and Nägeli traveled to Naples and Messina to collect and study the flora and fauna of the Mediterranean.[67]

By then both had gained their first appointments. Nägeli worked with Schleiden for a year and a half, and the two academic botanists soon thereafter founded and published together in 1844–1846 the short-lived *Zeitschrift für Wissenschaftliche Botanik*. In the meantime, Kölliker became Henle's assistant and then prosector when the latter acquired a position in Zurich. Once established as Ordinarius for Physiology and Comparative Anatomy in Würzburg, Kölliker joined forces with Theodor Siebold to found the more successful *Zeitschrift für wissenschaftliche Zoologie* in 1849. Thus, not only were Kölliker and Nägeli close friends, but they had achieved scientific standing at a time when the cell theory appeared on the scene and when their mentors, Müller, Henle, and Remak for Kölliker, and Schleiden for Nägeli, had directed their research in the new morphological directions opening up. As the titles of their new journals implied, both saw the need to raise their respective specialty above the mere collection of observations to render it more "scientific" by subsuming it under the basic mechanistic principles offered by the cell theory.[68]

As his basic textbook on human and other vertebrate species testifies, Kölliker had only a peripheral interest in the significance of nuclear cytology until 1884. Then he added a special note to the second edition of his *Grundriss* that emphasized the importance of the "karyokinetic" process, the hermaphroditic nature of the first and subsequent cleavage nuclei, and the significance of indirect cell division for the process of heredity. There is even a suggestion that he adopted momentarily a version of van Beneden's ontological meaning of hermaphroditism.[69]

The full impact of the new situation became clearer the next year, by which time Kölliker had assimilated the implications of the idioplasm described in Nägeli's *Abstammungslehre*. In Kölliker's mind his long-time friend was the first to present a clear and empirically based hypothesis about the substance of heredity. In a short report to the Physikalisch-medizinische Gesellschaft in Würzburg (a society he had helped found thirty-seven years earlier), Kölliker began developing a theory of his own.[70] Although cautious at the outset, he presented a new understanding that amplified and significantly altered Nägeli's sketch of the idioplasm. First, should it turn out that the nucleus is the mediator of reproduction and transmits the idioplasm to the offspring, it would be "simpler" to consider this "the single formative factor" rather than consider any contribution of the nutritive plasma. Second, he thought the idioplasm, as both a chemical and morphological structure, needed to be identified with the chromatic material in all nuclei. He never understood it to be an integral part of the whole cell.

Kölliker also expected that the process of heredity could only be grasped through reproduction; that is, the intergenerational transmission of the idioplasm explained the form of the offspring. The idioplasm was to be found in the germinal vesicle of the egg and in the spermatozoa, each of which had the value of nuclei and manifested itself in the chemical and morphological nature of the observable chromatin. The first nucleus of the new organism, he continued, consisted of the union of these male and female nuclear structures and was to be seen, so to speak, as a hermaphroditic structure. He meant by this that it bore both paternal and maternal characters. Addressing development, Kölliker asserted more confidently that the adult body was derived "in an unbroken sequence" from this first embryonic nucleus. Through the activity of the smallest particles, the nuclei thus brought about the reproduction of cells, their growth, and their qualities.[71] Finally, Kölliker concluded in an emphatic manner, "the typical formations of the organs and the entire organism are the result of specific combinations of cell divisions and processes of cell growth and thus to the extent their typical, propagation preserving forces allow, the nuclei control the entire formative process of the organism or the heredity."[72] Despite the fact that cellular growth and propagation were essential aspects of transmission, Kölliker had moved far beyond heredity through a simple overgrowth. Nuclear control and nuclein structures comprised his mechanism for heredity.

Kölliker expanded upon this theme in far greater detail later in the year.[73] By then he was confident that the nucleus had become central to understanding heredity. His recognition of his debt to van Beneden, Strasburger, Hertwig, and above all Nägeli was unequivocal. When considering the maturation of the gametes at the other end of the organism's cycle, Kölliker argued that germ cells were derived from embryonic cells that had experienced a minimum of differentiation and resided in many organs and tissues in both animals and plants. Thus, by claiming that an early embryonic sequestration from the maelstrom of embryonic differentiation was unnecessary, he distinguished his views from those of Weismann and Nussbaum. For Kölliker, the intragenerational continuity was to be found in the undifferentiated nuclei and chromatin of all developing and mature structures rather than in a designated line of isolated germ cells or germ-plasm.

The reader readily senses that in the background to Kölliker's opinions lay the experience of a seasoned microscopical anatomist and pathologist who had spent a lifetime teasing apart and identifying tissue and cell types on a grand scale. According to Kölliker, Weismann's examination of the migration of germ cells in hydromedusae, as "beautiful" as the study had been, demonstrated only an intergenerational migration not a sequestration of a germinal line. Speaking for himself, Kölliker insisted that "male and female germ cells are consequently for me simply cells of embryonic character, which for this purpose have acquired special characters."[74]

After Weismann hypothesized that an isolation of stable germ-plasm, rather than germ cells, occurred in the nucleus and that only the somatoplasm changed during development, Kölliker pointed out that his Freiburg colleague had produced no evidence that embryonic differentiation affected a deep-rooted change in the germ-plasm of the somatic cells. As Kölliker had repeatedly insisted, differentiation was easily understood as the consequence of different degrees of cell division, contrasting division rates, and differences in growth patterns of the tissues. These, after all, could be seen under the microscope and could be, Kölliker felt, ultimately explained in terms of "molecular forces and internal causes."[75]

Although Kölliker did not accept change in the idioplasm during development, he followed his friend Nägeli by intimating that "internal forces" ensured such changes in phylogeny. In retrospect, Weismann appeared to be altering the ground rules. Kölliker gladly accepted playing the role of an anatomist who had not been deeply involved with the theory of descent, but he felt justified in objecting to Weismann's recent claim that variations, introduced through sexual reproduction, altered the idioplasm at each generation and might transform a land animal into a whale. "So long as the followers of

Darwin's theory of descent do not comprehend that the first organism arose from internal causes and that internal causes brought about their further development . . . the chasm between the two positions will not be bridged."[76] Although Kölliker endorsed Weismann's rejection of the inheritance of acquired characters, as we will see an unbridgeable chasm was beginning to appear.[77]

Weismann and *Die Continuität des Keimplasmas* (1885)

To read Weismann's 1885 essay on the "Continuity of the Germ-Plasm" is to enter into a different world from his previous essays. No longer does one find a detailed discussion of the natural history of an organism. No longer are there discussions of life cycles, gross anatomy, the formation of eggs in a particular germ layer, or even life and death or a hypothetical phylogeny of the isolation of germ cell lineages from the soma. All of these subjects continued to be important to him, but at this moment Weismann focused on the cell nucleus and on the "loops" of chromatin material.

Only two years had passed since he had addressed his colleagues on heredity. As we have seen, Strasburger had written his *Zellbildung und Zelltheilung* in 1880 and two years later Flemming had crowned his important papers with his *Zellsubstanz, Kern und Zelltheilung.* These were enormously influential works that redefined the nature of nuclear and cell division. We have also noted that Weismann felt strongly enough about these to admonish Brandt for not being familiar with them. We have also seen that during the five-year period between 1880 and 1885, Nägeli, van Beneden, and Roux had sketched out important aspects of an idioplasmic model for the transmission process. Weismann was familiar with these major contributions too, and despite the fact that van Beneden interpreted his results within the framework of an ill-specified hermaphroditic theory of heredity material, Weismann gave him credit "for having constructed the foundation upon which a scientific theory of heredity could be built."[78]

As for fertilization, Weismann admired the works of Bütschli, Hertwig, Fol, and van Beneden. They had justifiably made much of the fusion of the two "pronuclei" in the fertilization process. Nevertheless, Weismann did not hesitate to point out that "it was only necessary to replace the terms male and female pronuclei [*Vorkernen*] by the terms nuclear substance of the male and female parents, in order to gain a starting point from which further advance became possible."[79] Weismann's ensuing critical comment reflected his personal belief that within a year of Fol's demonstration "my account of the sperm-cells of Daphnidae followed and this should have removed every doubt as to the cellular nature of the sperm-cells and as to their possession of an entirely normal nucleus, if only the authorities upon the subject had paid more attention to these statements."[80] Fertilization then comprised not a creation of a new kind of entity out of two preceding and different ones but simply an increase in bulk of the chromatic material for the offspring. This "nuclear substance" became the reality that Weismann had hoped for since he mentioned the "chemical and physical constitution" of heredity back at the time of his inaugural address as planmässiger Ordinarius in 1868.

Given this convergence of a variety of research endeavors, it is not surprising that following the 1885 Versammlung in Strassburg, Weismann took a fortnight overland trip through Metz and Luxemberg to visit the van Benedens "père et fils" in Louvain. Then, after meeting former colleagues and friends in Amsterdam, he returned to Freiburg via Bonn where he called on Strasburger. Unfortunately we have no details of his two visits.

In Weismann's mind, an understanding of heredity went far beyond an identification of the union of the nuclei of fertilization. Central to his thinking was the rise of the germ cells, the subject he had committed a decade to investigating in daphnids and hydroids. The problem, as he saw it, was how the germ cells circumvented the complexities of differentiation. How was it, he asked in 1885, that "a single cell out of the millions of diversely differentiated cells which compose the body, becomes specialized [*sondert sich*] as a sexual cell. . . . How is it that such a single cell can reproduce the *tout ensemble* of the parent with all the faithfulness of a portrait?"[81] Accordingly, Weismann was insisting that a theory of heredity also required a theory of development that went beyond simply the processes of assimilation and growth. He felt only Darwin had made a fair effort to do so in his theory of pangenesis, but in the end he had to resort to an undocumented circulation through the body and vague "elective affinities" to explain how the germ cells were anything more than a mélange of unorganized gemmules. There existed an alternative way of getting around this differentiation dilemma, a recycling of certain cells through a process of de-differentiation so that they matured into germ cells that were similar to the parental germ cells. Such a process, promoted by Strasburger, struck Weismann as highly improbable given the necessary complexity of the germ-plasm and the intricacies of the projected de-differentiation.[82]

This left him a third alternative, namely, a "continuity" from the parents' germ cells to the offspring's germ cells. Through an argument by exclusion and because recent embryological data showed that in most organisms the germ cells were not apparent until late in the cleavage process, or even beyond, Weismann returned to his position established on hydromedusae in 1883 but with a difference. Instead of a continuity of germ cells, as he and Nussbaum had previously championed, he envisioned a continuity of germ-plasm embedded in a lineage of nuclei. This resulted in a continuity of germ-plasm. Weismann failed at first to reckon with a fourth alternative, that is, differentiation of continually totipotent cells, which in fact would make all cells potential germ cells. He would not be able to overlook a variation of such a scheme once he read Kölliker's theory of embryonic cells. This he was unable to do until after he completed his own proposal.[83]

Weismann tells us that his resulting essay, "The Continuity of the Germ-Plasm," was developed in February and March 1885 as part of a lecture course on "The Theory of Descent" ("Descendenztheorie").[84] His comment provides us with a clue about its format and about the months during which he began to work out the implications of nuclear cytology. It was only the third time he had given this general course and the first time he had presented it in the winter semester. So when Gustav Fischer published Weismann's "Continuity of the Germ-Plasm" with its pages wrapped in his signature pale-orange paper covers, it had less the semblance of a unified monograph and more the structure of three separate lectures, each of which converged in its own way on a mechanical model for heredity.

The essay's full title, "The Continuity of the Germ-Plasm as the Foundation of a Theory of Heredity," promised much more. By highlighting "continuity," it declared itself to be an extension of the conclusions reached in his research on hydroids and his addresses on heredity and on life and death—but there was a difference. Weismann now wrote of germ-plasm rather than germ cells. His concerns were less on documenting the immortality of sequestered germ cells than on focusing on the role of their nuclei. In his effort to update his idea, he created a problem. It was possible to point to germ cells as they migrated and matured in the body, but it was quite another matter to specify in optical terms the germ-plasm.

The second phrase in Weismann's title, "Foundation of Heredity Theory" ("Grundlage einer Theorie der Vererbung") carried a double meaning. At first reading, it appears to be simply reinforcement to "continuity," the term that would soon become the hallmark of Weismann's biology. After going through the essay, the reader may well conclude that Weismann intended to stress the empirical nature of his discussion. It is clear that "foundations" alluded not to *a priori* first principles but to an empirical base for his discussion. Weismann demonstrated that he could build on the detailed cytological research of others. Finally, we have seen how he had enjoined Brandt to heed the material in the books of Strasburger and Flemming. Now he turned his thoughts to something materially smaller—first to the nucleus, then to the germ-plasm.

Weismann devoted the first substantive section of his essay to the importance of the nucleus and the concept and rationale for a substance called germ-plasm as a special entity. This required that he had not only digested what Nussbaum, Nägeli, and Strasburger claimed in their competing theories of germ cells, but that he had come up to speed with the discussions about indirect nuclear division pursued by Flemming and Strasburger and with the noncoalescence of parental nuclei at fertilization, first recognized by van Beneden. He understood that the outrageous experimentally based claims of the physiologist Eduard Pflüger were successfully countered by a younger generation, led by O. Hertwig, Gustav Born, and Roux.[85] They had made significant strides in demonstrating that the cell and nucleus rather than the physiological or external environment were the primary vehicles of inheritance. Weismann's own observations on daphnids and hydroids, the work at his institute performed by his brother-in-law on anucleated fragments of protozoa,[86] and the follow-up dissertations on hydroids by two of his students were also significant in this logical progression toward the conclusion that the germ cells were the result of sequestered germ-plasm of minimally differentiated cells.[87] His was not a textbook presentation but a series of discussions with his peers. It was a long argument for the transmission of a specified hereditary material through his familiar strategy of warping his theory through the reefs of observations and counter positions of the best work and theories available.

This emphasis on the nucleus did not change Weismann's attitude toward the development of the organism as a whole. After all, he was an embryologist by training and had followed his profession, which since Baer had been devoted to an epigenetic pattern of organic development. This is a point worth emphasizing because later in his career Weismann would be identified, even ridiculed, as a preformationist. Listen, however, to Weismann's elaborate metaphor about the differentiation of the idioplasm in somatic cells in the first section of his essay.

> The development of the nucleoplasm during ontogeny may be to some extent compared to an army composed of corps, which are made up of divisions, and these of brigades, and so on. The whole army may be taken

to represent the nucleoplasm of the germ-cell: the earliest cell-division (as into the first cells of the ectoderm and endoderm) may be represented by the separation of the two corps, similarly formed but with different duties: and the following cell-divisions by the successive detachment of divisions, brigades, regiments, battalions, companies, etc.; and as the groups become simpler so does their sphere of action become limited.

He then added some reservations:

It must be admitted that this metaphor is imperfect in two respects, first, because the quantity of the nucleoplasm is not diminished, but only its complexity, and secondly, because the strength of an army chiefly depends upon its numbers, not on the complexity of its constitution. And we must also guard against the supposition that unequal nuclear division simply means a separation of part of the molecular structure, like the detachment of a regiment from a brigade. *On the contrary, the molecular constitution of the mother-nucleus is certainly changed during division in such a way that one or both halves receive a new structure which did not exist before their formation.*[88]

As long as this metaphor was, it reveals a good deal about Weismann's early concept of the nucleus and the germ-plasm. It took a stab at the problem that would haunt the rest of the century: how to accommodate both change and permanence in development in a comprehensive theory of heredity. Weismann simply declared that the differentiation must be explained by "a gradual transformation of the nuclear substance," which in turn would be followed "by a gradual change in the character of the cell-bodies."[89] Unlike Nägeli, Weismann had not yet committed himself to a specific architecture for either his somatic or reproductive idioplasm. Nevertheless, like Nägeli, his argument rested on logic rather than facts: "Although in many cases the cell-bodies of such early embryonic cells fail to exhibit any visible differences, the idioplasm of their nuclei must undoubtedly differ, or else they could not develope [*sic*] in different directions."[90] The empirical evidence on the nuclear level for illuminating development was slim, but nuclear cytology had opened a window on a new dimension. For seven years, Weismann was to think in these epigenetic terms as he struggled, alongside his many biological colleagues, to explain development, heredity, and, of course, species evolution—all at the same time.

   It was logical that Weismann turned, albeit in a much shorter segment, to the polar bodies. At the time, their morphology, and hence function, was far more ambiguous than the events of fertilization. The polar bodies were clearly connected with the maturation of the egg. With the exception of a few problematic reports, however, they had not been identified with the production of spermatozoa. It was also unclear whether two, three, or even four polar bodies were involved in ovogenesis. Some zoologists were not even certain whether the polar bodies were basically cells or simply nuclei. Bütschli had argued that they represented a reduction in the mass of idioplasm in preparation for fertilization. Beneden and others had considered their expulsion as an elimination of the male component of a hermaphroditic nucleus. Still others spoke of rejuvenation of the nucleus or of equalization in nuclear mass of the female with the male nucleus. Weismann's explanation fitted elegantly into his own evolving theory of heredity. In its development, he explained, the egg required ovogenetic idioplasm in order to develop into a mature structure. Once formed, the egg could rid itself of this strictly histogenetic material. Unfortunately his explanation assumed that in time polar bodies would, for the same reason, be found in association with the production of spermatozoa as well.[91]

Claiming that his understanding of polar bodies also led him to a better understanding of fertilization, Weismann devoted the final section of his tripartite monograph to parthenogenesis. The subject, barely mentioned in modern textbooks, was an important issue throughout the second half of the nineteenth century. As we have already seen, Richard Owen coined the term in 1848; by the late 1850s, it had been stripped of Owen's hypothetical forces and incorporated into a spectrum of reproductive mechanisms, where it served a generation of biologists as the imaginary link between eggs needing fertilization and "pseudo-ova." Leuckart had described the changing understanding of the egg and sperm in a study of alternation of generations and parthenogenesis. "The history of generation," he pointed out to a male-dominated era, "shows in a clear way how the role which the sperm plays in fertilization has become even more restricted by the advance of science. Initially the sperm was seen as a young germ, which used the egg merely as a cradle and the yolk for sustenance; later it was seen as an element which had to a certain extent an equal status with the egg, which joined the egg and only with the union produced the germ; now the egg has become the germ even though something more is perhaps needed for development."[92]

The cytological discoveries of Hertwig and Fol put to rest the suggested diminutive role for the sperm nucleus, but neither one had explained the phenomenon of parthenogenesis. At the time it was well known that the eggs of some lepidopterans, other insects, and even some vertebrates could develop completely or partly without being fertilized. Bee eggs, studied intensively by both Siebold and Leuckart, revealed a situation where identical eggs could fully develop either sexually or parthenogenetically. For van Beneden, Minot, and Balfour's proposals that polar bodies signified the expulsion of sexually specified material that then was replenished by the fertilization process, parthenogenesis clearly presented an "inconvenient truth." With the case of *Moina,* the daphnid that laid sexual "winter eggs" and parthenogenetic "summer eggs," Weismann had earlier resorted to a utilitarian evolutionary explanation for this alternation of generations.[93] By 1885, he concluded for a variety of reasons that the chromatic material was the essential factor in fertilization, and its quantity might determine at what point eggs needed to supplement a quantitative deficiency through fertilization. In *Ascaris,* for example, he pointed out that there are generally four loops (*Schleifen,* but after 1888 *chromosomes*) in the adult, and so "it follows that an egg, which can only form two or three loops from its nuclear reticulum, would not be able to develop parthenogenetically."[94]

The quantity of chromatin material alone, however, was not sufficient to explain heredity. After further sifting through the contemporary understanding of parthenogenesis in bees and daphnids, Weismann added that the constitution of the involved nuclear material must play a determining role. "It is, I think, clear that these are obvious instances of the general conclusion that the direct causes determining the direction of development in each case are not to be looked for in external conditions, but in the constitution of the organs concerned."[95] Quantitative or constitutive loops? In either case, polar body expulsion appeared to be only indirectly related to the initiation of first segmentation. Although he did not prove it empirically, Weismann was convinced that polar bodies would be found in both summer and winter eggs.[96]

The central point, which can be easily overlooked by his warping through the contemporary evidence presented by polar bodies and parthenogenesis, is reflected in Weismann's insistence that "all these considerations depend upon the supposition that the egg-nucleus contains two kinds of idioplasm, viz. germplasm and ovogenetic nucleoplasm." The dilemma, however, remained. "I have not hitherto brought forward any direct evidence in favour of this assumption, but I believe that such proofs can be obtained."[97] Weismann's essay on the "Continuity of the Germ-Plasm" pressed forward the issue of the separation of the germ-plasm from the somatoplasm rather than proceeding to a model for the transmission of traits—that would come later. For the moment, the notion of "continuity" for Weismann provided the real "foundation" for heredity.

Conclusions and Responses

Standing back from the years of 1884 and 1885, we can see that it was an important period for initiating competing theories about hereditary transmission. At this time, a number of independent developments had converged to make this inevitable. The establishment of the near universality of indirect nuclear division, a general agreement that fertilization consisted of the unification of the nuclei of male and female gametes, and the demonstration that paternal and maternal nuclei did not simply fuse into a new unorganized blend—all these provided a common focus for further deliberations. Nägeli's highly theoretical discussion of the importance of a structured idioplasm and its component micellae added a structural perspective, which was soon corrected by Roux's equally theoretical suggestion that the chromatin elements might be structured on the molecular level and serve as the bearers of hereditary differences. Nussbaum and Weismann's conception of a continuity of germ cells and Weismann's more speculative revision of his continuity of the germ-plasm helped isolate the processes of transmission from development.

One might conclude that together these different theoretical positions not only challenged but eliminated the older view that nutrition, assimilation, and growth were the overriding physiological processes in reproduction and in the overgrowth view of heredity. In the past, I have pushed such an opinion, but the revision of a complex set of processes in science is rarely this unilinear.[98] With Nägeli, Roux, Kölliker, and Weismann, in particular, growth still played an important role in the machinery—any machinery—that made heredity work. Nussbaum, Hertwig, and Strasburger continued with a holistic view of the cell and organism. Together, these unresolved issues compromised the architectonic aspects of their heredity theories.

One thing that becomes evident in reviewing these two years is how important Nägeli's role was. From the perspective of the late twentieth and early twenty-first centuries, his substantial work on *Abstammungslehre* appears completely off the wall and outdated even before its appearance. Nevertheless, his sketch of the idioplasm and its component micellae and his unquestioned mechanistic credentials caught the attention of his younger contemporaries. The sheer mass of his work evidently did not faze his readers. It was not that he spoke on almost everything, as Ernst Mayr chided, but that he did so in a way that caught the spirit of the time: a commitment to mechanisms as the foundation to scientific explanation, a faith in an evolutionary process where life left nothing to chance or accident, a belief in a phylogeny that became ever more complex and proficient over time, and a view of organisms as individually formed structures that could pass on their acquisitions to their offspring.

In justifying his own contribution, Weismann considered that the "origin of germ-cells explains the phenomena of heredity very simply, inasmuch as heredity becomes thus a question of growth and of assimilation,—the most fundamental of all vital phenomena."[99] With these thoughts, Weismann had moved only part way to an architectonic explanation of heredity, development, and evolution. Nussbaum, Hertwig, and Strasburger, on the other hand, continued to maintain a holistic vision of the origin of the germ cells. Their reluctance to be more specific compromised any architectonic view of heredity they might have harbored.

13

Transmission of Adaptations and Evolution

1885–1890

Between 1880 and 1885 Weismann saw the need to shift his program of research to account for the advances in nuclear cytology. This did not mean he immediately adjusted the new findings into his own deliberations about heredity, development, and evolution. Instead, he wrote of continuity of the germ cells, then of the germ-plasm. Despite his conviction that the nucleus of the germ-plasm comprised in some way the hereditary constitution that he had so long assumed existed, despite his visits to Louvain and Bonn, despite his image of marching battalions changing formation during ontogeny, and his identification of the polar bodies and parthenogenesis as key phenomena for better understanding the structure of the germ-plasm, Weismann first paid attention to a closely related problem—the problem that ultimately defined his biology. This concerned the mechanism of evolution. Darwin had passed away in 1882. Weismann's generation, who came of age both in Germany and Great Britain with the publication of *Origin of Species,* had moved into leadership positions. Pushing for a hearing was now an even younger generation ready to use new tools, eager to make new distinctions, and confident that their standardized professional training in biology gave them standing not only, employing T. H. Huxley's words used in an earlier context, to hear "the scientific lions of the day roar" but to join in the chorus themselves.

Both Weismann and Haeckel were now recognized as leaders in German evolutionary biology. Their career trajectories, described in Chapter 4, however, were very different. This is symbolized by the fact that Haeckel delivered the last of three plenary addresses before the Versammlung the year after Weismann presented the first of three parallel addresses. In his three, Haeckel addressed in one way or another the importance of evolution as the mechanistic foundation of biology, which for him was summed up in his biogenetic law. In contrast, Weismann focused on the mechanism of heredity. Since early in 1885 he had argued that a germinal line remained undisturbed by the activity of development and isolated from direct environmental influences, but it remained incumbent upon him to explain the origin of the variations that were the grist so essential for the mill of Darwinian selection. Natural selection, rather than the more esoteric subject of the structure of the nucleus, arose from these particular concerns and spoke more directly to those in the broader scientific community. Of course, there was a connection between the two, and Weismann soon found himself balancing the mechanisms of both evolution and heredity against storms of opposition. It makes sense in this chapter to switch gears back to evolution theory and to pick up the further development of the nuclear structure in Chapter 14. In Weismann's mind, the two topics needed to be handled together, and over the next twenty-five years he struggled to do so.

Zoological Institute

The year of 1883 perhaps best represents Weismann's reestablishment into the local and national community of biologists. His major work on hydromedusae, which came out in May, reflected a deep and thorough scholarship of the particulars of a single major group of organisms.[1] In April, he was offered a position with Bonn University, which to his wife Mary's relief he declined with seemingly little anguish; soon thereafter, he assumed the prorectorship of Freiburg University, which at minimum required him to present a general address to the university in July. On that occasion, he spoke to his colleagues on the subject of "Vererbung"—heredity. This subject had become the dominant focus of his research and publications, up to his foremost publication *Das Keimplasma: Eine Theorie der Vererbung* of 1892. The mechanism of transmission from parent to offspring entailed a process that carried Weismann's thoughts and research almost, though not completely, to his discussions and publications on both the development of the individual and the evolution of species for the rest of his life.

As prorector and Geheimer Rat II Klasse, Weismann participated in the medical requirements (*ärztliche Vorprüfung*) in zoology, a requirement to which he contributed until his retirement. His two general courses, one in "Descendenztheorie" and the other in general "Zoologie," were undoubtedly responsible for his rising enrollments and eventually for the university's need to provide him with a new lecture hall. His lectures in 1902 (later editions were published in 1904 and 1913) indicate that he was a thorough and up-to-date lecturer; although he did not jump on the "genetic" bandwagon during the first decade of the twentieth century, he discussed Mendel's achievement and worked it into his broader theory of heredity and evolution of his *Vorträge*. (This relationship will be discussed in Chapter 20.)

Back in 1883, Weismann had begun a new series of experiments on seasonal dimorphism with *Vanessa levana* and *V. prorsa,* which he would pursue until 1889. This work became the starting point of his work "Neue Versuche zum Saison-Dimorphismus der Schmetterlinge."[2] His administrative duties beyond the running of his institute included guiding advanced students interested in zoology and providing zoology courses for premedical students. In addition, his colleagues elected Weismann to be the nominal chairman of the academic Krankenkasse. Beyond the local scene, Weismann delivered an address (*Festrede*) in July for the unveiling of a monument to Lorenz Oken in the latter's hometown of Offenburg. Not only was it logical for Weismann as a zoologist in Freiburg to present the homage, but he also found it a fascinating challenge to honor the leader of Naturphilosophie, and he succeeded admirably. We can imagine that even more stressful was his deep involvement in hosting the Versammlung in Freiburg between September 18 and 22. In addition all of these activities, including the acceptance of candidates for an advanced degree, demanded a deep commitment to his students; Weismann commented in retrospect that his eye troubles began again after reading the proofs for *Hydromedusen*. This may well have been, although his "Autobiographical Sketch" (1896) was more emphatic that his retinal decline became more pronounced after the death of his wife in 1886.[3]

Inheritance of Acquired Characters

There are few general works that attempt to analyze historically what has become to be known as the "inheritance of acquired characters."[4] For the historian there are obvious problems. The English expression "inheritance of acquired characters" is of post-Darwin origins, and its three operative words relate to nineteenth-century concerns. "Inheritance" and heredity assume their biological meanings only in the 1860s.[5] "Acquired" in its biological sense presents an ambiguity about when in the life of the individual—at fertilization, during embryonic development, or in adulthood—the acquisition occurred. The word "character" leaves two further confusions: one about whether it encompasses both physical and functional traits, and the other about whether, at a time before the distinction between "genotype" and "phenotype" was decisively made in the twentieth century, it referred to both constitutional (germinal) as well as anatomical and physiological traits. In the biological realm, inheritance of acquired characters as a conjectured phenomenon also embraced a number of different processes: for instance, the results of use and disuse, the effects of the external environment and nutrition, the consequence of habits and education, the outcome of physical mutilations, and the consequence of disease—all appearing on following generations. Thus, during the last twenty years of the century the ambiguities, including their combinations and permutations, were legion, and this only spurred debate and promoted discord among scientists who felt licensed by their own experience to promulgate on the matter: zoologists, botanists, psychologists, ethnologists, physiologists, anthropologists, and paleontologists. Barely below the surface of dissension lay a general confusion about the process of evolution itself. Thus, we find hues of neo-Darwinians, neo-Lamarckians, orthogeneticists, mutationists, progressionists, and even a few creationists, to use Peter Bowler's useful classification of the participants. Involved also were national interests, particularly among those who felt Darwinism was simply an updating of Lamarckism, which brings us back to Weismann.[6]

A turn-of-the-century photograph shows Weismann looking out from his desk at the viewer, still vigorous, brooking no compromise in standards, and sporting a full white beard. Books, writing material, a microscope, and an oil lamp adorn the crowded desk and above them all hangs a lithograph copy of Jacques Louis David's portrait of Lamarck. The juxtaposition of the two biologists, who flourished nearly a hundred years apart in history, may seem a paradox—the portrait within a portrait being of a strong advocate of the transmission of characters acquired through the use and disuse of structures, the other being of a fierce opponent of that process. This section is devoted to discussing the fortunes of this idea and to rationalizing how the paradox of this portrait within a portrait reveals a modern misreading.

In his prime, Lamarck had been a survivor of the worst years of the French revolution that remade many scientific institutions and claimed lives from all levels of society, including France's most famous chemist. Prior to the political upheavals, Lamarck had published extensively in botany and had been befriended by Buffon, who secured support for him at the Jardin et Cabinet du

Roi. With the death of his patron and the reorganization by the National Assembly of the royal gardens and cabinet into the Muséum d'Histoire Naturelle, Lamarck found himself installed as a professor of invertebrates with the obligation to organize and teach a subject he knew little about. The appointment appeared both a slight to his past achievements and a challenge to his own abilities to adapt, and adapt he did but in his own way.[7]

He immediately published a two-volume work, with a title of over a hundred words. Therein he presented a philosophical system of mineralogy, chemistry, and meteorology and argued for the organic origins of all basic minerals and elements. His system opposed the new pneumatic chemistry associated with Lavoisier and promoted by his own colleague at the museum, the professor of chemistry Antoine-François Fourcroy.[8] In addition, Lamarck was influenced by a geologist friend to think in terms of gradual rather than catastrophic geological changes in the earth, a position that brought him onto a collision course with another colleague, the young, talented, and politically astute professor of anatomy Georges Cuvier.

Notwithstanding, Lamarck concentrated on mollusks and was soon recognized as the authority in this area. It was his museum studies of the shells and internal anatomy of snails that led him to broach the notion of gradual organic advance from the most basic forms of life to mammals. After 1800, he became ever more involved in promoting classification of snails and then all invertebrates as representations of an overall increasing complexity. His best known work on this subject, his two-volume *Philosophie Zoologique* of 1809, not only presented a classification of the entire animal kingdom but provided a detailed physiology of the conjectured processes of organic change that justified it. As Burkhardt pointed out, the details of what later became called "transmutation" or "evolution" continued to develop through Lamarck's final years, with additional statements included in several volumes of his monumental seven-volume *Histoire Naturelle des Animaux sans Vertèbres* of 1815–1822.

Central to Lamarck's evolutionary system were two "laws" of organic change. A modern précis of his 1809 presentation might render them as follows: (1) the continued use of organs develops and enlarges them, and the continued disuse of organs diminishes them and ultimately causes them to disappear; and (2) the resulting modifications, when they occur in both parents under certain circumstances, are "generated" in their offspring. It is worth noting that in neither the French original nor in the Elliot translation did Lamarck himself speak of the heredity of acquired characters.[9] These "laws" pinpointed the obvious and smuggled in that which had generally been assumed to follow. Both had probably been part of received wisdom from time immemorial, for it was not difficult to recognize that the use and disuse of, let us say, muscles had a clear impact on structure, and as we have seen in an earlier chapter it was equally apparent that "like begat like." These laws became a source of great debate later in the nineteenth century, but they failed to do justice to the extraordinary "system" that Lamarck constructed.

The notion of organic evolution had been in the air prior to 1800. Burkhardt has pointed out that even several of Lamarck's colleagues at the National History Museum had entertained evolutionary ideas, but none of them had done so systematically. It was Lamarck's great achievement that he did so explicitly within the context of his curatorial work with invertebrates and comprehensively within a survey of both the animal and plant kingdoms. Moreover, and this is what may well have appealed to Weismann, Lamarck confined his explanation of the process to the mechanisms offered by the sciences as how Lamarck himself understood them. These mechanisms included the pervasive influences of subtle and imponderable fluids such as caloric and electricity, which aggregated then agitated mucilaginous (with plants) and gelatinous (with animals) materials to acquire internal functions such as the faculties of assimilation and motion. As the process continued, the faculties, organs, and the internal fluids for respiration, irritability, and intelligence arose. The consequence of the physical interactions between matter and subtle fluids was the rise of complexity, the inevitable "plan of nature" that Lamarck found reflected in his classifications.

To return to Lamarck's two laws of organic change, there was more to his system than simply the increasing complexity found in his museum cabinets— there was diversity at similar levels, a lateral diversification as well as an increasing vertical complexity due to nature's plan. Multitudinous external circumstances resulted in a multitude of different physiological responses, in different uses and disuses of bodily faculties, habits, and parts, and hence a diversity of observable characters. These, Lamarck assumed in his second law, may under certain circumstances be transmitted to later generations. Thus, there were two different laws in addition to two fundamentally "different causes" in Lamarck's theory of organic change, both of which were evident by 1809. Burkhardt speculates that it was easy for later evolutionists "to believe that Lamarck's theory was primarily concerned with change through the inherited effects of environmentally inspired use and disuse." For Lamarck, it was the eternal plan of nature that counted most.[10]

Unfortunately, Weismann never made it clear which aspects of Lamarck's evolution theory he found admirable enough to warrant hanging up a copy of David's portrait above his desk. Might it have been because he admired Lamarck's attention to the many sides of the evolutionary process, or was it his insistence on a thoroughly mechanistic explanation, or did he secretly identify with Lamarck's embattled position, or might he simply have liked the portrait of a fellow evolutionist? In his published work and letters, he scarcely mentions Lamarck, but we do know he owned an 1830 reprint of Lamarck's *Philosophie Zoologique*. Long before the scene was photographed, Weismann had begun his battle against Lamarck's generalization of inheritance. With his articulation of the contrast between the germ-plasm and the soma in the early 1880s, he had become the leader against the inheritance of acquired characters and even more of a champion for Darwin's natural selection. For his efforts, he would be designated a "neo-Darwinian" by Wallace, a man who knew Darwin and later Weismann and had the nominal authority to coin such a word.[11]

As mentioned in Chapter 4 the inheritance of acquired characters was a feature that Haeckel also integrated into his theory of evolution. He had dedicated the second volume of his *Generelle Morphologie* to Darwin, Goethe, and Lamarck—"the founders of the theory of descent." The third always and the first two occasionally had resorted to the inheritance of acquired characters as an integral part of their understanding of transmission. Five months after Darwin's death in 1882, Haeckel returned to these "three founders" in his third and final plenary address to the Versammlung, which had assembled that year in the Thuringian textile city of Eisenach.[12] Although largely a homage to Darwin and an invocation to the humanistic and monistic implications of the theory of descent, Haeckel took a moment to single out Lamarck as the most important of many pre-Darwinian evolutionists. It was the combination of Lamarck's "many sided" empirical researches and his *Philosophie Zoologique*, "one of the greatest products of the great literary period at the beginning of our century," that captured Haeckel's attention as well as Weismann's. Therein Haeckel found a classification that included fossil as well as extant species, that depended on living conditions and especially use and disuse of organs to bring about a gradual change in organic form "whose principal features are transmitted through inheritance from generation to generation." The result was a "natural ancestral lineage." Haeckel also supported Lamarck's physiology in which "an unnatural living force" was rejected and "all life's phenomena rested on mechanical processes." He finally applauded

Lamarck's claims that man descended from apes and that life began with spontaneous generation. There were indeed uncanny parallels between Haeckel and his French predecessor.[13]

As described in Chapters 4 and 9, Haeckel's biogenetic law, his repeated invocation of "overgrowth" as a way to understand transmission, and even his admittedly speculative theory of perigenesis only made evolutionary sense in the context of an inheritance of acquired characters. There was no way that Haeckel could throw away such inheritance without totally rethinking his biology. I have also indicated that Haeckel severely criticized Weismann in the 1890s for ruling out this key process of heredity with the tools of modern nuclear cytology.

In Haeckel and Weismann's day, the "inheritance of acquired characters" came to designate a number of phenomena. For Lamarck, as mentioned, the expression did not exist. His laws of development and hypothetical mechanisms, however, spoke of the generation of offspring bearing new parental traits acquired through use and disuse. Many of his younger contemporaries took a slightly different route. By focusing on the transmission of the impact of the environment on the morphology of organisms, they promoted his own view transmutation.

By the half-dozen years framing the *Generelle Morphologie* (1866) and Darwin's *Variations of Animals and Plants under Domestication* (1872), "inheritance of acquired characters" came into fashion as an expression and designated an array of phenomena, which included the transmission of mutilations, disease, habits, disease, and the products of use and disuse. It is an expression that soon became confounded with parallel induction, orthogenetic patterns of evolution, and even with the direct changes in the germinal constitution itself. The key for any meaningful modern discussion about the process must place the emphasis on an acquired change, that is, a change manifested in the soma of the first generation and traced through its transmission to subsequent generations. From the 1880s on, an increasing understanding of the nuclear dimension of heredity, an anatomically more sophisticated understanding of mutilations and regeneration, the irrefutable demonstrations supporting the germ theory of disease, and a greater awareness of the difference between habit and instinct joined the debate.[14]

"On the Significance of Sexual Reproduction"

Three years after Haeckel's final plenary address on the founders of the theory of descent, Weismann presented his second plenary address to the Versammlung in Strassburg on "The Significance of Sexual Reproduction."[15] The contrast between the two could not have been greater. Haeckel's address was backward looking, a historical reflection rather than a scientific contribution. Except for a number of papers and monographs on the Radiolaria and Siphonophorae from the HMS *Challenger* expedition, his creative scientific career appears in retrospect to have slowed down. Instead, he organized collecting expeditions and devoted most of his literary energy to new editions and new popular works, to essays on science and religion, and to promoting monism.

On the other hand, Weismann appears to have just reached his full stride. It was his turn to challenge the scientific community with new conceptions about heredity and evolution. Unlike the fragmented nature of his monograph on the "Continuity of the Germ-Plasm," Weismann's plenary address provided a unified message on a broader problem—one, to be sure, that better suited his general audience. The address acted as a double salvo against the enduring belief in the influence of the environment on the organism and any internally sustained evolutionary forces. It was the first time he joined his claim of germ-plasm continuity to his belief in the sufficiency of natural selection in evolution. The address might also be viewed as indirectly dismissing Lamarck's two causes of evolution or, within the context of the moment, as explicitly rejecting both Nägeli's internal perfecting mechanism and the generally held belief in the inheritance of acquired characters, so prized by Haeckel. In a roundabout way, Weismann's words signaled nothing short of emancipation from the widely held "overgrowth" concept of heredity, which promoted the proposition that like begat like. "But I believe," he continued, foreshadowing a break from that long-standing tradition, "it is possible to suggest that the origin of hereditary individual characters takes place in a manner quite different from any which has been as yet brought forward." He further pointed out, "The phenomena of heredity" had led "to the conclusion that each organism is capable of producing germs, from which, theoretically at least, exact copies of the parent may arise."[16] In the place of an inheritance of acquired characters, Weismann substituted the primacy of sexual reproduction and natural selection to explain the profusion of variations and adaptations to external circumstances.

The original argument was complex for a plenary address. When it appeared the following year in a separately published monograph with additions and appendices, it was swollen to more than twice its original size.[17] Weismann expressed a strong belief in evolution as a gradual process of accumulated variations over extended time. He explicitly related these variations to an imaginary molecular structure of the germ-plasm—yet had he burdened himself with a contradiction? On the one hand, the germ-plasm, as he had first conceived it, remained unchanged in the maelstrom of development; on the other hand, he now suggested the gradual accumulation of changes in the very same germ-plasm. Weismann responded that sexual reproduction brought about new combinations of germ-plasms, which in turn resulted in new hereditary individual characters appearing in each generation. "When we remember that, in the tenth generation, a single germ contains 1024 different germ-plasms, with their inherent hereditary tendencies, it is quite clear that continued sexual reproduction can never lead to the re-appearance of exactly the same combination, but that new ones must always arise."[18]

There were, of course, the "variations" caused by the environment, the actions of use and disuse, and the effects of other conditions of life, but these by his theory were simply transitory somatic differences. Because they did not influence the germ-plasm, their impact on the next generation was nil. By contrast, "individual hereditary characters" were commonplace to all individuals. Although at this time Weismann was not consistent or clear about "hereditary differences" being germinal or expressed traits, that was the direction in which his thoughts were clearly carrying him.[19] As such, the individual hereditary characters were numerous and subject to natural selection. They were, as he maintained, caused by the "minute alterations in the molecular structure of the germ-plasm, and as the latter is, according to our supposition, transmitted from one generation to another, it follows that such changes would be hereditary."[20]

At this time, Weismann also made it sound as though he considered larger variations, those deviations from "species characters" ("Species-charaktere"), but these he considered rare, too large, and too vulnerable to being rapidly blended out.[21] The contrast between individual characters and "species characters" was common at the time. Darwin had spoken of the same two building blocks. From our perspective, it seems to reflect a confounding of a taxonomic tool with a physiological process; in short, an essential contrast between individual and species characters seems to fly in the face of the neo-Darwinian direction, in which Weismann began pushing his biology. This is worth stating, for it emphasizes that the dramatic changes that were taking place in hereditary and evolution theory were not a simple matter but must be examined on many levels at the same time.

Aside from the essential but restricted phenomena from cytology, the purported cases of germinal continuity from Nussbaum and himself, and Nägeli's suggestive but highly speculative idioplasm, Weismann's strongest argument for the purported significance of sexual reproduction came from negative arguments. Here he had recourse to cases of complete parthenogenesis ("reine Parthenogenesis"). There were animal species known to reproduce solely asexually, such as the gall wasp *Cynips,* certain unnamed "lower Crustaceans," and some ferns described by his close friend Anton Bary. These might be well adapted to their environment and thrive, but according to Weismann, they would be unable to adapt to new circumstances because they were unable to accumulate variations and produce new characters. "If it could be shown," he challenged his audience with Darwin-like rhetoric, "that a purely parthenogenetic species had become transformed into a new one, such an observation would prove the existence of some force of transformation other than selective processes," an inherent mechanical perfecting force or the inheritance of acquired characters. Thus, he concluded, "monogonic reproduction can never cause hereditary individual variability, on the other hand, it is very likely to lead to its [i.e., the species'] entire suppression."[22]

In those heady days of phylogenies, Weismann's emphasis on sexual reproduction required that he suggest an origin for such a reproductive mode. That meant a hypothetical phylogeny for multiple variations and for individual characters. His earlier phylogenetic speculations about the origin of the germ-plasm seemed sufficient. Prior to the rise of Metazoa and Metaphyta, as he had previously insisted, there could be no distinction between body cells and germ cells. External conditions had directly brought about changes in the most primitive single-celled organisms. Consequently, changes could be directly transmitted to the next generation. "Here parent and offspring are still, in a certain sense, one and the same thing: the child is a part, and usually half, of the parent."[23] Thus, at the beginning of life traditional overgrowth prevailed! Once a system emerged to allow for the exchange of individual variations between two organisms, natural selection had a new source of differences upon which to operate. According to Weismann, recombinations of variations and their selection were part and parcel of the same momentous advance in evolution that created the independence of the germ-plasm. As a consequence, overgrowth became irrelevant, and the inheritance of acquired characters became impossible at the metazoan and metaphyte level.[24]

A Tangle with Rudolf Virchow

The Berlin pathologist and political reformer Rudolf Virchow was a commanding presence on the scientific scene in Germany during the Wilhelmine period. His early successful scientific career in Berlin was interrupted by his anti-Prussian evaluation of the state's lack of involvement in a typhus epidemic in Upper Silesia, and he became blunt in his criticism and active even at the barricades in the short-lived successful revolution of 1848. Virchow became a *persona non grata* in Berlin and accepted a Lehrstuhl of pathological anatomy at the smaller and far less prestigious University of Würzburg.[25]

The seven years he spent in Bavaria proved to be important ones that enhanced his scientific reputation. He had the opportunity to research and to rework his understanding of cell theory. He picked up again the investigation of cancer that he had begun in Berlin and came to the conclusion that cancers were best understood as functional deviations from normal development rather than as ontological disease entities. He also began to challenge Schwann's doctrine of *de novo* cell production from a generalized cytoblastema, and by 1855 he had pronounced his famous aphorism that "all cells were derived from cells" (*Omnis cellula e cellula*). By this, he meant that cells multiplied only by direct cell division. In addition, he began to interpret cells as the basic unit of life, below the morphological level of which occurred only unstructured chemical and physical reaction. He called his new cell theory a "new vitalism." Finally, and this flowed directly from his new vitalism and becomes critical for understanding his reaction to Weismann's germ-plasm theory, Virchow insisted that "if we concede to the agreement between pathological neoplasm and the embryological, so must we naturally regard the egg as the analog of the pathological mother cell and fertilization as the analog of the pathological stimulus."[26]

A year after he had returned in triumph to Berlin and had occupied his new pathological anatomy institute, he delivered a famous lecture series to an appreciative audience of academics and physicians. These were collectively published under the title of *Die Cellular Pathologie*.[27] A structural analysis of this classic in the history of medicine reveals that Virchow carefully constructed his extended argument to show the genetic relationship between normal and pathological cells and tissues.[28] Despite the contrary opinions of other pathologists, he explicitly maintained that "I will nevertheless endeavour in the course of these lectures to furnish you with proofs that every pathological structure has a physiological prototype, and that no form of morbid growth arises which cannot in its elements be traced back to some model which had previously maintained an independent existence in the economy."[29] Thus, Virchow's *Cellular Pathology* was based on his updated version of Schwann's cell theory, with its commitment to the continuity of all cells through direct cell division, to his new vitalism, and to Müller's analogy. With these principles in hand, he interpreted the infinite variety of neoplasms in terms of altered, misplaced, and exchanged developmental histories of the cellular parts of the organism. His explanations are drawn not only from a wealth of microscopic observations but from experiments on cell reactions to various stimuli. His demonstrations were drawn up to exclude any competing interpretation that neural or vascular stimuli might have caused the abnormal growth. Cells and their territories alone reacted to external stimuli, "From which I draw the conclusion that these active processes [new formations] have their foundation in the special action of the elementary parts."[30]

This brings us back to Weismann and the 1885 Versammlung in Strassburg. Virchow most certainly attended Weismann's plenary talk on the significance of sexual reproduction. It became his turn to address the same audience the following Tuesday at the second and final plenary session.[31] Ostensibly he was concerned about Bismarck's intentions to push Germany into founding overseas colonies to compete with England, France, and the Netherlands, but the internal dynamics of his talk swung between illustrating the effects of climate and geography on human health and inheritance and the explicit faulting of Weismann and all academic biologists for ignoring the general lessons of pathology. Proud and cantankerous, he insisted that pathology was not simply an ancillary discipline, a "Nebenfach" that could be ignored when the discussion turned to acclimatization and evolution. It was the pathologist who understood best that all changes were initially induced by external causes, and that subsequently all such changes could be transmitted by the mechanics of cellular reproduction. Deviations from the type produced variations, and these variations in turn led to varieties and new species; so Virchow could readily conclude that "Earth had formed humankind" ("Die Erde hat die Menschen gebildet").[32]

Virchow closed his somewhat rambling address with a long litany of pathological changes wrought during human colonization in foreign climes. Acclimatization might gradually succeed, but it was a slow process, and necessarily took decades instead of a few months of vacation. At the same time, he dismissed Weismann's sexual theory of variations, and he warned Germans and Alsatians against embarking upon colony building.[33]

Weismann was permitted two short rejoinders in which he first explained his understanding of the difference between externally and internally induced variations. The former, he declared, provided unacceptable candidates for inheritance of acquired characters; the latter, when transmitted, were not examples of it at all. The second rejoinder came from Weismann's heart. He alluded to the death in Veracruz of his brother Julius, which had reinforced for him the potential dangers of foreign emigration. The selective action of yellow fever rather than a failure to acclimatize, he insisted, was the better way of explaining his family tragedy. That Weismann felt confident about his confrontation with the famous Berlin pathologist is indicated in a letter to Fritz Müller: "He had, however, misunderstood the entire point; so it was easy for me to reply."[34]

In neither speaker's case were the issues clearly developed. Each turned, however, to rejoinders in professional publications. Virchow naturally enough mobilized his *Archiv* to publish a three-part response.[35] Weismann arranged with Gustav Fischer to bring out his talk with emendations and a supplement of six appendices. Neither would have seen the other's additions before their publication. After the new year, their arguments reappeared in the *Biologisches Centralblatt,* which had become an acceptable format to broadcast one's controversial ideas to a wider audience of biologists. The Basel anatomist Julius Kollmann started things off in January by repeating the two positions in a review of the two addresses in Strassburg.[36] His sympathies clearly lay with Virchow. Weismann responded in March by pointing out that Kollmann and he had used the term "adaptation" ("Anpassung") in different senses. The first sense was highlighted in a quotation taken directly from Kollmann: "One must understand adaptation as nothing other than the acquisition of a certain trait during the life of the individual under the pressure of external agents." "Adaptation in the sense of Darwin's selection theory," Weismann retorted, "should not be understood with respect to the life of the single being, but only in the life of the species, that means, in a sequence of generations."[37] Thus, Kollmann used the concept in an embryological sense, which explained his sympathies with Virchow's developmental concept of individual variations, whereas Weismann envisioned "adaptation" in a statistical sense, which allowed him to see it as the product of selection and gradual evolution.[38]

Virchow's view was again presented in April, when the Erlangen physiologist Isidor Rosenthal, as chief editor of the *Biologisches Centralblatt,* reprinted Virchow's article that had appeared in the *Archiv* three months earlier.[39] Rosenthal goes on to suggest that there were not only conceptual differences that needed to be explained but Virchow's recent complaint that biologists no longer paid attention to the pathologist's perspective needed clarification. Thus, reproduction, heredity, the nature of variations, adaptation, and a mechanism for evolution were all thrown into the mix at the same time. It was no wonder that the inheritance of acquired characters became such a heated and confusing topic over the next twenty years.

Although the Weismann-Virchow confrontation had not been the first public scrutiny of the inheritance of acquired characters, it resulted in more than simple restatements by the principal participants. Very quickly, the relevant scholarly journals, both biological and medical, were pressed into publishing case studies and discussions on the matter.[40] The general press aired the issue as well.[41] Interest in the subject was great enough for the organizers to schedule a rerun of the standoff between Weismann and Virchow in two plenary sessions of the 1888 Versammlung in Cologne.[42] Virchow was scheduled to speak on Thursday morning during the traditional third week of September. (The Versammlung met from Tuesday, September 18, to Sunday, September 23.) Because of overcrowding of the large hall, and perhaps because a larger than expected audience wanted to hear Europe's most celebrated politician/scientist, Weismann willingly changed places with him. So Virchow spoke at a prime time on the following Saturday.

Whatever the reason for the shuffling of the speaking order, Weismann again went first and, as was his style, again offered a carefully constructed professional argument. As he had three years earlier, he filled in some details in the republication in monograph form of his talk.[43] This time he focused exclusively on the question of the inheritance of induced mutilations.[44] In particular, he examined reports about the inheritance of a "bobtailed" condition in given lineages of domestic cats—a condition that, according to tradition, was reputably traced to the accidental removal of the tail of the mother cat of the population. There were many such claims throughout Europe, the most famous being the Manx cats of the Isle of Man.[45] Also examined in anatomical detail and illustrated in the Fischer publication was the case of the scarred earlobe of a woman whose earring had been torn off her as a child. The injury, as reported, reappeared many years later in the ears of her sons. Careful anatomical discussion of the ear by Weismann and anatomical examinations of the lower spinal column of bobtailed cats by the young anatomist Robert Bonnet, however, indicated that the histological details of initial injuries and of parallel scars in offspring failed to match. Because other key pieces of precise information about the initial injury and the parentage were missing, and proponents of the inheritance of acquired characters recognized that bobtailed cats were common, Weismann assigned such cases to a grab bag of insufficiently documented myths. "Insufficient" meant that the documentation could not meet the scientific standards of the time for a positive proof of what he called "Lamarck's principle."

It was in this context that Weismann famously reported on his recent experiments with amputation of the tails of mice. Between October 17, 1887, and August 25, 1888, he had raised over 700 mice, including the parent and four offspring generations. He had cropped the tails immediately after the young opened their eyes and began growing hair. He segregated the generations into separately numbered cages, with the parent generation continuing to reproduce in their initial cage. He measured the length of the tails, and either returned the mice to their designated "generational" cage for further reproduction or killed and preserved the surplus mice. He also kept controlled populations. In not a single mouse, Weismann reported to his audience, did the tails deviate in length from the norm of between 11 and 12 mm.[46]

There could be much to criticize about his procedure, but Weismann argued not that he had disproved the inheritance of the induced bobtailed condition in mice, but that it was impossible to prove such inheritance could *not* take place. Had he performed the amputations on enough generations? Were any of his procedures at fault? The objections were unlimited. This was a dilemma he and other critics of the inheritance of acquired characters faced. It was a burden that had been placed on the issue in the first half of the nineteenth century when Friedrich Blumenbach was willing to reject the environmentally induced origin of human races if, and only if, it were shown that the inheritance of acquired characters could not occur. The newly appointed professor of comparative anatomy, embryology, and histology at Dorpat, Johannes Brock, alerted Weismann and his peers to both Kant's and Blumenbach's ambivalence over the inheritance of acquired characters in human races. "Up to now (i.e., 1795)," Blumenbach had written and Brock had translated into German, "I side with neither of the two parties, neither the supporters nor the opponents of this theory of heredity, though I would gladly make the conclusions of the latter mine, if they could prove to me why such peculiar developmental failures, which first have arisen through purpose or accident, can in no way be able to be transmitted to the offspring."[47]

For Weismann, Blumenbach's eighty-year-old demand ("Forderung") had been rendered obsolete by the advance of science. By the 1880s, the hereditary theory had demonstrated the complexities in sexual reproduction. If one accepted as part of the process a continuous and isolated germ-plasm, a matter that many did not accept, the demand came from the other side. The complexities in the mechanisms for transmission now required that the supporters of the inheritance of acquired characters had to propose an even more complex mechanism to explain how mutilations could be introduced into a lineage of offspring. Weismann and his supporters moved the argument from Blumenbach's demand to show the impossible to the recognition of the inordinately more complex requirements of the possible. Those who still wanted to push for the Lamarckian principle as part of the mechanism of evolution had to turn to experimentation or move to inferences drawn from far more complex observations than anecdotal tales. Weismann responded by placing the burden of proof on the other side: "Only if the phenomena presented by the progress of organic evolution are proved to be inexplicable without the hypothesis of the transmission of acquired characters, shall we be justified in retaining such a hypothesis."[48]

When it became Virchow's turn on Saturday, he could only have disappointed those in the audience who were eager for a direct confrontation. He acknowledged the great public interest in the subject and the fact that the Berlin Academy had offered a prize for a decisive verdict on whether artificial disfigurements could or could not be inherited.[49] He implied that until such a work appeared, the issue would remain in flux.

The bulk of his talk drew upon Virchow's growing interest in physical and cultural anthropology. The worldwide practice of head compression of the newly born and foot-binding of young girls left lasting physical abnormalities in the individual. Whether the process was accidental or purposeful did not matter. The question was whether these deformities were transmitted to offspring whose head or feet were not bound as children. After some levity at the expense of insect-shaped women in their corsets, Virchow turned to Weismann's words and agreed with his conclusions about the limitations of many of the purported bobtailed cats. Virchow, indeed, had made an effort to collect and histologically examine many. What he could not agree with was Weismann's identification of two fundamentally different types of causes: one "fetal," the other postpartum. An infection, he urged, at either stage may cause the same distortion. It was no use for Weismann to respond that Virchow had distorted his argument about changes in the germ-plasm and in the soma. Virchow clearly had his own well-forged take on the disagreement. "More than that, we anthropologists and physicians are somewhat more used to investigating reasons or causes. That Herr Weissmann [*sic*] as zoologist needs not do. He frames a theory, and if this satisfies, then the affair is good. He will not recognize that a malformation can also be an acquired disruption."[50] It was a remark that in retrospect seems quite irrelevant and was bound to infuriate his Freiburg opponent.

Reaction to Comments in General

The double-barreled exchanges with Virchow and Kollmann were not the only scientific clashes Weismann experienced concerning the inheritance of acquired characters. During these years, he also became deeply involved in warding off priority claims concerning the idea of a continuity of germinal material. There had been no question that Nussbaum had anticipated his theory of the continuity of germ cells (not idioplasm). Weismann never questioned it, but within short order Julius Sachs and August Rauber criticized Weismann's theory while at the same time they claimed to have come up with the idea of continuity themselves. These claims were already addressed in Chapter 10. It is relevant to point out here, however, the toll this contentious opposition evidently took on Weismann. In responding to his old music friend from Frankfurt, who had noticed his recent publications, Weismann unburdened himself:

> Just as valuable to me at the moment is your support. My purpose is naturally to follow through with these ideas and where possible by means of observation and experiments to plant them more firmly— where necessary—to improve them and to clarify them. That naturally happens only in the course of years. In the meantime I have enough to do, in order, on the one hand, to defend myself against attacks and, on the other hand, against priority claims. With respect to the first I sent you a few days ago an attempt, which was minted chiefly against Virchow's involvement; as for Kollmann I would have scarcely needed to have said anything. Two other polemical writings have already been written and will shortly appear, both against prio[rity] claims. Now where these ideas are evident to many, there is one who comes, who at one time has said something similar, and claims, reasoning where possible with the reproach, that he had been purposely ignored by me. These are not pleasant outcomes, which one must however accept. Time and voice will be wasted and benefits no one.[51]

It is important to recognize that many private and public reactions were also supportive of Weismann's efforts to elaborate upon his theory of intragenerational continuity and the resulting negative consequence for a belief in the inheritance of acquired characters. Literary reviews and professional papers both indicate that Weismann had opened up important new research issues with the new perspective. A brief look at his correspondence in 1886 indicates some of these possibilities.

(1) The most complex of these issues was the relationship between the germplasm and the environment. We have seen that Weismann had been focused on the interaction of the environment on his postulated soma and had barely

suggested that the isolated germ-plasm might also be influenced. Carlo Emery, a specialist in ichthyology and entomology and at the time professor of zoology at the university in Bologna, evidently wrote him a letter of extensive comments that culminated with this issue. Weismann was pleased by Emery's constructive queries and dealt in detail with this one. He recognized the problem and argued that "the small individual" differences were the product of a fluctuating environment. "However, I believe that its direct change of the structure of the germ-plasm through external influence must be accepted; but only for such cases, in which these external influences remain similar throughout a long time."[52] To Jacob van Rees, his colleague and companion during his search for hydromedusae, Weismann further indicated that environmentally induced changes in the germ-plasm was a long-term process that influenced the entire species.[53] Despite his efforts to explain the distinctions, Weismann was finding it hard to forge boundaries between individual and species differences and between inherited and transitory variations.

(2) A question apparently highlighted by the same communication from van Rees also concerned variations and the evolution of single-celled organisms, which seemed to be excluded from Weismann's theories of continuity, sexual reproduction, and the dominance of natural selection. Weismann balked, but in his response he promised more details in a published form.

(3) An important matter for the issue of the inheritance of acquired characters, one that had already been explored by Alexis Jordan and Nägeli, involved the transplanting of plants to different locations, observing the environmentally induced changes, and later returning their offspring to the original habitat to see whether the changes endured. With these experiments in mind, Weismann proposed to Fritz Müller, who had written to him about the apparent permanent changes in a member of a California poppy genus transplanted to Brazil, that the two of them exchange German and Brazilian seeds of the same species.[54] As far as the record indicates, the experiment was not carried out.

(4) Although it had less to do with the inheritance of acquired characters than with mechanisms of fertilization, an important issue emerged that was given new impetus by Weismann's germ-plasm theory. This concerned the origin of parthenogenesis and its relationship to sexual reproduction. Weismann had examined parthenogenesis at length in his daphnid work, had focused a part of his treatise on the subject of continuity, and had devoted an appendix to parthenogenesis in his treatise on the "Significance of Sexual Reproduction." Gustav Adolf Ernst of Caracas, Venezuela, communicated the possibility

of parthenogenesis in a species of moonseed. If true, it would have been the first example of parthenogenesis in phanerogams. Weismann remained politely skeptical, for "as unbelieving the botanists always shake their heads, when parthenogenesis in phanerogams is mentioned." Nevertheless he understood the possibilities that this regulative, and hence this overtly utilitarian botanical example might parallel his own claims about absolute and relative parthenogenesis in insects.[55]

All four phenomena reinforced Weismann's contention, made throughout his career, that theories are to be tested—even at their own expense. "Maybe my ideas are also false, they must nevertheless be considered because they first pose the problem from which they are derived."[56]

Weismann and English Science

It is sometimes asserted that Weismann was better received in England than in Germany. There is some evidence to support this casual claim, but so far no one has attempted to explain or even wrestle with the historical details of the matter. It is tempting to point to Weismann as the most persistent and elaborate defender of natural selection—as the ultra-Darwinian of the day. In the land of Darwin, the argument continues, Weismann's work would have been appreciated and promoted. This certainly was part of the story, but as Weismann discovered to his own discomfort there were plenty of English critics who pointed out that Darwin did not rely exclusively on the natural selection of chance variations for the operation of evolution.

Weismann was not an unknown figure on the zoological scene during the last dozen years of Darwin's life. He himself corresponded with Darwin in a perfunctory manner when he sent the famous English evolutionist copies of his own works. Darwin appreciated their solid contents and encouraged the young professional chemist and avid entomologist Raphael Meldola to bring out a translation of the full text of Weismann's *Studien*.[57] Darwin even agreed to write a short prefatory notice in which he praised in particular Weismann's study of the utility of caterpillar markings. Generalizing further, Darwin pointed out that the overall "merit of [Weismann's] work . . . mainly consists in the light thrown on the laws of variation and inheritance by the facts given and discussed." This modest but astute endorsement turned out to be Darwin's last published statement written before his death.[58]

Alfred Russel Wallace reviewed the three installments of Meldola's translation for *Nature* as they appeared separately over a two-year span.[59] His assessment followed the lines of praise that Darwin had suggested. Weismann's research, Wallace insisted, consisted of a carefully executed zoological analysis along with simple experimentation, putting an emphasis on "the physical constitution, which causes different species, varieties, or sexes to respond differently." Wallace also extolled Weismann's dismissal of explanatory internal phyletic forces and laws of growth. He recognized that Weismann was a mechanist in his explanations but not an ontological materialist. He also noted, probably with some satisfaction, that Weismann did not resort to sexual selection as a separate mechanism as Darwin had continued to maintain. Wallace closed his three reviews with praise but a realistic caution to the reader: "To students of evolution, Dr. Weismann's volume will be both instructive and interesting, but it is a work that requires not merely reading, but study, since its copious facts and elaborate chains of argument are not to be mastered by a hasty perusal."[60]

The following year Henry Nottidge Moseley, professor of zoology and comparative anatomy at Oxford, wrote a lengthy review of Weismann's monograph on hydromedusae.[61] The challenge suited Moseley, for he had been a naturalist, specializing particularly in the invertebrates, while on the HMS *Challenger* voyage. He, too, had a very high regard for the details of Weismann's studies and for Weismann's ability to draw generalizations from the particulars. "Prof. Weismann of Freiburg is most highly skilled and most indefatigable in research, and all the memoirs which he publishes are of extreme scientific importance, and abound in original views and suggestions which render them of peculiar and widely spread interest." He continued his praise by suggesting that Weismann's monographs, from those of his *Studien* to his contributions on daphnids, were already known to English "naturalists." He then sketched the complex argument of the hydromedusae through nine of *Nature*'s closely packed columns. He included on the way four illustrations from Weismann's text and constructed a table to help orient English readers to the German vocabulary of germ cell migration. As an uncommonly thorough discussion, coming on the heels of Meldola's translation of the *Studien*, Moseley's review certainly highlighted Weismann's research for the English biological community. During the next three years, Moseley followed with two other strong reviews of Weismann's *Continuität des Keimplasma*s and "Bedeutung der geschlechtlichen Fortplanzung."[62]

By the Easter holiday of 1886, Weismann undertook a long-planned three-week trip (April 4–26, 1886) to England where he visited Cambridge, Oxford, and London. He met a number of the younger biologists in Cambridge, who were destined to become leaders in the discipline. These included Adam Sedgwick, Frank Raphael Weldon, William Bateson, Hans Friedrich Gadow, and Francis Darwin. Arthur Everett Shipley, the demonstrator of comparative anatomy in Cambridge at the time and the author of a short discussion in *Nineteenth Century* of Weismann's theory of death and immortality of unicellular organisms and germ cells, evidently had invited Weismann to spend a weekend at his family's home of Datchet in Berkshire.[63] It was Shipley who also made arrangements for Weismann to purchase a Cambridge rocking microtome. This serial sectioning machine became a highly prized instrument in Weismann's new institute in Freiburg.[64]

From Cambridge, Weismann visited Oxford where professional colleagues also treated him warmly. Gilbert Bourne, who had studied in Freiburg under Weismann's supervision, greeted him at the railroad station. Weismann later wrote with enthusiasm about spending time with John Obadiah Westwood, the elderly Hope professor of zoology, taxonomist, and antiquarian. When it came to his large insect collection, it apparently made little difference to the two insect enthusiasts that Westwood was one of the few holdouts against evolution theory of any kind. Unfortunately, Moseley was out of town the day Weismann visited.[65] From Oxford, Weismann traveled to London. He visited the Crystal Palace, Kew Gardens, and met Mr. and Mrs. Thiselton-Dyer, the former having just replaced Joseph Dalton Hooker as director of Kew Gardens. Armed with a Baedeker guide gifted to him by Shipley, Weismann undoubtedly saw many London sights. The entire trip took three weeks and had brought Weismann into direct contact with many English scientific notables, including those of the upcoming generation of biologists. In turn, it gave English biologists a firsthand chance to interact with and assess the German naturalist and evolutionist who was beginning to be known for his very strong pro-Darwinian views.

The success of his trip for both sides as well as the detailed reviews of Weismann's research undoubtedly encouraged Lankester to organize a session at the British Association for the Advancement of Science (BAAS), which in 1887 met in Manchester during the first week of September. Weismann was invited to participate in what turned out to be a discussion session and to present in a second session a paper on his current research. He traveled to Great Britain with two of his closest colleagues, his brother-in-law and Freiburg comparative anatomist Robert Wiedersheim, and the former Freiburg (1859–1872) and then-current Strassburg botanist Anton de Bary. It was an impressive trio of "biologists" from southwest Germany that crossed the English Channel. Traveling together, his colleagues must have provided Weismann with two trusted comrades with whom he could exchange a few words of encouragement during the meetings.

These annual national gatherings of scientists were patterned years earlier after the German *Versammlung*, but their monumental yearly *Reports* tell of a more complex set of activities than the latter's *Tagesblätter*. They opened with a hundred pages listing the BAAS rules, the former presidents, vice-presidents, and secretaries, the multiple scientific committees and their members, and the newly appointed committees. In addition they gave full details of BAAS expenditures and provided reports on the state of the various sciences and commitments of grants for new scientific projects. Following these "business details," as we might call them today, came the text of the annual presidential address, which in 1887 was given by the chemist Sir Henry Enfield Roscoe. The next 500 pages of the *Report* were consumed by specific committee accounts and their endless "communications" ordered previously by the general committee. Only then did the "transactions" of the year for the eight scientific sections come into focus, occupying another 400 pages.

It is here we can read that Wiedersheim contributed two short papers in English on his research with the olfactory organs of fish and with *Protopterus*. (Both of these talks appeared in German publications the same year.) In the Biological Section we read of Weismann's participation in a discussion session entitled "Are Acquired Characters Hereditary?" This session was led by Lankester and included M. Hartog, Hubrecht, P. Geddes, and E. B. Poulton. As for the last day of the meetings, we discover that Weismann read the principal paper in a session "On Polar Bodies" and presumably participated in the ensuing discussion.[66] As we will see in Chapter 14, this subject was the focus of his laboratory research of the moment. Participants again included Lankester and Hartog with the addition of Krause, Gardiner, Sedgwick, H. M. Ward, and Carnoy. Later on the same day, E. B. Poulton presented a paper on "Further Experiments on the Protective Value of Colour and Markings." Whether he attended the session or not, the subject would have deeply interested Weismann. Overall, then, Weismann and his ideas were carefully scrutinized by the biological community in Great Britain. In turn, Weismann learned firsthand about the BAAS and the "parade" before a mixed audience of specialists and a well-informed English public of frontier science in action.

After the meetings and in the company of Wiedersheim, Weismann traveled to Scotland to see the Firth of Forth, Edinburgh, and the Highlands. For two days, he enjoyed Scotland's major city, but he was clearly disappointed that fog and clouds prevented him from seeing the rugged northern landscape. It would be nice to know more about Weismann's second trip in August and September 1887 to Great Britain. Aside from a few letters, however, we learn little about his experience in Manchester, his reactions to the other scientists he met, or his travels from Oxford, Manchester, Liverpool, and Edinburgh and back to London in the "Flying Scotchman." He made a final foray to the Isle of Wight before returning to Freiburg on September 22, just eight days short of a year after the death of Mary. A passage in a later letter to de Bary added some color to the professional visit to a country where the language and customs were imperfectly understood:

> The Manchester Versammlung, was certainly very interesting and stimulating, but to be sure terribly stormy. With the Naturforscher-Versammlungen I have always found now and then a small quiet half hour in order to speak with someone about something important, there however this did not happen; one goes from one hand to the next—genuinely very inviting hands—but at the end my head was so full of the thing, that I could not grasp a comprehensible idea any longer. The foreign language also does much. It strains one unduly.[67]

Two results of these trips stand out. First, Weismann clearly felt appreciated; in turn, he was engaged in serious discussions over the mechanism of evolution that brought forth substantive issues. Second, by the time of his visits, English biologists were busily engaged in trying to fill in the gaps of Darwin's theory of evolution by natural selection. After all, Darwin himself had increasingly resorted to the evolutionary effects of "use and disuse" where the utility of a trait or a genetic sequence of such traits could not be imagined. It was generally agreed that besides natural selection, other factors such as "use and disuse," the correlation of parts, geographic and behavioral isolation, and various external forces might in one way or another contribute to the evolutionary process. Were they causes or conditions? Were they equal to or secondary to the impact of natural selection?

Despite the many alternatives, there existed a coterie of "pure" Darwinians who indefatigably championed the sufficiency of natural selection to form new species out of old. In the 1860s and 1870s, Wallace had out-Darwined Darwin. With the exception of cases of male combat, Wallace rejected Darwin's "sexual selection" of ornaments, colors, and even some behaviors, which he felt could be better explained in terms of natural selection than by invoking a separate cause. However, he had weakened his mechanistic explanation of evolution when it came to the origin of mankind. Furthermore, he appeared to be moving away from his interests in natural history and theoretical biology and devoting more attention to social and political issues that deeply concerned him. Finally, the codiscoverer of natural selection had compromised the mechanistic thrust of modern science altogether by becoming seriously involved with the nation's contemporary fascination with spiritualism. To be sure, Wallace made up for this questionable distraction by publishing at the same time an important work on *The Geographical Distribution of Animals* and a highly professional, though popular "Exposition of the Theory of Natural Selection" under the general title of *Darwinism*. The latter appeared at the end of the 1880s in the same year that Weismann's *Essays* were translated and published by the Clarendon Press. We have seen in an earlier chapter that Weismann had been influenced by Wallace's understanding of nonblending forms of seasonal dimorphism.[68] In turn, Wallace strongly supported Weismann's germ-plasm theory and pointed out how it (and Galton's theory of stirps) eliminated the need for the inheritance of acquired characters. "The names of Galton and Weismann," Wallace opined in a footnote, "should therefore be associated as discoverers of what may be considered (if finally established) the most important contribution to the evolution theory since the appearance of the *Origin of Species*."[69] Nevertheless, Wallace's status as a reliable stalwart of science had been shaken by the time Weismann traveled to England. Ross Slotten, a recent biographer, has poignantly summed up the situation: "By the late 1870's Wallace occupied an anomalous, Janus-like position. In 1859 he had spearheaded a scientific revolution. Twenty years later, he was in the vanguard of a counterrevolution—a paradox to everyone but himself."[70]

By way of contrast, other "pure" Darwinians remained steadfast in their defense of natural selection. Meldola, Moseley, Shipley, and Poulton stand out in their efforts. A perusal of letters to the editor in *Nature* between Darwin's death and the end of the decade reveals how active these four were in defending the sufficiency of natural selection. It is no accident that all four of these "biologists" promoted Weismann's ideas through translations of his works, and that three of them (perhaps Meldola, too) were involved in one way or another with Weismann's visits.

Relevant to the English scene on the eve of Weismann's two visits was the rise of systematic challenges to the dominance of natural selection. One was championed by the financially independent physiologist George J. Romanes; the second was presented by the newly appointed professor of botany at Oxford Sidney Howard Vines; the third was by the autodidact, philosopher, and science expositor Herbert Spencer. All three relied on the inheritance of acquired characters to complete the evolutionary process as they saw it. Because Spencer became deeply embroiled in a public jousting match with Weismann in the 1890s, I will develop his position in Chapter 17. The other two challenges, both published in 1886, would have been on his mind after the next summer when Weismann once again crossed the channel for his participation in the BAAS.

George John Romanes

George John Romanes is often identified as a comparative psychologist and successful experimenter on the rhythmic motions of medusae. After successfully concentrating on the natural sciences at Cambridge, Romanes worked under the guidance of the physiologist Michael Foster and later in the physiological laboratory of John Burdon-Sanderson at University College. Beginning in 1874 and continuing until Darwin's death in 1882, he played a strong supportive role to the elderly founder of modern evolution theory.[71] It was his studies and writing in the comparative aspects of animal behavior and intelligence that first brought Romanes into contact with Darwin. Despite Darwin's initial reserve, Romanes appears to have maneuvered into Darwin's good graces and into the latter's reliance on his suggestions and experiments. Central to their interaction was Darwin's theory of pangenesis, which had been cast into doubt by the experiments of Francis Galton. His talented cousin notwithstanding, Darwin would not give up his "provisional hypothesis," and Romanes presented him with the opportunity, as it turned out a chimerical one, to demonstrate experimentally the truth of his invention. Both appeared intent upon saving a theory of heredity that also served as a mechanism for the inheritance of acquired characters.

After Darwin's death, Romanes pursued further public evolutionary discussions by publishing three observational and theoretical works on comparative psychology, which involved him in debates with C. Lloyd Morgan and other more rigorous psychologists. Romanes's most controversial involvement with evolution, however, came when he hurled himself into the fray with an elaborate proposal about the existence of a new factor for the origin of species. He christened his child "physiological selection," and without special supportive evidence tossed this new mechanism out into the world, as though it were a can of oil onto a fire. His seventy-five-page essay in a prominent science journal, despite its originality, was repetitive, diffuse in language, and cumbersome to follow.[72] In short, Romanes claimed that Darwin, with his mechanisms of natural and sexual selection, had provided an explanation for adaptation of organisms rather than for the origin of species. Furthermore, Darwin and his followers had not satisfactorily responded to Fleeming Jenkin's "swamping effect," which had emphasized that a new variation was bound to be overwhelmed by successive generations of cross breeding between the original and newly varying individuals. Geographic and behavioral isolation, an increase in the rate of appearance of new variations in each generation, and a reduction in the size of the interbreeding group were a number of ways that Darwinians and non-Darwinians alike had wrestled with the problem. Weismann and Wagner's exchange over isolation in the 1870s represented some of the same general effort.

To remedy what he obviously viewed as conceptual failures, Romanes argued that a variation in the reproductive system of incipient species caused a relative sterility between those individuals who possessed it and those that did not. Natural selection over time, Romanes continued, would eliminate those individuals who wasted their efforts by attempting to cross breed. "The facts of variation have been known; and the facts of specific sterility have been known," Romanes repeated as he closed the postscript to his lengthy presentation, "but hitherto it has not been suggested that the former may stand to the latter in the relation of cause to effect, or that when a particular kind of variation occurs in the reproductive system a new species must necessarily ensue."[73] The physiological, indeed morphological, variation "caused" an effective isolation of the two groups and allowed them to diverge into separate species. Unfortunately, Romanes's argument relied largely on deductions, and when it presented documentary evidence, it exclusively relied on correlations rather than an identification of the actual morphological changes, which presumably appeared in reproductive systems.

There appeared swift and somewhat annoyed reactions in the letters to the editor of *Nature* as well as a full-scale rebuttal launched by Wallace in the *Fortnightly Review*.[74] Some of these reactions addressed minor points, but Wallace and his close friend Meldola captured the heart of the arguments when they pointed out that where Romanes had written of "independent variations," Weismann more than a dozen years earlier had "very ably" discussed the same evolutionary condition, which he had called "amixie" ("Amixia"). Directly

confronting the significance of "physiological selection," Meldola reasoned
through several paragraphs that natural selection, as presented by Darwin
and Wallace, had the capacity to bring about divergence by seizing upon the
relative physiological and morphological isolating traits. "In other words,
'physiological' appears as one particular phase of natural selection, and as far
as we can see there is no reason why there should not be other modes of variation leading to the same result by acting indirectly upon the reproductive
system. But all such modes of variation would still be subject to development
or suppression by natural selection."[75] When Romanes in turn peppered the
editor with letters, it was clear he was not going to give in easily.[76]

What is fascinating about the affair of Romanes and "physiological selection" is that it burst onto the English scene the year of Weismann's first visit.
It was a very English style of disagreement: letters to the editor of *Nature,* a
focus on the adequacy of natural selection to do the job Darwin and Wallace
had variously assigned it, and above all, questions concerning geographic isolation and distribution. All of the participants considered themselves to be
followers of Darwin. Although Darwin had allowed it, and Romanes at the
time required it, the matter of the inheritance of acquired characters remained
in the background. By the time of Weismann's second trip, however, Lamarckian inheritance had become the focus of a scheduled discussion at the
meetings of the BAAS and helped galvanize additional letters to the editor in
Nature. Within two years, the attention of English biologists had abruptly
turned away from Romanes and his self-promotion and had become focused
on Weismann and the full range of his research. The first indicator to appear
was a complimentary letter to the editor on Weismann's paper on "polar
bodies."[77] This was followed by acquiescent letters summarizing Weismann's
1882 theory of death and 1883 essay on heredity.[78]

That these essays by Weismann needed abstracting five and six years after
their original appearance in Germany further suggests a need on the part of the
English scientific community to come to terms with how Weismann's mul-
tisided research fit in with the ongoing struggle to supplement evolution
theory. This need was amply demonstrated by the translation and publication
of eight of Weismann's recent essays, all of which together left the impression of
the extraordinary development in Weismann's thoughts. The translations
were shepherded through the collection and publication process by the acknowledged expert on the utility of coloration and markings in insects, Edward Bagnall Poulton. Poulton quickly became one of Weismann's strongest
supporters in England. His compact navy blue volume of translated essays

emphasized the anti-Lamarckian side of Weismann's Darwinism by concluding with two essays dedicated completely to the issue of the inheritance of acquired characters.[79] As a collection, these translations were unique and an enormously useful benchmark for the history of heredity in England. Only after a second volume of further essays were issued by Poulton did a Parisian publisher bring out a French translation, and only after becoming convinced of its utility, did Gustav Fischer, who after all held the copyrights to the original essays, publish a German collection that mirrored Clarendon Press's original venture.[80]

Sidney Howard Vines

A year younger than Romanes, Sydney Howard Vines had not yet celebrated his tenth birthday when *The Origin of Species* first appeared. His education as a boy was varied, as he grew up in Paraguay, received his first formal education in Moravia, Germany, and completed his schooling in England.[81] He toyed with the possibility of going into medicine until he became captivated by physiology and then with botany, the first as a student in medicine at Guy's in London and as a demonstrator for Thomas H. Huxley at the Royal College of Science in South Kensington, the second when assisting William Thiselton-Dyer in a course of botany also in South Kensington and in the Jodrell Laboratory at Kew Gardens. Having earned a science degree in London and a first in the natural science tripos at Christ's College, Cambridge, Vines was awarded in 1876 a fellowship and lectureship in botany at his college. To better prepare for his teaching duties, Vines spent the summer term of 1877 working at Julius Sachs's institute in Würzburg. In 1879–1880, he sojourned again to Germany to work with Anton de Bary in Strassburg and again in Würzburg. Enormously impressed, particularly by Sachs and his approach to plant physiology, Vines later reflected in retrospect that "I sought and found in Germany what was unobtainable in my own country."[82]

Back in Cambridge, his courses in practical botany, emphasizing laboratory experience and plant physiology, turned out to be enormously successful. Vines trained a legion of students, many of whom found paying positions in botany and found time to pursue research on the influence of light on growth of plants and on the metabolism of protein particles (aleurons) in protoplasm. Nevertheless, throughout his Cambridge period Vines constantly had to struggle to improve laboratory conditions and equipment over the resistance of the elderly university professor of botany and an old-style plant taxonomist Charles Cardale Babington.[83] It was perhaps for this reason that when the Sherardian professorship in botany at Oxford became vacant, Vines applied and won the appointment and directorship over the botanical gardens and herbarium. It was just prior to and during this transition that Weismann would have interacted with Vines at the BAAS in Manchester and through a critical exchange on each other's views.

Vines's perspective on botany and life in general is quickly revealed with a glance at his published *Lectures on the Physiology of Plants* (1886).[84] Vines pointed out in the preface that this was a work that had remained in gestation for ten years. These were the years when Vines first worked in Würzburg with Sachs, expended considerable time in producing exciting lectures, and struggled to establish a respectable botanical laboratory. It was also a time when illness intervened, and Vines simply began to wear down from overwork. When it finally appeared, coincidentally in the year of Weismann's first visit to England, his *Lectures* turned out to be far more modest in scope than Julius Sachs's *Vorlesungen,* which had supplanted his celebrated *Lehrbuch* in 1882.[85] Nevertheless, Vines, producing a genuine physiology textbook, followed closely Sachs's model. Both were devoted to exploring the bodily functions of plants rather than displaying the traditional botanical systematics based on the form of leaves and flowers and focused on geographical ranges. The starting point of both consisted in lectures on the "Structure and Properties of Plant-Cells"; they moved on to basic functions, and metabolism and irritability in plants, before they both closed with discussions of plant reproduction.

Both Vines and Sachs presented a plant physiology that was cell based. They referred to some of the most recent literature in nuclear cytology—particularly to Strasburger's *Zellbildung und Zelltheilung* (1880) and to an advanced monograph by Flemming (1880). In a short chapter on nuclear and cell division, Sachs closely tracked Strasburger's work in presenting the movement of chromatin and spindle fibers during nuclear division, but as with Strasburger in 1880, Sachs seemed unconcerned about any special significance that might be attributed to such patterns other than to suggest a relationship between cellular and nuclear division, and conjugation of single-celled algae and fertilization.[86] In his turn, Vines provided even scantier details of nuclear division and generalized only by implying that the nucleus itself might be important for the nutrition of the cell. Sachs and Vines, both following Strasburger (1880), continued to balance the activities of the fluid protoplasm and chromatin. What was missing from both texts was any indication that Flemming's indirect nuclear division, his "mitosis" of 1882, had radically changed the discourse. "The fact that a nucleus has now been found in almost all living cells," Vines cautioned, "seems to shew [*sic*] that the presence of such a body is of importance to the life of the cell, but it is not yet possible to ascribe to it any definite function."[87]

It was within this framework that Vines, but not Sachs, described and critiqued a number of contemporary theories of reproduction and evolution in his final chapter. As a young English botanist he, to be sure, gladly began with Darwin's theory of pangenesis. Gemmules, for Vines, provided explanations for both vegetative and sexual reproduction and for the difference between the asexual formation of spores and sexual formation of gametes. Their different origins clarified why the offspring of vegetative and spore reproduction were similar to the parents, while the products of crosses and hybrids tended to be intermediate in form between two parents. The theory of pangenesis accounted for deviations from the hybrid average by invoking "prepotent" parents and postulating contrasting vigor and affinities among gemmules. Phenomena of reversion and alternation of generations could easily be encompassed by the differences in the activities of gemmules, while their accumulated bulk clarified how the production of a polar body prepared the way for fertilization and appeared to thwart accidental parthenogenesis. Finally, following Darwin's lead, Vines explained that "the increased variability which is induced by changed conditions, cultivation for instance, is ascribed to an influence on the reproductive organs which leads to an irregular aggregation of the gemmules in them, some being in excess and others deficient." Thus, according to his interpretation of Darwin, new variations, acquired through external influences, could be transmitted as they produced new gemmules for the next generation.[88]

Vines recognized the hypothetical nature of Darwin's gemmules. He also felt it was incomprehensible that spores could bear the total number of different gemmules to account for the variety of phenomena over a span of generations. Nevertheless, there was unquestioned acceptance on Vines's part of the external origins of new gemmules and new variations. The inheritance of acquired characters was an essential element, along with natural selection, of Darwin's evolutionary process. According to Vines, this feature also remained central to Brooks's revision of pangenesis. When Vines turned his attention to Nägeli and Strasburger, he addressed the difference between the former's belief in an extracellular idioplasm and the latter's recognition of indirect nuclear division. Both indicated the location of the hereditary material. In contrast to Darwin, however, Vines recognized that both Nägeli and Strasburger were set "in assuming inherent variability, and in denying the inheritance of acquired characters."[89] This Vines did not accept. Was he then ready for a new theory by Weismann?

There were three aspects to Vines's concept of the organism and his consequent rejection of the germ-plasm theory. First, Vines appeared inclined to a holistic interpretation of heredity and variations. Drawing upon his discussion of spore production among algae and fungi and among gamete productions in higher plants and animals, he found the postulation of gemmules or germ-plasm to be "altogether unnecessary."[90] He did not deny the importance of nuclear eliminations in these associated phenomena. They were for him, however, not an argument for centering an explanation of reproduction on the nucleus. They were simply an indication that the cells involved returning to an embryonic state before reproduction could begin. In this context, Vines rejected Weismann's claim that the first polar body simply contained the "ovogenetic" germ-plasm needed for the secondary features of the egg. Vines interpreted in a similar fashion Strasburger's differentiation of the generative and vegetative nuclei found in the pollen tube. It was the protoplasm as a whole, not the nucleus alone, that specified heredity and development for Vines.

Second, Vines took issue with both Strasburger's and Weismann's insistences that the nuclear contribution of the male and female gametes were in essence equivalent. Instead, he subscribed to the hermaphroditic views of Minot, van Beneden, Balfour, and many others of the reproductive elements. As we have seen they held that the gametes were functionally complementary. For this group, the polar bodies represented an elimination of the "essence of maleness" prior to fertilization. The mutual attraction of the opposite pronuclei during fertilization, the overwhelming predominance of female over male examples of parthenogenesis, and the commonplace contrasting behaviors of the male and female gametes provided further arguments to fortify his hermaphroditic claims. Finally, while comparing various theories of variation, Vines supported Darwin's contention that the conditions of life, not a special character, determined the nature of variations. This claim placed him in opposition to Weismann's apparent contention that all variations in phylogeny must be derived from the earliest stages acquired by single-celled organisms.[91]

His dismissal of the nucleus as the operative structure of heredity placed Vines in opposition to Strasburger's (and Flemming's) concept of the nucleus and to Weismann's germ-plasm theory. Instead, he tendered his allegiance to another botanist, Nägeli, when he accepted the more holistic concept of the latter's idioplasm and a belief in the organism's "inherent tendency to a higher organization" in phylogeny. "Evolution," Vines explained in closing his text, "is no longer a matter of chance, but is the inevitable outcome of a fundamental property of living matter."[92]

Weismann must have met Vines in Manchester when he spoke and participated in the session on his polar body research. It is likely that they had some discussion on the subject at the time, for Vines later sent him a copy of his *Lectures,* which was unknown to Weismann prior to his visit.[93] Weismann sent a cordial acknowledgment in return but also requested comments about his interpretation of polar bodies as they might apply to plants. By 1888, Weismann had become convinced that at least with animals the first and second polar bodies held different biological meanings. There is no record that either Weismann or Vines continued the informal exchange, but after the publication of Weismann's *Essays* a year later, such a scenario would have been unlikely.

It would have been unlikely because Vines wrote in *Nature* a lengthy but curious critical review of the *Essays.*[94] He chose selections from the eight essays to document his complaints, which in itself was fair enough, but he failed to be sensitive to the fact that Weismann's ideas had developed over the eight years covered by the essays. Vines wrote as though the theory of the germ-plasm was already a formal rather than a heuristic theory for further research.[95] Central to Vines's critique, however, was Weismann's apparent illogical notion that through the division of "immortal" single-celled organisms, such as protozoa and algae, "mortal" cells could arise. The fallacy, according to him, hung on a negation of the word "immortal." In other words, how could the substance of cells, which were phylogenetically immortal, be transformed into cell lineages that were not? Weismann, on the other hand, saw this as a linguistic misunderstanding, a confusion of "immortal" with "eternal."

However, in retrospect the disagreement between the two appears deeper. On the one hand, Vines had demanded an explanation of how an immortal substance could transform itself into a mortal substance. On the other hand, Weismann insisted he was thinking in terms of cyclical processes, which can change when conditions change. As a consequence, with the emerging phylogenetic state of multiple-cell organisms, Weismann's argument continued, cell divisions may become qualitatively unequal—one daughter cell continuing the persistent cycle of divisions as before, while the other daughter cell led to nuclear changes, differentiation, and a definitive conclusion. Ultimately, the key to the concepts of Vines and Weismann differed with their respective understanding of cell division. Indirect cell division had introduced a transforming process that entailed much more than a simple halving of substance.

Cell division placed some nuclear lineages onto limited paths while the germ-plasm lineage remained on the traditional, potentially immortal path of its single-celled ancestors. "An immortal unalterable living substance does not exist, but only immortal forms of activity of organized matter."[96]

From such a basic misunderstanding flowed Vines's other disagreements with Weismann: over the relationship between germ-plasm and the differentiating soma; over the presence of germ-plasm in, let us say, begonia leaves, which can readily regenerate the entire plant; over the relationship between the cytoplasm and germ-plasm of the ovum; over the apparent contradiction between a continuous stable germ-plasm, on the one hand, and a phylogenetically, evolving germ-plasm, on the other; and over the development of variations within certain families of multicellular fungi where sexual reproduction appeared never to have taken place. These issues, in Vines's opinion, had already been addressed in the theory of pangenesis by Darwin, "who held that it is not the sudden variations, due to altered external conditions, which become permanent, but those slowly produced by what he termed the accumulative action of changed conditions of life."[97] The mechanics of what Darwin, Vines, and many others had in mind when they argued that exposure to external conditions over many generations provided transmittable adaptations was rarely made clear. In retrospect, we can imagine the need for a critical mass of similar gemmules for trait expression. Even Weismann would propose something along these lines when his germ-plasm theory came under further attack. As long as a holistic perspective prevailed, the debate over the inheritance of acquired characters would continue.

It is interesting to note that both Romanes and Vines presented their views as derivative of Darwin's. Neither wanted to repudiate natural selection, but they saw beyond this fingerpost of Darwin's theory by extending the Lamarckian side, made possible by the pangenesis of gemmules. This brought them onto a collision course with Weismann and many of the "pure" Darwinians in England. Although two do not make a crowd, it is worth noting that both Romanes and Vines were physiologists by training and inclination. They both were an indirect beneficiary of Huxley's and Foster's new program for biology, which was both Darwinian and functional. The integration of organic processes of a single body was uppermost in their view of life. It was thoroughly appropriate from their perspective that heredity be described by holistic rather than architectonic modes. As his trips to England made clear, Weismann was steeped in the cytology of polar body production and the passage of variations from one generation to the next.

14

Polar Body Research

1887–1891

As biology moved into the 1880s, it entered into an exciting period not only for the maturing and often contentious conceptions of evolution but also for nuclear cytology. Unlike research on evolution, cytology required institutionalization in a way the more descriptive studies of natural history did not. The German zoological and botanical institutes, now often located in the philosophical faculties, also provided a home for technological innovations in a way that had not appeared so imperative a decade earlier. These institutes provided space not only for microscopes, microtomes, and related apparatus and for storage space for chemical reagents and stains, but also for an associated library for the many new books and journals, for separate spaces for the instruction of a new wave of medical students and doctoral research projects, for aquaria and collections, and for advanced studies and shared projects. The generation of Haeckel, Weismann, Bütschli, Sachs, Strasburger, and van Beneden represented the creators of many of these institutes, and by the 1880s they had attained the status of leaders of independent domains and the directors of common research goals.

It should be emphasized that the new zoological institutes and the newly reoriented research institutes of botany were not alone in investigating the general occurrence of and variations in karyokinesis. Histologists, embryologists, cellular anatomists and physiologists, and pathologists, who normally worked in the medical faculties, were equally absorbed in investigating the new details of indirect cell division. By tradition and practice, these professions were as involved in developing cytology and pondering the implications for their own specialties and for understanding heredity and development as were the zoologists and botanists. Just as the cell theories of Schleiden and Schwann, of Henle and Kölliker, and of Remak and Virchow had unified the sciences of life in the 1840s and 1850s, nuclear cytology of the 1880s and 1890s, following in the footsteps of Flemming and Strasburger, riveted attention on the mechanics and purposes of germ-cell maturation and fertilization.

This required cytologists coming to terms with polar bodies and parthenogenesis and recognizing the radical difference between Flemming's "indirect cell division," more formally named by him "mitosis," and the reduction division during gamete maturation. Two decades would have to pass before enough was understood that Farmer and Moore felt compelled to coin the word "meiosis" in addition to mitosis. As we shall see in a later chapter, that term denoted a very specific "chromosomal" process that resulted in the halving of the number of chromosomes during gamete maturation. Cytologists needed to discover the nature and significance of those chromatin loops, which only in 1888 were designated chromosomes by Waldeyer, and to recognize and understand tetrad formation and the synapsis of chromatids. They needed to follow up on van Beneden's report of 1883, confirmed soon thereafter by J. B. Carnoy, that the number of chromosomes in the gametes of *Ascaris* was halved during germ-cell maturation, and generalizing the phenomenon into what Weismann called a "Reductionstheilung" (reduction division).[1] Cytologists had to demonstrate the continuity and individuality of chromosomes through microscopy and experimentation, a collective accomplishment that was achieved through the research of van Beneden, Oscar Hertwig, Rabl, and many others, but above all Theodor Boveri.[2] Ultimately cytologists had to recognize that the complexities of the maturation of germ cells and their conjugation in sexual reproduction must be joined with Mendel's recently rediscovered laws. This was independently worked out in 1902 by Theodor Boveri and Walter Sutton. At the time, Boveri had been deeply steeped in the nuclear cytology of the preceding two decades; Sutton, a young medical student, was benefiting from a year of tutelage in Columbia University's department of biology under Edmund B. Wilson, one of the great practitioners of cytology.

However, presented in this fashion the trajectory of nuclear cytology seems too preordained and backward looking. What happened between 1882, when Flemming published his classic book on indirect nuclear division, and the Boveri-Sutton synthesis of 1902 appeared at the time far more complicated and far more uncertain in its course. What strikes this historian, as he surveys the published literature, is the large number of investigators deeply in-

Figure 14.1. Weismann and members of his Zoological Institute, 1886.

Klaus Sander, ed. *August Weismann (1834–1914) und die theoretische Biologie des 19. Jahrhunderts. Urkunden, Berichte und Analysen.* Freiburger Universitätsblätter, vols. 87, 88. Freiburg: Rombach Verlag, 1985. Helmut Risler, "August Weismanns Leben und Wirken nach Dokumenten aus seinem Nachlass," p. 34.

volved in exploring germ-cell maturation from a variety of angles. These investigators represented all levels of achievement, from well-recognized authorities to apprentice cytologists, working in myriad countries practicing modern biology not only in western Europe but also in the United States, imperial Russia, and even Japan. Collectively, their research examined a wide variety of organisms ranging across the spectrum of the animal and plant kingdoms. Many of these organisms revealed new secrets about the cytology of reproduction, which, when detailed, might or might not be variants of a more general pattern. This retrospective survey also reveals how generally it was believed that the nuclear cytology of germ-cell maturation, in keeping with

Flemming's prediction, held the key to heredity. If it was plant hybridization at the end of the century that put biologists in touch with Mendel's classic work of the 1860s, it was the nuclear cytologists, studying the mechanics of maturation and fertilization in both plants and animals, who quickly recognized how their work on animals could explain the hybridization patterns that Mendel and his followers had discovered.

Waldeyer's Review of 1888

Weismann's career extended fourteen years into the twentieth century. From 1885 to 1914, he remained steadfastly committed to developing a model to unite his views about heredity with his maturing defense of natural selection. It is only practical to present as separate chapters his work in the context of the intersection between two domains of biology. In later chapters, we will follow Weismann as he presented the two together. In this chapter, however, we will follow Weismann's attention to the nucleus while he examined the production of polar bodies and speculated on the significance of parthenogenesis. This will take us in his career from his essay on "Continuity of the Germ-plasm" to his extensive monograph on "Amphimixis" (1891). We will continue this side of his career with later discussions of *Das Keimplasma* of 1892. Significantly, Weismann subtitled this monumental work *Eine Theorie der Vererbung,* "a theory of heredity." We begin, however, with a general account of the history of polar bodies, reduction division, chromosomes, and parthenogenesis, all phenomena that drew cytologists' attention during this period.

For reasons of economy, it makes sense to examine a single review paper on the subject, 122 pages in length. This was written in 1888 by the Berlin anatomist Wilhelm Waldeyer in the midst of the elations and uncertainties over the discovery of indirect cell division. Today, this essay is known to the history of biology simply as the first time Waldeyer coined "Chromosomen" as a "terminus technicus" for the less technical terms of "Schleifen," "chromatische Elemente," "primäre Schleifen," "Karyosomen," "Fäden," "anses chromatiques," "bâtonets," "loops," and "rods," all of which had proliferated throughout the literature since the late 1870s. By the time of his writing, Waldeyer felt that such terms were either too cumbersome or already too theory-laden to be serviceable. As a technical term, "Chromosomen" was simply descriptive and neutral. "If the one I propose [i.e., this new term] is practically applicable it will become familiar," wrote Waldeyer, though he cautiously added, "otherwise it will soon sink into oblivion." It is interesting to note that Waldeyer himself did not consistently use the term in his review.[3]

There was much more to Waldeyer's review than a successful neologism. As a survey of over 200 contributions to nuclear cytology, the vast majority having been written in the previous fifteen years, it presented the reader with Waldeyer's assessment of the cytological evidence for and speculations about the function of indirect cell division.[4] A broad descriptive outline of the phases of karyokinesis had been generally accepted since Flemming and Strasburger had joined in agreement, but the finer details had to be worked out. The cell network during the resting stage remained controversial. Was it a single skein or a composite of primary and secondary threads? The indirect division of plant cells appeared somewhat different from that of animal cells in many respects. The staining pattern and relationship between the nucleoli and the network remained unclear. The origin of the nuclear membrane, whether coming from the general protoplasm of the cell or from a delimiting layer of protoplasm with a morphological status, was hotly contested. As descriptions of cell multiplication from more organisms became available, the contrast between achromatic threads, spireme threads, loops, and nucleoli became more obvious—but could they be interpreted as a range of different staining reactions as well as separate morphological entities? The polar rays found by van Beneden in *Ascaris* eggs were not found by Strasburger in plants; only later would they be detected there, too. The apparent disagreement over the origin of the nuclear spindle, whether from the cytoplasm, from the achromatic threads, or from the chromatic nuclear constituents, heightened the confusion about whether these structures might play a role in transmission. And so it went.[5]

That indirect cell division occurred in most situations of growth and repair appeared to be a given, and Flemming's important discovery of longitudinal division of the chromosomes, assumed to achieve their equal distribution to daughter cells, was in Waldeyer's mind a significant achievement. But not all would agree. Taking issue with Flemming's generalization about the lawlike regularity of the mitotic process, the Belgian cytologist J. B. Carnoy insisted that "all the phenomena of cell division are variable; none of them seemed essential."[6] Waldeyer agreed that "the karyokinetic process presented so many abnormalities that it is as yet impossible to lay down any sound general rule."[7] Thus, Hermann Fol was among a minority of cytologists who "would deny to the chromatic nuclear figure any essential significance."[8] Because of the uncertainties of many details, because the basic events of spermatogenesis were not yet determined, and because there were well-documented cases of direct cell and nuclear division, Waldeyer seemed to be at a loss to explain the relationship between direct and indirect division. At several points, he expressed frustration over the matter and took a nonessentialist approach in distinguishing between the two. "I must confess," he admitted frankly, "I cannot rid myself of the idea that nuclear division is a single process, with Remak's simple amitotic division as the fundamental form."[9] The bottom line of Waldeyer's paper was that the unique events of indirect cell division were still uncertain.

Fertilization

Waldeyer's review was not confined to the recognized and controversial details of karyokinesis. He directed the second half of his essay to an elaborate discussion of how indirect nuclear division affected contemporary views about fertilization and heredity. Thus, the relationship between the sperm and the germinal vesicle continued to be questioned. From the early observations of Purkinye and Baer, to Goette's more recent study *Entwicklungsgeschichte der Unke* (1879) on the development of the fire toad, there prevailed the general feeling that the germinal vesicle ("Keimbläschen," i.e., the nucleus of the developing egg) and its germinal spot ("Keimfleck," i.e., the nucleolus) simply vanished as a cytological entity before fertilization. For Waldeyer, this belief precluded a morphologically rooted explanation of this important process.

An alternative view, held earlier by Johannes Müller, then Carl Gegenbaur and van Benenden (1870), considered that the germinal vesicle underwent change before fertilization; whereas Haeckel, in addition to agreeing with both views, maintained that the germinal spot remained intact. Bischoff and Fol believed that the germinal vesicle continued whole but that the germinal spot disintegrated. It was a great achievement then when in 1875 Bütschli, van Beneden, and, most notably, Oscar Hertwig reported that the germinal vesicle persisted as the egg's "pronucleus"—this expression, first coined by van Beneden, was useful at the time, but even it was heavily theory-laden. All three investigators recognized that this female pronucleus upon fertilization joined with the male pronucleus, which most likely was derived from the head of the sperm.

The recognition of copulation by the two pronuclei was in itself not a prelude to Flemming's discovery of indirect cell division, but the question remained what this copulation entailed. Waldeyer suggested that this important demonstration of nuclear continuity highlighted the need to understand the various nuclear structures involved, such as the chromatic threads, the microsomes, pole spots, astral rays, and the spindle. Did the copulation of the pronuclei consist of a fusion ("Verschmelzung") of some or all of these elements? Or did it simply consist of an intermingling of enduring parts? Did fusion mean that fertilization was identifiable as a single event? Did intermingling imply that fertilization was really the initiation of first cleavage of the next generation?

Polar Bodies

Part of the story of coming to terms with fertilization involves understanding the preparatory process of egg maturation. As with so many cellular phenomena, the first observations of polar bodies were made years before there were enough details to generate speculations about their significance. Korschelt and Heider record that L. G. Carus made the first observations of these bodies. His sighting appeared in an embryological study of the pond snail ("Teichhornschnecke") four years before the appearance of Baer's first volume of *Entwickelungsgeschichte* (1828) and fifteen years before Schleiden and Schwann simultaneously published their papers on cell theory (1839).[10] Although maturing gastropod eggs and their polar bodies tend to be comparatively large, it is hard to discover how Carus had interpreted them.[11] Pierre-Joseph van Beneden and Fritz Müller, also working with gastropods, commented on these structures in the 1840s, and Charles Robin identified them as "globules polaires de l'ovule" in a treatise in 1862. Müller, who first recognized that these bodies appeared close to the point where the plane of first cleavage occurred, gave them their identification as "Richtungsbläschen." The accepted term soon became "Richtungskörper." Robin's term and the English expressions "polar bodies" and "polocytes" reflect the same coincidence. The 1870s brought renewed interest in these structures as cytologists began putting together the details of nuclear division.[12]

It was left to Bütschli (1876) and Oscar Hertwig (1877 and 1878) to ascertain with a large degree of certainty that polar bodies were the outcome of two cellular divisions of the germinal vesicle absent an intervening resting stage in the egg and before the union of the female and male pronuclei. The apparent "expulsions" implied that both the chromatic and achromatic structures should be observable in a continuous process of two consecutive divisions. Confusing the matter, however, was the fact that their expulsion and the sperm's entry happen in different sequences in different species. Thus, with the lamprey eel the sperm enters the egg between the two expulsions; in *Ascaris*, the sperm enters after both expulsions. During the late 1870s, it was also recognized by many that the first polar body, undergoing an additional division of its own, might produce a secondary—that is, a third—polar body. Although the daughter cells of these unique divisions were dramatically different in size compared with the egg, expulsion in these terms led cytologists such as Edward L. Mark (1881) to describe polar bodies as "abortive eggs," which encouraged others to place their production in a phylogenetic as well as embryological context. Beneden's outstanding work on *Ascaris* of 1883–1884, as described in Chapter 11, set the stage for Waldeyer's evaluation of first expulsion. "It is evident," he concluded, "that the whole process may in fact be regarded as a karyokinetic division of the whole egg-cell with very unequal products."[13]

The somewhat earlier work of Salvatore Trinchese, the newly appointed professor of anatomy and comparative physiology at the University of Naples, appeared to complement the impression that the polar bodies were abortive eggs produced by indirect cell divisions of the maturing egg. In 1880, he demonstrated that the large polar bodies of naked land snails (Gymnobranchiata) might occasionally be fertilized by additional sperm cells and even undergo some cleavage of their own. When Weismann became deeply involved in the developmental significance of polar bodies a few years later, he made a special effort to acquire the species of naked snails used by Trinchese.[14]

Waldeyer's Goal

The goal of Waldeyer's detailed discussions emerged when he turned to the relevance that cytology bore for theories of fertilization. He felt that the first theories, which recognized the penetration of the spermatozoon, obscurely referred to a dynamic action on the germinal vesicle. Others adopted a "chemical" theory that seemed equally vague. Once it was clear that the head of the entering sperm became the male pronucleus and possessed a complex nuclear structure that became intimately associated with the equally complex female pronucleus, there emerged two other ways of envisioning fertilization. Oscar Hertwig and his followers adopted assorted fusion or "Verschmelzung" theories, which held that a complete merger of the structures of both pronuclei into a new nucleus constituted the essence of fertilization. Accordingly, polar body expulsions entailed the halving of the nuclear mass before its doubling upon fertilization.

A fusion theory presented a problem for Waldeyer, however, who criticized it throughout his review. In particular, Hertwig, by this time Waldeyer's colleague at the university in Berlin, had been unclear about how and what nuclear structures fused. Then there was the "nuclear replacement theories" most recently championed by van Beneden, who was following the lead of Minot and Balfour. He argued that the organism was hermaphroditic in composition, that the nuclei of its cells contained both female and male material, and that the latter was expelled in the form of polar bodies only to be replaced upon fertilization by new male material from the sperm. Hence, his term of "pronuclei" identified both sperm and egg nuclei that had not yet joined and so had not regained the missing part of the organism's hermaphroditic state. Again, Waldeyer was not impressed. To start with, how did the traits of the paternal grandfather become transmitted to the grandson? Was there a morphological difference between the male and female material? Finally, Waldeyer mentions a "pure nuclear theory" of fertilization presented in the very recent publications of N. K. Kultschitzky, who remained unclear about the role of the polar bodies and spoke of a union, but not fusion, of the nuclear parts.[15]

According to Waldeyer, the multiple types of organism played a role in the confusion. On the one hand, in the echinoderm *Strongylocentrotus* studied by Hertwig and the plants studied by Strasburger a union of the "threads" of the pronuclei appears to occur during the resting stage before first cleavage. In *Ascaris,* studied by van Beneden and many others, fusion appears to occur only at a late stage of first cleavage itself, and even then the "nuclear threads do not fuse in *Ascaris megalocelphala.* They certainly come close together temporarily in a division-figure." Fertilization might occur with the union of the two pronuclei, Waldeyer reasoned, but fusion of their essence—that is, their chromosomes—did not follow.

Chromosomes and Their Reduction

Up to this point, if we accept Waldeyer's review of 1888 as setting out the important issues about fertilization, there appeared a singular lack of interest in the number of chromosomes involved in the process. To be sure, van Beneden and others had easily noted the different numbers of chromosomes in the first and second polar bodies and the related halving of the number in the mature unfertilized egg cell in *Ascaris.* Nussbaum had noticed a similar phenomenon in the daphnid *Leptodora nigrovenosa.* Moreover, Boveri had noted that there were two varieties of *Ascaris,* the univalens form with two chromosomes and the bivalens form with four chromosomes in the early developmental stages.

All these zoologists had also recognized the importance of the reduction of chromosome numbers for understanding heredity. At this moment in the understanding of maturation and fertilization, there was a minimum of concern about what such chromosomal numbers might mean. Comparing this with the chapter on fertilization and its table of organisms and their numbers of chromosomes in the first edition of Wilson's *The Cell* of 1896, one realizes that the next ten years would bring a widening perspective of nuclear division and heredity theory.[16]

A Fruitful Collaboration between Weismann and Chiyomatsu Ishikawa on Polar Bodies

In 1925, E. B. Wilson, by then the foremost cytologist in biology, in retrospect praised Weismann's significant essay on polar bodies and parthenogenesis of 1887.[17] "The first fruitful attempt to analyze the internal phenomena of reduction," Wilson explained, "was made on purely theoretical grounds by Weismann (1887) in one of those brilliant essays on heredity which contributed in so important a way to the enlargement of our views concerning cytological research." Wilson packed a good deal into this sentence and into the accompanying two paragraphs that explained the grounds for his strong endorsement.[18]

It is clear that Wilson was focusing on the theoretical dimension of Weismann's postulated "reduction division," and that this appeared to contrast with the equational division, which characterized Flemming's longitudinal division of chromosomes by mitosis. (According to Wilson, Weismann coined the expression "Reducktionstheilung.") In a general way, Wilson's interpretation is indeed right, as shown by his essay on "Continuity." Weismann had been thinking in terms of the chromosome as comprising the material elements of heredity. Wilson, however, was writing thirty-eight years later at the zenith of classic chromosomal genetics, and the process of meiosis had long since been firmly established. In 1887, Weismann was focusing far more on the existence of a single polar body in parthenogenetic reproduction and linking his findings with the contrast between his theoretical "continuous germ-plasm" and his equally theoretical "ovogenetic nucleoplasm." He, too, was thinking in terms of a chromosomal theory of heredity, but he had not yet developed a model for it in a way that would be later familiar to Wilson. The "brilliance" of Weismann's essay of 1887 and of a whole series of his related papers on polar bodies and parthenogenesis, I find, emerges from his practice and articulation of the interplay between facts and theories. What gets left out of Wilson's evaluation is the nature of the empirical basis upon which Weismann was building his case.

Through a series of eleven statements and papers, ranging from the "Note" at the end of his "Continuity" of 1885 to a paper on what he called "Para-copulation" in 1889, Weismann reported on the occurrence of polar bodies in the parthenogenetic eggs of daphnids, rotifers, and ostracods.[19] He was assisted through much of the investigation by the Japanese student Chiyomatsu Ishikawa, who had joined the Zoological Institute for the summer semester of 1885 and remained until he completed his dissertation in the winter semester of 1888–1889.[20]

Together they discovered single polar bodies in the parthenogenetic eggs of the rotifer *Callidina bidens* in November 1886.[21] Exactly a year later, Ishikawa spent a two-week stint on the Bodensee with Weismann while they collected more winter eggs of cladocerans. The Japanese scholar also traveled to the Naples Zoological Station with Weismann to continue microscopical studies on polar bodies during the first three months of 1888. Weismann described Ishikawa as serving "as his eyes," and it is evident that the latter was extraordinarily gifted in the manual techniques of nuclear cytology and at working through Weismann's research goals—so successfully, in fact, that he became coauthor with Weismann on six of the last papers in the series. This was a privilege, incidentally, not accorded to any of Weismann's other sixty-seven doctoral students. Ishikawa's close collaboration with Weismann reveals not only his personal accomplishments but is relevant to the issue of Weismann's training of students. Despite the extensive collaboration on polar bodies, Ishikawa wrote his dissertation on a totally different subject, on the manipulation experiments with *Hydra*.[22]

The monograph, so extolled by Wilson, spelled out the evidence the two had collected to date. Weismann's talk before the British Association for the Advancement of Science (BAAS) on polar bodies and an up-to-date report on his and Ishikawa's findings in Naples, both also published in 1887, gilded the lily.[23] First, after his initial discovery in 1885, Weismann extended their examinations of the summer eggs to include eight genera of cladocerans. He and Ishikawa examined a single species of ostracod, in which they could follow the polar body and its secondary derivative as far as the sixteen blastomere stage. They also found a single polar body with two species of parthenogenetic rotifers. Second, not only did they demonstrate that a single polar body alone was associated with parthenogenetic eggs, but they clinched their argument by showing that the subsequent karyokinetic division in such eggs led directly to first cleavage. In other words, doubters could not claim that Weismann and Ishikawa had simply missed the usual second polar body found in fertilized eggs.[24]

Parthenogenesis

Significantly, the study of parthenogenesis became Weismann's back door to explaining sexual reproduction and heredity. What especially caught Wilson's eye, when he praised Weismann's thirty-eight-year-old monograph, was embedded in the long and detailed section on the second polar body. Granting his ill-conceived argument that the first polar body expelled the ovogenetic nucleoplasm, Weismann turned to the hereditary material. We have already seen that a decade of cytological research culminating in the work of Flemming, Strasburger, van Beneden, and others had focused attention on the chromosomes during egg maturation. Picking up from van Beneden and Roux, Weismann concentrated on the loops, soon to be baptized by Waldeyer as chromosomes. Condensed from the spireme, these possessed structure, and Weismann assumed them to be qualitatively different in both a phylogenetic and developmental sense. He described and illustrated each loop as a linear accumulation of ancestral germ-plasms. Although he did not invoke Haeckel's biogenetic law, the same historical and material considerations informed his perception. Each chromosome represented a lineal arrangement of an organism's ancestral past. It faithfully duplicated itself with each longitudinal splitting during each indirect cell division of embryonic development—but this pattern had its limitations.

The beauty of Flemming's discovery of the longitudinal splitting of chromosomes and description of karyokinesis was that it guaranteed a precise mechanism for the exact distribution of the nuclear parts to each daughter cell. By 1887, most cytologists agreed that these nuclear structures must be reduced by half during the maturation of the egg to make way for the parallel hereditary material of the sperm. Putting this reduction into chromosomal terms, Weismann spoke of "another kind of karyokinesis, in which the primary equatorial loops are not split longitudinally."[25] Hypothetically this "other kind" might occur in one of two ways, he reasoned: (1) the chromosomes might not split but instead be segregated as undivided structures to the two daughter cells, in this case to a polar body and the maturing egg, or (2) the chromosomes might split transversely ("if such division ever occurs").[26] Either way opened the possibility of a reduction in the number of chromosomes and consequently in a halving of the collective ancestral germ-plasm and their unique hereditary qualities.

What Weismann constructed, to be sure, remained theoretical, but it was not without recent empirical input.[27] Both van Beneden and Carnoy had described the plane for the karyokinetic division in the formation of the polar bodies in *Ascaris* (i.e., the bivalens form with four chromosomes in the "diploid" state rendering eight half-chromosomes) to be at right angles to the standard plane of traditional karyokinesis. Carnoy had further recognized that the first polar body expulsion began with four chromosomes, which at the equatorial plate became eight. The process ended with only four chromosomes apiece in the first polar body and the egg. He went on to observe that the four chromosomes, still visible in the egg, were further reduced by half with the expulsion of the second polar body. In modern terms, he had seen a 4–8–4–2 sequence in chromosome numbers in *Ascaris* during the two maturation divisions.

For Weismann, the first step from 4 to 8 to 4 represented a normal karyokinesis with a longitudinal division of each chromosome before or at the equatorial plate. The second step from 4 to 2 depicted the true reduction of chromosome number and hence entailed, for Weismann, a different kind of karyokinesis. By interpreting Carnoy's observed pattern in this way, Weismann again reinforced his germ-plasm theory. The first division repeated the pattern of both qualitative and quantitative embryonic divisions that had already been sanctioned by Roux. During first maturation division, the karyokinetic division was different from the standard indirect nuclear division only in that it involved the separation and expulsion of the "qualitatively different" ovogenetic germ-plasm. In short, it consisted of a late developmental differentiation on the part of the parent. On the other hand, the second division was truly unique in that it empirically established the required reduction of ancestral germ-plasm in the form of halving of chromosome number. Utilizing recent findings by Carnoy on arthropods and even more recent comments by Flemming, Weismann recognized that in spermatogenesis only the standard karyokinesis had been observed.[28] Reduction of the ancestral germ-plasm contained in the male gamete remained an unexplained phenomenon. Nonetheless, it must be noted that at this time Weismann continued to be cautious about his conclusions. He admitted that there were alternative views

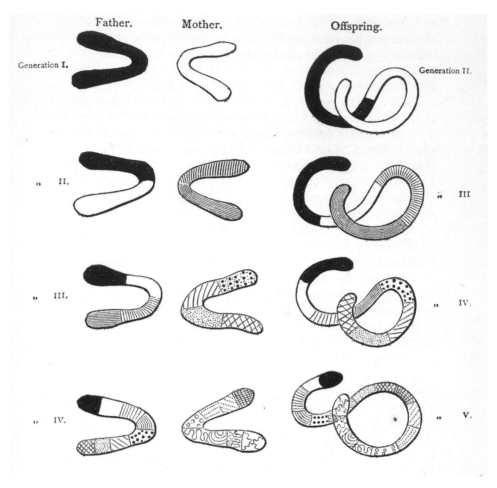

Figure 14.2. Illustration by Weismann of the reconstitution of father and mother chromosomes during four successive generations.

August Weismann, "Über die Zahl der Richtungskörper . . . ," *Aufsätze über Vererbung und verwandte biologische Fragen.* Verlag von Gustav Fischer, Jena (1892), p. 428.

and different ways of explaining the same phenomena, so he felt compelled to caution his readers that "a secure basis of facts is only very gradually obtained, and there are still many conflicting opinions upon the details of this process."[29]

Besides invoking chromosome numbers as support for his presumed contrast between the first and second polar body expulsion, Weismann became aware of the importance of the physical orientation of chromosomes during

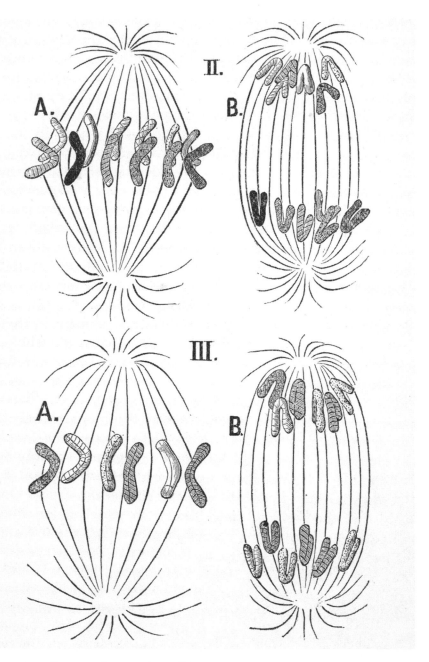

Figure 14.3. Illustration by Weismann of chromosomal action during reducing division number 2 and equational division number 3.

August Weismann, "Über die Zahl der Richtungskörper . . . ," *Aufsätze über Vererbung und verwandte biologische Fragen*. Verlag von Gustav Fischer, Jena (1892), p. 437.

division. He soon was forced to deal with the problem in *Ascaris* in which the spindle alignment during first polar body formation assumed a 90° orientation to the normal direction of expulsion. The consequence, Boveri had argued, should result in half of Weismann's hypothesized ovogenetic nucleoplasm ending up in the embryo rather than in the first polar body.[30] In response, Weismann diagramed how this would not be the case if the chromosomes, as well as the spindle, assumed the abnormal orientation.[31] He was thus inspired to elaborate that many developmental faults might arise through errors in faulty distributions of chromosomes. "It may perhaps be pointed out by this example that we will have really seen in this and in other imperfections of the mechanism of nuclear division one of the most significant sources of deformities, that a way opens here, which leads to the explanation and to the understanding of the origin of many teratological appearances."[32] Chromosome numbers and distribution mattered in both development and heredity. We will find in Chapter 20 that Boveri, although not a student of Weismann nor an exponent of Weismann's germ-plasm theory, followed in his footsteps by expanding and perfecting a chromosomal theory of heredity.

In a way that recapitulated his recent work far more faithfully than Wilson or other historians have been able to do by looking backward, Weismann brought together his accomplishments in the study of polar bodies. The law of polar body numbers demanded an explanation, which he gave by allocating a different function to the first and second polar bodies. Parthenogenesis and reproduction by fertilization played different roles with respect to his postulated ancestral germ-plasm. The first preserved the parental hereditary constitution of a single parent; the second moved toward the rebuilding of a new combination out of half the germ-plasm of each new parent. Weismann also insisted that the quantity of the germ-plasm in the mature egg determined whether it would produce parthenogenetically or through fertilization. In this respect Weismann echoed the tradition of invoking biological growth as part of the explanation of heredity. He continued the break from that tradition, however, by insisting that no two offspring of the same parents would be endowed with the same hereditary tendencies. "Thus sexual reproduction is to be explained as an arrangement which ensures an ever-varying supply of individual differences."[33] From the perspective of the cytology of microscopic cladoceran polar bodies, Weismann throughout kept his mind's eye on the neo-Darwinian process.

15

Protozoa and Amphimixis

Starting in 1825 with the reign of King Ludwig, Munich grew throughout the nineteenth century in population, wealth, and prestige. A growing city, ten times the size of Freiburg, Munich sits on the Bavarian plateau north of the Austrian Alps and receives the waters of the Isar River, which runs through the center of the city. Ludwig and his successors put a premium on building a fine city center of architectural reflections of Italian style. Two large boulevards converged near the imposing Ludwigskirche, which still dominates the skyline today and from which on a good day one may see the Starnbergersee to the southwest, the Chiemsee to the east, and the Bavarian Alps to the south. Renaissance-style buildings, palaces of Bavarian kings, and a special concert hall adorned with frescoes highlight the city center. Public fine arts and private holdings helped shape the future of this capital city of the Bavarian kingdom. The old residence of royalty and Italian rococo palaces sit close to the center of activity. Architecture, art, the Academy of Science founded in the middle of the eighteenth century, and a royal library, which by the end of the nineteenth century contained over 1 million printed volumes, reflected the high premium placed on learning. It is estimated that the university, founded in the fifteenth century and moved to the Bavarian capital in 1826, had 130 professors and lecturers and over 2,100 students by 1882, a decade before Weismann was offered a professorship there. Munich was also a crossroads to Italy, the Austrian empire, and Germany. Art, culture, architecture, mountains, and lakes were at the city's disposal.

As we have seen, Weismann had earlier turned down the opportunity to move with the youngest members of his family to this exciting city. In the winter of 1891–1892, however, he took leave from Freiburg to write in Munich, while Bertha and Julius attended school in the Bavarian capital. As he reported to Frau Antoine Lüroth back in Freiburg, his semester visit was for work, not sightseeing. During this leave of absence he finished *Amphimixis,* edited the German edition of his *Aufsätze,* and all but completed his best-known book. Work—yes, the evidence is in print. "It is to be sure a beautiful and stimulating winter for us. . . . I have seen to be sure that life here would be somewhat different here than at home. It is however curious how little one uses what one daily can have. I have come into no gallery and since the Mozart-performances came to an end also into no theater! The entire day work, and evenings, here and there, something entertaining."[1]

Weismann did visit the Zoological Institute, at the time led by Richard Hertwig; he was introduced to Theodor Boveri, who was working there as a research scholar, and on occasion he mentions other Munich colleagues such as Friedrich Wilhelm Baeyer, professor of chemistry, and Carl Maximillian von Bauernfeind, professor of geology and paleontology at the University of Munich.

In his correspondence, however, his focus appears to have been on completing his manuscript of *Das Keimplasma.* Weismann originally hoped that the German and English texts would appear at the same time, but the translation took longer than expected. Weismann wrote the preface and signed the German edition May 19, 1892. The English edition, translated by W. N. Parker and Harriet Rönfeldt, was completed on November 28, 1892, and was published in 1893. Curiously, G. H. Parker of Harvard revised chapters 13 and 14. Weismann dedicated the German volume to Rudolf Leuckart on his seventieth birthday and the English translation "To the memory of Charles Darwin."

Amphimixis, 1891

Chronologically Weismann's monograph *Amphimixis* (Plate 2) lies squeezed between the chromosomal studies carried out with the assistance of Ishikawa and his monumental statement about heredity in his *Keimplasma.* During this four-year period from 1889 to 1892, the publishing house of Gustav Fischer had become deeply involved with a decade's worth of Weismann's individual essays. It had been indirectly involved with the first and second editions of the English translations and with a French version of Weismann's *Essays upon Heredity.* Because the firm of Gustav Fischer held the copyrights to the original German essays, it followed suit and brought out a German edition of the same collection. A reprinting of *Amphimixis* appeared as the last essay in both the second English and the only German editions, but not in the French—freshly off the press from the previous year.[2] In both the English and German prefaces, Weismann expressed the importance of *Amphimixis,* which elaborated upon the themes of eleven of his most important essays of the 1880s:

I shall attempt to explain, as clearly as possible the close connection existing between certain apparently isolated problems and the subject of this essay, which although mainly concerned with so-called "sexual reproduction," forms in reality *the keystone* for the whole edifice of all previous essays. My object is to express more fully than before, the thought that the process, which we are accustomed to regard as reproduction, is not reproduction only, but contains something *sui generis,* something which *may* be connected with reproduction proper, and in the higher plants and animals *is* so connected, but which is entirely separate in the lower organisms.[3]

In Weismann's words, *Amphimixis* was not a mere capstone (*Kappe*) but a keystone ("Schlussstein"). It not only concluded but bound into a unified whole ten years of intensive research and writing. It drew together Weismann's essays on "Duration of Life," and "Heredity," on "Continuity of the Germ-Plasm" and "The Significance of Sexual Reproduction," and various statements about the sufficiency of natural selection into a central theme. It capitalized on his microscopic research on hydromedusae, daphnids, gall wasps, and ostracods. It made the most of nuclear cytology, the recent understanding of spermatogenesis, his own significant researches on polar bodies, and the rapidly expanding knowledge of parthenogenesis. Finally, it exploited the recent advances in the developing subdiscipline of protozoology. Weismann brought all these phenomena and more under a single arch to emphasize that reproduction, chromosomal reduction, and chromosomal doubling are distinct, albeit often closely connected, processes. They all helped him shape a model for heredity.

There was no specific point during this decade of research when Weismann explicitly broke from Haeckel's view of heredity that reproduction was basically an "overgrowth" of parent(s) to offspring. With the conviction that conjugation and sexual reproduction were not simply processes for "the maintenance of life" but reflected a complex and selective "mingling of individualities" (i.e., ancestral germ-plasms or ids), Weismann had irrevocably abandoned that older tradition.[4] Laboring with a more immediate concern, however, Weismann explicitly challenged the contemporary view promoted by a handful of influential biologists that chromatin reduction entailed an elimination of a male substance or essence in order to prepare for its replacement through fertilization. This view received support with the recognition that polar body expulsion accompanied egg maturation. Referred to variously as "rejuvenation," "vitalization," "compensation," or "hermaphroditic" theories of heredity, this view of maturation and fertilization was advanced in one form or another by Charles Sedgwick Minot, Francis Maitland Balfour, Eduard van Beneden, Viktor Hensen, Thomas Wilhelm Engelmann, Otto Bütschli, and Émile Maupas. Weismann had rejected this viewpoint earlier in 1887 as he postulated the function of the second polar body. A hermaphroditic theory of heredity, "which was quite justified at the time when it originated," he had at that time insisted "must be finally abandoned."[5] He would return in depth to this subject throughout his *Amphimixis*. His germ-plasm theory supplied an alternative, and as it developed in print both here and later, the theory served a double role by providing the arena for an intricate interaction with contemporaries over factual events and details and as a lofty space for the presentation of his general theory of heredity, development, and evolution. As he entered the last decade of the century and as he found it increasingly difficult to carry out his own microscopic research, it became harder for Weismann both to present and to factually support his ideas at the same time. As he accommodated to the most recent research on spermatogenesis, he complained in *Amphimixis* that "my impaired eyesight, which has so often put a stop to microscopic research, has again rendered the continuation of this research impossible."[6]

Maturation of Germ Cells

The structure of *Amphimixis* is logical, although given Weismann's "warping" style, it requires patience and an understanding of contemporary research to winnow the essence of his theoretical conclusions from the contemporary matters of biological details. The first section dealt particularly with the complex chromosomal events of maturation in both ova and spermatozoa. At stake was the increasing recognition that the changing number of the chromosomes was critical to understanding gametogenesis. Focusing on the polar bodies, van Beneden had demonstrated that both maturation divisions of the ova of *Ascaris* involved a reduction by half of the number of chromosomes from the number found in the oocyte, but he failed to distinguish between the *univalens* and *bivalens* forms. In 1890, Oscar Hertwig elegantly demonstrated that

spermatogenesis in both forms of *Ascaris* entailed a process of two maturation divisions which resulted in the ultimate halving of the number of chromosomes. Where in ovogenesis three of the four daughter "cells" became polar
bodies, in spermatogenesis all four cells ended up as functioning sperm cells.
The two processes were homologous and resulted in a formation of the
respective gametes by means of a double "mitotic" division. Hertwig's demonstration made it absolutely clear to Weismann that he must abandon his cherished claim that the first polar body represented the expulsion of ovogenetic
or histogenetic germ-plasm. "My previous interpretation of the first polar body
as the removal of ovogenetic nucleoplasm from the egg must fall to the ground."[7]
It was a low-key *mea culpa,* but he was ready to move on.

By 1890, the details of these two maturation divisions were recognized as
more complex than normal somatic divisions. In the first place, the number
of those chromatin strands, which were considered to be complete chromosomes in prophase of the first maturation division, appeared to be twice the
number of those found in the preceding oocytes and spermatocytes. It meant
that in the numerically simple case of *Ascaris bivalens* there appeared to be
eight instead of four "chromosomes."[8] By the end of the first maturation division, however, these eight chromosomes were reduced to four in number.
Furthermore, the nuclei of the resulting daughter cells did not disappear into a
normal resting stage but immediately initiated the second maturation division. This division, in turn, resulted in a second halving of chromosome
number, that is, from four to two. The whole sequence of maturation with *A.
bivalens* coursed through a pattern of 4–8–4–2 chromosomes. After being
supplemented upon fertilization by the sperm's two chromosomes, the
number of chromosomes in the egg was restored to the normal number of
four. Less certain, however, was a consistent longitudinal division of the
chromosomes, so heralded by Flemming at the end of the 1870s.

At first blush, this seemed to be an unnecessary step in the maturation process. As it became clear that the gametocytes of both sexes of *Ascaris* underwent a doubling of their chromosomes at the outset of the double reduction
divisions (i.e., 4–8–4–2), Weismann grabbed at a ready explanation, which
his evolutionary commitment provided. "To my mind the doubling of the
idants before the 'reducing division' possesses this very significance," he insisted, for "it renders possible an almost infinite number of different kinds of
germ-plasms, so that every individual must be different from all the rest. And
the meaning of this endless variety is to afford the material for the operation
of natural selection."[9]

With only its four chromosomes, this aspect of gamete maturation became readily obvious in *Ascaris bivalens.* Weismann's colleague in mathematics, Jacob Lüroth, persuaded him that a doubling of chromosomes would be advantageous, and with organisms that had many more chromosomes the advantage increased. Weismann quickly recognized that an organism with twelve chromosomes could provide without prior doubling just 924 combinations, but with a prior doubling could produce 8,074 combinations. Following similar calculations, Weismann saw that an organism containing twenty chromosomes would without a preliminary doubling produce merely 184,756 combinations.[10] On the other hand, with the initial doubling of chromosomes they would have the possibility of producing an extraordinary 8,533,606 possible combinations. To be sure, these were only abstract mathematical calculations, but they reinforced Weismann's neo-Darwinian message. Such calculations also reveal the evolutionary thrust to his increasing interest in the nuclei of gametes.[11]

There were limits in linking nuclear cytology to his evolution theory, but Weismann seemed prepared to accept them. At this point, however, his evidential arguments about the number of nuclear rods shifted to model building. He argued that the idants were made of ids, an assortment of ancestral germplasms of a hereditary lineage. Given the longitudinal division of the idants and the requirement that two maturation divisions yield half the number of ids for reproduction and the next generation, Weismann recognized that these ids must be arranged in a single line in order to achieve a halving rather than a quartering of the number of ids. This, at least, must be the case with both forms of *Ascaris,* with a butterfly and snail investigated by Platner, and with the salamander studied by Flemming.

A reprint sent to him by a young privatdozent in Göttingen, Hermann Henking, however, threw a monkey wrench into Weismann's conclusions. Working on spermatogenesis and ovogenesis in the red bug ("Feuerwanze") *Pyrrhocoris apterus,* Henking described and diagrammed a wreathlike pattern for the chromosomes at the equatorial plate before the first maturation division. No doubling of the chromosomes appeared to take place at that moment, so there was not the expected 24–48–24–12 doubling and halving following the 4–8–4–2 pattern of *Ascaris;* instead, the Feuerwanze exhibited a 24–24–12–12 pattern.[12]

Henking presented a serious problem because he had argued that the fault lay in attributing a doubling of the chromosomes prior to the first maturation division. Weismann had proudly assumed that the doubling had occurred at the outset of first division in order to expand dramatically the number of vari-

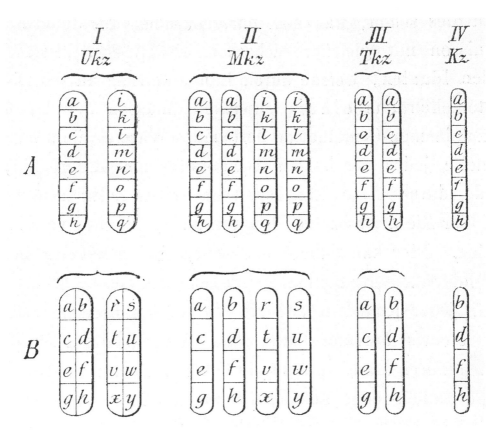

Figure 15.1. Diagram of two theoretical combinations of ids in the separate idants (chromosomes). During maturation division in row A, the single arrangement of ids results in their halving. Row B reduction division results in a quartering of the number of different ids.

August Weismann, *Amphimixis oder: Die Vermischung der Individuen*. Verlag von Gustav Fischer, Jena, 1891, p. 49.

ations for natural selection. Henking's solution presented an awkward situation. "If this explanation [Henking's] be valid," Weismann wrote in retrospect, "the interpretation offered above of the doubling of the idants in the mother-cells of *Ascaris* must fail, and I doubt whether any other feasible explanation is to be found."[13] He thanked Henking for a preliminary review of the forthcoming publication and asked for figures and preparations if possible. The subject was critical enough for Weismann to set Arnold Spuler, a gifted doctoral student at his institute, to double-check Hertwig's work. Having been assured that the chromosomal work of both Hertwig and Henking were dependable, Weismann came to the conclusion that there must be an additional wrinkle to reduction division. His conjecture, drawn from a closer examination of Henking's illustrations, held that the wreath form found with the idants of *Pyrrhocoris* at the first equatorial plate was actually double idants. This insight saved the original 4–8–4–2 pattern of reduction division. It also reinforced his belief that the first division revealed a qualitative reduction in the number of ids and that the second division concerned an equational division where the idants were longitudinally halved in preparation for fertilization.

Just as important, the qualitative reduction occurred through a transverse division across the wreath, and because this wreath could be divided by an arbitrary diameter across the circle, the process increased even further the combination of ids.[14] There was nothing sacrosanct in a given pattern. Evolution, Weismann figured, could develop several ways of multiplying potential variations. The *Ascaris* pattern of reduction entailed two selections of whole idants with different combinations of ids. The *Pyrrhocoris* pattern entailed a transverse cleavage of double circular idants followed by their longitudinal halving so as to provide even further combinations of ids.[15]

Heredity in Parthenogenesis among Protozoa

For investigating gamete formation, Weismann had to rely on correspondence, advanced students, and assistants at his institute. He could no longer perform sustained studies of chromosomes. When he turned to the second section of *Amphimixis,* he turned to his specialty, parthenogenesis. Ishikawa, his "eyes" in Naples, had left for Japan after completing his dissertation in 1889, but Weismann did not lack for support at the institute for a discussion of inheritance in parthenogenesis. Ishikawa's place was soon taken by Valentin Häcker, who became the second assistant at the institute in 1890. The place of first assistant was held by H. Ernst Ziegler, who had joined the institute in 1887 as a privatdozent and in early 1890 had been granted an unsalaried außerordentlicher professorship. He had a broad background in evolution and microscopy. In addition, there was Weismann's brother-in-law, August Gruber, whose work on protozoa had contributed substantially to Weismann's understanding of conjugation. A volunteer assistant, Otto vom Rath, who had completed his dissertation in Strassburg, also joined the institute in Freiburg to pursue research on germ-cell production in arthropods. It is evident that vom Rath not only produced some fine studies on germ maturation but that his affable nature set a collegial tone for all.[16]

When Weismann and other members of the institute turned to parthenogenetic reproduction, they had to resolve a double problem. The subject of parthenogenesis changed dramatically in 1890 with the work of Platner and O. Hertwig. Both had demonstrated the parallel between spermatogenesis and ovogenesis in that the two processes both contained two maturation divisions and produced two daughter and four granddaughter cells. Both spermatogenesis and ovogenesis, as described previously, appeared to result in a reduction of chromatin material in the form of chromosomes. To be resolved was whether there were two kinds of reduction: one of mass, and one of "types" of chromatin—one of quantity, the other of quality. Logic and the counting of chromosomes suggested that the first division must take place to prepare for the later increase in mass at fertilization.

At the time, however, inspection could not distinguish between either chromatin particles or chromosomes. So how was one to assign the two types of reductions to the two maturation divisions in gametogenesis? With its single maturation division and with no fertilization to complicate the issue, parthenogenesis appeared an ideal process to solve the matter, at least for Weismann. If Weismann accepted that a doubling of idants occurred before the formation of the equatorial plate and their random distribution to the daughter cells occurred at the completion of the first and only maturation division, he could claim that a visual qualitative reduction of the idants (i.e., chromosomes) and a hypothetical quantitative reduction in the diversity of ids could occur at the same time. Reviewing Weismann's old slides of both sexual and parthenogenetic reproduction in the brine shrimp *Artemia,* vom Rath found support for Weismann's calculated conclusions.

To argue his point further, Weismann drew upon parthenogenetic lineages of the ostracod *Cypris reptans,* an organism he had cultured in freshwater aquaria since 1884. There were two varieties of the same species, and these he had meticulously kept separate in isolated cultures. One variety possessed a consistently "clay" yellow shell marked by a few dark green spots; the other variety displayed a dominant dark green foreground of spots leaving visible only a little of the clay yellow background. The coloring and spots remained constant through seven years and forty generations of cultivation and selected isolations at the start of each new culture. Thousands of individuals were examined, and it was apparent that descendants of each variety remained constant to type with only "minute differences." In 1887 and again in 1890–1891, however, single individuals of the opposite variety suddenly appeared in an aquarium holding the other type. The transformation could happen in both directions. Weismann felt confident in his isolation procedures and the constancy of external conditions. Instead, "these remarkable phenomena," he presciently commented, "must certainly be ascribed to internal causes, viz. to changes in the composition of the germ plasm."[17]

His explanation of these results depended on two propositions. First, that the two varieties shared a not-too-distant common ancestry that sexually multiplied. Second, with the natural selection of the parthenogenetic mode of reproduction, two varieties of different germinal composition thrived, each containing some of the ancestral germ-plasm (the ids) common to the other variety. The conclusion that followed validated his belief that the first maturation division must be the qualitative reduction in ovogenesis. If a few ancestral ids of one variety persisted in the germ-plasm of the other, a doubling of the idants through equational division followed by their random halving at the time of the single polar body expulsion would occur (the 4–8–4 portion of the *Ascaris* pattern). Statistically, over a number of generations that ancestral residue could prevail. Parthenogenetic lineages of *Cypris reptans* provided a demonstration of the combinatorial possibilities of the germ-plasm. Future research by others would challenge this conclusion about the roles of first and second reduction division, but Weismann's general view was that "we can safely affirm that in parthenogenesis individual variation exists, which, as in bisexual reproduction, has its foundation in the composition of the germ-plasm itself, and thus depends on the heredity, and is itself inheritable."[18] Qualitative and quantitative divisions and recombinations of the chromosomes during maturation divisions became essential concerns for the nuclear cytology of gametes during the next thirty years.

Conjugation and Fertilization in Protozoa

But then there were the protozoa! First seen by Leeuwenhoek in the seventeenth century and collected together with many other microscopic organisms under the informal category of "animalcules" in the eighteenth century, these motile, unique, and *inter se* widely different organisms entertained and puzzled viewers from the outset. From the point of view of taxonomists and physiologists, the diversity of structures and complexity in behavior expanded exponentially as we enter the nineteenth century and beyond. A reflection of the state of affairs with regard to nuclear division alone is captured in a passage in the first quarter of the twentieth century by Gary N. Calkins, a colleague of E. B. Wilson at Columbia University and one of the foremost protozoologists of his day. While commenting upon the success that zoologists had achieved in describing the similarity in mitosis throughout the metazoan and plant worlds, Calkins threw up his hands when he addressed the same phenomenon in single-celled organisms: "In Protozoa, on the other hand, there is no one type of nuclear division common to all forms. Here we find gradation, in association of constituent nuclear and cytoplasmic kinetic elements during division resulting in an enormous variety of division types." As he continued, the reader can sense the frustration seething beneath his formal, somewhat awkward style:

> These vary in complexity from a simple dividing granule to mitotic figures as elaborate as in the tissue cells of higher animals and plants. Some observers see in these diverse types a possible evolution of the mitotic figure of Metazoa and use them as one would use the separate pieces of a picture puzzle to reconstruct its past history in development. Terms like "promitosis" (Naegler), "mesomitosis (Chatton) and "metamitosis" (Chatton) may serve a useful purpose to indicate general types of association of nuclear and cytoplasmic elements during division, but when an effort is made to give a specific name to each step in an increasingly complex series the result is a confusion of terms which defeats the useful purpose intended.[19]

In the 1880s, it was protozoan nuclear division that also interested Weismann and many protozoologists. The trajectory of understanding about protozoa reproduction, however, was not a simple curve. Elsewhere colleagues and I have described the tortuous route from Siebold's convincing demonstration that conjugation involved an act of nuclear exchange between two ciliates of the same species.[20] During the intervening years, developments in cell theory and the advent of Darwin's theory of evolution had done much to change the intellectual landscape. By Bütschli's time "Protozoa" had come into their own as an identifiable phylum of single-celled organisms and taken their place below the simplest of metazoans. Views about their reproduction, however, still appeared guided by the spirit, if not the facts, of organ-based description of infusion animals. In the 1850s, it was widely accepted that protozoa reproduced by either fission or gemmation. So when Johannes Müller, whose interests, as we have seen, had veered from physiology and embryology to cell theory and finally to natural history, announced to the Berlin Academy and later to the Versammlung in Vienna that he and his students had found spermatozoa-like structures in *Paramecium,* and later in *Stentor* and *Kolpoda,* he threw open the possibility that that which Siebold had earlier identified as the nucleus and nucleolus might instead be seen as gonads.[21]

Ernst Haeckel and Friedrich Stein both produced a number of major works on protozoa and embryogenetic origins of protozoa. The former, emphasizing phylogeny, argued that protozoa had evolved from anucleated Moneren; while the latter accepted Müller's argument that the nucleus and nucleoli were either sexual organs, eggs, or embryos of parasites. So prior to the 1870s there was still much to be understood about protozoan maturation and conjugation and consequently of protozoan morphology and taxonomy.

Otto Bütschli and Protozoa

In 1859, when Heinrich Bronn published the first volume of his *Klassen und Ordnungen der Thier-Reichs,* he surveyed the simplest of animals mixed together with an assortment of sponges, radiolarians, mollusks, and infusorians. Collectively these were the "Urthiere" and represented what Bronn called the "formless animals" or Amorphozoa. His was a short volume of 122 pages and twelve plates, which represented a modest achievement in the natural history and zoology at that. After significant later pruning and other adjustments, it would become known by Siebold's term "Protozoa." Compared to this short work, Bronn's second volume in the series was devoted to the echinoderms and contained 434 pages and forty-eight plates; the third volume, embracing mollusks, contained a whopping 1,500 pages (exactly!) and 136 plates. The comparison may be taken as an informal measure of the state of knowledge in the natural history and zoology of the first three phyla of the animal kingdom. This comparison is further enhanced in realizing that it took Otto Bütschli the entire decade of the 1880s to rewrite the same first volume. Now forthrightly entitled *Protozoa,* Bütschli's second edition came in three parts, together comprising over 2,035 pages and seventy-nine plates. In twenty years, the study of protozoa had expanded many fold and become focused on a domain that had grown from an ancillary zoological novelty to an independent field of active investigation.[22]

Bütschli did not simply emphasize the classification of the protozoa but generalized across groups. This is especially noteworthy in his 600-page examination of the Ciliata within the part on Infusoria.[23] The Ciliata are the first protozoa examined today in beginning biology classes as Paramecium, Stentor, and Vorticella are readily found in hay infusions. Bütschli discussed in detail the many special internal and external structures of ciliates. He included sections on the contractile vacuoles, locomotion by means of cilia, the internal gelatinous mass and its chemical makeup, the formation of colonies, encystment, many physiological aspects, and above all, the behavior of micro- and macronuclei and the contrasting acts of copulation and conjugation. He ended his survey of the Infusoria with a bibliography of nearly 900 citations, two-thirds of the items written after 1859. There could be no doubt by this time that the Ciliata in particular were an important animal group deserving of intense investigation.

Now we have already recognized that Bütschli had described the disintegration of the micronuclei in the reproduction of ciliates in his Habilitationsschrift of 1876. A few additional words about his early career are worth interjecting at this point. Like Weismann a decade and a half earlier, Bütschli was a native son of Frankfurt. Like Weismann, he had been captivated by science instruction at the Senkenbergische naturforschende Gesellschaft. Bütschli's first university studies took place at the Karlsruhe Polytechnic Institute where he studied mineralogy, zoology, and chemistry and where in 1865–1866 he assisted Karl von Zittel in paleontology just before the latter's move to Munich. Bütschli's earliest publications came from this period and concerned mineralogical and geological subjects. A year of military service interrupted further studies in this direction, and upon his release from service, he sought out Rudolf Leuckart for guidance in becoming a zoologist.

At the time, Leuckart had just been appointed professor in zoology and zootomy at Leipzig and was willing to provide Bütschli with working space. Just as he had done earlier with Weismann in Giessen, Leuckart encouraged Bütschli to examine the microscopic events of fertilization and early cleavage. After a semester of advanced studies, Bütschli's career was again interrupted by military service—this time by the brief Franco-Prussian War. When he returned to civilian life for the second time, Bütschli secured an assistantship in zoology with Karl Möbius in Kiel. This proved to be an unhappy experience, so he assumed the life of an independent investigator in his native city of Frankfurt.

As described in Chapter 11, during these unsettled years in Bütschli's life he presented in 1873 his first important descriptions of the fusion in nematode eggs of the male and female pronuclei at fertilization. At the same time, he followed the subsequent metamorphosis of the combined nucleus during first cleavage.[24] It was during this freelance period that Bütschli also completed the research that gained him his *venia legendi* and a position at the

Technische Hochschule in Karlsruhe. There is a fascinating parallel between Bütschli's early career and Weismann's: both were natives of the free city of Frankfurt, both had formative experiences with the Senkenburg, both studied chemistry (Weismann for his MD, Bütschli for his PhD in mineralogy), both had unsatisfactory initial experiences at a Baltic university, both resorted to an interlude period of freelancing in their home city of Frankfurt, and both turned for a semester of retooling under Leuckart's tutelage before each completed an outstanding Habilitationsschrift.

Bütschli was more than fourteen years younger than Weismann and his entry into zoology focused on nuclear rather than gross cellular developmental phenomena, but soon after the appearance of his Habilitationsschrift, referred to here as his *Studien,* Bütschli became Weismann's junior colleague as professor of zoology and director of the Zoological Institute at the older and better-funded Baden University in Heidelberg.[25] After the two briefly crossed swords over the matter of life and death in the 1880s, their personal interaction remained minimal.[26]

Bütschli's lengthy monograph not only extended his nematode studies on fertilization but allowed him to compare somatic nuclear divisions in the cells of other organisms, such as snails, certain insects, and the chick. The work was microscopically comparable to Oscar Hertwig's studies of the same years, both of which, as we have seen, appeared prior to the definitive works on indirect nuclear division of Flemming and Strasburger at the turn of the decade. Unique to Bütschli's *Studien* was its extended third chapter, which occupied more than twice the space of the two previous chapters put together. It was here that he delved into the microscopic realm of the protozoa, which for nearly fifteen years became his obsession. Up to this time copulation, that is a fusion, had been seen in several groups of protozoa. In contrast conjugation, a temporary lateral joining, or syzergy, of two individual organisms had been observed in many ciliates and possibly in flagellates and amoebas. This event appeared episodically, sometimes in "epidemics." Little was understood, however, about the behavior, and it was even less clear whether it was a product of external conditions or internal forces. Because of his earlier examination of fertilization and nuclear division in metazoans, it was natural for Bütschli to focus on the role of the nuclei.

It was recognized by this time that ciliates possess a large nucleus, soon to be identified as the macronucleus. This appeared to fragment and then disappear during the conjugation process. In contrast a single or, depending upon the species, several nucleoli, soon to be identified as the micronuclei, divided and some of its fragments appeared to be exchanged with the other conjugant. The micronucleus or its parts seemed to be the necessary structure in conjugation. Bütschli, too, was uncertain of the initiating cause of the process, but he was convinced that conjugation represented an act of rejuvenation analogous to fertilization with metazoans.

There were differences, to be sure. Bütschli pointed out that "the fertilization act with higher animals, as far as we know, amounts to a complete fusion of spermatozoon and egg cell. The usual form of conjugation among the Infusoria differs from this because a complete fusion does not take place." In his view, this difference boiled down to a matter of degree. Colonial *Vorticella* provided an intermediate step, for in this organism the small and larger individuals often fused together. Furthermore, where the process of egg maturation in metazoans included the expulsion of polar bodies, conjugation consisted of a disintegration of the secondary nucleus, the macronucleus. The reformation of the nucleus in the zygote of metazoans also appeared comparable to the reformation of the secondary nucleus in ciliates.

As problematic as these comparisons seem in retrospect, Bütschli insisted that his work on fertilization, cell replication, and conjugation firmly established the analogy between metazoan cells and individual protozoans. "Siebold's daring analogy of 1845 was vindicated by Bütschli in 1876."[27]

As he turned to surveying the protozoa for the second edition of Bronn's *Klassen und Ordnungen,* Bütschli had no difficulty accepting the primary feature of the whole group as single-celled organisms with one or more nuclei. Thoroughly committed to the group's phylogenetic location, he could not escape the Janus-like question of whether, when facing into the future, protozoa represented forerunners of both animals and plants, or whether, when looking toward the past, the group arose from non-nucleated "Moneren," as Haeckel would have it. Bütschli rejected out of hand the value of Moneren as a concept and begged off deciding whether protozoa, as a taxon, contained ancestors of both animals and plants or just animals alone. The first option suggested that the taxon was polyphyletic in composition; the second that it might be monophyletic. Polyphyly appeared a more practical consideration in the short run, but Bütschli preferred monophyly as providing a more natural grouping. As for his 2,000-page revision for Bronn's *Klassen und Ordnungen,* Bütschli divided the world of single-celled organisms into four classes. The last one, the Infusoria, contained the "Unterklasse" Ciliata, which by that time had received by far the widest attention from zoologists and would continue to do so in the future. This group contained material and activities that not only concerned protozoologists but were relevant to Weismann's quest to explain heredity.[28]

When in 1887–1889 he brought out the final part of his revision, Bütschli focused on the ciliates and radiolarians. This obligated him to review the significance of conjugation. Although the ciliates possessed several easily identifiable means of reproduction and metamorphosis, which included copulation (fusion), division (fission), and encystment, it occasionally happened that two individuals temporarily joined in what was at the time described as a syzegy. This behavior, referred to technically as conjugation, appeared to be neither a fusion nor fission. In fact, was it even correct to consider it a unique form of reproduction at all since a multiplication of organisms did not directly follow? Complicating the matter was the fact that it had been clear for some time that ciliates possessed at least two differently sized "nuclei." The German terms "Hauptkern" and "Nebenkern" for the two are particularly descriptive. One was large, sometimes lumpy, and it seemed to break up during the conjugation process. The other, often coming in multiples, was much smaller and, depending upon the species, was located in various internal places including infolds of the large nucleus. By the mid-1870s, it was widely accepted that the small nucleus regenerated, at least in part, the large nucleus after the latter had broken up.[29]

This left the dilemmas of whether the small nuclei were simply nucleoli and what the functions of such diverse structures with such different metamorphic histories could be. Previously Stein, one of the foremost students of ciliates, and Balbiani had interpreted the spindle formation of the small nuclei as spermatozoa and had maintained that conjugation represented a form of fertilization. Their view suggested that the larger nucleus was either an egg or an embryo. In the mid-1870s, Oscar Hertwig and Thomas Wilhelm Engelmann had proposed that the large nucleus, or its bearers, were in essence female and that the small nuclei or nucleoli, or their bearers, were in essence male. Forging a similar analogy, Bütschli had compared the fractionation and elimination of the large nucleus to the expulsion of polar bodies during egg maturation of higher organisms and supported previous suggestions that the smaller nucleolus was a cluster of spermatozoa. In retrospect, Bütschli commented on his own earlier view: "Bütschli," he impersonally reported, "sought to make parallel, even in their finer details, conjugation and fertilization."[30]

The key to unraveling the nuclear changes taking place during conjugation came only after Flemming and Strasburger had worked out the stages of indirect nuclear division at the end of the 1870s. After much uncertainty about the nature of the smaller nucleoli, its true nuclear nature was finally recognized by the formation of the mitotic spindle of indirect cell division. It was soon designated the "micronucleus." By this time it was also generally accepted that in the process of conjugation the micronucleus fashioned anew part of or the entire "macronucleus." The latter was assigned a developmental and metabolic function, and the former seemed to play a hereditary role.

Despite this general morphological advance, confusion still reigned among zoologists about the purpose of conjugation and why it occurred at all. Thus, in one way or another, leading microscopists initially interpreted conjugation as an elaborate form of protozoan copulation and sought to bring it more into line with the sexual reproduction of metazoans. The sexual analogy between protozoans and metazoans, however, only served as a bridge between a phylogenetic and a physiological explanation.

Like most of his contemporaries in the 1870s and 1880s, Bütschli believed that conjugation extended the life of the lineage by restoring the energy needed for division. In short, conjugation became an act of physiological rejuvenation.[31]

Émile Maupas and Conjugation

I have already discussed elsewhere Émile Maupas's extensive studies of protozoa.[32] What follows is a brief statement of Maupas's contributions to protozoa studies. An archivist by training and a microscopist of protozoa by avocation, Maupas spent most of his life in Algiers between the 1870s and his death in 1916, where he worked as government city librarian and studied microscopic life on his own time. Never married, he spent his free time in his bedroom where the natural lighting was good, working with and writing about microscopic life—particularly, at first, protozoa. It was a simple avocation, but Maupas took his studies as an independent scholar seriously, and he published professional-quality papers that were precise and replete with carefully collected data. A man of strong opinions about evolution, Maupas would publish two papers that built on Bütschli's work and brought him into direct conflict with Weismann's use of protozoa in his evolutionary accounts.[33]

Maupas's work involved studies of conjugation, and he published about his isolation experiments with ciliates and observations about the colonies that developed and then expired from these single organisms. He came to the conclusion that only through conjugation could a colony rejuvenate itself and continue to live and expand. Because of these results, he felt there was a life-endowing property brought about by conjugation (and by analogy by fertilization among metazoans).[34] Of the two papers Maupas published in the 1888 and 1889 volumes of the *Archives de Zoologie Experimentale,* the first described his experiments with building and examining colonies; his conclusion was summed up in his remark that "the final goal of conjugation is a rejuvenation, which manifests itself most especially through the development and reorganization of a new nucleus."[35]

As with his previous paper, the bulk of his much larger monograph the next year detailed conjugation. Fifteen elaborate plates, comprising hundreds of figures, portrayed the complex stories of nuclear metamorphosis, growth, and degeneration, exchange, union, and cleavage, which differed among them in their fine details, but all told the same general story about the nature of the macro- and micronuclei of protozoa during and immediately after the conjugation process. Maupus particularly observed the fragmentation and disappearance of the macronuclei during the process, the contrasting divisions of the micronuclei, and their preservation in what appeared to be analysis of mitotic divisions. He then noted the passage of half of these micronuclei to the opposite partner of the conjugating pair and their union with the micronuclei left behind. This exchange of portions of the micronuclei encouraged Maupas to analogize the process with a bifertilization among lower metazoans. What he felt he had discovered was the phylogenetic roots of sexual reproduction, which was then followed by a production of multiple offspring. The maleness and femaleness of the two nuclei seemed to him morphologically clear; only the delayed production of offspring after the conjugating "couple" disengaged seemed somewhat arbitrary. The nuclear exchange and reproduction were in protozoa demonstrably separate processes.

There could be no question that he was familiar with the recent accomplishments of Flemming and Strasburger and with Waldeyer's review of fertilization. Maupas associated his experimental and cytological work with the contemporary discussions about evolution and fertilization. Furthermore, he attacked in strong terms Weismann's contention that single-celled organisms were potentially immortal. In this context, he claimed to have refuted the "immortality theory," which had served as Weismann's phylogenetic justification for his theory about continuity of the germ-plasm. Maupas's claim was laced with strong criticism as well as insulting innuendos:

> The theory, as one sees it, is very simple; even too simple, I will say. It
> expands with the allure of geometrical and metaphysical conceptions,

the axioms of which and the propositions succeed themselves and are fatally linked, so to say. But it rarely goes so well with biological studies and applying these geometrical methods to them, one exposes oneself into taking a distorted path. It is this that has happened with Weismann. He would have done better to have returned to the experimental method, or, if he had neither the time nor the will, to let the questions remain undecided, the certain and definitive solution of which was unable to be obtained in another way.[36]

The above passage is only a sample of an unrestrained diatribe Maupas leveled against Weismann.

A year later, after detailing the process of nuclear transformation and after summarizing in twelve points the general picture that he could now draw from his many observations and pages of contemporary literature on nuclear cytology, Maupas returned to the attack.[37] By this time, he had read Weismann's essay on the "Meaning of Sexual Reproduction," and reserved his choicest comments for Weismann's belief in the importance of material and morphological exchange in both fertilization and conjugation. Weismann's comparison between the two was, for Maupas, "inexact" and "confused." He accused Weismann of identifying protozoan conjugation with metazoan fertilization, which was the result of a false identification of the individual protozoan with the compound metazoan. He saw it differently: complete fusion of two protozoans was equivalent to the fusion of two metazoan pronuclei. This new analogy left him room to assign a very different function to conjugation. When Weismann had offered the amphimixis of chromatin material as the goal of hereditary transmission in metazoans and had envisioned conjugation in ciliates as the phylogenetic precursor in this process, Maupas, although rejecting Weismann's germ-plasm theory, did not question a material explanation for the phenomenon of heredity. Heredity for him, however, was merely of "accessory and secondary" concern. The "primary and essential function" of fecundation "was in providing the capacity for the perpetuation of the species."[38]

Thus, Maupas reasoned from a very different direction. Conjugation and fertilization were primarily processes of a necessary rejuvenation. His cultures of senescent lineages of protozoa provided an experimental foundation for this claim. Here Maupas joined ranks with Bütschli, Engelmann, Hensen, and van Beneden. The latter even provided Maupas with the quintessential element in the analogy: "La fécundation est la condition de la continuité de la vie." ("Fertilization is a condition of the continuity of life.")[39] He was speaking from a physiological perspective. He may not have been able to explain rejuvenation in its ultimate details, but he considered the process to be in full accord with the demands of contemporary physicomechanical science. When Weismann argued that there was enough evidence from some uncommon species of crustaceans and insects to warrant the claim that parthenogenetic reproduction could continue indefinitely, Maupas demanded further proof.[40] Echoing van Beneden, Maupas deflected Weismann's comments. "As for me, I consider the alternation of agamic generations with the karyogamic fecundation as a primordial law of life, assuring its maintenance and its perpetuity. This law derives from its intimate and necessary relationship with the grand physico-mechanical factors which have preceded its appearance and are always the source from which it is going to draw its special energies."[41] Maupas was a mechanist and evolutionist believing in the inheritance of acquired characters, but he cared more for the operation of the cyclical reproductive process than for the phylogeny and mechanisms of variations.

This was the last of Maupas's substantial publications on protozoa. He soon turned his bedroom laboratory into a place for the study of rotifers and of parasitic and free-living nematodes. Maupas may have been an amateur in science in the sense of earning his living in a totally different career, but he was hardly a dilettante in his chosen avocation.

Weismann on Protozoa and Amphimixis (1891)

Weismann did not personally participate in the extraordinary advances of protozoology detailed by Bütschli and Maupas, but his first assistant surely kept him abreast of important developments. Weismann's brother-in-law August Gruber had joined the institute in 1878 after completing a dissertation with Leuckart, which according to Weismann contained "very interesting" observations on the spermatophors in freshwater copepods.[42] Gruber's *venia legendi* in zoology at Freiburg was awarded "in specie" after the approval of eleven of twelve votes of the academic senate who reviewed Gruber's previous research, his trial lecture, and his colloquium.[43] The implied preferential treatment notwithstanding, Gruber contributed in important ways to research at the institute. Having already done some research with protozoa before he arrived in Freiburg, he concentrated fully on the subject after his appointment.[44]

On relevant occasions, Bütschli and Maupas discussed Gruber's protozoa publications, and it seems probable that Weismann depended upon his brother-in-law's research and growing experience in this area. It cannot be doubted that the presence of Gruber at his institute helped focus Weismann in his quest to understand the relationship between conjugation and fertilization and reinforced his conviction that there was a difference between somatic and reproductive germ-plasm. In turn, Weismann certainly influenced Gruber in forming the questions he addressed with his research. It is, of course, relevant that Gruber adopted Weismann's germ-plasm theory and saw his work with protozoa contributing to its scientific status.

With advancement to a salaried außerordentlicher Professor in 1885, Gruber assumed many of the teaching and guidance activities at the institute. During Weismann's leaves of absence, he became responsible for its routine administrative duties. He became particularly acclaimed for demonstrating that the ciliate *Stentor* was unable to regenerate excised structures after its nucleus was removed. Although his scientific activities appear to have dwindled after the mid-1890s, he remained the nominal second in command at the institute until 1913 when he retired shortly after Weismann did. Gruber's own assessment of his career seven years after his retirement is encapsulated with a wry quip that reads far better in German than English: "Der bin ich geblieben—als Mensch ordentlich, als Professor außerordentlich" ("There I remained—as an ordinary human being, as an associate professor").[45] Nevertheless, his work was frequently cited by Bütschli and other protozoologists during the 1880s and early 1890s.

In addition to Gruber, Eugen Korschelt and Ernst Ziegler, both privatdozenten and assistants at the institute, published minor papers on protozoa during the period. Soon after he returned to Japan in 1890 or early 1891, Ishikawa published some noteworthy research on the nucleus during conjugation in the flagellate *Noctiluca*, which he communicated to Weismann.[46]

Thus, there were institutional resources at Weismann's beckoning as he devoted the last third of his monograph on amphimixis to conjugation. The relationship between conjugation and fertilization was a compelling subject. It was the academic duty of any German professor of zoology to incorporate the recent advances in the subject into his curriculum. Weismann had, however, two additional motives for closely following protozoan research. First, he had a decade earlier provided a speculative phylogeny for the separation of the germ cells (later, germ-plasm) from the soma, which ran back in time to single-celled organisms. He had argued that the indefinite continuation of protozoa lineages by means of fission justified his claim that the lineages of germ cells (later germ-plasm) in metazoans must be continuous and potentially "immortal" as well. The metazoan germ cells were simply the totipotent left-overs from the specialization process of metazoans. It was an argument worthy of Ernst Haeckel's *Generelle Morphologie*. Second, as he became increasingly focused on the heredity of individual variations, Weismann found it rewarding to extend the variations he found implicit in the chromosomal amphimixis of sexual reproduction among metazoans.[47]

Weismann made sure to spell out for his uninformed readers the nuclear details of ciliate conjugation as he introduced the subject at the outset of the third part of *Amphimixis*.[48] He detailed the two reduction and single equational divisions of the micronucleus, the exchange of one of two resulting micronuclei between the conjugants, and the reconstitution of the macro- and micronuclei from the cleavage nucleus before the conjugants parted and began their separate lives of reproduction through fission. His account followed the "wonderful investigations" of Maupas, and by borrowing an illustration and an enhanced diagram from Maupas's text, Weismann emphasized how important Maupas's research was for his own understanding of evolutionary events. Despite these compliments and attributions, Weismann did not let Maupas's earlier brusque remarks pass unnoticed.[49]

Substantively, Weismann was disinclined to homologize the two processes of conjugation and fertilization, but he leaped at the similarity in the behavior of the chromosomes, that is, Weismann's idants. In both cases, they showed distinctiveness and growth during the onset of reduction division. One feature of the sequence, which Maupas ignored but was close to Weismann's immediate concern, involved the pattern in the change in their numbers during the reduction divisions. Weismann surmised from Maupas's illustrations, which unfortunately were not drawn with an eye toward the number of chromosomes in question, that there might well be a doubling then a double halving as the micronuclei advanced through two reduction divisions. This pointed to a close parallel with what he had noted during the maturation of *Ascaris* and cladoceran egg and sperm cells. The pattern, he insisted, would "lead to a fresh grouping of the idants, just as in the analogous 'reducing divisions' of the egg- and sperm cell." Alas, the available details were insufficient, and dropping the matter, Weismann had left it as simply a "justified" assumption. On the other hand, the unique third "maturation division" of ciliate micronuclei was evidently an equational one that prepared the "male" and "female" micronuclei for interconjugant fertilization.[50]

The bottom line of the entire process of conjugation for Weismann underscored that "*variety in the individual nature of the hereditary substance is thus brought about by means of these divisions.*"[51] In this context, Weismann suggested that sometime between the appearance in evolution of bacteria and other non-nucleated monera and the rise of nucleated protozoa, the micronucleus had developed as a specialized "organ" for assuring amphimixis and producing variety. One might also add that Gruber's recent regeneration experiments with *Stentor* showed how formative that organ had become in ciliates. Throughout the animal and plant kingdoms, this organ of heredity, the nucleus, represented a conserving structure; as the organ of amphimixis—that is, introducing variations—it became a progressive structure. His was a neo-Darwinian view of the nucleus.[52]

Up to this point, Weismann had drawn upon Maupas's detailed descriptions of mitosis of the micronucleus and found it an important validation of his own discussions of reduction divisions in metazoans.[53] Weismann devoted the rest of the essay to addressing the other side of Maupas's work. This side directly attacked Weismann's phylogenetic views of the "immortality of the germ plasm." For Maupas, all organisms or parthenogenetic lineages suffered a natural death. This fate could be overcome in a number of ways. In metazoans, the act of fertilization was an act of rejuvenation. A new, pristine generation would come into being as the parent generation continued to degenerate and eventually to die through the wear and tear of existence. When carried out in a timely fashion, conjugation served as the counterpart of fertilization. It produced a rejuvenation of two lineages of fissile ciliates. Rejuvenation was a physiological claim about both reproduction and natural death.

Weismann focused on another level of the phenomena. To him, fertilization and conjugation were acts of amphimixis: mechanisms for increasing variations and indirectly promoting evolution through natural selection. Of course, these were not the only ways some organisms could reproduce. According to Weismann, fission, gemmation, spore formation, and budding were all part of an evolving arsenal for the multiplication of life. Amphimixis, however, from an evolutionary standpoint, accelerated possibilities. Amphimixis came to dominate reproduction in ever more complex organisms. To be sure, agamic modes of reproduction were likewise subject to the conditions of the life of certain higher organisms under certain circumstances—Weismann had argued as much in his studies of seasonal parthenogenesis with Cladocera and in closely related hydroids.[54] We may easily interpret these agamic modes of reproduction as following the spirit of a physical overgrowth concept of heredity.[55] In contrast, as amphimixis became dominant, according to Weismann's view of the evolution of higher metazoans, the potential immortality

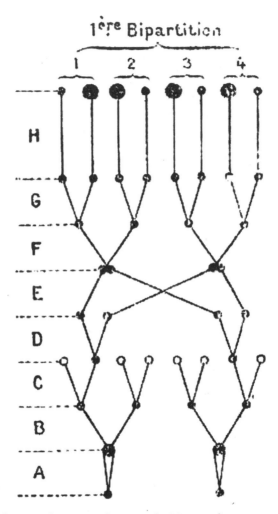

Figure 15.2. Schematic illustrations from 1889 by Maupas of maturation division as he interpreted the process in the ciliate *Colpidium colpoda.* Weismann was very much influenced by Maupas's two papers on the subject.

Émile Maupas, "La rejeunissement karyogamique chez les cilies," *Archives de Zoologie experimentale et generale* 17 (1889): 241, as reproduced in F. B. Churchill, "August Weismann Embraces the Protozoa," *Journal of the History of Biology,* 2010, 43:767–800.

characteristic of agamogony became subject to the principle of panmixia: if it was not useful, it vanished under the pressure of natural selection.[56]

From this evolutionary perspective, Weismann viewed death as an adaptation—a dramatic but unfortunately easy way of confounding the process. By describing death as an adaptation Weismann did not mean that a

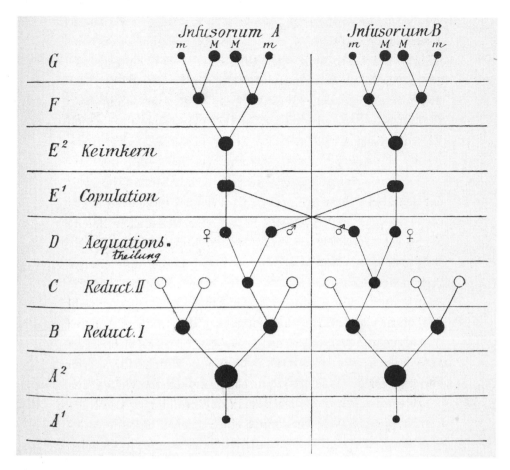

Figure 15.3. Diagram of the pattern of movement undergone by the micronuclei during the conjugation of a ciliate infusorian.

Émile Maupas, "La rejeunissement karyogamique chez les cilies," *Archives de Zoologie experimentale et generale* 17 (1889): 149–517; and "Recherches experimentales sur la multiplication des infusoires cilies," *Archives de Zoologie experimentale et generale* 16 (1888): 165–277.

specific constitutional or germinal determiner for death arose early in evolutionary history. He simply maintained, as he had done since 1881, that potential immortality became increasingly restricted to germ cells and then to the germ-plasm. In other words, an indefinite combination of life's conditions rendered it convenient for species to allow the nonreproducing and worn-out elements of the body to die. A division of labor or specialization led to their death. Continuity, on the other hand, led to potential immortality of the germ-plasm. Amphimixis guaranteed an abundance of variations to move evolution more rapidly onward.

Maupas included four such diagrams of the divisions of the micronuclei during conjugation in his 1889 monograph. All of them illustrated eight identifiable stages, A to H. Stage G or the "cleavage" stage for *Paramecium caudatum* and *P. Aurelia* pictured two divisions. The diagram for *Colpidium colpoda* showed a simpler postnuclear fusion pattern. Stage H illustrated the reformation of the macro- and micronucleus from the cleavage nucleus.

As stated earlier, Weismann's illustration of the micronucleus was taken from the simpler pattern in Maupas's diagram of *Colpidium truncatum*. Weismann was clearly interested in identifying two reduction divisions and an equational division in stages B to D. He interpreted the separation of the macro- (M) and micronucleus (m) to take place during stages F and G.

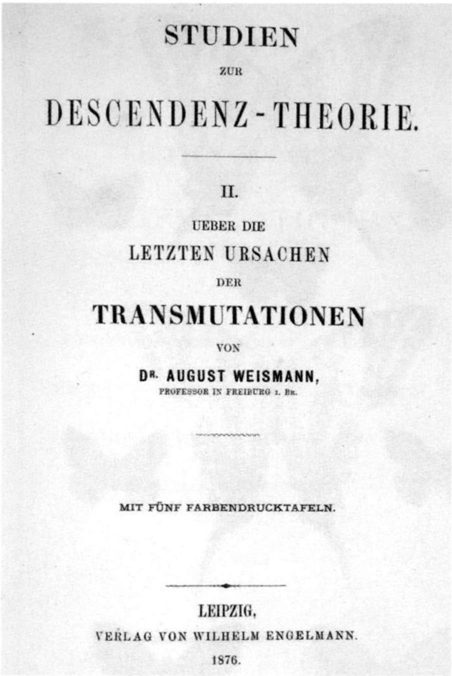

STUDIEN

ZUR

DESCENDENZ-THEORIE.

II.

UEBER DIE
LETZTEN URSACHEN
DER
TRANSMUTATIONEN

VON

DR. AUGUST WEISMANN,
PROFESSOR IN FREIBURG i. Br.

MIT FÜNF FARBENDRUCKTAFELN.

LEIPZIG,
VERLAG VON WILHELM ENGELMANN.
1876.

Plate 1. Title page of Weismann's *Studien zur Descendenz-Theorie*, Part II (Leipzig: Wilhelm Engelmann, 1876).

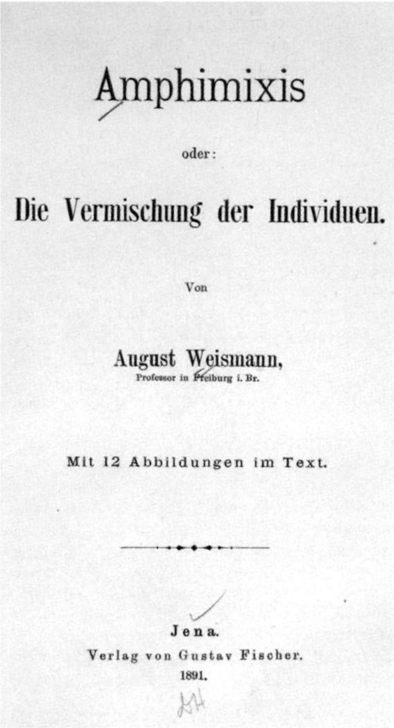

Amphimixis

oder:

Die Vermischung der Individuen.

Von

August Weismann,
Professor in Freiburg i. Br.

Mit 12 Abbildungen im Text.

Jena.
Verlag von Gustav Fischer.
1891.

Plate 2. Title page of Weismann's *Amphimixis oder: Die Vermischung der Individuen* (Jena: Gustav Fischer, 1891).

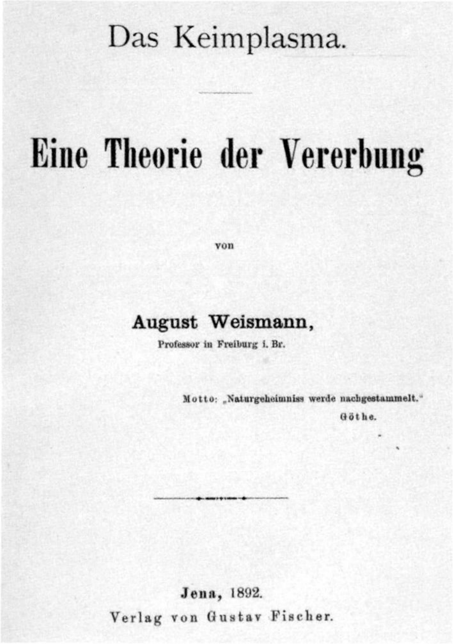

Das Keimplasma.

Eine Theorie der Vererbung

von

August Weismann,
Professor in Freiburg i. Br.

Motto: „Naturgeheimniss werde nachgestammelt."
Göthe.

Jena, 1892.
Verlag von Gustav Fischer.

Plate 3. Title page of Weismann's *Das Keimplasma: Eine Theorie der Vererbung* (Jena: Gustav Fischer, 1892).

Plate 4. The first plate at the end of Part I of the *Studien*, showing winter and summer forms of *Vanessa levana*. It was Weismann's tradition to capitalize both the genus and species names. (From Weismann, *Studien*, Part I, 1875, after p. 94.) The accompanying caption read as follows:

Winter and summer forms of *Vanessa Levana*:
1. Winter form of male
2. Winter form of female
3. Artificially raised intermediate form
4. Female of *Levana*, derived from an artificially raised intermediate form (porima)
5. Male of *Levana*, summer form (*prorsa*)
6. Female of *Levana*, summer form (*prorsa*)
7–9. Of the first summer generation artificially raised from intermediate forms (*porima*)
10–11. Male and female of *Pieris Napi* winter form raised artificially from the summer form; the yellow basic color of the undersides of the hind wings of the summer form is livelier than the natural winter form
12–13. Male and female of *Pieris Napi*, summer form
14–15. *Pieris Napi* var. *Byroniae*, male and female raised from eggs

Plate 5. The second plate at the end of Part I of the *Studien*, showing winter and summer Freiburg and Italian forms of *Papilio Ajax* (numbers 16 and 17), *Lycaena Agestis* (numbers 18–20), and *Polyommatus Phlaeas* (numbers 21–22), and *Pararaga Egeria and Paraga Meione* (numbers 23–24). (From Weismann, *Studien*, Part I, 1875, after p. 94.) The accompanying caption read as follows:

Descriptions: Comparisons between winter and summer forms of:

16–17	*Papilio Ajax* winter and summer
18–20	*Lycaena Agestis* winter and summer
21	*Polyommatus Phlaeas* winter (Sardinia)
	In Germany winter and summer forms are the same
22	*Polyommatus Phlaeas* (Genua)
23	*Pararga Egeria* (Freiburg)
24	*Pararga Meione* (Sardinia)

Weismann, (1863) *Senkenberg Nat. Ges.* 4:227–260.

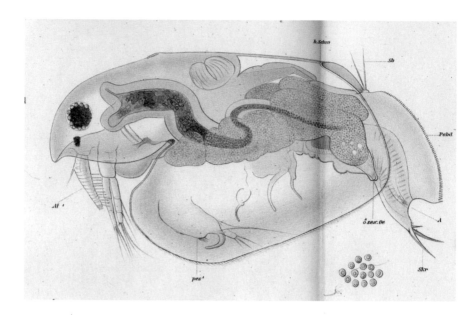

Plate 6. Left side view of a male *Eurycercus lamellatus* with the left testicle lying on the side of the gut with its opening at "oe." Beneath it are shown the sperm cells spread out in water. (From Weismann, "Zur Naturgeschichte der Daphniden, Part VII, Die Entstehung der cyclisehem Fortpflanzung bei den Daphnoiden," figs. 29A and B.)

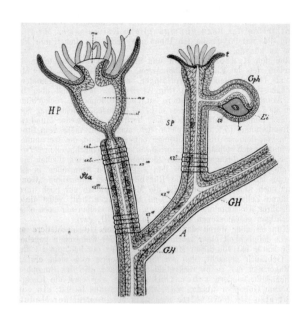

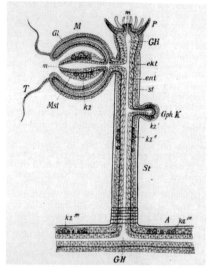

Plates 7 and 8. Composite presentations of the phylogenetic displacement of the germ cells in medusa and hydroids from their place of origin in the ectoderm of the gastric cavity. Illustrated is the root of the germ cells from their origin to the endoderm of the gonophore. Plate 7 (top): From Weismann, *Vorträge uber Deszendenztheorie* (Jena: Gustav Fischer Verlag, 1904), vol. 1, p. 338, showing migration of germ cell of *Eudendrium* by A. Petrunkewitsch (1902), taken from Weismann scheme in English translation. Plate 8 (bottom): From Weismann, *Vorträge uber Deszendenztheorie* (Jena: Gustav Fischer Verlag, 1904), vol. 1, p. 336, concerning A. Petrunkevitch's schema of the migration of germ-plasm in medusen and polyps (1902).

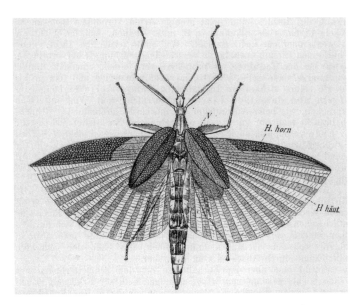

Plate 9. Tropidoderus. An excellent but not unique example of the complex actions of natural selection is revealed in Weismann's attention to the complex coloration of the Australian stick bug, *Tropidoderus childreni*, as illustrated in Weismann's published lecture. Here he pointed out an unusual green and yellow combination of the coloring of forewings and part of the hindwings. Weismann concluded that the yellow must reflect the action of the stick bug flying in the sun under the pressure of predation from birds while the green formed an adaptation for the resting position. The insect possessed the coloring of two contrary actions. For Weismann, two antagonistic life processes, resting and flight, were involved during the stick bug's evolution. Two contrasting colors could only arise through parallel though contrary actions. For Weismann, only a double natural selection rather than sympathetic colorations would explain the evolution of such contrary coloration. From Weismann, *Vorträge uber Deszendenztheorie* (Jena: Gustav Fischer Verlag, 1904), vol. 1, p. 65, showing *Tropidoderus* hindwings with protective coloration.

Verlag von Gustav Fischer in Jena.

Plate 10. Weismann illustrates mimicry in different genera and species of butterflies. From Weismann, *Vorträge uber Deszendenztheorie*, 3rd ed. (Jena: Gustav Fischer, 1913), vol. 1, plate I. The original caption read as follows:

1. *Papilio merope*, male, Africa.
2. The same species, one form of mimetic female.
3. *Danais chrysippus*, Africa, immune model of Fig. 2.
4. *Papilio merope*, second form of mimetic female, S. Africa.
5. *Amauris niavius*, S. Africa, immune model of Fig. 4.
6. *Papilio merope*, third form of mimetic female, S. Africa.
7. *Amauris echeria*, S. Africa, immune model of Fig. 6.
8. *Danais erippus*, immune model of Fig. 9, Central N. America.
9. *Limenitis archippus*, Central N. America, mimics the foregoing species.
10. *Danais erippus*, (a) caterpillar, (b) pupa.
11. *Limenitis archippus*, (a) caterpillar, (b) pupa.

Plate 11. Weismann illustrates "mimicry rings" in various species of butterflies. From Weismann, *Vorträge uber Deszendenztheorie*, 3rd ed. (Jena: Gustav Fischer, 1913), vol. 1, plate II. The original caption read as follows:

12–15. Represent a "mimicry-ring" composed of four immune species belonging to three different families and four different genera.

12. *Heliconius eucrate*, Bahia.

13. *Lycorea halia*, Bahia.

14. *Mechanitis lysimnia*, Bahia.

15. *Melinaea ethra*, Bahia

16, 17. *Perhybris pyrrha*, male and female, S. American "Whites" (Pieridae). The female mimics an immune Heliconiid, while the male shows only an indication of the mimetic coloring on the under surface.

18, 19. Dismorphia Astynome, male and female, also belonging to the family of "Whites," and mimicking immune Heliconiids; a white spot on the posterior wing of the male is all that remains of the original "White" coloration.

20. *Elymnias phegea*, W. Africa, of the family Satyrides, mimics the foregoing species.

21. *Acraea gea*, an immune W. African species.

22. *Danais genutia*, an immune Danaid from Ceylon.

23. *Plymnias undularis*, female, one of the mimics of Fig. 22.

16

A Model for Heredity

Das Keimplasma, *1892*

Emphasizing the nuclear implications of amphimixis constituted one of the real achievements of Weismann's science. His was an achievement that hinted at the way the next dozen years of heredity theory, classic hybridization experiments, and wholesale exploration into the mechanics of transmission from one generation to the next helped shaped the theory of modern science of nuclear genetics. Weismann, however, had reached the point in his science when in 1892 he naturally felt that transmission, development, and above all evolution needed to be brought together into a general synthesis. His tragedy was that the biology he employed, the scope of his presentation, and the depth of his analysis all reflected the exciting post-Darwinian period in the second half of the century, not the science to come.

This notwithstanding, Weismann's germ-plasm theory, as presented in *Das Keimplasma* (Plate 3), is worth understanding even though it may easily be interpreted as the end rather than the beginning of an era in biology. There is no better way of understanding the goal of so many biologists of that period or the revolution that was about to come.

Particulate Theories of Heredity: From Brücke to de Vries

In 1895, Yves Delage, professor of science at the University of Paris and director of the Zoological Station of Roscoff, published his monumental work on heredity, in which he described and critiqued in detail some of the major theories of heredity of the day.[1] These included pages of analyses focused on the ideas of Buffon, Darwin, Haeckel, Spencer, Altmann, Wiesner, Galton, de Vries, Jäger, and O. Hertwig. He, of course, included Weismann, whose germ-plasm theory had appeared in two versions, the first Delage described as the "Théorie des plasmas ancestraux" of 1882–1888 and the second he identified as the "Théorie des determinants" of 1892. The theories of all these authors were discussed by Delage under the general category of "micromérisme" (literally "microparts"). According to him, these theories postulated chemical or morphological molecules or particles that passed from parent(s) to offspring during reproduction. They physically transmitted traits from one generation to the next and singly or collectively became involved in the expression of those traits. Delage's own words reveal how difficult it was to define what exactly the process of micromerism entailed:

> The idea attributed to the life and formation of organisms as the reunion of very small particles endowed with the properties depending on their constitution, combined in immense numbers and grouped in a particular fashion in each species of being and in each organ of the individual, is certainly very fruitful.
>
> The multiplicity of the systems established by this conception, the high value of each one of them, shows us how fertile it is; the chemistry and histology show us that it is probable, because they independently confirm this, on the one hand, to establish that on the other hand by induction, that all substance is dead or living, in effect, formed from units of diverse orders associated in hierarchical groups: the atoms and the molecules more or less complex from one and another part, the cells and the intracellular organites.

Delage's kicker came in the next paragraph:

> But what are these particles? It is here that imagination has free reign, and we will see that there is a wealth of conceptions—a wealth, alas, proportional to the poverty of facts.[2]

Weismann had met Delage in Roscoff in 1880 when he was collecting hydromedusae, and in retrospect he reported having had a favorable impression of him.[3] Delage reviewed Weismann's *Keimplasma* when it came out twenty-two years later, and two years thereafter Weismann claimed to have read the comments in the first edition of Delage's *Hérédité*.[4] Although they disagreed,

Weismann expressed the desire, should the occasion again arise, to interact verbally with Delage. Beyond that, Weismann was impressed by Delage's treatment of the theories of many contemporaries. "Your presentation of the views of others," he wrote, "has often filled me with admiration, where you have sometimes even presented them more elegantly and more clearly than they themselves have been able to do."[5]

There is no need to follow Delage's many analyses. Some proponents, such as Spencer, Darwin, and Brooks, have already been surveyed here in Chapter 9. We will examine the ideas of Hertwig and Galton in Chapter 17. It is, however, worth briefly presenting three theories in the tradition of micromérisme, that is, of particulate theories of heredity. These include the ideas of Ernst Brücke, Johannes Hanstein, and Hugo de Vries, all of whom contributed directly to Weismann's theoretical understanding of the minute particles of the cell and persuaded him to move from an epigenetic to a predetermined theory of heredity.

Ernst Brücke, 1861

Brücke is best known as a student of Johannes P. Müller and a member of the cohort of students who admired Müller but rebelled against their mentor's quasi-reliance on vital forces for explaining physiological processes. Four students—Emile du Bois-Reymond, Carl Ludwig, Hermann Helmholtz, and Brücke—went on to brilliant careers in physiology in Berlin, Leipzig, and Heidelberg.[6] It is curious but understandable that of all of Brücke's publications the best known to today's historians of science is an address presented to the Viennese Academy of Mathematics and Science in 1861.[7] Twenty-five printed pages in length, the address belonged to a body of literature of the day that reevaluated and amended the Schleiden-Schwann twenty-year-old cell theory. In itself, Brücke's talk was not a surprising undertaking, for as we have seen Robert Remak and Rudolf Virchow had in the 1850s promoted direct cell division as the near-exclusive mode of cell multiplication, and Max Schultze and Ernst Haeckel in the 1860s had pictured the protoplasm as the essence of the cell itself, thus rendering nuclei, nucleolus, the cell, and vacuole walls as derivatives of the protoplasm. Brücke's slant was different. He challenged Schwann's understanding that animal and plant cells, no matter whether as single units, as communities, or as building blocks of multicellular organisms, represented the basic elementary unit of life. Anticipating his conclusion at the outset, Brücke gently warned his audience, "We should be inclined to conclude that those cells do not at all represent the often cited elementary organisms, but rather we must speak of the molecules, through which they arise, as such."[8]

As his opening remarks suggested, Brücke hammered away at the view that the cell as a whole—"das Schema," he called Schwann's position—was the unit of life. By this he meant the cell could not be the source of independent motions. He also contested that cellular structures, such as membranes, the nuclei and nucleoli ("Kernkörperchen"), and protoplasm, could not be considered in themselves the bearers of "life." He pointed out moreover that spermatozoa and stinging nettles, which were far more complex than the whole cells from which they were derived and were independently self-propellant, should not be considered units of life. The absorption qualities and refractive indices of pigment bodies in cells were so different from the commonly accepted structures of cells that a dynamic correlation between pigment particles and "das Schema" did not exist. Because of these implied contradictions, Brücke turned to organic molecules in the protoplasm. When visually associated with the motion of the corpuscles that made up various muscle fibrils, sheaves of Bowman's capsule, and plant protoplasm, it was clear to Brücke that such corpuscles must be considered independent. When isolated beyond visibility even with the microscope, Brücke argued that these entities must still possess their own capacities for contractile motions. When detected in polarized light, one had to assume that they were complicated and flexible. In conclusion, Brücke insisted that whole cells—from the perspective of Schwann's cell theory—should no longer be considered elementary organisms. As a consequence, these complicated and motile molecules within the protoplasm must instead be designated the "Die Elementarorganismen."

Besides being discursive and more poetic than precise, the address revealed two important aspects of midcentury German biology. First, as a physiologist, Brücke may have been unhappy with the state of the cell theory, even though developments in the 1860s to the 1880s of the germ-layer theory of animal development, cellular pathology, and nuclear cytology returned the cell theory to a dominant interpretive position in the eyes of histologists and cell morphologists. Second, Brücke revealed the assumption that life must be an attribute of a complex molecular form possessing internal motion. This aspect provided a convenient starting point for future "particulate" theories of heredity. As a German "reductionist physiologist," Brücke explicitly rejected the possibility that life might be an emergent quality that arose from cellular structure and function. So the question remained as to how and where to locate life in contrast to nonlife.

Johannes von Hanstein, 1880

Little is known about Johannes von Hanstein or his contributions to botany and cell theory.[9] Born in Potsdam in 1822, three years after Brücke was born in Berlin, von Hanstein studied gardening as a teenager before attending the University of Berlin, where he was encouraged by both botanists and zoologists. After completing a dissertation [*Plantarum vascularium folis, caulis, radix, utrum organa sint origine distincta, an ejusdem organi diversae tantum partes* (1848)] that focused on the morphology of plants, and after passing the state's teacher examination, von Hanstein taught in Berlin's secondary schools for five and a half years. Encouraged by the botanist Alexander Braun and the zoologist Hinrich Martin Lichtenstein, both professors at Berlin, von Hanstein habilitated and six years later became curator at Berlin's botanical museum. In 1865, the University of Bonn invited him to become a full professor and the director of their botanical gardens. Jahn describes von Hanstein as one of the "pioneers" of plant anatomy and embryology, and the Royal Society Catalogue of Scientific Papers appears to document this judgment by listing thirty-one papers, two of which were coauthored.

In the long run, von Hanstein presents a contextual problem for the historian. He left behind him after his death of a long and incapacitating illness two partial manuscripts on the subject of the intersection between plant physiology and morphology. The first consisted of an extended three-lecture study, designed for a lay audience as well as the specialist, collectively entitled *Das Protoplasma*.[10] This was a semipopular synthesis of his physiological and morphological work and his growing interest in the molecular mechanics of heredity. In the twentieth century, this work briefly caught the eyes of E. B. Wilson, Thomas S. Hall, and Ilse Jahn. Hanstein's second posthumous work was assembled, modified, and published a year and a half after his death by his former student and colleague Fr. Schmitz.[11] It is scientifically more detailed than the popular lectures, but both conclude with a general, nontechnical assessment of whether the protoplasm can provide the molecular platform for heredity and development. The second manuscript served as the focus of Delage's useful two-page exposition and critique of von Hanstein's theory of heredity.[12] In the late 1880s, de Vries (whom we will be discussing later) refers to both von Hanstein's *Protoplasma* and his "Beiträge." Together, these closely related publications reveal that their author had become deeply interested in whether it was possible to explain heredity in particulate terms.

For our purposes, a discussion of *Das Protoplasma* will have to be sufficient. Ostensibly this work concerned the structure of the cell, its formation of tissues, and the contrasting roles of protoplasm and the nucleus in the life of cells and multicellular organisms. The importance of these closely related subjects was enhanced by the molecular side of the question of what constituted life. Hanstein believed a living organism comprised an active mechanical system that was associated with nutrition, growth, motion, and reproduction. "Thus all labor, course and fine," he speculated, "will be brought about in the organism through small movements, by means of which they in actuality exert an influence one way or another on the molecules and atoms within and between them."[13] By 1880, such a statement represented a view of life that reflected the general trend toward the mechanization of organisms, shared by Brücke, other reductionist physiologists, and many other mechanistically inclined biologists. Although he recognized the limits to what physicists and chemists knew about complex molecules, von Hanstein picked up the challenge to verbalize how the atoms and molecules might come together to form the rudiments of an organism. He pictured three levels of organization, extending from atoms and molecules to what he called "Micellen." Grouped together through assorted and unspecified adhesions, these levels together formed the building blocks of an organism.

Hanstein invoked heat as the underlining physical principle that generated the appropriate interactions and motions between the three levels. He then spent page after page describing how these basic building blocks might interact under their given motions and the motions around them of water, carbon, salts, and other simple chemicals. He imagined how the more complex molecules and Micellen might construct protoplasm, walls, and the more solid structures of nuclear consistencies. It was a fantasy of thirty pages; he himself referred to it in one place as a "biochemischen Phantasie," adorned with scenarios of organic motions and textures based on his long experience with the microscopic details of plant cell structures and tissues.[14] The majority of the Micellen soon became identified by von Hanstein as "Protoplastin-Micellen," that is, the building blocks of the protoplasm. Baptizing them "protoplasts," he described them as growing by intussusceptions and envisioned them leading to the construction, according to a "specific plan" ("nach bestimmten Bauplan") of whole cells and the entire organism.[15]

Hanstein recognized the limitations to his vision: the sheer number of different protoplasts involved, the space necessary for them in the gametes, the question of whether superior molecules ("Obermolekeln") were necessary to organize the different protoplasts, and the matter of explaining regeneration and new adaptations. Compared to the construction of a simple chemical crystal, the building of an organism necessitated millions upon millions of "Urmicellen" to anticipate differentiation, new forms, and novel adjustments. "By itself," von Hanstein added in a resigned way, "should we want seriously to maintain this whole Urmolekel-Hypothesis as possible for the normal run of things, we shall not succeed."[16]

This discouraging conclusion provided the justification for viewing the protoplasts in particular and life in general in teleological terms. Ilse Jahn rightly pointed out in her brief sketch of his life that von Hanstein was aligned with the finalistic teleological philosophy of his Berlin mentor Alexander von Braun.[17] His teleological arguments were based on the premise that the instincts of animals and a simpler level of physiological responses of plants to their environment revealed purposeful reactions built into all living organisms. Differentiation and the range of morphological complexities in organisms also expressed this purposefulness. Life itself, in contrast with inorganic chemical processes of crystallization, was teleological. Such purposeful acts expressed an unconscious drive built into the protoplasts and expressed in the protoplasm of the cells they in turn constructed.

How did this purposefulness begin? How did the first organic seeds arise? Hanstein was critical of contemporary opinions, exemplified in Haeckel's Bathybius or in the suggestion that the first "organic seed" came to Earth in a meteorite. He rejected origins and diversification of life through spontaneous generation, Darwinian natural selection, or a gradual mechanical evolution from the inorganic to the most advanced organism. The science of the day, von Hanstein insisted, simply did not have the necessary facts to support such solutions.[18] He had shown, however, that there was purposefulness in the entire system of life and that this attribute resided in the first building blocks of the protoplasm. "The Protoplasts are the artists, the instruments and adaptable material all at the same time," he explained with an air of certainty.[19] The implied finalism was emphasized by von Hanstein in his closing remarks after the first two lectures on cell and tissue formation: "Thus the final form of the development of an organism is not the result but the cause of atomic motions, which dispose molecule by molecule to put together and develop cell by cell. This is already evident with the previous superficial observation of the processes of tissue formation."[20]

There is an irony in von Hanstein's efforts to explore a molecular theory of heredity. One of the first and influential particulate theories was fundamentally teleological rather than purely mechanistic. The grounds for this are clear. How does the biologist travel on an uninterrupted pathway from matter and motion to the complex processes and apparent purposes of life?

Hugo de Vries

In the opening pages of his *Keimplasma,* Weismann mentioned Hugo de Vries with respect but caution: "De Vries has so far been my most powerful opponent as regards the ancestral germ-plasm," he forthrightly declared. Continuing, Weismann made clear that he disagreed with de Vries's recently published theory of heredity.[21] We will return in a moment to the specifics of their differences, for their respective stories tell much about the development of particulate theories of heredity during the last two decades of the century.

De Vries came from a lineage of Dutch scholars and academics, but he appears to have been the first to have turned the family's scholarly inclinations toward the natural sciences.[22] He avidly collected and identified most of the plants of the Netherlands while still in grammar school, and when he advanced to the gymnasium at the Hague, he was soon recruited to assist in the classification of plants housed at the Netherlands Botanical Society. His success enhanced his reputation as an accomplished taxonomist, and by the time he entered the university in Leiden, he quite easily expanded his interests to plant physiology and evolution theory. Unable to find reliable support from the faculty or satisfactory local research facilities, de Vries pursued research in his attic; working under these circumstances, he completed a dissertation in 1870 on the influence of temperature on the physiology of plants.[23]

More a review of contemporary literature than a report on a focused research project of his own, his initial publication surveyed the research of others on how changes in temperature affected plant functions. Even the most casual reader cannot help but notice how completely physiological in focus his discussion was. Prominent among the works de Vries reviewed were the writings of Julius von Sachs, so it is not surprising that after de Vries became a newly minted doctor, he quickly found his way to Würzburg where he worked under the guidance of Europe's foremost plant physiologist. Although de Vries soon returned to the Netherlands to begin a less than successful teaching career—he had a similar unhappy experience at the end of the 1870s as a privatdozent teaching plant physiology at the University of Halle—he returned to Würzburg during the summers where he continued to investigate physiological problems under Sachs's tutelage. With his mentor's support, he secured a position in botany at the new University of Amsterdam, where he worked until his retirement in 1918.

For the first twenty years of his career, de Vries continued his research into the effects of physical forces on plants, such as the effects of temperature, osmosis, and plasmolysis, the influence of turgor pressure on the shape of cells, the control that indirect pressures exercised upon the shape of annual rings in wood, the transport of fluids, and a variety of other mechanical forces that influenced plants. During the same period, de Vries also investigated the germination of domesticated agricultural plants. This latter subject prompted de Vries to explore the relationship between the lineages of cell formations and their different functions. The research he published during this period, especially his studies of plasmolysis and turgor pressure, was highly regarded. In no way, however, did his published work appear to incorporate the recent advances in nuclear cytology.

Thus, it came as an apparent break in his research trajectory when in June 1889 de Vries published an extensive theoretical discussion on heredity.[24] This sudden reorientation of his research interests is worth noting, for being steeped in mechanical physiology rather than the equally mechanistic behavior of cell division, de Vries provided a very different perspective on the subject of heredity from van Beneden, Flemming, Strasburger, the early germ-plasm theory of Weismann, or the other German nuclear cytologists. It is also important to note that at the time of writing *Intracellular Pangenesis,* de Vries had not yet embarked upon the single-minded hybridization experiments that became the hallmark of his botany in the 1890s and beyond.

Because of his physiological orientation, it is understandable that de Vries viewed heredity from another side of the process. Instead of mining deeply into the cell's nucleus with the new tools of cytology, he approached the subject having been mentally immersed in the makeup and movement of protoplasm, in the constitution of the walls between cells, and the growth and shapes of cells themselves. The immediate inspiration for considering heredity, at least as told in the introduction to his *Intracellular Pangenesis,* was to follow Darwin's lead. "We shall not try to explain the morphological details of those processes," he explained at the outset, as "our knowledge is yet too limited for that. But following the method of Darwin, to find in the special cases the material substratum [das stoffliche Substrat] of the physiological processes, that is our problem."[25] De Vries, of course, was responding to Darwin's provisional hypothesis of pangenesis, but rather than considering Darwin's focus on the transportation of gemmules from the cells of the body to the sexual glands from whence they could be delivered to the next generation, de Vries was concerned about how this material substratum, starting as molecules in the nucleus, might construct the protoplasm of the cell. To honor Darwin and yet distance himself from Darwin's questionable transport of "gemmules" within the parent, de Vries renamed his particles "pangenes" and speculated about how they might determine the nature of each cell.

His focus was not on the maturation of germ cells, as it roughly was in Darwin's theory, but on a speculative particulate physiology of each cell. His was an "intracellular" rather than "intercellular" pangenesis. If Darwin's theory had inspired a name, it was the continental particulate tradition beginning with Brücke and von Hanstein that helped reify the particles involved.[26] Dismissing the special teleological dimension of von Hanstein's protoplasts and deemphasizing the primacy of protoplasm over the nucleus as the vehicle in which the particles are conveyed during reproduction and development, de Vries explicitly argued that his trait-endowing pangenes began in the nucleus, migrated at the appropriate time into the cytoplasm, and collectively determined the cell's nature.[27] As with both Darwin's gemmules and all other particulate theories of the era, these pangenes were conceived of as large molecules that assimilated nutrition, grew, and reproduced.

De Vries was also impressed by the continental particulate tradition beginning with Brücke and von Hanstein. He adopted the latter's notion of specific individual protoplasts that determined histological and morphological character. Through their physiological actions or inactions, they determined the differentiation of embryonic cells and established the differences between species, the variations of individuals, the aging processes, and the regeneration capacity. According to him, *"All living protoplasm consists of pangenes; they form the only living elements in it."*[28] Finally, de Vries depicted protoplasm, and hence the living nature of cells themselves, as a mosaic of independent, miscible, and varied pangenes.

The simplicity yet comprehensiveness of this picture introduced a potential problem to de Vries. Because both the protoplasm and the nuclei of dividing cells consisted of pangenes, the possibility existed of new variant pangenes appearing in the protoplasm of somatic cells and becoming the determiners of acquired characters during the process of development. De Vries, however, emphatically rejected such "Lamarckian-type mechanisms" to explain evolution. Up to a point, he subscribed to Weismann's claim that the germ-plasm destined for gametes separated early in development from the soma. This, of course, was the core of Weismann's neo-Darwinian crusade. No matter whether the variant pangenes were passed on from cell to cell during growth and development, and no matter whether they became active or remained in an inactive state, "we can join Weismann in quietly answering, no" to the matter of the inheritance of acquired traits.[29]

His model for heredity occupied only the final pages of de Vries's *Intracellular Pangenesis*. He devoted the bulk of his 200-page booklet to a somewhat digressive presentation of what he considered must be the basics of any theory of heredity. First, as a botanist, he felt he had certain advantages over his zoological colleagues, for he could control reproduction better through the self-fertilization of plants than they could in animals. He could carry out innumerable selective hybridization experiments and anticipate the appearance of given variations. Foremost, as a botanist he was in the position to identify the combinations and permutations of individual "characters," that is, pangenes, whether they expressed themselves or remained latent on the cellular level. As a result, de Vries felt, "it is clear that every thorough consideration of the character of the species and every comparison with other characters, leads us to conceive of the first as a composite picture, the component parts of which are miscible in the most varied of ways."[30]

Second, in a roundabout way, de Vries had intellectually traveled from an epigenetic belief—from a commitment to *de novo* formations of generating factors—to a panmeristic view, a particulate understanding of differentiation and growth.[31] His early epigenetic view had held that the generation of new structures or physiological molecules arose from the domain of atoms and elements interacting within a continuum of chemical and physical forces. By 1889, his panmeristic attributions applied to the smallest organic molecule that could reproduce itself through physical division. De Vries described his pangenes within the protoplasm of the cell as reproducing themselves by the accrual of new material, growth, and division. Conceived of as omnipresent in the nucleus and surrounding protoplasm, they explained the innumerable combinations and permutations of individual characters of organisms.

De Vries's panmeristic conclusions were explicitly derived from Brücke's *Elementarorganism* and from von Hanstein's concept of the protoplast.[32] In addition, the organic chemist Jacob van't Hoff, his colleague in Amsterdam, provided him with a mechanistic and reductionistic justification on a deeper level. "From the chemical properties of carbon," van't Hoff had written and de Vries quoted, "it appears that this element is able, with the help of two or three others, to form the numerous bodies which are necessary for the manifold needs of a living being. . . . Therefore, one does not go too far, in assuming that the existence of the vegetable and animal world is the enormous expression of the chemical properties which the carbon-atom has at the temperature of our earth."[33] Most of van't Hoff's textbook was written for students of organic chemistry, but this statement, repeated by de Vries and later picked up by Weismann, served to anchor the heredity speculations of both to the domain of physics and chemistry.[34]

The stuff of heredity became one of the innumerable carbon-based molecules that organic chemists were set on identifying and chemically describing. De Vries's pangenes, however, had to be more than just one more carbon-based molecule. They must be flexible and variable enough to bear in their chemical interstices the hereditary details of hundreds of thousands of different plant and animal species. They must assimilate material, grow, and divide as each organism came into being with its peculiar combination and permutation of pangenes. Statistically, enough organisms must survive long enough to pass on their pangenes. Assimilation, growth, and reproduction became a common definition of life on both the molecular and organismal levels.

Ontologically mechanistic and now particulate in design, there was a dilemma in such a theory. De Vries called attention to the problem by alluding to the foremost particulate theories of heredity to date, those constructed by Darwin, Spencer, Weismann, and Nägeli. The organic molecules conceived thus far addressed heredity either as the transmission phenomenon or as the developmental process that led to a visible differentiation. These tasks seemed to him contradictory. On the transmission side of the equation, each hereditary unit must act together as a cohesive organic whole, which bore the totality of individual traits of a given offspring. On the differentiation side, the hereditary molecules must act in such a manner as to specify the character of the new and developing individual. De Vries correctly identified Spencer's theory of physiological units, Weismann's first 1883–1887 theory of the ancestral germ-plasm, and Nägeli's conjecture of an extended idioplasm as theories that sought to solve the transmission side of the cycle of life. By contrast, he argued that Darwin's hypothesis of pangenesis focused on the uniqueness of individual gemmules as they collected in the gametes at the end of differentiation. In these examples, neither transmission nor differentiation dovetailed with the other in a fully mechanistic way.[35] Transmission and differentiation apparently demanded different kinds of solutions. De Vries sought to solve this dilemma physiologically by insisting that all nuclei of the individual were totipotent and then calling upon some pangenes to leave their inactive state in the nuclei and migrate into the protoplasm at appropriate moments to become active and determinative. The nature of subsequent cell lineages resulted in the relative numbers, and hence strength, of the activated pangenes or variations in these pangenes. By using the same hypothetical physiological mechanism, de Vries explained not only inherited variations but organized sequential differentiation and regeneration.[36]

In a short section, de Vries focused on what he considered an important weak point in Weismann's theory, specifically: the germ-plasm theory appeared to be making two conflicting claims. On the one hand, Weismann seemed to be arguing that in the process of segregation and eventual transmission the germ-plasm remained an integrated whole, bound together in its original form. On the other hand, during development the germ-plasm of the soma (the somatoplasm) must break up into basic parts to fashion the different organs and traits of the offspring. De Vries felt that Weismann's solution to these two contradictory processes was to posit an "ancillary hypothesis" that a reduction division, eliminating half the ancestral plasms, must take place before fertilization. The connection, however, between unified ancestral plasms and reduction division is not easy to see, despite de Vries's claim. Nevertheless, it is clear that Weismann's germ-plasm theory at the end of the 1880s attempted to solve the transmission side of heredity but failed to provide a compatible scheme for development. De Vries clearly recognized a problem, but he repeatedly confused Weismann's ancestral plasms as trait-determining entities.[37]

Again on the attack, de Vries lay great stress on the ability of botanists to determine what he called "cell pedigrees" ("Zellularstammbäumen").[38] This biological principle rested on Schwann's cell theory and its midcentury corollary "omnes cellula e cellula." In de Vries's day, Sachs had impressed upon his followers the importance of observing each plant cell as it diverged from the apical meristem or any other point of development in order to document the entire pedigree of each cell, cell by cell, from meristem to maturity. De Vries found in his mentor's demonstrations the justification for regarding such pedigrees as empirically documentable, not merely hypothetical.[39]

The study of cell lineages demonstrated the rigidity in Weismann's system, which maintained an absolute separation of germ-plasm and somatic cells. Instead of denying Weismann's sequestration doctrine out of hand, de Vries simply loosened it up. After exploring the well-known capacity of begonias to grow whole plants from fragments of leaves, after pointing out the "entire and very rich doctrine of adventitious buds," and after describing the varied and complex oak galls induced by the gall wasp larvae of cynipids, de Vries insisted there must exist a spectrum of secondary germinal tracks and totipotent cells in these and other organisms. He boldly concluded by stating a problem and offering a conjectured solution: "If anyone is ever successful in growing in this way an entire willow from a gall, it will be clear that in the latter, all the hereditary characters of the willow are present in a latent state." "This," he insisted, "would obviously be much more useless than their presence on any given normal somatic track. We may even regard the result as perfectly clear, however, *that germ-plasm is by no means limited to those cells which need it for their own development, or for their progeny.*"[40]

It must be remembered, however, as we enter this period of conflicting theories of heredity in the competing disciplines of botany and zoology as well as the contrasting approaches of physiology and microscopy that, at the writing of his *Intracellular Pangenesis,* de Vries knew of Weismann's germ-plasm theory only through the latter's writings up to 1887; this included Weismann's important works on the meaning of sexual reproduction and the number of polar bodies. Even so, de Vries devoted a short section and then a chapter to Weismann's research, more than he had done for any other biologist with the exception of Darwin. This reflected his high regard for Weismann's work. That Weismann carefully read *Intercellular Pangenesis* indicates the high regard he held in return for de Vries's achievement.

Weismann was impressed enough with de Vries's work that he wrote in his copybook two drafts of a letter of praise to Amsterdam. It is likely that he sent only the second, which was somewhat longer. It is particularly interesting to note Weismann's explicit contrast between de Vries's "evolutionistic" view and his own "epigenetic" theory. That would soon change.

Herrn Prof. *de* Vries   Freiburg i.Br.
                                          Amsterdam   6 March 1889
Dear Sir!

Permit me to tell you with how lively an interest I have now read your work. I have desired for a long time that a professional among the botanists take up from the botanical point of view the heredity question. The reading of your book was for me a real pleasure, even though your views stand to a certain extent in fundamental opposition to mine. You are, however, completely right, that Darwinian Pangenesis can on this basis be made fully viable again. I also willingly grant you that in this way a large range of phenomena becomes relatively simple and easily understandable. It will nevertheless surprise you if I cannot join you without further joining you in every way; yet each of us sees things from another standpoint which is confirmed by *true* and *living* experiences, which are compelling for him. For these moreover govern our conceptions and these are for botanists necessarily other than for zoologists. Whether finally you with your evolutionistic or I with my epigenetic view will be proved correct, Who can say today? It is also finally irrelevant, if only on the way to justifying one and the other conception, the truth will be discovered.

For progress is also reached on the error of much.

I will not go into single points, which then I could only stop with difficulty. Discussion by correspondence of comprehensive questions is scarcely possible and in the long run [scarcely] beneficial. In any case, however, you have my hearty thanks for the pleasure and multiple stimulation that you have created for me through your work. It is a shame that you do not live here, so that we might exchange our views in verbal intercourse. Actually the problem of heredity should be worked on by a botanist and zoologist.

> With sincere salutation yours faithfully
> *August Weismann.*[41]

If Darwin had been his personal inspiration and the theory of pangenesis the model for his theory of heredity of pangenes, de Vries saw Weismann as his principal opponent and the germ-plasm theory as the most important means for its studied refutation.

Das Keimplasma, 1892

Weismann closed his monumental work on *The Germ-Plasm* with an affirmation of his general assumption that heredity must be grounded in a substance that was extensively complex with minute particles. It was a tall order that strained the credibility of many of his contemporaries, but as Weismann stood back from his volume, his last sentence concluded with a cosmic allusion. "We are thus reminded afresh, having to deal not only with the infinitely great but also with the infinitely small, that the idea of size is a purely relative one, and that we ourselves stand in the middle while on both sides stretches infinity.[42] Despite the imprecision in his use of "unendlich" ("infinite") and "Unendlichkeit" ("infinity"), Weismann was poetically reasserting that this world was an architectonic one of molecules, varying biological units, and worldly and cosmic structures beyond.

The work that elicited such terminal poetry was, in fact, a solid work of biological information drawn together to bolster Weismann's germ-plasm theory in fourteen chapters, sorted into four parts. The first part presented in detail his model of heredity material: the germ-plasm in its internested hierarchy of heredity particles and cytological structures. The second examined the action of this material within the context of various forms of "monogonic" or asexual reproduction. The third performed the same service with respect to sexual reproduction and amphimixis. The final focused on two widely opposed perspectives on evolution, which pitted the inheritance of acquired characters against the selection of variations. The last three of the four parts were filled with biological information drawn from a lifetime of work in the field and in the institute's laboratory and collections. The outstanding question about Weismann's strategy was the degree to which his rich biological experience justified the model he so carefully articulated at the outset.

Das Keimplasma, Part I: The Material Basis of Heredity

First a word is in order concerning the hereditary stuff that formed the foundation of Weismann's mature panmeristic germ-plasm theory. In his introduction, Weismann openly expressed both his debt to and his rejection of de Vries's *Intracellular Pangenesis.* He applauded de Vries's denial of the "transport" side of Darwin's provisional hypothesis of pangenesis. As discussed previously, de Vries, among many others, had found this hypothesis implausible, but in its stead he had theorized that hereditary particles existed in the nuclei of all cells, and, by migrating into the cytoplasm, they would affect its histological nature and collectively the characters of the descendants. Weismann was strongly influenced by this conjectured action of "pangenes" on the developmental side of heredity, but his Dutch colleague had failed to explain how the supracellular organization of the organism could be derived from the nuclear architecture of the fertilized egg. De Vries, he pointed out, holds that the pangenes were "independent and perfectly and freely miscible, and, in fact, postulates a germ-mechanism which admits of their separation in any manner required."[43] In short, Weismann adopted de Vries's living, cell-controlling pangenes, renaming them "biophors," but he incorporated them into a hierarchy of higher living units, which collectively served the necessary cell determination, body organization, and parental and ancestral contributions required in a heredity theory.

The biophors were, in Weismann's view, living physiological units. They were imagined to be biological molecules that were considerably larger than the molecules of the chemist, and as such they possessed the distinctive living property of being able to accrue material, grow in size, and divide into replicas. Once discharged from the nucleus into the cytoplasm, the biophors multiplied by divisions and controlled the expression of the character of a formerly undifferentiated cell.[44]

Following his model further up the line of organization, we read that the biophors initially resided in a second level of vital units. Again characterizing these units by their ability to accrue material, grow, and divide, Weismann named them "determinants." These vital units controlled the differentiation and diversity of cells in multicellular organisms. They were the primary constituents of the nucleus of a developing or developed cell or cell group. Their distribution, as the organism developed, reflected the mechanics of mitotic divisions during ontogeny. Similar determinants controlled the release of specific biophors from the nucleus of cells, which were destined to possess the same histological character. Slight differences of homologous determinants in similar cells reflected independent variations within homologous cells of organisms of the same species or in comparable cell groups of similar parts of the same organism. Where the biophors were the physiological units of character expression on the cellular level, the determinants were the hereditary units of differentiation and variation on the histological level.

The determinants, too, resided in the chromatin of a third-level vital unit. When he had first enunciated his germ-plasm theory, Weismann had referred to this unit as ancestral germ-plasm. Later and more explicitly this became the "id," or ids in the plural. (The name was to honor the idioplasm of Carl Nägeli's theory.) Given the continuity of the germ-plasm between generations, given the unity of species over time, and even given the branching phylogeny of all life, the ids reflected the germinal past of a given organism—but only in part, because the reduction divisions during germ cell maturation led to half of the germ-plasm being eliminated with each generation. Each id or ancestral germ-plasm embedded within the chromatin of the nucleus contained a single copy of all the necessary determinants for a complete organism. It was the interaction of the homologous determinants in the many ids that in fact determined the physiological action of the biophors.[45]

Finally, the chromosomes of the nucleus, for which Weismann preferred his term "idants," constituted the fourth level of nuclear organization. It was widely accepted by now that these structures multiplied by longitudinal and perhaps also transverse division. We find that the focus of Weismann's research after 1885 had zeroed in on the behavior of chromosomes in germ cell maturation and fertilization, in parthenogenesis, and in conjugation. By 1892, they had become the visual anchor for his germ-plasm model.

The hypothetical hierarchical nesting of these four levels of vital units not only helped Weismann explain what occurred in reduction division, fertilization, parthenogenesis, and differentiation, but this particular architecture of the germ-plasm gave him the opportunity to explore further in a hypothetical way the appearance of individual somatic variations. Given that natural selection would promote some individuals and eliminate others, and given that fertilization would randomly advance in number certain ids and that reduction division would randomly reduce in number, even eliminate, other ids, here was a biological mechanism to ensure the appearance of new determinants and hence new somatic variations. Was there also a mechanism for the change or disappearance of determinants and hence the alteration or disappearance of a somatic trait? To answer such questions, the reader must bear in mind what Weismann suggested happened in the germ-plasm to the determinants and biophors, for it was only here that changes played a phylogenetic role. The reader must also keep in mind that in Weismann's mechanistic world changes might appear independently in the somatic idioplasm (i.e., somatoplasm), but these would have no evolutionary impact unless a parallel change also occurred in the germ-plasm.

Dealing sometimes with the germ-plasm and sometimes with the somatoplasm in ontogeny, Weismann wrestled with these critical questions for the next decade. The rudiments of his solution, however, emerged in 1892, for here he claimed that the different rates of multiplication of homologous determinants of different ids told the story. Again, it was assimilation of nutrients, growth, and division of vital units that provided his mechanistic scenario. Slight differences in nutrients available to different homologous determinants in different ids, his hypothesis continued, not only ensured a difference in their molecular makeup but resulted in "stronger" and "weaker" homologous determinants during development. His anthropomorphic implication aside, Weismann simply held that the stronger determinants must grow more rapidly and their biophors must outnumber the biophors of weaker determinants. As a result, their expression in ontogeny would come to dominate the determination of the relevant cells, and a new somatic variation would be expressed. With this momentous step, natural selection of the variant individual would come into play. Weismann was confident enough in his mechanism to suggest that it helped explain not only new traits but also the degeneration and correlation of parts.[46]

In retrospect, two conclusions seemed to follow—one taken for granted by Weismann, the other explicitly discussed by him. Both revealed his debt to the past, including the influence of his old friend Ernst Haeckel. First, the very idea of a biological molecule possessing the properties of accrual, growth, and division was a common perspective for the period from Brücke's famous discussion of the units of life on through to the end of the century. Such a mechanism characterized one side of Schwann's cell theory, which envisioned, among other mechanisms, a direct cell division. As we saw in Chapter 13, this was a feeling shared by the proponents of the exclusive role of direct division in multiplication, championed by Remak and Virchow and maintained up to the time of Flemming's and Strasburger's demonstrations of the overwhelming presence of indirect cell division at the beginning of the 1880s. The continuity of living material by means of accrual, growth, and direct division was such a logical and mechanistically satisfying way of describing the uniqueness of life and the perpetuation of form that it was hard for Weismann to move beyond it without plunging into mystery or vitalism. One might argue that it satisfied the same logical requirements as the earlier overgrowth principle of reproduction, except for the fact that the latter, as we have seen with Baer and Müller, was not explicitly mechanistic.[47] Although by 1885 Weismann followed Flemming and Roux, he held on to a residue of miniature "overgrowth" in explaining the multiplication of his four vital units.

Second, Weismann felt compelled to speculate, with reservations to be sure, about the parallel between phylogeny and the ontogenetic order of determinant and biophor activation. The argument is fascinating enough to repeat here a key passage, which drew upon the biogenetic law. In addition, Weismann assumed that new traits are often, but not always, formed by the doubling of determinants and a nutritionally induced divergence of the duplicates. "This may be expressed in terms of the idioplasm as follows":

the determinants of the id of the germ-plasm become endowed with a greater power of multiplication, so that each one of them causes the addition of one or more cell-generations to the end of the ontogeny. At the same time, the determinants in the germ-plasm increase in number, and each of them becomes differentiated in a fresh manner. As, however, every two new determinants always follow the same course from the id of germ-plasm to the final stage in ontogeny as was taken by the single original determinant, they will pass through the same determinant figures as before, and only lead to the formation of new structures in the final stages, when they become separated from one another.[48]

It is doubtful that Weismann believed that new traits arose only as terminal additions. Nevertheless, the architecture of the germ-plasm expressed the past and determined in the future the mechanically ordained unraveling of the expression of new determinants and biophors. Weismann might have broken the grip of a strong version of the biogenetic law by recognizing that reduction division constantly eliminated ancestral ids, but the architecture of the remaining and future ids still reflected a historical order of determinant expression.[49]

Das Keimplasma, Part II: Heredity and Monogonic Reproduction

After laying down his basic model explaining heredity in particulate and mechanical terms, Weismann turned to the known biological phenomenon of reproduction. It is immediately clear to the modern reader that there is diversity in these phenomena that would not be found in the standard account of heredity to appear twenty or thirty years into the twentieth century. Heredity, that is, "Vererbung" for Weismann, meant not only transmission of traits but their multifaceted expression. As an embryologist and naturalist by training and as the holder of a Lehrstuhl in zoology at a German university, Weismann was not only prepared for a larger canvas, but he was obligated to fill in the many blank spaces with what he claimed would be a comprehensive model. It is as though his entire professional career of study and research now spilled out on that canvas before him. It was his genius to be able to fill in such a large portrait with the paint he had available and remain true to his earlier mechanistic and evolutionary commitments. It was part of his fate that such an all-embracing model turned out to be hopelessly premature for the science he represented.

The material to be covered conveniently sorted out into three basic subjects: asexual reproduction, sexual reproduction, and the evolution of species. Although it presented one long argument for his model, the volume might have served as a digest of contemporary zoology focused on reproduction, development, and evolution. In the second part of *Das Keimplasma,* Weismann dealt with asexual phenomena, processes that ran the gamut from regeneration, fission, and gemmation to the alternation of generations and even the formation of the germ cells. Each of these processes challenged his model of a simple mechanical sorting of determinants producing a predictable organic form out of a fertilized egg. The alternatives seemed to Weismann to be few.

Either his model itself was deficient and one had to have recourse to some style of *nisus formativo,* that is, to some teleological force popularized by Blumenbach at the beginning of the century, or one had to rely on simple repeatable crystallizations at the chemical level, which according to Herbert Spencer would reflect and duplicate a given organic form; otherwise, Weismann had to add to his model further biological units to address the complexities of the real world as they arose. He chose the latter option. In a telling passage, Weismann emphasized the importance of thinking of development in terms of complex sequences of dividing cells and a parallel segregation of determinants that, when active, determined the nature of the differentiating cell with their biophors. His was a cellular embryology governed by each cell arriving at a specific location at a specific time. "The harmony of the whole," he explained, "is primarily brought about by the variation and increase of the cells, the kind and rhythm of which respectively, is prescribed by the idioplasm of each individual cell, rather than by the mutual influence of the cells during their differentiation."[50]

Front and center Weismann placed regeneration, the discussion of which comes at the outset of part II and occupies by far the longest chapter. This is not surprising because by this time experiments on the process of regeneration had become popularized by Wilhelm Roux and Oscar Hertwig and had entered a period of contentious counterdemonstrations by Laurent Chabry and Hans Driesch.[51] (More on that phase of the phenomenon will be discussed momentarily.)

Regeneration came in many guises, from the normal continual growth of hair and nails, to the complex replacement of limbs, tails, and eyes commonly found in some amphibians and many invertebrates. The problem was emphasized by Weismann in a figure illustrating the hypothetical distribution of determinants in the normal development of the bones in the forelimb of a species of *Triton*.[52] His diagram assumed a segregation and activation of different determinants for the individual bones during normal ontogeny. When mapped on to this diagram of the regeneration of the limb, all the structures removed are envisioned to be replaced from the proximal cells, which must contain accessory idioplasm with supplementary determinants for the missing parts. These must be activated in a sequence, in a specific "rhythm" to reestablish the order, proportions, and size of the missing part. Weismann also employed supplementary determinants to explain how adult regeneration might introduce a different developmental sequence from the original developmental pattern. A case in point was the replacement of a lizard's tail. Embryologically the tails developed in conjunction with the notochord,[53] but in adult lizards, where the notochord had long since degenerated, regeneration of the tail must take a different course. Thus, a new or coenogenetic regenerative process replaces the standard or palingenetic embryonic process. As a consequence, the supplementary determinants must be significantly different from their predecessors.[54]

In experiments on the whole adult organism, the results were often found to be conditional in the sense that an arbitrary cross section of flatworms, annelids, and *Hydra* produced a head at one side of the slice and a "tail" end at the other side.[55] Recent efforts had revealed an association between the phylogenetic standing of an organisms and the vulnerability of the structure to loss. Weismann himself later performed a single extirpation experiment on a lung in *Triton,* and when, as predicted, this organ failed to regenerate despite the animal's extended and otherwise normal life at the institute, Weismann emphasized the importance of utility in determining a capacity for regeneration. There would be no point, his argument went, for *Triton* to evolve the germinal apparatus for lung regeneration when a natural scenario of its loss could scarcely be imagined. Once again, Weismann pointed out the role of natural selection in establishing the form and function, the capacities and forms of the organic world.[56]

Part II labored through many additional asexual phenomena, such as regeneration in plants, fission, gemmation, alternation of generations, and the production of the specialized germ cells proper. All these phenomena became understandable in his terms of additional determinants that remained latent until circumstances required. The notion of an "accessory idioplasm" containing "supplementary determinants" and "double determinants" provided Weismann with the ad hoc solution compatible with the rest of his germ-plasm theory. It made sense that he had discussed in part I the size limitations and projected numbers of possible organic molecules in the nucleus. His calculations on the sizes and quantities of determinants and biophors, however, were totally speculative, and his overall model of heredity, as we will see in Chapter 17, strained the credulity of his critics.[57]

Regeneration of lost parts lent itself particularly to experimentation. Many earlier investigators who had sliced and diced *Hydra,* snails, and salamanders, such as Lazzaro Spallanzani (1729–1799), Abraham Trembley (1710–1784), Tweedy John Todd (1789–1840), Dominique-Auguste Lereboullet (1804–1896), and Camille Dareste (1822–1899), have been duly recognized by recent historians.[58] As innovative as their regeneration experiments were—and many of their ingenious accomplishments were well known in Weismann's day—they may be collectively classified as discoveries in natural history and designed to establish the diversity and order of organic nature. This sweeping characterization describes the six years of experiments by Privatdozent Paul Fraisse (1851–1909), which appeared in 1885 and served as a source on the subject for Weismann.[59] Fraisse worked in the post-Flemming era of the cell theory, and like Flemming he recognized that the cells at the onset of regeneration did not conform to the standard pattern of indirect cell division. His regeneration studies, he argued, helped identify the origin of tissues and supported the notion of a specificity of germ layers and tissues. Aptly enough, one of his interests was in the relationship between ontogeny and phylogeny.[60] Another talented experimenter on embryos whose work appears to fall into the same category was Laurent Marie Chabry—but more on Chabry in a moment.

In contrast to these extended and *inter se* generally independent series of experiments, a related movement took zoology by storm in the 1880s. It sprung to prominence at the hands of a younger generation.[61] It was engineered in part by anatomists who were impatient with the reigning descriptive mode of their profession and who turned to asking causal questions that were free of the endless phylogenies associated with zoology and natural history. It was also a reaction to simplistic experiments on the embryo with physical and chemical models of development, which were associated with the embryologist Wilhelm His and the physiologist Wilhelm Preyer.[62]

Eduard Pflüger

The immediate stimulus for the work of the new movement came in 1883 at the hands of the Bonn physiologist Eduard Pflüger. A student of Emil du Bois-Reymond, well established by 1859 in a Lehrstuhl at the age of thirty, founder and editor of the *Archiv für gesammte Physiologie*,[63] and known for his work in respiration physiology, Pflüger had tried his skills at exploring the relationship between the first three cleavage planes of frog eggs and the primary axis of the ensuing larvae. He found a way to immobilize the fertilized eggs within their perivitelline membranes prior to first cleavage so that he was able to rotate the egg to any given position during the first three cleavage events. He discovered that the third cleavage alone always followed the horizontal and that it alone must determine the point of the appearance of the blastopore, that is, of the first organized embryonic structure. This horizontal cleavage consequently determined the dorsoventral axis of the tadpole and of the adult frog. Gravity, Pflüger proclaimed, must organize the material in what appeared initially as an isotropic egg. In a dramatic image, which became well known in embryological literature, he compared the contents of the egg to snowflakes in an avalanche:

> I imagine that the fertilized egg bears no more relation to the later organization of the animal than the snowflakes bear to the size and shape of the avalanche which develops under certain situations from them. From a germ there always arises the same structure because the external circumstances remain the same. From the small snowflake there will also always arise an avalanche of the same size and shape if it regularly moves through the same track with the same quality of snow.[64]

For a trained embryologist this conclusion appeared far too simple and detached from the diversity and details of development. The next generation, including Wilhelm Roux, Gustav Born, August Rauber, and Oscar Hertwig, leaped at the chance to pick up the challenge thrown down by a well-established physiologist.

Gustav Born

As a student, Gustav Born had been influenced by the physiologist Rudolf Heidenheim in Breslau and Pflüger in Bonn before he had joined Carl Hasse's anatomical institute in Breslau. He was Hasse's prosector at the time of Roux's arrival in Breslau in 1879.[65] Born appears to have been the first to publish his doubts about the validity of Pflüger's experiments. He did this initially in a paper he submitted to Pflüger's *Archiv* on the subject of hybridization between local species of frogs.[66] There he noted that Pflüger had paid no heed to the division of the nucleus, which was known to have marked the plane of first cleavage, and he promised a closer examination of the problem—presumably when the new spring breeding season for *Rana fusca* arrived. When his chance came to test Plüger's work and comment on recent papers by Roux, Hertwig, and Rauber, Born presented his experimental results to the April 1884 sitting of the medical section of the Silesian Society for National Culture. His suspicions were solidly confirmed in his full report: "Nothing more remains than to accept that the specific hereditary structure belongs only to the nucleus, which suffers no visible changes because of gravity."[67] Born undoubtedly pleased his friend and colleague Roux, who sat supportively in the audience of his oral presentation.

Wilhelm Roux

We already followed the early career of Roux in Chapter 9. We left him as he published a speculative but highly suggestive paper in which he wrestled with the implications of Flemming's and Strasburger's descriptions of indirect nuclear division. Roux's suggestion was to consider mitosis as a mechanism for dividing the hereditary elements (which by hypothesis lay along the chromosome) into equal and sometimes unequal parts and which were distributed to the ensuing daughter cells. We have seen, too, that this mechanism soon became an integral part of Weismann's germ-plasm theory of 1885. Roux's career, however, was just getting started. For ten years he remained in Breslau, beginning as a privatdozent in Hasse's anatomical institute and rising to außerordentlich Professor in his own institute for embryology. As mentioned previously, Born was his colleague in Hasse's institute, and he joined Roux in the new embryology institute when it was set up. Thus, during a formative decade in his career, Roux was associated with a friend and contemporary who shared a keen interest in experimentation and the mechanization of biology.

When Pflüger's rotation experiments appeared in 1883, Roux had already sent to press descriptions of a puncturing technique (known in the German plural as *Anstichversuche*) for the purpose of identifying the location of blastomeres. This monograph described the correlation between the first two planes of cleavage and the medial axis in frog eggs.[68] Despite addressing similar questions about the early determination of axes, it is clear that Pflüger's work represented not only a manual manipulation but an attempt to analyze the causal processes. Roux's first work was descriptive enhanced by his *Anstichversuche*.

Within the year, Roux quickly met Pflüger's challenge with a set of refuting experiments that involved the construction of a vertical carousel containing vials with loose but living, fertilized frog eggs.[69] The speed of rotation and the position of the vials along the radius of the carousel allowed *him* to counteract the force of gravity with a controlled centrifugal force. The results showed that cleavage and the three primary axes were independent of gravity (and for that matter of light, heat, and magnetism as well). It was a successful attempt to isolate and test hypothetical external forces on cleavage. The exercise was, in Roux's mode of speaking, a causal analysis. Roux argued that the egg was a self-contained unit and that Pflüger's simile of a snowflake in an avalanche was absurd. Over the next ten years, Roux published six additional *Beiträge* on this new experimental approach to embryology, in what he began calling "Entwickelungsmechanik."[70]

But there was much more to Roux's achievements. The early results of Roux's most famous set of experiments were demonstrated at the Wiesbaden Versammlung in 1887. The official report described this contribution in a manner that captured only a portion of Roux's work and none of the excitement it must have generated. It also missed the real point of why Roux went after the nucleus with a hot needle. "Herr W. Roux (Breslau)," the scribe unemotionally recorded, "spoke on self-differentiation of the blastomeres. After the destruction of one of the first two blastomeres of the frog's egg he pursued the fate of the other surviving cell. This cleaved, formed a Semimorula, then a Semiblastula, a Semigastrula lateralis and finally a Hemiembryo lateralis. In the further progress of the development a regeneration often commenced with the half of the body, which up till then was missing, so that finally a normal embryo resulted."[71]

Apparent half-embryos were a dramatic outcome. However, in his published account, Roux revealed that his real goal was to understand the roles of the cytoplasm and nucleus in early cleavage:

First of all, there must be substances that are not essential for development and, secondly, those whose disruption or loss in very slight quantities from the blastomere destroys its ability to develop. At the present stage of our knowledge we shall consider the latter substances preferably as nuclear components. I attempted when operating with the needle to disrupt the arrangements of the nuclear parts by manifold movements within the egg; as mentioned above, I was so rarely successful that I preferred to make use of heat as a destructive agent. This then accomplished the desired effect.[72]

Roux had achieved the goal of testing the role of the nucleus in differentiation by destroying it at the two- or four-celled stage with a heated needle. In doing so he was supporting his suggestive essay of 1883. It must not be forgotten, however, that a regeneration, or what Roux call a "post-generation" of the lost portion of the organism, be it a half or a quarter of the blastomeres, often took place as nuclei from the developing part invaded and reconstructed the protoplasm of the still-attached dormant part.[73] This would have been significant news in Freiburg.

As mentioned earlier, there is no record that Weismann attended the Wiesbaden Versammlung, which took place between September 18 and 24 of 1887, but this is immaterial. Sometime in the spring or early summer of the same year, at least two months prior to Wiesbaden, Roux likely visited Freiburg and Weismann's institute. His visit may have been connected with an informal exploration of a prosector's position in the neighboring anatomical institute. Weismann would surely have heard from Roux directly at that time about his results.[74] Two years after his successful demonstration, Roux became a professor of anatomy at Innsbruck, where he proudly explained his experiments to Emperor Francis Joseph I. His "half-embryo" experiments had captured public as well as professional attention. Generalizing his results in 1893 in his seventh "Beiträge zur Entwickelungsmechanik," Roux designated the pattern of cleavage, which his experiments purported to demonstrate, "mosaic development."[75] Two years thereafter, Roux moved back to Prussia where at the University of Halle he assumed the Lehrstuhl of the university's institute for anatomy.

Laurent Marie Chabry

Coincidentally, Chabry, who we have already discussed, performed his experiments on cleaving ascidian eggs, which produced the same ambiguous results as Roux's. Chabry undertook his experiments as part of a much larger study on ascidian development for his dissertation.[76] Employing a mechanized stylus and holding the ascidian egg fast in a capillary tube, he was able to destroy designated blastomeres under ×300 power magnification.[77] After the numerous distortions created by his mechanical assault, Chabry calculated he could create in theory 65,535 different developmental "monsters" by the eight-cell stage.[78] The initial distortions aside, Chabry showed that as development continued the embryo was often able to reestablish a nearly normal, though smaller larval stage. This "regulation," as it was later called, contrasted with the "post-generation" described by Roux in that Chabry did not trace a nuclear reconstruction of the disabled blastomeres. The publication of Chabry's very clever experiments anticipated the publication of Roux's cruder heated-needle experiments.[79]

I have maintained that Chabry's work represented the French teratological tradition, which was primarily concerned with classification of normal and artificially created monstrosities.[80] Reinhard Mocek has more recently argued that Chabry was also inspired by a materialistic social philosophy absorbed during his student days in Russia, which promoted a causal analysis in his later science. This may have been so, but it is also clear from his published results that Chabry was not concerned about the role of the nucleus in development nor the perplexing antithesis between dependent and independent development, which motivated Roux and many of his fellow Germans.[81] No matter where Chabry's motivation and priority may have lain, Weismann recognized the importance of his work but pointed out that he "has drawn no theoretical conclusions from his observations."[82] Unfortunately, after developing his elegant microtechnique, Chabry died in 1893, so he had little if any impact on subsequent embryology.

Oscar Hertwig

As described in earlier chapters, Oscar Hertwig was the author of several seminal works in zoology. In 1875–1876, he published as part of his Habilitationsschrift the most cogent demonstration of the fertilization of the animal egg. By showing that a male "pro-nucleus," assumed to be derived from the head of a sperm cell, fused with the female "pro-nucleus," which was clearly derived from the germinal vesicle of the egg, Hertwig had challenged Haeckel's claim that all cells began in a non-nucleated monera stage. Shortly thereafter, Oscar and his brother Richard published a series of innovative monographs on the formation of germ layers, which in a more direct way challenged Haeckel's biogenetic law. In addition, as we have seen, he and Gustav Platner had between 1888 and 1890 independently argued beyond question the parallel between the formation of a tetrad of spermatozoa derived from a single spermatocyte and the formation of the mature egg and three polar bodies. As argued in Chapter 11, Hertwig's outstanding demonstration set the stage for Weismann's work on amphimixis.

Before writing his influential monograph, Hertwig had picked up Pflüger's challenge as well.[83] He devoted some of his Easter holiday of 1884 on the Mediterranean to the exploration of the relationship between gravity and early cleavage. His technique was to allow the fertilized egg of an unidentified species of sea urchin to undergo first cleavage in a drop of seawater hanging from a microscope slide. There were enormous advantages to his arrangement, for the eggs were small, transparent, and unencumbered by yolk inclusions or fat droplets. Furthermore, the nucleus floated in the middle of the egg mass. Within an hour and a half at a given temperature, cleavage began; Hertwig could claim that the spindle of the dividing nucleus, and hence the line of first cleavage, might occur at any angle. Although it may have had an indirect influence on the distribution of the protoplasm, gravity did not influence the nucleus and thus could not be considered the formative factor.

August Rauber

Like Born, Roux, and Hertwig before him, Rauber argued that the formative factors must be internal to the cleaving egg itself. Halfway in age between Weismann and Haeckel on the one hand, and Born, Roux, and O. Hertwig on the other, August Rauber hailed from the Bavarian Pfalz. He received his medical education in Munich and Vienna, habilitated in anatomy back in Munich, and then served two years as a military surgeon. In 1872, he became the prosector for Wilhelm His in Basel; before the end of the year, he would move with His when the latter accepted a call to the anatomical institute in Leipzig.[84] Within three years of arriving in Leipzig, according to his biographer, Rauber and His had a falling out. As a result, Rauber—like Remak, Auerbach, and Maupas before him—set himself up in a private laboratory.

Only in 1886, a few years after the common reaction of younger embryologists to Pflüger, did Rauber secure a Lehrstuhl in anatomy in the Estonian city of Dorpat.[85] There, he spent the rest of his life in apparent exile on the fringes of German academic life, teaching Russian medical students who frequently could not understand his German lectures.[86] In a now classic work, *Formbildung und Formstörung,* Rauber demonstrated in 1880 his enthusiasm for many of the same traditions that were to coalesce some nine years later in Roux's program of Entwicklungsmechanik.[87]

In the work's 150 pages, Rauber reviewed the extensive literature on experimental teratology. It is clear that he saw the same reservoir of information hidden in the results of abnormal development. "Most valuable," he insisted, "is the investigation of the connections that exist between the normal and disturbed courses of development and that not infrequently place those obscure relationships of the normal in better light than could be done in other ways."[88] It is obvious, too, that Rauber was familiar with some of the physiological work that had been done on embryos; although his knowledge was not profound in this area, he recognized the importance of physiology for cellular mechanics.[89] Finally, Rauber was in touch with the discoveries that were current in histology, and he definitely saw that this work had a direct bearing on the meaning of early embryonic cleavage.

Cognizant of a great number of research traditions, Rauber strove to bring them together into a discipline that he baptized "Cellularmechanik." This purported discipline was to explain ontogeny by observing the basic cellular events of development, among which Rauber listed the numerical increase of cells, changes in their size and shape, cell migration, and cell differentiation.

"Cellularmechanik," he said, was the study of these events not only in normal development ("Formbildung") but in distorted development ("Formstörung") as well. Guided by "Cellularmechanik," Rauber foresaw that embryology was to launch itself into a new and rich era of discovery and understanding.[90]

Thus, it is thoroughly understandable that three years later Rauber also picked up the gauntlet laid down by Pflüger. By fixing fertilized trout eggs with silvered clamps and turning them upside down or placing them in a centrifugal apparatus, Rauber observed that the yolk attempted physically to compensate for its dislocation. When it did so successfully, normal cleavage began; when it was unsuccessful, cleavage failed to occur.[91] In a general way, Rauber argued that Hertwig may have unwittingly disturbed the protoplasm and thus misjudged the isotropy of the cleaving egg. He was prepared to argue, along with Pflüger, that gravity did after all have some direct influence on the orientation of first cleavage.

Hans Driesch

One last investigator who made a perplexing discovery associated with blastomere regeneration prior to the publication of *Das Keimplasma* was Hans Driesch. Born in 1867 to a wholesale merchant and his wife in Hamburg, Driesch graduated from the city's famous Johanneums gymnasium with highest honors. Having been inspired by Haeckel's writings, he was determined to go into science, but before embarking for Jena, he spent his first two university semesters, the summer of 1886 to winter of 1886–1887, in Freiburg. There he attended lectures by Weismann and Korschelt.[92] Spending the next seven semesters in Jena, Driesch associated with a galaxy of established and rising stars in zoology, including Wilhelm Preyer, Oscar Hertwig, and Curt Herbst. He attended Haeckel's lectures, which incidentally he thought were boring. He also spent a semester in Munich at Richard Hertwig's zoological institute where he had the chance to interact not only with Oscar's younger brother but with Boveri and Gustav Wolff. During this period, he was introduced to the delights of marine fauna, first in Helegoland, then along the Adriatic coast. The experience led him to a morphological dissertation on hydroid polyps, which he submitted to Haeckel and successfully defended in the summer of 1889. Haeckel offered him an assistantship, but having inherited a considerable fortune from his deceased mother, Driesch turned down the opportunity to climb the next rung on a formal academic ladder. Instead, a trip to Ceylon, temporary relocation to Zürich, and multiple research visits to the marine laboratories in Plymouth, Trieste, and Naples reinforced Driesch's determination to pursue zoology seriously.[93]

The scientific context of Driesch's research in the early spring of 1891 determined his experimental career. He was impressed by Roux's frog half-embryo experiments. He was inspired by the advantage of using sea urchin eggs from the first experiments of Oscar Hertwig and the subsequent collaborative experiments by both Hertwig brothers. So with a university lecture by Roux devoted to the half-embryo frog experiments in hand, Driesch set out for the marine laboratory in Trieste to confirm the results with other organisms.[94] Previously, the Hertwig brothers and Boveri had used shaking techniques to damage the blastomeres of sea urchin eggs after first cleavage. Driesch learned it was not an easy manipulation to shake fifty to a hundred eggs in 4-cm-long, roughly 0.6-cm-diameter vials for five minutes. The timing of fertilization and thereafter of cleavage turned out to be critical. The individual isolation of the few separated blastomeres was essential because dehydration and bacterial contamination was always a danger, and containment of blastomeres by the vitelline membrane caused complications. Nonetheless, Driesch succeeded in a number of bouts and duly reported an example where after the first five and a half hours and five further cell divisions the isolated blastomeres appeared to be headed toward half-embryos, as Roux's results predicted. By the second day, an invagination at the vegetative pole of fifteen blastulae began, and by the third day Driesch found normal but small gastrulas swimming about in their dishes. Development continued until active larvae in a few specimens became evident. "These experiments," Driesch commented, "therefore show that under certain circumstances each of the two first segmentation-cells of *Echinus microtuberculatus* can give rise to a larva of the normal form, which is *entire* as regards its shape; and that a partial formation, and not a semiformation, occurs in this case."[95] The expression "partial formation" ("Theilbildung") was important for Driesch, for it contrasted his results with Roux's "post-generation." The significance of his work, Driesch felt, refuted His's and Roux's claims of "organ-forming germ-areas" ("organbildene Keimbezirke").[96] The experiments led to results that dramatically altered the discussion of early development.

Driesch's paper appeared six months before Weismann signed the preface to his *Keimplasma,* but Weismann had time to include the results and fashion a response before sending his completed manuscript to Gustav Fischer. It was important for him to address the work of both Chabry and Driesch (he did not mention Hertwig's experiments in this regard) because they both presented a serious challenge to his germ-plasm account of development. Weismann's response was twofold. First, he endorsed Roux's notion of "an unfolding of multiplicity, i.e., evolution," in ontogeny. This in turn was validated by the assumption that "the primary constituents of the germ-plasm are distributed by means of the processes which can *actually* be observed in nuclear division."[97] Consequently, the surprising production of whole larvae from isolated blastomeres in ascidians and sea urchins must, in Weismann's mind, be due to the same mechanism of accessory and supplementary determinants that he had earlier employed to explain regeneration of amputated and damaged adult parts.[98]

When exactly Driesch sent an offprint of his paper accompanied by a letter to Weismann is not known, but it is clear from Weismann's reply that Driesch had read the section on experimental embryology in *Das Keimplasma.* Weismann's reply in return was thoughtful but defensive. The letter is worth quoting extensively, for it poignantly demonstrates the bind he soon found himself in:

> My best thanks for your work and your letter, both of which have interested me to a high degree. They are very remarkable results, which you relate to me and I gladly admit that I do not yet see how they can be ordered into our, that is, into my currently developed point of view. Just as little, however, am I able to find, how by giving up of accumulated achievements on the basis of these new experiences one can gain any degree of insight whatsoever into the processes of development. These are both only first impressions, which I do not take as definitive. I am used to allow the facts, with which at first I do not know what to do with, to toss for a while quietly about in my head, in order gradually to obtain greater clarity. . . .
>
> Even though it should turn out, however, that it should be as you believe, that the character of the single cell should be determined by external conditions, evolution theory [i.e., "an unfolding multiplicity" in ontogeny] would still remain the single possibility, because every true epigenetic theory is not compatible with the fundamental phenomena of variation, as far at least as I am able to see.
>
> As for your pragmatic nod with respect to the duplication of your experiments I thank you most kindly; it would naturally be very important for me to be able to repeat them myself, however, that would be hard to do, but perhaps one of my students can do it and that would be desired by me. . . .

I wish you success in your research and hope that you yourself will discover still more facts, which will permit the current affairs to appear in yet another light, more favorable for an understanding of the events of development.

Respectfully yours,

August Weismann.[99]

The reply to Driesch and its antecedent source was not decisive, but it was prophetic. At the very moment when Weismann seemed to have fashioned an extraordinary mechanical model buttressed by his very productive research in natural history, embryology, and nuclear cytology, the practice of a younger generation hurdled over his accomplishments with new techniques, less explicit models, and new ways of explaining biological phenomena. In the same year that Driesch carried out his initial researches, Dietrich Barfurth, a newly minted Ordinarius in comparative anatomy, histology, and embryology at Dorpat University, wrote the first of twenty-three reviews covering the contemporary work on "regeneration" for *Ergebnisse der Anatomie und Entwicklungsgeschichte*.[100] The list of works by investigators and commentators of regeneration examined in Barfurth's first review included, among many others, Balbiani, Boveri, Gruber, Ishikawa, Nussbaum, Wolff, and Barfurth himself. Some, like Driesch and Wolff, would turn toward vitalism; others, like Boveri, T. H. Morgan, Hans Spemann, and Ross G. Harrison, would find new ways to lead embryology without violating the general precept of a mechanistic view of life.[101]

Weismann was perfectly right. In 1892, the experiments on blastomeres were far from definitive. The different techniques employed and the different species involved rendered the experimental results contradictory and confusing. Was he, however, simply stubborn? It is also worth remembering that in the past Weismann had given up cherished beliefs when the evidence seemed strongly against them. He had, after all, just relinquished the notion that polar bodies contained spent ovogenetic idioplasm when Hertwig and Gustav Platner had revealed a cytological parallel between ovogenesis and spermatogenesis. By 1892, there were several anchors that held him fast to a carefully considered logical construct, regardless of what experimental evidence threatened to pull embryology in another direction. It was certainly relevant that he had just articulated a complex germ-plasm model for heredity, development, and regeneration at the outset of his *Keimplasma*. His model had been premised on the conviction that determinants were fixed to the ids of each idant (chromosome) and were distributed, either equally or unequally, at each indirect cell division during development. It must be difficult for any scientist to give up a carefully articulated model in the face of uncertain, contradictory evidence.

The second anchor was imbedded in the mechanistic imperative that shaped his theory of life. This did not consist of a reduction to the bedrock of physics and chemistry but to a hypothetical "pre-molecular biology," which projected complex, replicating, biological molecules to satisfy the multiple demands of heredity, development, nuclear cytology, and evolution theory. Thus "biophors" physiologically differentiated all cells; "determinants" transcribed individual traits; "ids" represented selected ancestors; and "idants" were those demonstrable nuclear chromosomes. All those supplementary accessory and supplementary units of the germ-plasm served, however, as *dei ex machina*.

Das Keimplasma, Part III: Heredity and Sexual Development

The second part of *Das Keimplasma* focused on the application of the model in various kinds of asexual reproduction: regeneration, cleavage experiments, fission, gemmation, and the alternation of generations. To meet contemporary embryological and cytological challenges, Weismann invoked reserve and supplementary ids and determinants. It was an unwarranted castle built on sand, to be sure, but it was a model that framed research and debate for the next dozen years.

The third part of *Das Keimplasma* served as an extension of his *Amphimixis* and its promotion of the sexual production of variations. The nature of his model forced him to address this issue on two levels. The first consisted of an explanation of the origin of changes in the germ-plasm of newly conceived individuals; the second addressed the appearance of recognized variations in the soma of individuals. In the ambiguous language of classical genetics, this would be considered the difference between variations of the genotype and phenotype.[102]

On the germ-plasm level, Weismann had to face a number of empirical questions. Were his idants continuous and enduring through the cycle of cell divisions of reduction, fertilization, and ontogeny? This remained an unresolved question at the cell's resting stage, when the chromosomes seemed to disappear into a reticulum of chromatic material and reemerge as a spireme, which then condensed into the chromosomes of the offspring cells. This was a problem he had faced with his first microscopic studies in 1859–1860, and thirty

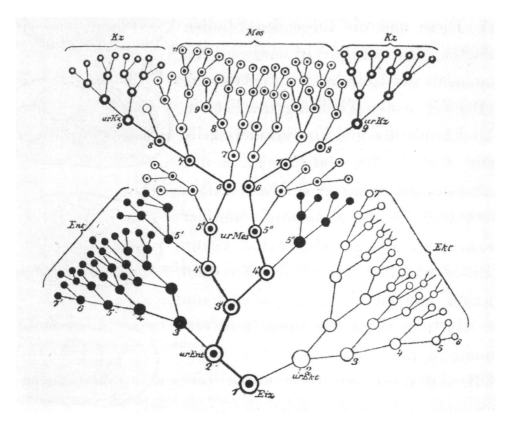

Figure 16.1. Presentation of the theoretical passage of the "Keimbahn" of twelve germ cell generations (*Keimzellen*) in *Ascaris*. Its passage contrasts with the cells of the ectoderm (*Ektobast*) and mesoderm (*Mesobast*) during initial development. The cell generations are identified by Arabic numerals. The first generation begins with the egg cell, "Eiz."

August Weismann, *Das Keimplasma: Eine Theorie der Vererbung. Drittes Buch, Capitel VI: Der Bildung von Keimbahn* (Jena: Verlag von Gustav Fischer, 1892), p. 258.

years later no one had yet found a way to solve it. Another issue concerned the geometry of chromosomal divisions during both maturation divisions— were such divisions always longitudinal, as described by Flemming, or did transverse divisions also take place? During which of the maturation divisions did a qualitative reduction actually occur? How consistent and how significant were the number of chromosomes of various species? Finally, what was the significance of the appearance of the so-called tetrads of chromosomes during maturation divisions? As nuclear cytology developed in the decade following Flemming's and Strasburger's first support of indirect nuclear divisions, these issues became increasing concerns. These issues will be explored in greater detail in Chapter 17, but reference to one of them is unavoidable here. It must be remembered that in 1892 there were no recognized physical markers for individual chromosomes, which might have led Weismann from the development to the transmission of traits. This was soon to change, but not in Freiburg.[103]

Weismann was certainly aware of these problems. He introduced some of these issues by pointing out once again that reduction division, as van Beneden had first discovered in *Ascaris,* eliminates half the number of idants (i.e., chromosomes), and consequently half of the theoretical ids of each parent. Fertilization brings together new combinations of idants with each subsequent sexual union of the gametes. The advantage of two maturation divisions, explained in his earlier essays, was revisited. "All these phenomena of heredity," he emphasized in bold print, "which are spoken of as the intermingling of the characters of ancestors, such as degeneration or atavism of all kinds and degrees, depend, I believe, on this complicated structure of the germ-plasm."[104]

Given the fact that they disappear from view at the resting stage after each mitotic division, it was essential that Weismann take a stand as to whether the idants were continuous and permanent entities throughout ontogeny. We have seen that Roux had implied this in his provocative paper of 1883. We have seen that Weismann insisted upon it two years later when he first presented his germ-plasm theory. Inspired by Flemming, Carl Rabl, the prosector of anatomy in Vienna at the time, presented a lengthy discussion followed by detailed illustrations of indirect cell division in the epithelium layer of the floor of the mouth and gills of salamanders. Although the resting stage was not the focus of his studies, he described the reticulum and portrayed the chromosomes remaining intact throughout resting stage.[105] Working with *Ascaris,* Boveri provided an experimental endorsement of Rabl's conclusions by demonstrating eggs with an abnormal number of chromosomes continue to develop with a corresponding abnormal number. He also ratified van Beneden's claim that the paternally and maternally derived chromosomes retained their individuality through early cleavages.[106]

Boveri's work must have been echoing in his ears when Weismann wrote his *Keimplasma* in Munich. Back in his institute in Freiburg, Otto vom Rath produced evidence that the ids (i.e., microsomes) were bound together by threads of achromatic nuclear reticulum or "linin." Häcker supported vom Rath with his own studies on copepods, and at the same time demonstrated that the maternal and paternal idants persisted through the resting stage after fertilization. Together their work gave Weismann all the evidence needed to build his model on the hypothesis that the idants persisted through indirect cell division.[107]

The halving of the idants, that is, the permutation of ids at reduction division and the doubling of the idants during fertilization, provided for germinal variations—hundreds of thousands of them. As his critics pointed out, however, the two in tandem, reduction and fertilization, would not insert new determinants into the germ-plasm line. This dilemma had to be solved if Weismann's model was to be relevant to the contemporary debates over evolution. Furthermore, it could not be directly solved by recourse to external forces creating new variations, for this would introduce a Lamarckian solution and render his model obsolete.

On the second level, the level of empirically verifiable somatic variations, Weismann as an accomplished naturalist and someone who had discussed identifiable variations since the early 1870s was no stranger to these features. By 1892, he must have devoted considerable thought to how this might happen in detail and in connection with his germ-plasm model. It is clear that he began consulting experts and literature about different types of variations. He relied extensively on Darwin's *The Variation of Animals and Plants under Domestication* (1868). He structured much of his discussion on a threefold classification of characters constructed by Wilhelm Focke and drew upon Focke's *Die Pflanzen-Mischlinge* (1881) for many of his examples.[108] Weismann also consulted with his botanical colleague Adolf von Hildebrandt about the latter's breeding experiments with two species of *Oxalis*. Through Wiedersheim, Weismann had gotten to know personally Otto Ammon, a retired engineer in Karlsruhe, and the two had discussed the results of Ammon's investigation of the physical traits of the citizenry of the Grand Duchy of Baden.[109]

Rather than a concerted effort to understand the patterns of somatic variations, however, Weismann retreated to the germinal level and to the appearance of variations during ontogeny. It is as though the spirit of his early studies of caterpillars took over from his earlier examination of seasonal dimorphism in *Araschnia*. His interest in the changing, developmental forms of an individual prevailed over the more quantitative comparison of predetermined varieties. Twenty years later and steeped with the importance of chromosomes in the explanation of heredity, he reached back to the more basic level for a germinal explanation of the production of variations. This was not an unwarranted focus for a dedicated mechanist, and in the next century the spirit, if not the details, of his model of heredity would be vindicated. In 1892, however, there was too little known about reduction division and about those chromosomes to lead him very far. The fundamental processes Weismann did comprehend were assimilation, growth, and a natural competition—an overgrowth and selection, if you will. Back he turned to apply these nineteenth-century principles to a lower level after he had denied the overgrowth principle in somatic heredity as a whole. Drawing his cue once again from Roux, he reached this time for the latter's youthful tract on *Kampf der Theile* (1881). What better than a natural competition among ids, determinants, and even biophors to provide a mechanism for the origin of variations?[110]

From the architecture of his germ-plasm Weismann turned to the union of two parental germ-plasms in ontogeny. "The nature of the combination of the parental ids, which takes place during fertilization," he explained in bold type, "predetermines the whole subsequent ontogeny."[111] With this claim in hand, Weismann turned to Focke's three classes of parental contribution: the mean between the two parents, the domination of either the father or the mother, and the varied influence of each parent at different moments in ontogeny. During development, a continuous competition among the homologous determinants of the multiple ids contributed by both parents resulted in a changing mix of the influence of each parent during the expression of any given character—an "ontogenetic shifting of the hereditary outcome," as Weismann expressed it. Incidentally, this provided even more patterns of variations for the species.[112] Homologous determinants moreover competed for nutrition, and slight molecular differences between them could lead to their different rates of assimilation, growth, and reproduction. "Stronger" determinants, inherited from one or both parents, would dominate during differentiation, and acting together they would produce the trait to be expressed. Thus, the process of this "ontogenetic shifting of the hereditary outcome" explained Focke's three breeding patterns. Competition and the same shifting introduced new determinants and hence new traits into the phylogenetic lineage where Darwin's natural selection singled out the more fit under the given circumstances.

Latencies of all kinds, reversions, atavisms, symmetry, bilateralism, and metamerism, in theory, arose as the result of amphimixis, germinal competition, and natural selection. Polymorphism and sexual dimorphism of many types were the result of competition between double, even triple determinants. Degeneration and loss of given characters became a matter of panmixia or the relaxation of natural selection. The distinction between "individual" and "species" traits, commonly used by taxonomists and Weismann himself, was explained through the calculus of amphimixis, germinal competition, and natural selection over thousands of generations. Those traits that survived the germinal competition and natural selection would be those that would be phylogenetically older, physiologically stronger, and consequently, in Weismann's vocabulary, not only homologous but homodynamic determinants. The traits they produced would prevail together in the germ-plasm of all parents of a given species. Taxonomists would see in them a permanence that individual traits did not have.

The accessory mechanism of germinal selection in all its elaborated detail was necessary if Weismann was going to serve his primary mission of supporting Darwinian evolution. The standard trio of required mechanisms for natural selection, heredity, and variations had expanded in his mind to include the additional mechanisms of reduction division, amphimixis, and germinal selection. It is in keeping with his goal that he devoted the last two chapters of his tome to two major mechanistic theories of evolution: the inheritance of acquired characters and his updated version of the Darwinian theory of natural selection.

Das Keimplasma, Part IV: Transformation of Species

The two closing chapters of *Das Keimplasma* not only presented familiar themes but added new supporting materials to the ongoing debate between neo-Lamarckians and neo-Darwinians. At the outset, Weismann defined "acquired characters" as characters "which are not preformed in the germ, but which arise only through special influences affecting the body or individual parts of it."[113] With respect to claims about the "supposed transmission of acquired characters," Weismann made clear that his sequestration of the germ-plasm from the soma was not simply a counterclaim dependent upon perceived ontogenetic tracks of germ cells and a conjecture about the phylogenetic origins of metazoans, but that the histology of germ-layer formation and the cytology of indirect cell division rendered obsolete all theories of transmission dependent upon "neural excitation" or pangenesis. So important were the cytological developments of the 1880s that in respect to Darwin's theory of pangenesis Weismann could emphatically assert that "*the process of the fission of the idioplasm in nuclear and cell-division seems to me directly and conclusively to refute the whole idea of the circulation of gemmules.*"[114]

His claim appears clear enough until one explores, theoretically, the action of germinal selection at the microlevel. Weismann made a distinction between (1) acquired traits on the somatic level caused by an ontogenetic shifting in the ids and (2) determinants and induced changes in the germ-plasm itself. The former included injuries, functional changes, and environmentally (including nutritionally) induced changes in the idioplasm of the soma. These were, in Weismann's terminology, somatogenic characters ("somatogene Eigenschaften"), and in accordance with the mechanics of the germ-plasm theory they could not be transferred to the germ-plasm. The latter were the changes wrought by a germinal selection in the ids and determinants of the germ-plasm. He designated these blastogenic characters ("blastogene Eigenschaften"). The somatogenic changes were properties appearing in an individual reacting to the vagaries of life such as nutrition, activities, and environmental stimuli. The blastogenic changes could be transmitted to the next generation but might remain latent, or sooner or later might be expressed. To Weismann's consternation, much confusion resulted from the distinction, but it was necessary to follow the details of his mature germ-plasm theory if one was to follow Weismann's conclusions about evolution.

When it came to empirical support for his adamant rejection of an inheritance of acquired characters, Weismann mentioned two of his ongoing research projects. He reported once more on his experiments with severing the tails of newborn mice. A certain fatigue in the matter is apparent: he was now on his nineteenth generation without the results of his amputations appearing spontaneously in subsequent generations. In this case, a somatic injury failed to reappear as a transmitted trait.[115] More challenging was his debate with believers of Lamarckian inheritance. Weismann turned briefly to a description of his renewed study of seasonal and geographic dimorphism in butterflies. His focus was on the effects of temperature on the *Chrysophanus phlaeas/eleus* complex, which he had mentioned earlier in his *Studien* of 1875. Common throughout the northern hemisphere, *Chrysophanus* was known in Europe as the small copper or fire butterfly (*Feuerfalter*) because of the brilliant orange color on the upper surface of its forewings and an orange marginal band on its hindwings. In southern climes, the orange becomes dusky, which might appear to be an unambiguous example of the influence of external temperature on heredity. The American social and biological scientist Lester Ward had just published a lecture claiming as much. Weismann would have nothing of it, and he proposed that somatic and germ-plasm determinants could be affected by similar influences through parallel germinal selection. Because I will be examining Weismann's new butterfly work in the context of work done by his peers in Chapter 17, I will leave the matter here. The general subject and Weismann's intense involvement suggests how Weismann was developing his mature germ-plasm theory to reflect the experimental work at Freiburg and elsewhere.

When in the final chapter of his *Keimplasma* Weismann turned to a defense of his neo-Darwinian convictions, one does not find an elaborate justification of natural selection. Instead, he presented a detailed examination of variations. Evolution and natural selection were assumed. At the end of the nineteenth century, the challenge for the believer in natural selection was the rise and transmission of new variations—a counterrepost, if you will, to the neo-Lamarckian explanation he had dismissed in the previous chapter.

His starting point was framed by his conclusions on amphimixis. "I am convinced," he underscored, "that the two forms of amphimixis—namely, the conjugation of unicellular, and the sexual reproduction of multicellular organisms—are means of producing variation."[116] A means for mixing and recombining ids and their determinants, yes, but now he had to dig deeper. The variations brought about by the innumerable combinations of ids and determinants produced by amphimixis were nevertheless limited—that is, they were bound by what already existed in the collective germ-plasms of the two parents of a given species. Weismann had already alluded in earlier essays to the necessity of escaping from this phylogenetic trap and had indicated the solution in his treatment of regeneration and ontogeny earlier in *Das Keimplasma*. Now was the time to exploit his capital idea: germinal selection within the blastogenic germ-plasm itself. This rendered the involved determinants different in kind. As with the selection of homologous determinants from different ids during ontogeny, Weismann reckoned on the assimilation of nutrition and growth within the blastogenic germ-plasm.[117] The resulting germinal selection unavoidably produced minor molecular variations between homologous determinants. After generations and through the processes of reduction divisions and amphimixis, the variant homologous determinants could accumulate; when numerous enough, they could, according to his argument, result in a phylogenetically new somatic trait. Becoming operative, the new trait would contribute to the natural selection of the whole individual.

The germ-plasm model, because Weismann so consistently segregated the molecular events on the somatic and blastogenic levels, saved the Darwinian process as he understood it. In contrast, new variations introduced by environmental fluctuations of ontogeny were irrelevant in evolution. On the other hand, blastogenic variations, which altered the germ-plasm, could have a direct influence on phylogeny through natural selection. It is as though Weismann sensed the need for "genomic mutations," but explained the phenomenon in terms of a struggle for nutrition rather than as random physical events. The very core of his germ-plasm theory seemed to void the architectonic advantage of the rest of his theory of heredity.

The evidence from natural history that might document this blastogenic process of change was hard to come by. Discussions of sports and the difference between large and small variations, which had plagued an earlier generation of evolutionists, were no longer useful. Weismann once again called upon his butterflies to illustrate the origin of blastogenic variations. He had renewed his studies of geographic differences in *Polyommatus phlaeas,* where a gradual increase in dusky wing scales from the center to the margins of the wings is found as the collector moves in a southerly direction from northern Europe to central Italy. The butterflies from Naples, whether raised there or in Freiburg, suggested that homologous determinants for certain wing scales had diverged from one another over time. His experiments suggested that the variation was not due to natural selection and definitely not to the inheritance of acquired characters. It is an interesting feature of the goals in Weismann's agenda that he considered this particular example as "the only carefully observed cases of blastogenic variation, due to the direct influence of external conditions." Furthermore, as he emphasized the flux in determinant changes, he reasoned that "it is evident that the influence of the temperature was not quite uniform," even when broods were raised under the same conditions.[118] In short, a visible variation arose through a combinatorial algorithm involving nutritional differences within the germ-plasm, germinal selection, reduction division, and amphimixis. *"A hereditary individual variation,"* he concluded in bold print, *"will therefore arise if many of the homologous determinants vary in the same way."*[119]

The sheer "force" of numbers as the altered homologous determinants prevailed over the original determinants represented a mechanism for germinal change. Weismann nevertheless found a second mechanism in the formation of double determinants. This was a process he had described earlier in his chapter on regeneration. With a nod to Darwin, Weismann attributed the increase in the number of tail feathers of fantail pigeons from about twelve to "about forty" to a doubling of determinants in the ids at a critical time during germ-plasm development. The lengthened tail of many hummingbirds provided an additional example of sexual dimorphic variations of size resulting from determinant doubling. Although neither of these cases reflected Weismann's personal research, he felt comfortable in concluding that "all really new structures are not merely the result of transmission, but are due to the variation and frequent multiplication of determinants."[120]

At this point Weismann was stuck. His interpretations of the rise of certain somatic variations were easily reinterpreted. In accordance with his own standards, he had to find visible germinal changes, but these were not easy to come by. By the 1890s, it became increasingly clear that the number of chromosomes was constant within a species.[121] As described earlier, Weismann identified his ids with discrete, visibly stained units making up the chromosomes, but hard and fast optical evidence of the different determinants within the ids simply did not exist. Their presence was premised on Weismann's theoretical claims about development and the indirect evidence from natural history, such as his detailed examination of colored ornamental flecks (*Schmuckfarben*) on the appendages of cladocerans.

According to theory, the ids from each parent contained a single determinant for each somatic character. The cells of every somatic character, as they developed and differentiated, involved an interaction between the homologous determinants. As mentioned previously, Weismann added to his mix: supplementary determinants for regeneration, double determinants for sexual dimorphism, alternation of generations, and polymorphisms. None of this could be microscopically seen or chemically detected. To explain inheritable somatic variations, Weismann added more "double, triple, and even quadruple" determinants, which during extended lineages might deviate so far from one another as to become entirely different—no longer homologous determinants. These served as the root explanation of new heritable variations.

Weismann's demonstration of these claims could not be cytological but depended on phylogenetic lineages of variant traits, such as the scales of the wings of seasonally dimorphic butterflies. Each differentiating cell became the battleground between competing homologous determinants, and each generation of individuals contributed to how this battle fared cell by cell and embryonic stage by stage, and how the gains and losses for the germ-plasm of the next generation reflected the random nature of reduction division and amphimixis. The lineages themselves reflected the fitness of a given pattern in the wild over time. What more could he do in 1892?

Weismann closes his lengthy discussion of the germ-plasm by going back to where he started with the subject over a decade earlier.[122] Instead of simply attributing the indefinite extension of life to a gradual specialization and isolation of body cells from the germ cells and presenting a phylogenetic history of the process from amoebae to metazoans, Weismann now had a model of a highly elaborate germ-plasm that explained many of the common phenomena known in natural history, embryology, and evolution. Besides riding piggyback on the limited cytological evidence mentioned earlier, his model provided detailed hypothetical explanations for regeneration, fission, alternation of generations, gemmation, sexual reproduction, differentiation, reversion, atavism, sexual dimorphism, polymorphism in insect colonies, the relationship between the environment and hereditary variations, degeneration, and bud variations in both animals and plants. All these phenomena, discussed widely in the literature of the day, were ascribed to "the power of transmission." Growth might alone explain reproduction in the "very lowest conceivable organisms with which we are not acquainted; while in all forms which have already undergone differentiation, it results from the possession of *a special apparatus for transmission*."[123] It made sense that the subtitle to Weismann's *Keimplasma* announced "Eine Theorie der Vererbung."

17

Controversies and Adjustments

1893–1896

It is easy to overlook the supportive reactions to Weismann's *Keimplasma*. After all, it is tempting to see the history of science, as most areas of intellectual history, thriving more vigorously with contrasts and arguments about specific issues than to parade forth agreements and sanctions. The immediate professional response to Weismann's major work, though muted and cautious, seems respectful even though cautious. Within a few years the likes of Theodor Boveri, Wilhelm Johannsen, and Hans Spemann found his work valuable and challenging though few adopted the elaborate details of his germ-plasm theory. In retrospect, his fully developed germ-plasm theory appeared to be driven by mechanisms, and an integrated reductionist strategy attempted to solve the major issues of morphology, that is, heredity, development, and evolution. This strategy certainly appealed to the next generation despite the fact that they were not attracted to the details of the theory itself. Weismann certainly would have agreed. It was his outwardly mechanistic and suggested reductionism that counted rather than allowing a finely disguised or elaborately justified teleological element or framework to shape the final enterprise. Although many half-friendly critics of Weismann zeroed in on the enormous deficiencies of the germ-plasm theory, many generalists such as Francis Galton, Otto Ammon, H. E. Ziegler, and Edward B. Poulton appeared to honor, if not follow, the general approach broached by Weismann.

Weismann, however, also had serious detractors who took issue with his theoretical assumptions, and it is to these that I turn to explore what deeper issues were involved. Wilhelm Haacke, Oscar Hertwig, Herbert Spencer, and

George John Romanes serve as exemplars of this reaction with whom Weismann directly interacted.

Wilhelm Haacke and a Rival Theory

History has not treated Wilhelm Haacke kindly—or more accurately, it has hardly treated him at all.[1] When he is favorably recognized, it is for coining the expression "Orthogenesis," for independently discovering that Echidna lays eggs, and for carrying out hybridization experiments between mice. We will deal more with this last enterprise later.

Haacke had an unorthodox career for a professional zoologist. Born in the town of Lüchow in Lower Saxony, he studied with Haeckel and wrote an inaugural dissertation in 1879 on the "Blastologie der Korallen," which won muted praise from his mentor.[2] After serving two years as an assistant, first in Jena and then in Kiel, Haacke migrated with Haeckel's support to New Zealand where he became an assistant at the museum in Christchurch. Between 1882 and 1884, he served as director of the Public Library, Museum, and Art Gallery of South Australia in Adelaide. He returned to Germany in 1886 and within two years was appointed director of the Frankfurt Zoological Society, where he worked between 1888 and 1893. His reintegration into German zoological circles gave him the opportunity in 1890 to habilitate at the Technische Hochschule in Darmstadt. After completing the basics for a university career and after serving as a privatdozent in Darmstadt until 1897, Haacke worked as a private teacher and finally as an assistant teacher at the gymnasium level. Haeckel appears to have supported him, despite the latter's failure to mention either the *Generelle Morphologie* or *Natürliche Schöpfungsgeschichte* in his early writings. Haeckel, however, became discouraged at recommending him for positions only to have Haacke turn them down or relinquish them after a few years. By the end of the century, Haeckel frankly advised his former student to work more "thoroughly and scrupulously" and "less superficially in the handling of difficult general problems."[3]

Within the context of this irregular career path, Haacke published a number of criticisms of Weismann's germ-plasm theory. By the time *Das Keimplasma* appeared, he was primed for an all-out assault. His most telling point remained the same, so it is worth noting his earlier, brief critique directed at Weismann's essay "On the Number of Polar Bodies and Their Significance in Heredity."[4] It may be remembered that E. B. Wilson praised this essay as "one of those brilliant essays on heredity which contributed in so important a way to the enlargement of our views concerning cytological research." It should also be recalled that at this time Weismann still believed that only the second polar body was involved in the formation of the gamete. Haacke, however, was not concerned about the cytology; for the most part he praised Weismann's empirical work. In fact, at the outset he singled out Weismann's "extraordinary clarity" and "above all the strong logic" in his papers. It was Weismann's conclusion about the multiplication of traits, which resulted from the processes of reduction division and sexual fertilization—the process that Weismann would soon associate with "Amphimixis"—that upset Haacke. His criticism was simple and justified in the context of a corpuscular-trait relationship. He explained that reduction division and sexual reproduction in Weismann's theory may produce new combinations of traits, but sexual reproduction could not by itself introduce new traits into the phylogeny. Furthermore, sexual phylogeny, begun according to Weismann in the Cambrian period, was limited to the traits available in the germ-plasm derived from the asexual ancestors of the pre-Cambrian, at a time when the influence of the environment and the inheritance of acquired characters were acceptable to Weismann. According to Haacke's view, no matter how extensive the number of traits in the germ-plasm might initially have been, it was subject to a constant reduction in number by the elimination of unfavorable traits through natural selection. "With justice it remains to Weismann to take into account this supposition and its ensuing point of view [*Stellungnahme*]."[5]

When *Das Keimplasma* appeared, Haacke was primed with more extensive arguments. Before turning, however, to his elaborate critique of amphimixis, Haacke spelled out some new concerns. He first made it clear that the problem of heredity was to be solved within the context of a choice between preformation and epigenesis. His was not a searching analysis of these antinomies, but he found in Haeckel's discussion of protoplasm (Haeckel's *Plankton-Studien* of 1890) a modern way of viewing the two. The theory of preformation maintained that the protoplasm of the germ was "polymiktic," made up of many different substances that were ordered in a definite way. On the other hand, the theory of epigenesis held that the protoplasm of the germ was "monoton," made up of a mixture of similar elements.[6] It was not a formal definition of either position or, with the exception of the reference to Haeckel, was it in any sense historical, but it served as a convenient way of contrasting Weismann's germ-plasm theory from the one he would propose.

Turning directly to an obvious problem in Weismann's mature germ-plasm theory, Haacke criticized not just the elaborate architecture of biophors, determinants, and ids, but the arrangement of qualitative divisions, which during differentiation according to Weismann's presentation, mechanically distribute the determinants to the right place at the right time. How could the ids divide during ontogeny to segregate the determinants of different traits and then heal their broken edges in preparation for the next qualitative division? "If we toss away an epigenetic explanation," Haacke sarcastically asserted, "the increase of the ids by division remains an insoluble riddle, and nothing further remains than either to accept repeatedly new acts of creation by God or to let the germs of all the future representatives of a species be initially inserted by God into the first created individuals of this species."[7] It was not that Haacke was arguing against a complex hereditary material, but that Weismann had postulated an inviable architecture made up with his ids, determinants, and biophors that could not accommodate to the developmental task assigned them. "If they [the bearers of heredity] are architectonically ordered, as Weismann accepts and must accept, the Ids which form them, may only multiply in the manner of Epigenesis."[8]

As if that were not enough, Haacke also ran through a litany of developmental and evolutionary events that he insisted were not explainable in terms of a "preformationist" model of heredity. For example, how can development and evolution based on "preformationism" produce asymmetrical types from symmetrical forebears as one finds with the mollusks? The brown rat must have replaced the house rat in European towns and cities so rapidly because it possessed independently varying adaptations that must arise epigenetically rather than be anticipated through a preformed heredity. Haacke applied the same argument to the appearance of human eyes and ears. "Epigenetic [theories of heredity] assume a reciprocal interplay, a correlation, and where correlation rules, the organs are dependent on one another and can therefore not adapt themselves to the external world in an arbitrary way." With the term "Orthogenesis," Haacke captured what he felt was the continued and directional evolution of given types. Even the development of parasites bespoke a directional evolution, albeit through increasing simplicity. Again, Haacke argued, this systematic degeneration could only be explained through an interactive and flexible epigenesis, not a static and prophetic preformation. In these and other examples, Haacke associated preformation with Weismann's inflexible, complex germ-plasm of independent, hierarchical units of heredity. He insisted that epigenesis, with its simple, monotone germ-plasm (Haacke was comfortable using Weismann's term throughout), could produce mechanistically the developmental and evolutionary patterns he described.[9]

In Haacke's opinion, parallel distortions of the opposing parts in bisymmetrical organisms refuted Weismann's germinal system of the symmetric segregation of ids and their respective determinants throughout differentiation. Similarly distorted fingers on opposing hands, similarly but abnormally colored feathers of opposite bird wings, or the unpredictable number of accessory toes on the opposite hind feet of guinea pigs all presented the same dilemma for Weismann. "Examples such as these may be brought forward a thousand times," Haacke insisted.[10] Moving on from problems of symmetry to the influence of the environment; from the contrast between the plasma-centrosome complex on the one hand, to chromosomes on the other; from meaningful organs to nonfunctioning ones; from rudimentary traits; from directly useful to indirectly useful traits, and from supportive to nonsupportive traits, Haacke drummed forth his message a dozen times.[11] "With these facts," he repeated, "the Preformation theory is incompatible with independent variations of the single biophors of the germ-plasm that correspond with the ensuing selection; only the Epigenesis theory finds itself in agreement with them."[12]

Haacke's *Gestaltung und Vererbung* is a broad study that carried its author into an examination of different kinds of selection. He insisted on distinguishing personal and racial selection and endowed them with his specialized nomenclature of "konstitutioneller" and "dotationeller Auslese." He claimed that only an epigenetic theory could explain the latter.[13] The distinction soon became central in his refutation of Weismann's reliance on the individual selection of variant determinants in the evolutionary process.[14] Haacke also returned to the subject that had been the focus of his earlier critique of Weismann's study of polar bodies and their implications for sexual reproduction. He recognized that Weismann by 1892 had revised his views; besides expressing some anguish over the changing ideas of his opponent, Haacke settled down to some new calculations of the number of combinations available through amphimixis in sexual reproduction. These were more complicated than the earlier calculations because chromosomal reduction, according to Weismann, must now occur during the first maturation division and thus provide an additional division for the mixing of ids during two maturation divisions. Haacke's conclusion remained the same: "Is nature really as productive as it must be if Weismann's theory of Amphimixis were correct? I don't believe it!"[15]

Haacke did both less and more than provide abstract quantifications of trait accumulations during reduction divisions and fertilizations. On the negative side, he failed to take into serious consideration Weismann's short discussion of slight changes in the structure of determinants, and hence the introduction of determinant variants during germ-plasm growth and amphimixis. On the positive side, Haacke topped his discussion by introducing experimental hybridization studies between two races of mice to prove his point. It was here that biologist and historian Hans Stubbe discussed Haacke as a forerunner of Mendel's rediscovery. Haacke, Stubbe claimed, had failed to keep quantitative records of his experiments and had failed to perform back-crosses with the third generation.[16] True enough on both counts! But this is reading history backward. When Haacke chose two races of mice, the gray Japanese dancing mouse ("Tanzmaus") and the checkered climbing mouse ("Klettermaus"), and carried out their interbreeding through 3,000 hybrids, he purposefully chose contrasting behavioral traits and may have used the associated color traits as a means to ensure the purity of the offspring. All he wanted to show was that the offspring [F2 generation] assumed only one of the four possible combinations; that is:

1. ♂ Tanzmaus × ♀ Tanzmaus → all Tanzmäusen.
2. ♂ Tanzmaus × ♀ Klettermaus → all Tanzmäusen.
3. ♂ Klettermaus × ♀ Tanzmaus → all Tanzmäusen.
4. ♂ Klettermaus × ♀ Klettermaus → all Klettermausen.

Further crossing of the subsequent generation [F3] revealed the reappearance of pure ("reine") dancing and climbing mice and, most important for Haacke, the apparent fact that the dancing and climbing traits did not blend. "Reduction division in no way brings about Amphimixis, i.e., the mixture of different plasmas, but instead Apomixis, i.e., the lack of mixing of two not correlated germ plasms."[17]

Only in recognizing that there would be sixteen combinations including both sexes did he provide an unimportant anticipation of Mendel's approach popularized seven years later. Although Haacke claimed to have carried out this and many other hybridization experiments on 3,000 mice, he generally failed to pursue the experiments to the F2 generation or to publish his experimental records. He brushed off any results he may have noticed concerning dominance or recessiveness of the two pairs of traits, white-gray and climbing-dancing. His focus was on the various combinations between the mice, white versus gray and climbing versus dancing. It was not a pre-Mendelian but an anti-Weismannian exercise. It is highly possible, however, that his results scat-

tered throughout his lengthy monograph devoted to Weismann's *Keimplasma*
prompted an attempt by Georg von Guaita, a doctoral student then voluntary assistant in Freiburg, to hybridize the same species of mice.

Haacke ended his formal critique of *Das Keimplasma* with summary lessons. In his mind, Weismann had produced the best and most detailed preformation theory of the day, but its completeness made clear to Haacke a
number of unacceptable premises inherent in this kind of solution to heredity
and development. Primary among these was what Haacke described as the
dualism found in all preformation theories, which necessarily contrasted the
creator and the created. Haacke did not explicitly accuse Weismann of introducing God into science, but he felt it was unscientific for him to sketch out
a prearranged, asexual reproductive structure to the germ-plasm. After all,
preformation must start out with the assumption of the near equivalency of
forms and let the Darwinian principle of natural selection differentiate between them. In contrast, epigenesis "remained true to the principles of the
investigation of nature." By recognizing natural hierarchies, Haacke felt he
allowed for the intrinsic patterns of nature to establish both an embryological and evolutionary development in specific, generally progressive, directions.[18] This dimension to Haacke's argument is reminiscent of eighteenth-
century debates between preformationists and epigenesists in that the former
had recourse to God while the latter confined themselves to natural laws.[19]
In addition to these arguments against Weismann, Haacke found that his
own quantitative and experimental demonstrations against amphimixis were
a strong argument against the primacy of natural selection.

His resort to epigenesis, however, rested on an entirely different foundation. "What makes an organism an organism," Haacke declared in bold print,
"is the possession of *acquired* characters."[20] Further along he compared it to
a fundamental law of physics: "Consequently the inheritance of acquired characters is proven to be a process that must exist with absolute necessity. To deny
the inheritance of acquired characters means to negate the law of the conservation of energy."[21] There were those who recently demanded experimental
demonstrations of such a process, but the proof for Haacke was evident
throughout nature. For example, according to epigenesis 10,000 years of
nonuse had reduced the index finger of the East Asian Lemur ("Plumplori")
half a centimeter in contrast to its functional middle finger. By comparing
the evolution over an extensive period for such a gradual anatomical diminution with the mere decade of lopping off of the tails of generations of mice,
one might understand Haacke's feeling of futility with the rather casual

experiments in Freiburg. Haacke even suggested the possibility that the mouse tails might nevertheless have already diminished as much as 1/1,000 of a millimeter—way below the margin in Weismann's measurements![22] Haacke completed his summary of the germ-plasm theory with a touch of irony. Extending the collective thanks of his profession to Weismann for his challenging preformation theory, Haacke threw Weismann's often cited words back at him: "Weismann believes correctly not 'to have worked in vain; for error, as long as it is based on correct conclusions [observations], must also lead to the truth.'"[23]

Haacke would not forget Weismann's *Keimplasma* and its failures, but he devoted two-thirds of the remaining 300-plus pages of his *Gestaltung und Vererbung* to the context and justification of his own theory of organic form and its inheritance. The exercise consisted of endless chapters on various aspects of morphology. The text traversed matters of material assimilation, the rise of basic forms, organ development, the details of external developments, the forms of sexual, asexual reproduction, regeneration, hybridization degeneration, alternation of generations and polymorphism, and the reproduction of various types and patterns of heredity. The sequence might be interpreted as a synopsis in a university course in biology except for the fact that it was a loosely constructed text without exacting examples. It would also have been a course without reviews of detailed anatomy, histology, or cytology, which formed the bedrock of university zoology and botany texts of the day. Haacke's basic message was that at all levels were found the operation of the inheritance of acquired characters intertwined with the actions of the selection of individuals and collectively of races. The glue that bound his discussion together was his own theory of heredity.

The doctrine of Gemmarians ("Gemmarienlehre") was intentionally an epigenetic theory and explicitly antithetical to the assorted particulate theories of the day, ranging from Darwin's gemmules, Haeckel's plastidules, de Vries's pangenes, and O. Hertwig's idioblasts to Weismann's biophors. Haacke considered all such particulate theories expressions of preformation. The differences between organisms were to be found in the differences between the original particles. Weismann's theory represented the ultimate and most completely constructed example of preformed particles. In contrast Haacke's epigenetic theory was based on an "Urform," the "Gemmen," which possessed a universal structure or specific molecule but which in combinations formed numerous configurations to form the building blocks for the limitless contrasting structures of animal and plant anatomy. Haacke, of course, was far less specific than Weismann had been in depicting his germ-plasm and provided far fewer details about the processes of development. There was one basic molecular urform, the gemmen. In different structural combinations and configurations this urform produced different molecular compounds or "Gemmarien." In each organism, in all the new races and species, and throughout every stage in development and evolution Haacke reckoned upon different Gemmarien producing the variety in living nature. The combinations and permutations would be open ended over time. It was an interesting strategy, and if it had been structured in a different format, articulated with better focus and a less contentious text, it might have had more sympathizers.

As it stood, *Gestaltung und Vererbung* had little impact. In enquiring of Gustav Fischer about reviews of his own *Keimplasma,* Weismann described recent books by Romanes and Haacke as "explosive" ("losplatzenden") and added that his own book "appears to have settled heavily in the gentlemen's stomachs."[24] Toward the end of the year, Weismann mentioned to Franz von Wagner that he had read two-thirds of Haacke's book but felt it would be inappropriate and far too time consuming for him to respond in print. He enquired whether von Wagner might not instead write such a review.

With the turn of the year, Weismann commented on epigenetic theories in general when writing to Wilhelm Roux. In the process, he indicated that Haacke's "amiable" book had brilliantly demonstrated how hard it was to think through a "somewhat satisfying" epigenetic theory. At the same time, he derisively dismissed Haacke, who boastingly indicated he had dictated his entire book to two stenographers.[25]

Oscar Hertwig Takes on Weismann

While Weismann was putting the final touches on his *Keimplasma* in Munich, Oscar Hertwig was occupied at his anatomical institute in Berlin with the first volume of a major textbook entitled *Die Zelle und die Gewebe.*[26] Weismann's work represented the climax, but not the conclusion, of his long and fruitful career, whereas Hertwig's was the first half of a second textbook in an equally successful career that would extend into 1922. These works are quite different in that Weismann's, as we saw in Chapter 16, was a 600-page monograph devoted to the author's elaborate theory of heredity. Hertwig's was a genuine textbook for students that surveyed the known anatomy and physiology of the cell. Both authors revealed a commitment to a mechanistic explanation of life and to the centrality of contemporary cell theory for understanding the processes of transmission, development, and evolution. They each could draw upon extensive personal achievements to further that goal. Hertwig's text is relevant to this biography because it leads us into his practice of science, which began to collide with the labyrinth of Weismann's germ-plasm theory.

Oscar Hertwig has already played an important role in providing a common disciplinary context for Weismann's work. We have seen him and his brother, Richard, pushing beyond Haeckel's monistic understanding of animal germ layers in embryology. He was an important participant in the 1870s in pursuing the rapidly developing field of nuclear cytology and in providing, along with Hermann Fol, a clear understanding of the identity of the head of the spermatozoa and with the structure that van Beneden had baptized the male "pro-nucleus." Hertwig appears for a third time in our narrative as one of two discoverers of the parallel processes of egg and sperm during gametogenesis.[27] Finally, contemporaneously with Roux, Born, and Rauber, he experimented with the early cleavage patterns of fertilized eggs in response to Pflüger's exaggerated claims about the influence of external forces on early development. Despite being fifteen years Weismann's junior, Hertwig was already an authority on microscopic and experimental embryology, nuclear cytology, and fertilization—all of which he interpreted in terms of his understanding of cells. He had already written a valuable textbook on vertebrate embryology and had been appointed, after much institutional and disciplinary wrangling, to a specially created "Second Chair of Anatomy" in Berlin.[28] His perspective and experience have to be carefully considered when evaluating his reaction to Weismann's mature germ-plasm theory.

Fortunately, Paul Weindling has provided us with a detailed study of Hertwig's career. This study has multiple messages concerning the biological and social implications of Hertwig's work. Although integrated into the chapters on the broader development of his science, three of Weindling's chapters—those dealing with Hertwig's research on fertilization, his experiments on cleavage and regeneration, and his comparison of polar bodies and concept of heredity—cover subjects where Hertwig's outstanding accomplishments interfaced with Weismann's evolving germ-plasm theory.[29] If his textbook on the cell and tissues is a fair indication, Hertwig accepted Weismann's microscopic accomplishments as a valuable part of the legitimate and rapidly developing understanding of cells, organisms, and theory building in biology. He respected Ishikawa's microscopic work on polar body formation; at the same time, Hertwig's extended monograph on maturation division in nematodes motivated Weismann, two years before he had begun writing *Das Keim-plasma,* to change his interpretation of polar bodies. Here then were two accomplished German zoologists whose scholarly work intercepted each other down to the fine microscopic details of cellular mechanics, yet whose interpretation of cellular and nuclear parts and whose vision of their functions in development and evolution widely diverged after 1892. This sort of divergence from a shared empirical foundation occurs frequently in the history of biology. It is also the case that in the Wilhelmine period such differences often descended into polemics.[30]

As a general text published in 1893, the first volume of Hertwig's *Die Zelle und die Gewebe,* as indicated by its subtitle, was committed to reviewing the general anatomy and physiology of the cell.[31] This meant that out of nine chapters Hertwig devoted the first five to what biologists knew about the chemical, physical, and vital properties of cells. Beginning with the lengthy sixth chapter on cell and nuclear divisions and followed by a chapter on fertilization, Hertwig covered material to which he had contributed in substantial ways. In a very short eighth chapter, he turned to the interactions between the protoplasm and the nucleus. Its brevity reflected how little was yet known on this subject. He tailored the ninth and final chapter to covering the elementary "germ" of the organism and theories of heredity. It was a chapter of conjectures based on recent cellular research in biology.

The summary of the book's contents reveals that Hertwig's focus was much broader than that of either Weismann or Haacke on the interpretation of the cell and its properties. In contrast to Weismann's nucleus-oriented *Keimplasma,* Hertwig was also concerned with organic functions and the protoplasm. In contrast to Haacke's *Gestaltung und Vererbung,* Hertwig paid equal attention to the overall functions and structures of cells attending to both the protoplasm and nucleus. Discounting the obvious differences between these very different texts with different objectives but written within a year and a half of each other, one might say, when one turns to Hertwig's general perspective, that it lay halfway between Weismann's focus on the nucleus and Haacke's attention to the protoplasm. Given the fact that Hertwig could not comment on either Weismann's or Haacke's books still in progress, we find him addressing matters that dealt specifically with Weismann's earlier germ-plasm theory and other contemporary theories of heredity from Darwin to de Vries.

Long before he had read *Das Keimplasma,* Hertwig had headed off on a very different interpretive path than Weismann's.[32] Cleavage and all cell divisions were assumed with good experimental reason to be equational while at the same time he considered that differentiation was not a consequence of a predetermined architecture in the idioplasm. When it appeared, the extensively elaborated new version of the germ-plasm theory gave Hertwig reason to turn polemical. In the first issue of a series entitled *Zeit- und Streitfragen der Biologie,* he examined the question of *Präformation oder Epigenesis?*[33] Gustav Fischer must have known what he was doing when he advertised ten of Weismann's essays and *Das Keimplasma* on the back cover of Hertwig's *Streitfragen.* Polemics were also good for the publishing business.

Wilhelm Roux: Preformation and Epigenesis

Because the first issue of Hertwig's *Streitfragen* focused not only on Weismann's *Keimplasma* but on the contrast between *Präformation oder Epigenesis?*, it is worth briefly sketching out the history of these terms. "Epigenesis" was coined by Wilhelm Harvey in his innovative *Exercitiones de generatione animalium* of 1651. It reflected his conviction that the generation of a visible chick or deer embryo initially emerged from a clear, homogeneous fluid in the egg or uterus. Although he did not endorse the age-old claim that organic development arose only after the formal, efficient, and final causes delivered by the male joined with the material cause contained within the female, Harvey's Aristotelian orientation and his epoch-making biological examinations confirmed Aristotle's observations. Until that moment, according to Harvey, the reproductive matter of the egg or of the uterus possessed no structure. Rather than unquestioningly adopting Aristotle's causes, Harvey remained neutral about a scientific explanation. He appeared inclined instead to invoke an act of God as the prime generative agent.

With the discovery of the simple and compound microscope, Antony van Leeuwenhoek, Marcello Malpighi, Jan Swammerdam, and above all, Charles Bonnet and Albrecht von Haller endorsed another kind of explanation: the embryo was preformed before fertilization. According to one's observations and proclivities, this preformed individual resided in either the male sperm or the female egg, hence the rival claims by "ovists" and "spermatocysts" in reference to one's belief in the origin of form before its development in the egg or womb. In either case, a preformation, most likely achieved through an act of God, appeared to many naturalists to be a satisfactory reply to Aristotle and Harvey. There were those, such as Bonnet and von Haller, who postulated that all generations had been preformed at one stroke and that the entire human race had been enclosed within the loins of Eve at the creation.

No matter how seriously scientists and natural philosophers might have taken such claims, by the sixth decade of the eighteenth century the German embryologist Caspar Friedrich Wolff destroyed the preformationist perspective by carefully examining the early development of the chick and arguing that the visible, physical interplay of tissues created anew the forms of each organism. Following Wolff's demonstrations, exacting microscopic studies of Lorenz Oken, Karl von Baer, Heinrich Rathke, and Johannes Müller emphasized that the early dynamics of the embryo shaped the basics of each organism. A new epigenesis reigned through the first two-thirds of the nineteenth century.

By the early nineteenth century, the controversy over epigenesis and preformation appeared settled. More important, developmental processes were constantly being examined and amplified; the cells' omnipresence, cell formation into germ layers, tissue folding into organs, and organs forming whole embryos supported a general epigenetic mode of development. One outcome of this success was that embryology joined comparative anatomy and natural history as a guide for taxonomic relationships and later for phylogenetic evolution.

Science, however, has an uncanny way of producing unanticipated outcomes. The new cell theory of the 1880s, with its emphasis on nuclear cytology and the possibility of explaining heredity in terms of nuclein particles, provided the groundwork for a new kind of controversy about the overall mode of development, whether by epigenesis or preformation. The complexities of this antinomy between epigenesis and preformation had already been recognized by Wilhelm Roux soon after he published his influential paper on the significance of indirect cell division. Still a young privatdozent at the anatomical institute in Breslau, Roux submitted in December 1884 a lengthy paper to the *Zeitschrift für Biologie*. The paper's introduction addressed the contrast between epigenesis and preformation.[34] The introduction also turned out to be a justification for Roux's new approach for studying the embryo: "Entwicklungsmechanik" as he called it throughout his life. This self-proclaimed science turned out to be more than simply experimental embryology. During the last twenty years of the century, it amounted to a crusade by Roux and other young contemporary embryologists against the descriptive embryology of earlier generations.[35] In Roux's vision, entwicklungsmechanik necessitated an understanding of the general principles of development and a recognition of the need for a causal analysis of specific moments of this development. Thus, it was essential to appreciate the difference between the kinematics and the kinetics of development, distinguishing the descriptive from the mechanical, chemical, and energetics of development. To explore the latter, it was necessary to distinguish the moments of independent development from the moments of dependent development ("Selbstdifferenzirung and abhängige Differen-zirung"), for this seemed to Roux to be the only way for embryologists to enter into a true causal analysis of development.

This second distinction became the focus and hallmark of Roux's embryo-logical research; however, as a result, interpreting development had to be considered a subjective act in "the perception of developing complexity" ("Entstehen von wahrnehmbarer Mannigfaltigkeit"). Roux recognized that thinking in this way about development reintroduced the old contrast be-tween epigenesis and evolution. (It was more common at this time to refer to the opposite of "epigenesis" as "evolution" rather than "preformation.") The former, epigenesis, denoted the "new formation of complexity in the strongest sense, the actual increase in the existence of diversity." In contrast, "evolution," Roux insisted, "is hereafter the mere perception of preexisting latent differences."[36]

These two modes of development were not easy to distinguish in practice, and Roux's categorization foretold problems biologists would have in assigning one or the other in given instances. Epigenesis simply indicated the new de-velopment of complexity out of less or no complexity, whereas developmental evolution could entail an increase in size, an unperceived complexity made perceptible through metamorphosis, or the two acting together. He wisely lim-ited his discussion to unambiguous physical examples of the emergence of per-ceptible from the nonperceptible, such as the responding pattern of iron fil-ings spread on a glass over a magnet or the "development" of a photographic film. Roux included not just unperceived configurations of matter but the ef-fects of energy and chemical reactions as elements in the unperceived com-plexity. The lesson, he advised his readers, indicated that "the deeper we pen-etrate into an observed developmental event . . . the more as a rule we recog-nize that a greater part of that which appears to us upon first consideration to be the observation of new formed complexity, owes its richness of observ-able complexity to a metamorphosis of preexisting differences."[37] The argu-ment applied to the undetectable limits of nuclear cytology. The bottom line for Roux was that there existed three possible patterns in development: epi-genesis, evolution, and a mixture of the two. "It will then be our problem, in the interpretation of our observations to show double care and double acute-ness in order to separate appropriately the role of each one of the principles from the other."[38] In a footnote to his reprint, Roux criticized Weismann for espousing in his *Keimplasma,* "almost pure Evolution," and Hertwig for espousing in his *Präformation oder Epigenesis?* "almost pure Epigenesis."[39] Developmental circumstances, not philosophical principles, must rule.

Hertwig on Weismann's *Keimplasma*

Acknowledging Roux's discussion, Hertwig informs us that he had personally resisted preformationist strategies even before Weismann had developed his early germ-plasm theory.[40] With the appearance of *Das Keimplasma* in 1892, Hertwig had a strong motive both to deal on a theoretical level with the contrast between epigenesis and preformation, and to attack Weismann's work as the foremost exemplar of preformation "wrought with the greatest care and acuteness, and totally irreconcilable with my conclusions."[41] Hertwig elaborated on two principal criticisms of Weismann's preformation theory. The first was Weismann's insistence that cleavage and further differentiating divisions were the results of nuclear divisions, serving to segregate from each other the different kinds of determinants. Weismann, Hertwig rightly claimed, had failed to bring forth any proof of this process other than the argument that the resulting correspondence between germ-plasm and character would be more efficient. Hertwig then enumerates a series of known phenomena that demonstrate the reverse process, that is, equational division: (1) the routine divisions of single-celled organisms; (2) the divisions of the germ-plasm prior to maturation; (3) reproduction through budding and regeneration; (4) heteromorphosis or the production of whole organisms from budding in somatic parts; (5) the division experiments by Driesch, Chabry, and others on blastomeres; (6) a wide range of grafting, transplantation, and transfusion experiments that demonstrate the ubiquitous existence of a great variety of latent germinal materials. Hertwig's argument drew directly from modern experimental biology, which was building up a wealth of facts that militated against unequal nuclear divisions during embryogenesis.[42]

The second area of criticism Hertwig leveled concerned Weismann's notion of determinants and biophors. Despite his own invocation of idioblasts, Hertwig rejected Weismann's particular use of hypothetical particles as a "false use of the conception of causality." This was not an uncommon fallacy, particularly among zoologists, Hertwig noted, which relied on the assumption that the somatic traits were the direct products of the anlagen dictating "the visible complexity of the final stage of the development" correlated with the "invisible complexity of the first stage."[43] By assuming such an unequivocal correspondence, Weismann missed all the correlative features in development resulting from the interaction of cells with one another. Hertwig considered this error to be the outcome of a lack of understanding of development on the cellular level. There "are secondary formations that can arise only after the multiplication of cells, and from the varied combination of cell-characters that accompanies the multiplication of cells."[44]

Finally, Hertwig carried the same argument to the supracellular level where recent experimental embryology had shown that developing blastospheres and gastrulae of amphibia, reptiles, and birds, among others, were known to be reshaped and that the positions of their blastomeres could be exchanged experimentally without causing distortions in the resulting embryo. "Here we have epigenesis—the appearance of a new formation, not the becoming visible of preexisting complexity."[45]

After taking Weismann's *Keimplasma* to task—from a modern perspective, his criticisms were valid—Hertwig presented his own theory of development. Here a comparison between Weismann's and Hertwig's theories shows the dilemma between all serious competing theories of science. In certain ways, one fails and the other succeeds, and vice versa. Viewing this generalization at the end of the nineteenth century, we find that epigenesis as a view of development succeeded and preformation failed at one point, and at another the reverse occurred. Despite his elegant and fundamental investigations into germ layers, cell theory, fertilization, nuclear division, and early development, Hertwig avoided detailing any comprehensive theory of development besides writing in a most general way about cellular and supracellular interactions. By the 1880s, his approach to development became experimental and physiological rather than simply microscopic and morphological. Nevertheless, as have seen, he also included in this nonspecific perspective particulate elements, the idioblasts, to transmit character differences. In Hertwig's opinion, Weismann assumed that "every cell *must* have become what it is, because it was provided only with the definite Anlage assigned to it beforehand, according to the plan of the development of the germplasm [*sic*]." In contrast, Hertwig held that "the causes we recognize are first, the continual changes in mutual relations that the cells undergo as they increase in number by division, and second, the influence of surrounding things upon the organism." Although he could not describe a single causal link from stimulus to its reaction, he insisted that "this principle indicates the path along which explanation of the differentiation of cells is to be sought."[46]

In his conclusion, Hertwig revealed that the distinction between preformation and epigenesis was more complicated and subtle than the provocative subtitle of his essay suggested. Endorsing an earlier criticism by Roux, Haacke had in his *Gestaltung und Vererbung* already accused Hertwig of being a preformationist simply because of his employment of idioblasts as part of his theory of heredity and development. "For the concept of preformation," Haacke explained, "it is not necessary that one glimpses in the germ a microscopic copy of the finished organism, but one need only to accept, as Hertwig has done, the presence of a prearrangement of qualitatively different idioblasts, to steer into the harbor of preformation with all sails set."[47] The key to Haacke's criticism of Hertwig was his expression "prearrangement of qualitatively different idioblasts." Weismann's germ-plasm theory certainly hypothesized a complex "prearrangement" of determinants and biophors. It is unclear, however, that Hertwig did. In the end, Hertwig saw himself offering a theory with aspects of both preformation and epigenesis. In his view, however, Weismann "merely transfers to an invisible region the solution of a problem that we are trying to solve . . . by investigation of visible characters; and in the invisible region it is impossible to apply the methods of science."[48] By emphasizing the word "preformation," Hertwig tarred Weismann with the brush of eighteenth-century theories of "emboîtment."

Hertwig and Weismann on Reduction Division

One last aspect of his interaction with Hertwig continued to dog Weismann until the end of the century. It concerned the significance of chromosome division during the two maturation divisions. Both Weismann and Hertwig agreed that a quantitative reduction by one-half of the chromatin material occurred during maturation. Accepting this, however, brought two key issues to the fore: first, whether and how a qualitative reduction of hereditary units took place at the same time; second, how the chromosomes, as bearers of the particles of heredity, were significant units both during transmission and development. Both Weismann and Hertwig accepted Roux's suggestion that the hereditary particles lay longitudinally along the length of the chromosomes. On the one hand, it became the core of Weismann's mature germ-plasm theory that the ids, those ancestral germ-plasms first envisioned by Weismann in the early 1880s, were distributed lineally along the chromosomes (i.e., his idants), and that the separate determinants and biophors were systematically arranged within the ids. Fertilization consisted of a physical mixing ("Vermischung") of the halves of two parental germ-plasms—an "amphimixis," as he called it in 1891.[49] As described in Chapter 16, Weismann's chromosomes contained a highly organized architecture, which explained both a quantitative and qualitative reduction at germ maturation and an ordered, sequential distribution of determinants and their biophors during development. Somatic variations simply reflected the different combination of homologous determinants and biophors of the different ids. It was this explicit, complex architecture that Hertwig attacked as a preformation.

On the other hand, Hertwig considered chromosomes to be transitory elements that appeared in the nucleus as a spireme of nuclein at the onset of cell division. This spireme condensed into chromosomes, which distributed the nuclein particles throughout the four basic phases of indirect nuclear division but disappeared again with the formation of resting daughter nuclei. For Hertwig, fertilization consisted of a physical fusion ("Verschmelzung") of the nuclei of two parents, and the important hereditary structures were the different idioblasts that interacted with the protoplasm of the cells and the extracellular environment to cause cellular differentiation. Variations reflected the slightly different physiological reactions of different homologous idioblasts during the developmental process. Where Weismann considered somatic divisions to bring about the orderly, morphological separation of different determinants, Hertwig considered the physiological activation of specific idioblasts in totipotent nuclei to bring about differentiation at a specific place and time.

There were ways to mediate between these rival claims both experimentally and microscopically.[50] Weismann's advanced associates in Freiburg, Valentin Häcker, a privatdozent and first assistant since 1892, and Otto vom Rath, a nationally recognized microscopist and volunteer assistant, both pursued the issue. Their respective studies of spermatogenesis in copepods and in crickets emphasized the importance of the chromosomes, but instead of recognizing four chromatids during first maturation, they envisioned a transverse division of chromosomal pairs. The strategy, suggested by Weismann in his 4–8–4–2 calculations of chromosome numbers during reduction division, allowed them to preserve the same number of chromosomes in successive generations.

Hertwig had empirical support as well. August Brauer had completed his doctorate in Bonn where he studied with Richard Hertwig and the classical microscopist Franz von Leydig. He habilitated in zoology at Marburg in 1890. He focused on development in invertebrates, but when he felt that Oscar Hertwig had used pathological material for his epoch-making monograph on sperm production in Nematodes, Brauer turned his microscope onto spermatogenesis in *Ascaris*. After a detailed study of the whole process, he confirmed that Hertwig had, indeed, used some pathological material, but he supported most of Hertwig's findings and adopted Hertwig's theoretical conclusions. Brauer summed up that "on the basis of these observations the splitting of the chromatin corpuscles may not only begin very early, but it must be at the first prophase of the new division." "Because this [chromatin] division, however, constitutes the essence of the entire Karyokinesis," he continued, "I consider all those phenomena which follow it, such as the assembling of corpuscles into a few threads, their union into large corpuscles, the disintegration of a thread into segments and finally into chromosomes, as less significant; they all merely follow up the purpose of making possible the transport of the divided halves to the daughter cells in the surest and simplest way."[51] His observations assumed the widely held belief that maturation divisions were unlike somatic divisions in that they consisted of a double division of the chromatin followed by two longitudinal divisions of the spireme and of the ensuing chromosomes. Because chromatin mass rather than chromosome number was important during maturation and fertilization, Brauer suggested that degeneration of one parental set of chromatin or fusion of both sets together prevented the doubling of chromatin mass upon the completion of the two processes. With a fusion theory of chromatin, chromosome numbers were not significant.

From Weismann through Roux, to Haacke and Hertwig, there lay a danger of using antiquated terms. Both "epigenesis" and "preformation" had strong previous implications, both of which, in fact, often evoked the Deity as a final cause. In the late nineteenth century, these terms needed to be forgotten, and better descriptors for theories of development needed to be established. Professional status, rivalry, and a rapidly advancing technology all played a role in the ensuing confusion.

Despite the criticisms and ignoring Haacke, we recognize the other three participants—Weismann, Roux, and Hertwig—all continued to pursue productive careers. At stake in this decade of long controversy were a number of key issues of the 1890s: (1) the significance of the chromosomes, (2) their relationship to the particles that were the basic conjectured units of life and their diversity, and (3) the significance of each maturation division. All three issues ultimately played into the more loosely articulated but more general controversies over the mechanism of evolution.

Herbert Spencer: A Philosopher's Response

Haacke and Hertwig were only two of the most prominent German critics of Weismann's *Keimplasma*. Others, such as Theodor Eimer, Erich Wasmann, and Haeckel, also responded critically to the work. Such negative attention suggests Weismann's germ-plasm theory was taken seriously and generated counter alternatives both on the level of germ cell production and in the explanation of evolution theory. When we cross the channel, where Weismann's work had been followed with eagerness even before his first visit, we find that his ideas had already generated considerable discussion. The first volume of his collected *Essays upon Heredity,* which appeared in 1889, had already elicited eight short abstracts, twenty-nine articles in English and American journals, and twenty-two excised letters to the editor of *Nature* by the time Weismann began writing his *Keimplasma* in Munich.[52] The English translation of *Das Keimplasma,* appearing only months after the publication of the original, had been carefully shepherded through translation and publication by Weismann's son-in-law, W. Newton Parker.[53] It was scheduled to hit the bookstores so soon after the Jena edition had appeared that its publisher Walter Scott was concerned that the German version might directly compete with its English counterpart. Trying to assure him that Gustav Fischer was an outstanding and honorable publisher, Weismann wrote of Fischer that "he had answered me today, saying that *not a single* copy of my book has been sent by him to England, and that he has informed the booksellers of the Leipzig market not to send copies to England or America before having received new information from him . . . [and] that he has not received orders from his ordinary English customers for the German book, probably because they are waiting for the English edition, which will be 3 times cheaper than the German one."[54]

Whether it was the more substantial German book or the more compact and cheaper English edition that was used by the English-speaking academics and public at large, Weismann's *Keimplasma* immediately stirred up controversy on the British Isles and in America. We will examine two of the English critics excited by Weismann: Herbert Spencer and George John Romanes. Both had commented upon Weismann's work prior to the publication of *Das Keimplasma,* but after its appearance they embarked on further critiques. Both commentators were active evolutionists, and both were well informed in certain areas of biology. Neither, however, represented the "professional" English zoologists at the time, men such as Edward Bagnall Poulton, Edwin Ray Lankester, Henry Nottidge Moseley, and Francis Galton, all of whom either were or became Weismann supporters before and after *Das Keimplasma* appeared.

As a philosopher of science—using an anachronistic but apt identification—Herbert Spencer was no stranger to contemporary studies of life.[55] As a philosopher he had garnered much of his biological information through his friendship with Thomas H. Huxley, Joseph Hooker, and John Lubbock. All three, in addition to Spencer, were members of the influential X-Club that unofficially guided the biological sections of the British Association for the Advancement of Science (BAAS) and steered the development of English science education in the immediate post-Darwin period.[56] Spencer had already incorporated organic evolution into his cosmology long before the 1880s when Weismann first visited England. The term "evolution," in fact, had been wrenched by Spencer from its embryological context and had been fashioned into a general appellation for cosmic and phylogenetic change over time.[57]

In the prolegomenon to his *First Principles,* Spencer had articulated the fundamental principle of universal cosmic change, which held that homogeneous systems tended to evolve into heterogeneous ones.[58] At the same time, he had described organic development, adaptation, and heredity simply as the moving equilibrium of a single life cycle. He referred to the first and second laws of thermodynamics but failed to understand their negating implications for his principle of increasing heterogeneity. How did the homogeneous germ develop into a complex entity? This, as we have already seen, would become one of the most basic issues for biologists during the second half of the century. To get more details on Spencer's answer to this ostensibly biological question, we must turn to his *Principles of Biology,* which appeared in serial form between 1863 and 1876.

In his classic biology text, which incidentally served as a standard reference for students of biology both in England and America, Spencer defined life as "the continuous adjustment of internal relations to external relations."[59] His was consequently a dynamic, physiological vision of organisms, which sharply differed from Weismann's morphological examinations and speculations. In addition, Spencer attended to the prevailing opinion of the 1860s that growth was simply a surplus of matter over expenditure and that development was simply an increase in the size and diversity of structure. In combination, these two processes comprised the physiological essence of biological evolution.[60] Spencer reinforced his reductionistic predilections by analogizing the selective assimilation of chemicals during the process of crystallization with the selective assimilation of "physiological" units from the body fluids during biological development.[61] Heredity, in keeping with this view, was understood by Spencer to be simply the inverse relationship between the growth and development of the parent and the production of the offspring. The relationship remained the same regardless of the mode of reproduction of the organism being discussed. Sexual reproduction, or "gamogenesis" in Spencer's terminology, entailed nothing unique; it denoted a mere "coalescence of a detached portion of one organism, with a more or less detached portion of another." He then proceeded to claim that gametes "have not been made by some elaboration, fundamentally different from other cells." Gametes were "unspecialized" and "have departed but little from the original and most general of types."[62]

Spencer was not one to bow out of a technical controversy involving evolution theory. In a general article in 1886, he had reviewed his accumulated ideas of the subject.[63] At that time, he seemed uninvolved with Weismann's emerging ideas; nor did he question the operation of the inheritance of acquired characters as an integral part of the evolutionary process. After Weismann visited England in 1887 and participated in a session on just this subject at the BAAS meetings, Spencer could not help but be aware of Weismann's dismissal of the subject. With the translation of Weismann's *Essays* and of *Das Keimplasma* in early 1893, Spencer found ample reason to reassert in detail his claims about "use inheritance" in a two-part series entitled "The Inadequacy of Natural Selection," written for the general educated public and appearing in the *Contemporary Review*.[64]

Spencer's cosmology as a whole and his biology in particular were products of the 1860s. They are not unlike those found in Haeckel's *Generelle Morphologie* of the same period. It is not coincidental that as committed evolutionists they both found their views justifying their intuitive beliefs in the inheritance of acquired characters. In the decades to come they cited standard examples of use and disuse of parts as demonstrations of this inclusive form of inheritance. They both continued to downplay the uniqueness of gamete production when it became such an important issue in the 1880s and 1890s.

Within this narrow scientific context it is worth emphasizing the accelerating interest in England in Weismann's research before the publication of his *Keimplasma*. Perhaps Raphael Meldola's full translation of Weismann's *Studien* with its "Prefatory Notice" by Darwin (this was Darwin's last scientific publication), which appeared in two impressive volumes in 1882, started the ball rolling. The editors of Weismann's translated *Essays* of 1889 (Edward B.

Poulton, Selmar Schönland, and Arthur E. Shipley) and P. C. Mitchell's favorable review of that volume capture some of the growing recognition in England of the broader implications of Weismann's research.[65] At the time, the editors of the *Essays* listed eight abstracts of Weismann's single essays, all of which appeared in *Nature* between the time of an 1885 article by Arthur E. Shipley and their assembled collection of complete essays four years later. Mitchell was of the opinion that "Since Mr. Shipley's article, entitled 'Death,' in *Nineteenth Century* in May, 1885, first called the attention of English biologists to Prof. Weismann's essays, the interest in that author's conclusions and arguments has become very general." Mitchell also referred to "various" articles in *Nature,* a lecture by E. Ray Lankester at the Royal Institution, and the BAAS meetings in Manchester—at which, as we saw earlier, Weismann made two presentations. To whet the appetite, the same year as the publication of Weismann's *Essays* and Mitchell's review, Alfred Wallace presented a popular synopsis and update of Darwin's theory. When dealing with heredity, he strongly supported Weismann's research and conclusions.

The major concern that Weismann had stirred up was the claim that his germ-plasm theory denied the traditional view that acquired characters could be inherited. Wallace, and Weismann after him, both realized that Francis Galton had presented similar views against "use" heredity twenty years earlier, but Wallace also recognized that they had not created a stir. In his opinion, Weismann, as an experienced biologist, "has worked it out more thoroughly, and had adduced embryological evidence in its support." Nevertheless, "the names of Galton and Weismann should therefore be associated as discoverers of what may be considered (if finally established) the most important contributions to the evolution theory since the appearance of the *Origin of Species.*"[66]

Although involved in other projects by this time, Spencer reviewed his own ideas on organic evolution for a general audience in 1886.[67] It was a becalmed weather vane of a paper that pointed backward to old ideas and away from the brewing storm.[68] Spencer did not mention Weismann, nor did he appear to be aware of the controversy then developing in Germany over the germ-plasm theory. The following year, when Weismann attended the Manchester meeting of the BAAS, he delivered, as we have seen, a formal paper on his work on polar body formation and introduced the audience to his newest ideas on reduction division. As significant as this research was in the long run for the germ-plasm theory, Weismann's participation on a panel the preceding day was unquestionably of more immediate interest. The topic for debate was "Are Acquired Characters Hereditary?" Weismann's copanelists included

Lankester and Poulton, who had become neo-Darwinians, and Geddes and Hartog, who decidedly had not. For the next six years, the columns of *Nature* were filled with claims and counterclaims about the adequacy of the "Weismann position." Romanes, Vines, Hartog, Poulton, Wallace, Cunningham, and Mitchell were among the most frequent contributors. The more popular and religiously oriented *Contemporary Review* and the *American Naturalist* across the Atlantic quickly picked up the controversy. Important books by Wallace, Ball, Romanes, and Poulton and a translation of Eimer's *Entstehung der Arten* established the issue in the English-speaking world in a more substantial form. A collected edition of authorized translations of Weismann's early essays on the germ-plasm, guided through the press by Poulton came out by the end of the decade. A second volume, including the important essay on *Amphimixis,* appeared in 1892—the same year as *Das Keimplasma.*

This surge of literature reflected more than simply a breakthrough on a recalcitrant scientific problem. The confusion and profusion of reactions could only indicate that deep assumptions had been misunderstood and shook. It was also the case that the issue of heredity, then as now, had implications that stretched beyond the scientific claims. But let us turn directly to the Weismann-Spencer controversy that followed the publication of *Das Keimplasma,* chiefly in the pages of the *Contemporary Review,* and which seemed more of a response to Weismann's recently published *Essays* than to the English translation of his *Keimplasma.* Spencer's opening attack came in the spring of 1893 in a two-part salvo entitled "The Inadequacy of Natural Selection."[69] Weismann responded that autumn with a two-part volley of his own on natural selection.[70] Spencer returned fire in December with "A Rejoinder."[71] The following spring, Weismann traveled to Oxford to give, as we have seen, the annual Romanes Lecture. It consisted of a broadside entitled "The Effect of External Influences upon Development."[72] The series' benefactor, George John Romanes, as we shall see, had contributed immensely to the broader controversy; it was unfortunately one of the last public affairs this disciple of Darwin attended before his untimely death.[73] Spencer's reply, "Weismannism Once More," thundered forth in October of the same year,[74] and Weismann's "Heredity Once More," followed twelve months later.[75] A final one-page discharge to the editor by Spencer, also entitled "Heredity Once More,"[76] ended the engagement.

That Weismann's responses appeared in the *Contemporary Review* and in English was appropriate because of the venue of Spencer's critiques. Probably to avoid a cross-language controversy, Weismann elicited William Newton Parker to assist in his response. It was a comfortable choice, for Parker had worked voluntarily in Wiedersheim's Anatomical Institute in Freiburg, had married Weismann's second daughter Hedwig, and subsequently would serve for the rest of his career as professor of zoology at University College in Cardiff (also formerly University College of Wales and Monmouthshire). Parker and Gregg Wilson, another former student in Wiedersheim's institute, executed the translations. At the same time, however, Weismann made arrangements with Gustav Fischer to publish simultaneously his original German versions.[77]

The details of the controversy involved four issues: (1) the capacity of natural selection to explain the accumulation of small variations appearing over an extended period of time, such as the refined sensitivity of the human feeling of touch in fingers; (2) the likelihood of natural selection to bring about the correlation of complex structures within the same organism, such as the multiple adaptations in associated structures to accommodate the increase of antler size of the prehistoric Irish elk; (3) degeneration and disappearance of structures due to panmixia, that is, the relaxation of natural selection, such as the eyes of cave-dwelling organisms; and (4) the evolution of contrasting structures of the neuter castes in social insects. In each case, Weismann cited inheritance through the germ-plasm with its different ids and determinants and the action and/or the relaxation of natural selection on the level of either the individual or the colony. Spencer implied that Weismann's hierarchical model was deficient and that the only way to explain these phenomena was through the processes of use and disuse and the consequent inheritance of acquired characters. Although Spencer's criticisms were informed, they were short on carefully recorded observations and personal experience. It is also evident that he was not fully acquainted even with Weismann's *Essays*. Not surprisingly Spencer's criticism rested on his ability to take apart the arguments of his opponent and to show that they led to neither necessary nor sufficient conclusions. He expected Weismann to respond in detail to his selected examples, and by his third paper in a note of triumph he pointed out Weismann's failure to comply. "No reply," Spencer hurled back at Weismann at least seven times to the latter's lack of response to his direct challenges. Spencer's were not papers of research or model building; they were instead papers against incompleteness and illogic. At the same time there was no effort on Spencer's part to get to the core of Weismann's model of heredity, which from the start had rested on the complex and evolving studies of chromosomal transmission.

Above all, he made no attempt to explain how acquired traits could pass from one generation to the next, which was the basic challenge laid down by Weismann's theory of continuity.

Weismann's papers were more informative, and relied on contemporary research by experts in specific fields of research. Instead of responding in detail to Spencer's example of sense discrimination in fingers and hands, he turned to a subject that had been recently and carefully investigated and was unambiguously relevant. The pertinent example concerned whether the deformed little toe of the human foot was the direct result of the use of boots and thus represented an inherited acquired trait, or whether it was due to the little toe's evolving unimportance in the human mode of walking and thus reflected a relaxation of natural selection, panmixia. When he turned to the problem of coadaptation, Weismann relied on the extensive investigations of other specialists such as Leuckart, who had studied parthenogenesis in bees and applied his conclusions, Lespès, who had recognized that the ovaries of ant neuters were a good guide for classification, and Auguste Forel, who more recently had examined both in nature and the laboratory variations in the egg-laying capacities of worker ants. He had also studied in detail the peculiar structures and behavior of neuter castes among ants. Both the retrogression and disappearance of sexual structures, such as the ovary and spermatheca of worker ants and the appearance of elaborate and specialized jaws, the increase in brain size in soldier castes, and the unique behaviors of all neuter castes could not be attributed to use and disuse of characters. "None of these changes," Weismann insisted, "can rest on the transmission of functional variations, as the workers do not at all, or only exceptionally, reproduce; they could only have arisen by a selection of the parent ants dependent on the fact that those parents which produced the best workers had always the best prospect of the persistence of their colony."[78]

Throughout this controversy Weismann's strategy was an argument of exclusion, not ocular proof. It was a foregone assumption that neither he nor Spencer accepted vitalistic or creationistic explanations of the phenomena. The remaining explanations offered by either participant were evolution alternatively through natural selection or through the inheritance of functionally acquired characters. Given his mature concept of the germ-plasm, as he had developed in *Das Keimplasma*, Weismann was able to describe the evolutionary process not only in terms of the selection of hypothetical variant structures but of variant active and inactive homologous determinants. It was not a question of proof, as Spencer had repeatedly demanded, but a question of logic about that which Weismann considered the most plausible alternative. Weismann was careful not to sound dogmatic about the general operation of natural selection, "but if, as in the case of the ants, the other possible explanation, that of the transmission of functional variation, can be excluded, *we have a demonstration, at least for the particular instance, of the actual occurrence of natural selection.*"[79] This was a temperate but firmly asserted response.

Darwin and the Degeneration of Parts

One aspect of the germ-plasm theory that was difficult for Spencer to comprehend was Weismann's recourse to panmixia. As we have earlier seen, this term was introduced into biology by Weismann in 1883 in his lecture "On Heredity" ("Über die Vererbung"), delivered in Freiburg on the occasion of his prorector address. At that time, he had been discussing common imperfections in human eyes. "Those fluctuations on either side of the average which we call myopia and hypermetropia," he explained, "occur in the same manner, and are due to the same causes, as those which operate in producing degeneration in the eyes of cave-dwelling animals." Such maladies as the degeneration of sight exhibit a deterioration due to a lack of natural selection. This entailed an unrestricted mixing of related variations, that is, a panmixia.

How evolved structures also devolve, that is, degenerate and disappear, was a challenge for a neo-Darwinian theory that maintained that natural selection was the sole mechanism driving evolution. One might explain the process by recourse to savings in nutrition and metabolic energy, which could then be devoted to other structures and functions, but this was hard to measure and unlikely under any but the most extreme conditions of nutritional need. The example of myopia moreover was not a good one for Weismann's purposes because according to his own account myopia may be either an effect of ontogeny or heredity. Nevertheless, panmixia denoted in Weismann's science the state of a relaxation of natural selection. "This suspension of natural selection may be termed Panmixia, for all individuals can reproduce themselves and thus stamp their characters upon the species, and not only those which are in all respects, or in respect to some single organ, the fittest." Weismann added that "in my opinion, the greater number of those variations which are usually attributed to the direct influence of external conditions of life, are to be ascribed to panmixia. For example, the great variability of most domesticated animals essentially depends upon this principle."[80]

From the way in which he introduced this concept, it seems clear that Weismann was thinking of variations as represented in a population of the whole. "Fluctuations" of a somatic trait about a mean was pictorially easy to envision, but how was he to put this evolutionary feature into the terms of his germ-plasm theory of ids and determinants? In *Das Keimplasma* and after, Weismann resorted to variant homologous determinants of different ids. If the least effective of such determinants possessed no ill effects, the germ-plasm with the least effective determinants would have as great a chance for survival as the most effective ones. It was simply a panmixia at the germinal level.

The problem Weismann was addressing was not new. While discussing rudimentary, atrophied, or aborted organs in the first edition of *Origin,* Darwin had described many natural and domestically induced cases of the inheritance of degenerated structures. The best known and most effective cases, he had mentioned, were those of the naturally degenerated or total absence of wings of beetles living on islands and the diminished wings of common farmyard fowls. As was Darwin's bent, he had also accounted for such phenomena, which intuitively might seem to play against his basic mechanism of natural selection. Four explanations, operating singly or in combinations, had come to his mind. First, he found that the continued inheritance of increasingly degenerated structures was not only plausible but self-evident. Here he explicitly accepted without apology the inheritance of acquired characters, and later such examples served as a justification for his provisional theory of pangenesis. Second, Darwin recognized that natural selection might play a role in the increasing degeneration of parts. With reference to the degenerated or vanished wings of beetles, he concluded that "in this case natural selection would continue slowly to reduce the organ, until it was rendered harmless and rudimentary." Third, he accepted a principle of nutritional economy "by which the materials forming any part or structure, if not useful to the possessor, will be saved as far as is possible, will probably often come into play; and this will tend to cause the entire obliteration of a rudimentary organ."[81] In his "provisional hypothesis of pangenesis," Darwin offered a fourth explanation by arguing that the gemmules of unused parts must eventually perish from the lineage.[82] Only the last of these explanations appeared to be original to Darwin's thesis.

Five years later, however, Darwin made his fifth and most unique contribution to the problem in a "Note" to *Nature.* While expanding upon Wyville Thomson's "interesting account of the rudimentary males of the barnacle *Scal-*

pellum regium," Darwin had drawn attention to Lambert A. J. Quetelet's deliberations about variations in the overall stature and size of other structures in a group of men, which fell into a bell-shaped curve.[83] Darwin had seen that under unfavorable living conditions Quetelet's symmetrical curve would become lopsided toward the smaller and less functional size. When these diminished individuals, he had argued, interbred with the less numerous individuals of larger size, they would produce "in the course of time, the steady diminution and ultimate disappearance of all such useless parts." In his "Note," Darwin addressed both aspects of the problem: the degeneration of parts and their final disappearance from the phylogenetic line. At the time, he had not done more than describe the phenomenon.

Weismann and Panmixia

The matter of degeneration of structures did not become a lively issue in the discussion of evolution until Weismann independently posited a similar process in his inaugural address as prorector of his university in June 1883.[84] At that time he enriched his famous declaration of a theoretical contrast between germ and somatic cells with an examination of the process of degeneration in human eyesight. His example was of Europeans who could nonbiologically compensate for their waning sight with man-made eyeglasses. One could not explain visual degeneration either through natural selection, nor did his new understanding of heredity permit him to invoke the inheritance of acquired characters. "Those fluctuations on either side of the average, which we call myopia and hypermetropia," Weismann elaborated, "occur in the same manner and due to the same causes, as those which operate in producing degeneration in the eyes of cave-dwelling animals." In short, degeneration was due to that "suspension of the preserving influence of natural selection [which] may be termed Panmixia, for all individuals can reproduce themselves and thus stamp their character upon the species."[85]

With one exception, Weismann seldom had reason to refer to this process through the rest of the decade. After all, he was concentrating on the new nuclear cytology in general and polar bodies in particular.[86] That exception came in a popular lecture he gave to a mixed audience at the Freiburg Academic Society in January 1886. The address in its original format was not well known; the Reports of the Academic Society at Freiburg was not a frontline science journal, and Edward B. Poulton and his coeditors of Weismann's *Essays* did not select Weismann's "Rückschritt in der Natur" for translation and inclusion in what later turned out to be only the first of a two-volume publication.[87]

By the time of his general address to the Versammlung in Cologne in September 1888, Weismann felt compelled to respond to critics who refused to dispense with the inheritance of acquired characters. He discussed in depth the illusion held by experts that mutilations of many kinds, such as bobtailed cats and dogs, were inherited. Science had only recently debunked the reports of the inheritance of maternal impressions, which earlier in the century had been supported by outstanding zoologists such as Karl Baer. Weismann's message was clear: like the belief in maternal impressions, the belief in the inheritance of acquired characters, despite its support by many experts, could now be explained with greater scientific understanding. It was only in passing, however, that the phenomena of degeneration, he argued, could now be understood through the lack of natural selection, panmixia.[88]

How evolved structures also devolve, that is, degenerate and disappear, was a challenge for all neo-Darwinian theories that maintained that natural selection was the sole mechanism driving evolution. One might explain the process by recourse to an economy in nutrition and metabolic energy, which could then be channeled to other structures and functions, but this was hard to measure and unlikely under any but the most extreme conditions of nutritional need for the organism. The example of myopia moreover was not a good one for Weismann's purposes because according to his own account myopia might be an effect of either ontogeny or heredity. Nevertheless, panmixia denoted in Weismann's science the state of a relaxation of natural selection. As he explained at the time, "this suspension of natural selection may be termed *Panmixia,* for all individuals can reproduce themselves and thus stamp their characters upon the species, and not only those which are in all respects, or in respect to some single organ, the fittest." Weismann added that "in my opinion, the greater number of those variations which are usually attributed to the direct influence of external conditions of life, are to be ascribed to panmixia. For example, the great variability of most domesticated animals essentially depends upon this principle."[89] From this statement it is evident that Weismann recognized that his chief rivals would be those biological commentators who attributed degeneration and disappearance of hereditary traits to the inheritance of acquired traits.

Weismann's concerns during this period were focused on nuclear cytology in general and polar bodies in particular. The word "panmixia" appears only four times in the first volume of his *Essays.* In his general address to the Versammlung in Cologne in September 1888, Weismann also felt compelled to respond to critics who refused to relinquish their beliefs in the inheritance of injuries, a common form of acquired characters.[90] At the time, he pointed out to his audience that well-established scientists, including Baer, had only recently abandoned their beliefs in the inheritance of maternal impressions. His allusion was clear. Like the belief in maternal impressions, direct links between degeneration and disappearance of parts and the inheritance of acquired characters were strongly held even into the 1890s. It was only in passing, however, that Weismann mentioned degeneration and argued that he could explain that phenomenon not through the customary recourse to the inheritance of acquired characters but through the lack of natural selection, that is, through panmixia.

Romanes on Degeneration

If Weismann had pretty much ignored the problem through the decade of the 1880s, George J. Romanes, whom we saw earlier, had quickly picked up Darwin's earlier discussion of degeneration.[91] This close friend of Darwin elaborated upon the first three of Darwin's explanations, neglected Darwin's reference to gemmules as the fourth, and went into more detail of Darwin's fifth cause, which he identified in a separate paper in *Nature* as a "Cessation of Selection." Romanes recognized this action as another cause "which co-operates with the [other] reducing causes in all cases, and which is of special importance as an accelerating agent when the influence of the latter becomes feeble."[92] Failing to find suitable quantitative ways to compare the actions of disuse and cessation of selection in a follow-up "Note," Romanes nevertheless affirmed his belief that disuse causes atrophy on the individual level, and that this and cessation both may cooperate to eliminate the structure on the species level.[93]

He let the matter drop from his discussions until 1890 after the first volume of Weismann's *Essays* had appeared and been favorably reviewed in *Nature*. The collection of essays within one cover provided a ready target for Sidney Vines and Spencer to attack the accumulating neo-Darwinian message found throughout Weismann's technical writings. Week by week in an unrelenting sequence notes by Vines, Spencer, Romanes, Wallace, Lankester, and others appeared in *Nature*'s letters to the editor section and commentaries. Uncharacteristically, Weismann entered the fray in *Nature* as well. The exchange eventually subsided, but after *Das Keimplasma* appeared in both German and English (late 1892 and early 1893) the exchange reignited—not only in *Nature* but also in the *Contemporary Review*, where it merged with the Spencer-Weismann controversy. Romanes, however, played the role of a tempered neo-Darwinian bulldog in Great Britain. By 1895, he had submitted fourteen letters and separate notes and wrote two books dealing with panmixia in particular and Weismannism in general. Romanes did not agree completely with Weismann's assignment of causation to the problem of degeneration and disappearance, but he willingly substituted Weismann's term panmixia for his older expression of "the cessation of natural selection." Despite Darwin's earlier inclination in the sixth edition of his *Origin* to accept the inheritance of acquired characters in cases of degeneration and disappearance of structures, Romanes became inclined to believe that the Lamarckian option appeared increasingly problematic.[94]

These multiple dialogues are difficult—perhaps even pointless—to organize topically. The key disagreement between Romanes and Weismann, both of whom must be described as followers, albeit different, of Darwin, appears to sort out in terms of the various causes involved in the general process of degeneration and disappearance. Romanes disagreed with Weismann's belief that panmixia, the cessation of selection, alone would be enough to explain both. He argued instead that Darwin's own comparison of the size of wing bones in wild and domesticated ducks showed that only about 30 percent of a decrease of size could be attributed to panmixia. After this point, Romanes argued, a "reverse selection" must intervene because the reduced organ would still be large enough to compete with other structures for nutrition and energy. When this reverse selection diminished the structure to a mere rudiment—when the structure had became about 5 percent of its original size—the rudiment no longer would compete with the rest of the body. At this point, the structure disappeared as the "forces of heredity," the Darwinian gemmules declined enough to make the structure in question, vanished altogether from the hereditary mechanism. Romanes felt that Weismann had rendered this last step impossible because of his insistence that the germ-plasm is immortal, unchangeable.[95]

Weismann barely took up the issue of degeneration and panmixia in *Das Keimplasma,* but it is clear that he had considered integrating the consequences of a relaxation of natural selection into his germ-plasm model. He alluded to certain determinants changing their ability to multiply in cases of degeneration.[96] In the language of his germ-plasm theory, he mirrored Darwin's explanation that gemmules underwent change and disappeared, and he unwittingly incorporated Romanes's claim that a change in the "forces of heredity" was needed at some point in the process. It is doubtful that Weismann drew directly from Romanes's pre-1892 commentaries on degeneration, for the notion of changing determinants had already become an integral part of his germ-plasm model. By burying his brief mention of degeneration in a penultimate section of his *Keimplasma*—to be more precise, in the final chapter on variation—it was unlikely that many biologists saw it. Weismann integrated the processes of competition for nutrition between homologous determinants, panmixia, and amphimixis—in a different sequence than Romanes had recited in 1890. Weismann insisted,

> *The process of degeneration of parts* must be attributed to the disappearance of the respective determinants from the germ-plasm. . . . The cause of the regression of a determinant is to be looked for in insufficient nutrition. If this occurs in the majority of the ids either directly, or in consequence of the accumulation produced by amphimixis, the character controlled by these determinants becomes regressive in that particular individual. If, however, it no longer has a physiological value, it becomes slowly but surely suppressed by panmixia in an ever-increasing number of individuals until it disappears.[97]

It is hard to determine whether the difference in emphasis between the two was significant or not. Romanes was interested in the degeneration of somatic characters. While partaking of a vacation in Madeira, he vigorously wrote of Spencer's failure to distinguish panmixia from the economy of nature and the action of reverse selection.[98] After a belated return from his holiday he was able to read the English edition of *Das Keimplasma* and became much more sympathetic to Weismann's point of view even though not to specific details. Weismann notwithstanding, Romanes continued to believe in the inheritance of acquired characters in the form of "use-inheritance," and so it must remain a tempered dimension to his evolutionary theory. "My position with regard to this question is one of suspended judgment."[99] In their understandings of degeneration, both Weismann and Romanes recognized the evolutionary importance of panmixia in a way that Spencer had not.[100] In another letter to the editors, again in *Nature,* and again on panmixia, Romanes mentioned in passing that Weismann would soon be giving the third Romanes Lecture in Oxford.[101]

Weismann and the Romanes Lecture, May 1894

The exchanges between Spencer, Romanes, and Weismann, among others, were unquestionably responsible in part for the invitation to Weismann to visit the United Kingdom a third time. To be more specific, it likely was Romanes's 1893 essay on Spencer in *Contemporary Review* that had initiated Romanes's inquiry about a visit and lecture by the author of *Das Keimplasma*. As Weismann recalled to his son-in-law, he had felt that the essay had been an effort on Romanes's part to take full credit for the concept of panmixia.[102]

From Weismann's response to the invitation it appears that Hermann von Helmholtz may also have been asked, but he apparently had declined, thus leaving Weismann the choice of giving a talk in nearly six months' time.[103] Weismann's acceptance followed those of two outstanding figures on the English scene: William Ewart Gladstone, who opened the series in 1892 with a lecture entitled *An Academic Sketch,* and Thomas Henry Huxley, who delivered the 1893 lecture on the subject of *Evolution and Ethics.*[104] Both had strong but conflicting views about the place of evolution in science and society. Weismann presented his lecture in the Sheldonian Theatre on May 2, 1894, but conceived it more narrowly. It was the last of the three lectures to be given before Romanes's unfortunate death eighteen days after Weismann's visit.

Weismann's lecture, entitled *The Effect of External Influences upon Development,* was in form and structure an elegant justification of his ideas as evidenced by recent findings in natural history. Weismann presented his lecture in English, which had been translated from the German manuscript by Gregg Wilson and Weismann's son-in-law William Newton Parker. The printed version represented an expansion of what he actually delivered in Oxford and contained an additional fifteen pages of explanatory notes that amplified Weismann's technical points. The German edition was immediately picked up by Gustav Fischer and published the same year.[105] The printed preface, written after Romanes's death became known in Freiburg, contained a salute to the series's founder. In it, Weismann noted that in spite of his poor health Romanes was present at the lecture and "followed it with a lively interest." He then penned a touching paragraph about Romanes's "valuable writings" and "ceaseless energy." This was a far cry from Weismann's initial reaction to Romanes and his commentaries on Weismann's works.[106] It was clear that not all evolutionary disputes need end on a contentious note.

The lecture was presented within the context of the heated international controversy over the inheritance of acquired characters, but because this was a general lecture, it was not the format in which to introduce new scientific ideas. While Weismann kept his discussion on a nontechnical level, he provided many examples from the professional literature taken from recent natural history. Thus, he focused on how all those external factors such as temperature, humidity, gravity, and nutrition traditionally invoked by believers of the inheritance of acquired characters could easily be transposed from formative influences into physiological stimuli. His claim was that morphological, physiological, even behavioral adaptations could be accommodated within the structure of his germ-plasm theory. It meant returning to 1881 and Wilhelm Roux's *Kampf der Theile,* in which the young Breslau anatomist had proposed that a competition for nutrition took place between the similar cytological units within an organism. With the more rapid multiplication of more successful cells during development, an intraselection between cells would naturally take place. In turn, this cellular struggle determined the size and shape of the cells and their comparative success during development. The results determined the success of variations of comparative organisms. Roux had composed his suggestive monograph at a time when an invocation of the inheritance of acquired characters was not controversial. Thus he assumed a direct inheritance within lineages of organisms that had acquired the more successful forms and shapes of vessels and other structures during the intraselection processes of development.

Later in the 1890s, Weismann borrowed the notion of intraselection but thought of the problem somewhat differently. "It is not the particular adaptive structures themselves that are transmitted, but only the quality of the material from which intraselection forms these structures anew in every individual life."[107] This meant ignoring the inheritance of acquired characters while taking the notion of intraselection from Roux's *Kampf der Theile* and attaching it to the structural details of his *Keimplasma.* There, as we have seen, it became a selection of competing homologous determinants from different ancestral ids as they multiplied in the germ-plasm during development. The successful constituents would then be passed on following the precepts of reduction divisions and amphimixis—and eventually the natural selection of individual organisms.

The principle of intraselection of germinal constituents became Weismann's anchor for explaining new variations. In his talk he interpreted many recent experiments, which at first blush might appear to confirm an influence of the environment: the widespread phenomenon of mimicry, dimorphism and polymorphism in general, Poulton's experiments on the impact of their surroundings on the color of caterpillars, and Frederic Merrifield's new investigation of seasonal dimorphism with Weismann's old lepidopteran friend, *Vanessa levana/prorsa*. All of these and more provided serviceable examples of the hypothetical rise of variations. The matter of dimorphism in sex, although not yet extensively studied, was particularly relevant. Experiments by Yung on sex determination in frogs, Maupas's work on the temperature influence on the sex of rotifers, and Siebold's and Leuckart's older studies of the caste system in bees all offered cases showing di- and polymorphism. Could they help distinguish between the impact of possible external stimuli and the internal efficient causes of intraselection? The challenge of the many cases revealing di- and polymorphism was to distinguish between the impact of possible external stimuli and internal efficient causes of intracellular selection.

Weismann had himself examined the influence of starvation on larvae of blow-flies (*Musca vomitoria*), which developed into smaller but anatomically complete and sexually active adults. His results were a challenge to those who wished to generalize from bees, where it had been shown that nutrition was a critical element in determining the difference between queens with their complete ovaries and workers that generally had rudimentary and rarely functional ovaries. In brief, the workers could not pass their morphological and physiological acquirements to the next generation. The starved blow-flies' ability to remain reproductive showed Weismann that *"the disappearance of a typical organ is not an ontogenetic but a phylogenetic process . . . is always due to variations of the primary constituents of the germ, which to all appearances can only come about in the course of numerous generations."*[108] The germ-plasm, after all, was not unchanging but subject to an internal competition during embryogenesis.

"Zusatz 16"

Between the time when he wrote his preface to the English edition on May 27 and when he delivered the final proofs of the German edition to Gustav Fischer on June 12, Weismann received from his publisher a copy of Hertwig's *Präformation oder Epigenesis*. Because Hertwig showed agreement with Spencer's strong position presented in 1893, Weismann took this opportunity to add a sixteenth note, "Zusatz 16," to meet both his opponents head on. The note this time resulted in a critique against the inheritance of acquired characters. Here, moreover, Weismann concentrated on the evolution of castes in social insects, which both Spencer and Hertwig had used as an illustration of their parallel causes but about which neither had firsthand experience.

Spencer had insisted that the numerous castes, that is, queen, drones, and the elaborate intermediary forms, represented a spectrum of both morphological and functional traits. Ants and termites best revealed complexities of what was at stake. How did these clearly distinct forms arise phylogenetically and ontogenetically? Spencer argued that nutrition was the overriding cause of intermediate forms, just as it had been demonstrated earlier during the production of queen and worker bees. Hertwig, as a cell biologist, could go further than Spencer's simple narrative. In each individual from the moment of first cleavage to the adult stage, he explained, the cells constantly interacted with one another—they were sensitive to external conditions (including nutritional differences), and the tissues and organs were correlatively adjusting to the whole.

In both accounts, the conclusion was that the inheritance of acquired characters explained the transmission of new traits to the next generation. Theirs was clearly a physiological perspective that contrasted with Weismann's reliance on the morphology of ids and determinants in the germ-plasm. Weismann's approach, Hertwig explained, "incorporates in the rudiment what really are stimuli coming from external conditions during the process of development." He concluded in a quotable précis that Weismann "makes a grave confusion between the rudiment and the conditions of its development."[109]

Weismann was indignant about Hertwig's accusation: "Das ist freilich sehr schlimm!" ("To be sure that is very bad!"). But he preferred to concentrate on the details of what was known about the castes themselves. It turns out that there had been a fair amount of recent research on the castes of social insects by experts such as Father Erich Wasmann, Carlo Emery, Sir John Lubbock, and above all Auguste Forel. The latter had been an important neuroanatomist and professor of psychiatry at the University of Zurich until his retirement in 1893. The freed-up time allowed him to concentrate on his avocational interest in the taxonomy and behavior of ants, upon which he had already published in a comprehensive volume.[110] As the title of this 550-page study implied, the work was crammed with all kinds of details, particularly those concerning the structural and behavioral differences between castes of many species, and it closed with a restrained review of Darwin's contribution. It was exactly the kind of work in natural history that impressed Weismann and to which he would naturally turn in his confrontations with Hertwig and Spencer—neither of whom were experienced naturalists.

Weismann had started corresponding with Forel shortly after the latter's retirement. That was also the year when he first tangled with Spencer and became deeply involved in defending his mature germ-plasm theory and natural selection as the primary factors in evolution. He enquired of Forel about reproduction in neuter castes, that is, parthenogenesis among ant workers, and he asked for specimens of the slave-making genus *Polyergus,* which Forel had used in his discussion of extraordinary inherited behaviors in worker ants. By the next year, Weismann was again soliciting help and other specimens as he sharpened his attack against Spencer and ultimately Hertwig.[111]

The issue regarding parthenogenetic reproduction among worker ants was one of trait transmission. How could Spencer reasonably explain such a reproductive pattern in terms of an irregular delivery of nutrition to these intermediate forms when they were larvae? One might imagine an irregular system of feeding, first with worker-induced nutrients, then with queen-induced nutrients. Such a scenario might explain a rare individual, but how to explain such a process involving 20 percent of the workers that Forel had found in certain nests?[112] Reduction and amphimixis of the germ-plasm would readily explain the necessary combinations of traits. It was easier for Weismann to imagine discrete units in the germ-plasm that brought about such conflicting results. Nests with so many intermediate forms might not last for long in the wild but long enough for Weismann to refute Spencer's and Hertwig's insistence that nutrition was the determining cause and that this demonstrated the inheritance of acquired characters. One must be impressed by the contrasting preparation and familiarity with the fine biological details of ant natural history that Weismann brought to the field when he jousted with Spencer and Hertwig. It was the difference between a deep versus a casual understanding of the natural history involved.

It was not reading alone that apprised Weismann of current research on the problem. His colleagues and students at his institute pursued relevant research. In this case, he entrusted a talented American student, Elizabeth E. Bickford, with a project to investigate the structural details of the ovaries of worker ants.[113] As Weismann described the project to his senate colleagues in his supporting "Gutachten" of Bickford's ensuing dissertation, "the work deals with material that at the present time has been selected to answer certain general questions and whose exact analysis therefore is desired and would be of particular importance." As her chosen title indicated, Bickford examined the microscopic morphological details of the ovaries of workers in nine species of ants in two major subfamilies. She explored the reproductive abilities of the workers of some of the species. She confirmed some experiments Weismann had previously done on workers, and she investigated the influence that the application of heat had on the larvae of additional species of ants. Her dissertation was a representative morphological survey of parthenogenesis among workers and a testing of a few reproductive influences. It was a modest but important conclusion when she reported "that under normal conditions the reproductive capacity of the worker ants has clear characteristic boundaries for the species, that is, different for different species." Her characterization supported Weismann's claim that spoke directly to the evolutionary degeneration of the reproductive system in neuter workers. At the close of her account she quoted with approval her mentor's statement that "the loss of a typical organ [that is, the number of ovarioles in the rudimentary ovary] is no ontogenetic process but a phylogenetic one."[114]

When he urged his senate colleagues to accept Bickford's dissertation, Weismann insisted that it provided factual relevance not just opinion to "certain general questions." Those unstated questions were, of course, the relative merits of the inheritance of acquired characters contra natural selection in explaining the appearance of parthenogenesis in normally sterile castes. As with the research of many other students at Weismann's zoological institute, Bickford both promoted her technical ability to solve a specific technical problem and fit her message into the larger framework of biological debate. After Freiburg, Bickford returned to the United States eventually to become a professor of biology at Vassar College.[115]

18

The Germ-Plasm and the Diversity of Living Phenomena

1890–1900

Not only did questions about the validity of the inheritance of acquired characters loom as a central topic in biology during the 1890s, but Weismann focused a good deal of his effort on responding to his critics and defending his *Keimplasma*. As we have seen, a significant amplification to his germ-plasm theory began in 1894 when in the Romanes Lecture he expanded upon his notion of germinal selection. The elaboration of this heredity theory, however, did not consume all of his professional attention. He had an institute to run, increasingly larger classes in his basic lecture courses to teach, and an ever-increasing number of doctoral students in zoology and a few in anatomy to guide.[1] He gave his core course in "Descendenztheorie," which he had tentatively initiated in the summer semester of 1883 and then transformed into a winter semester course in 1884–1885, as four and later five scheduled meetings per week including Saturdays from noon to one.[2] Gruber, as the salaried außerordentlicher Professor, assumed the routine business of running the institute during Weismann's leave of absence, as was the case when Weismann wrote *Das Keimplasma* during his stay in Munich in the winter of 1891–1892. The outcome of his teaching of evolution once a year for two decades resulted in the expansion of his lectures, which he assembled and published in 1902.

By this final decade of the century, Weismann was apparently still able to read and write some of his professional literature and mail, but he was continually searching for "readers" to assist him.[3] Often with the help of his students, he also continually pursued research projects. Some of these were of minor interest, but two projects were major concerns and resulted in a revisiting of his papers on seasonal dimorphism in butterflies and a challenge to the generally accepted claim that drones were derived from parthenogenetic eggs. Before we review either of these specific projects, it is important to examine the remainder of Weismann's major theoretical elaborations to his germ-plasm theory. This arose as a response to the fierce criticisms he had received from Haacke, Hertwig, Spencer, and Romanes. As we will see, Theodore Eimer, among others, found Weismann's germ-plasm theory deficient on many counts; as with other critics, he emphasized Weismann's inability to explain the rise of variations, which were necessary for natural selection to drive evolution.

When Ernst Mayr included a chapter of forty-five pages on variations and genetics in population in his 1963 work *Animal Species and Evolution,* his emphasis was well warranted.[4] The background of population genetics informed Mayr's entire work; going back sixty years before that, the subjects of species and evolution existed, of course, but they were poorly identified. Classical genetics was missing from the mix, and the techniques for exploring the behavior of individual variations in different generations hardly existed. Weismann was not the only one who was interested in the matter, but like most of his contemporaries he focused on the morphology and taxonomy of variations rather than trying to identify their relationships to one another or to find a causal pattern in their appearance.

The strategy of mathematizing the consequences of selection and the breeding individual variations was certainly not in Weismann's world of study. The results could have been informative though not conclusive for evolutionary and even developmental questions, but they were not as easy to generalize as the simple sets of patterns that characterize modern genetics. Weismann and his generation were basically naturalists, evolutionists, and embryologists. A different set of questions was required to focus on the hybridization of trait variations. What follows are two detailed examinations of Weismann's studies in the older framework of explaining the consequences of generational changes as they might enlighten the evolutionist.

Address on Germinal Selection

With the appearance of Darwin's *Origin of Species,* critics and supporters of evolution alike honed in on two immediately pressing deficiencies in his account: (1) how to explain heredity and (2) how to account for variations. The first of the two occupied the attention of Darwin and others from the outset. As we have seen, Weismann thought about the process within the context of the maturing of nuclear cytology, which led to his germ-plasm theory as a result. For those who believed in the theory of natural selection, it became imperative to explain the origin and nature of *de novo* variations.

When Darwin came up with his controversial "provisional hypothesis of pangenesis," he also made an effort to explain the origin of variations. Although he did not push the point in quite this way, he conceived of individual organic beings as consisting of two physical levels above the chemical processes at the organic base. One of these levels entailed the visible self, collected and studied by natural historians, anatomists, morphologists, and physiologists. This was the level on which naturalists traditionally identified and described variations. The other was a subvisible level, which for Darwin consisted of hypothetical gemmules, perhaps associated with differentiated cells. According to him, the gemmules circulated in the body and became the hereditary elements transmitted to the next generation.[5] These gemmules were the particulate determiners of traits from one generation to the next and set an undetermined stage for development. They were the material source for explaining transmission and development. Darwin had relegated the varied differentiation of specific traits to the physical nature of specific gemmules. The differentiation and variations in traits on the visible level were simply passed off as the results of different gemmules and their chance variation on the lower, subcellular level.

In keeping with the mechanistically dominated spirit at the end of the century, Weismann had to confront two radically different alternatives to Darwin's enigmatic conclusion of "chance." The first of these was the renewed enthusiasm for Lamarckian-inspired accounts of heredity and variations in terms of the inheritance of acquired characters. By the last quarter of the nineteenth century, this account normally involved a commitment to the physical and/or chemical origin of variations. New variations were acquired through either external physical influences or internal physical and chemical processes in the body. In either case, there occurred a direct physiological response to life's activities and interactions. Once they had appeared, variations were generally, but not always, conceived of as adaptive. Weismann had debunked this solution head on in 1883 with his visualization of separate germinal and somatic tracks; his condemnation of the inheritance of acquired characters would continue unabated for the rest of his life.

Over a period of a dozen years (1883–1895), Weismann sketched out and modified a hypothetical yet coherent account for transmission. He had gone far beyond Darwin's theory of pangenesis and the efforts of many others to join contemporary cytology and natural history into a model for heredity.[6] Whether Weismann had adequately explained evolution by natural selection, or took advantage of increasing experimental efforts to document hereditary patterns, or was premature, too speculative, or downright misguided in venturing into the unknown about variations prompted debate and complaint. There were those, however, who found Weismann's mechanistic account and command of the morphological solution sketched out in his germ-plasm theory inspiring. After the publication of *Das Keimplasma* through to the end of the century, no more thoughtfully articulated mechanical models of transmission between generations or for development and evolution would be produced.

The other explanation for variations may also be found in Lamarck's writings, but it was not always a device for furthering an evolutionary theory. It called upon the body of the organism simply to manufacture new ones. These were not necessarily of immediate function to the organism, but they might be transmitted to the next generation and might help engineer an evolutionary process. Burdach's "vital force," Nägeli's "Vervollkomnungsprinzip," and Jägers's system of chemical reactions appeared to satisfy the appearance and possible utility of such phenotypic variations. Weismann, however, was hostile to such variation-creating scenarios from the outset because he thought they might be construed as teleological; clearly they were foreign to his cytological experience, and they soon became contradictory to his idea of continuity.

So was Weismann's germ-plasm theory irrevocably squeezed between theories of individually acquired variations on the one side and vitally or chemically induced variations on the other? Considering his definition of "acquired characters" and his concept of vital processes, Weismann was convinced he had found a third option—germinal selection.

We have pointed out that in the closing chapters of *Das Keimplasma* Weismann said that he had been inspired by Roux's *Kampf der Theile* and its description of a competition between histological units during development. This came in an invitation from G. J. Romanes, who had not made an issue of it in 1892; however, being pressed by his critics, Weismann envisioned a competition within the germ-plasm to create new heritable variations. In his subsequent Romanes Lecture of 1894, he delivered the first full statement of the process that he began calling "Germinal Selection."[7] The talk came early in the congress, at the general session of September 16, to be exact. Three days later, Theodor Eimer, professor of zoology at Tübingen, delivered a general talk at the same congress also on the subject of variations. We will discuss his contrary position later. For the moment, let us examine Weismann's presentation.

By mid-November of the same year, Weismann had completed a more extensive manuscript of his address and shipped it off to the publisher Gustav Fischer. The same translator who had converted the German text of an earlier talk into English directly from Weismann's personal manuscript performed the same task for the full monograph.[8] When his extended essay appeared, Weismann made it clear in his preface and in an autobiographical sketch written for an American publisher that he had expanded the application of germinal selection in response to his critics, in particular to Spencer.

Weismann explained that with his *Keimplasma* completed and his energies focused on the ensuing disputes, he had become convinced that the Darwinian mechanism of "selection," if applied to the level of the germ-plasm, might play an essential role in the production of new variations. He had suggested as much in his 1895 response to Spencer's criticism; Weismann assured von Wagner that when he chose the lead to the German title of the same essay that he had indeed promised "Neue Gedanken."[9] These "new thoughts" now appeared as necessary to complete the evolutionary picture in neo-Darwinian terms. "It is quite obvious how exactly this concept meshes with my views of heredity," he assured his readers. "Should the germinal substance not really be assembled out of Anlagen, then germinal selection could not play a role." Weismann further explained to his American audience: "The one falls with the other, and the next problem will be to gain certainty about which of the two great parties is right, that of the Evolutionists or that of the Epigeneticists."[10] Having produced an elaborate model to explain heredity, he now more clearly saw the need for creating a related model for the production of variations.

The first thing to notice about Weismann's published paper was its title: *Über Germinal-Selection: Eine Quelle bestimmt gerichteter Variation.* The subtitle "A Source of Definitely Directed Variation" did not appear on his original Leiden address, and it was awkward on its Chicago translation. However, Weismann meant exactly what he promised: a theory that in principle may have appeared at first glance to be teleological but was in reality a mechanical model without a hidden purpose. It was Weismann's task to persuade his readers that he had not introduced a contradiction to his previous worldview; rather, the occurrence of directionality in the appearance of variations, as evident from the taxonomist's perspective, was fundamentally mechanical. Moreover, his use of the word "variation" in both languages was in the singular, indicating that Weismann was dealing with a general principle rather than specific cases.

Spencer's physiological purposefulness may have explicitly shaped Weismann's reply, but the language of his title clearly was responding to another opponent. First, some details of his new essay: Weismann made no apology about adding one more feature to his model of the germ-plasm. He simply recited in his preface the opinions of the physicists Maxwell, Hertz, Boltzmann, and Newton on the value of models and hypotheses in science to ratify the style of his own scientific explanations.[11] Deferring to his submicroscopic particles, he expressed in a somewhat convoluted sentence a possibility:

> At the moment the much maligned concept of the determinants may be such an important matter, that not only will the present essay support it but it will also defend it on new grounds, above all as merely a symbol for something, which we to be sure, do not yet know more exactly, but that we can count on it as being present, and with which we can figure on leaving to the future to decide the extent to which our construction agrees with nature.[12] Again: . . . In its foundation my theory would yet prove more than a mere work of imagination, and that the future would find in it some durable points which would outlive the mutations of opinions.[13]

The query, which he posed to his audience and left for future historians, remained on the boundary between the fruitful "foundation" and artful "imagination." Nevertheless, the concept of germinal selection was a serious extension to Weismann's germ-plasm model with the intention of meeting his critics.

Weismann's real forte had always been to resort to the biological details. So when events compelled him to defend his model of heredity, he called upon his own experience and the vast personal and institutional collection of butterflies at his disposal. Thus, he reminded his readers of a significant difference in coloration and behavior of diurnal butterflies. Certain widely spread families, such as that of the danaids (Danaidae, such as monarchs) and the heliconids (Heliconiidae, e.g., many tropical butterflies) display bright colors on both the upper and undersides of fore and back wings. These butterflies flit about in open spaces, and when they rest, they do so with their wings in open display. It was well known by Weismann's day that these colorful butterflies were able to behave so cavalierly because they were protected from predation by their offensive taste to avian predators. These families had also been shown to have prompted the evolution of color and pattern in more delectable species in unrelated families, mimicking the less palatable.[14] By way of contrast, many species of diurnal butterflies such as the nymphalids (Nymphalidae, i.e., the brush-footed butterflies) dwell in shaded areas and forests and possess concealing coloration on the undersurface of their wings. Their posture of repose is one of folded wings, during which the leaf-color of the undersurface is exposed to conceal the butterflies in their leafy environment. "I have noted fifty-three genera," Weismann said of the nymphalids, and "they belong to different continents and have probably for the most part acquired their protective colorings themselves. . . . They are all *forest-butterflies*." Weismann's message was clear. Through mimicry of bad-tasting coinhabitants, protective coloration of the underwings, or even a precise imitation of leaves, these more delectable butterflies had acquired an undersurface coloring that corresponded to their upright resting posture in forests. These butterflies, however, contrasted with the "nocturnal butterflies"—that is, moths—which rest with their wings spread out and reveal thereby protective colorations on the exposed upper surfaces of their forewings. In brief, "the conditions of life have wielded the brush."[15]

The use of butterflies as the primary exhibit of germinal selection was no mere whim on Weismann's part. Not only had he studied seasonal dimorphism in butterflies early in his career and amassed a large collection of butterflies from around the world for his museum, discussion of Lepidoptera brought him into direct confrontation with a former student and current critic. What Weismann was explaining through the varying forms and behaviors of butterflies was how his mature germ-plasm theory might provide a mechanical Darwinian explanation for specific and directed germinal and somatic variations.

Eimer and Orthogenesis

Throughout his career, Theodor Eimer also had variations in butterflies on his mind.[16] As Eimer emphasized in bold letters in the opening paragraphs of the first of his three volumes of his *Entstehung der Arten,* he immediately made clear his basic objection to neo-Darwinism. "The Darwinian utility principle, the selection of the useful in the struggle for existence," he insisted, "does not explain the initial rise of new characters. It only explains—and moreover in my opinion only partially—the increase and the growing dominance of these characters."[17] To his mind, the rise of species through natural selection was of only minor significance in the evolutionary process. Instead, Eimer devoted his entire career to showing how orthogenesis, the progressive development of characters in the rise of new species, was brought about by the external environment, the internal processes of growth in all organisms, and the inheritance of acquired characters.

Eimer was nine years Weismann's junior. He had spent his youth in various towns in the Black Forest, including Freiburg. The son of a physician, he started his university studies with perhaps a medical goal in mind. He first matriculated in Tübingen and then spent a predoctoral year in Freiburg, followed by another in Heidelberg.[18] Eimer earned a doctorate after a final year and a half in Berlin where he worked in Virchow's institute and received his medical degree in 1867. Although his Berlin studies took place nearly two decades before the Weismann-Virchow controversy, Eimer's later assistant and admirer Countess Marie von Linden insisted that Virchow had strongly influenced Eimer and his later career.[19] Eimer spent the next twelve months in Freiburg with a short stint in Paris as a postdoctoral student. In the latter place, he worked in zoology with Weismann, who by that time was teaching zoology and comparative anatomy under the auspices of both the medical and philosophical faculties. In later years, when the relationship between the two had broken down, Weismann still referred to Eimer as "my oldest student" and "my former student."[20] In 1869, Eimer again left Freiburg for Würzburg to become von Kölliker's prosector and to complete a second doctorate; this time he focused on histology and performing experimental work on fat absorption.

After habilitating in zoology and comparative anatomy in Würzburg, Eimer volunteered for and served actively in the Franco-Prussian War as a field surgeon. For reasons of health, however, he was forced to retire from military service. He and his new wife spent time in the Bay of Naples, where he investigated marine organisms. (Throughout their marriage, Anna Lutteroth Eimer served as his scientific illustrator.) On the Isle of Capri, he became fascinated with the common wall lizard, *Lacerta muralis*. A local form of this lizard possesses a darker and bluer coloration than its counterparts on the rest of Capri and throughout Europe. Eimer felt he had discovered a new race, isolated as they were on the rocky promontories known as the Fragalioni Cliffs at the southern end of the island. He reasoned that the lizards had resulted from a physiological interaction of the unusual local conditions and the organism's inherent growth process. These provided the terms in which he conceived of the unique differences between races and ultimately between different species. "The inherited characters form the indicated line of descent," Eimer claimed,

> whose direction prior to or after birth in the effecting external influences may alone bring forth a change. The resulting changes can be nothing other than the necessary product of crystallization from an altered composition in the organism. The same necessary crystallization product from a changed mixture of given materials results in any new race that we raise through the hybridization of different parents. And in exactly the same way must all so-called correlative appearances be explained.[21]

A touch of Haacke's vague chemical allusion colored Eimer's language in referring to "the products" of "crystallization." Similar to Haacke, too, was his lack of any direct appeal to nuclear cytology in an effort to understand heredity and evolution. As with Haacke, O. Hertwig, and many others, Eimer invoked but made no effort to explain, let alone in mechanical terms, the inheritance of acquired characters. He did envision the possibility of working out the hereditary process through hybridization experiments—though, unlike Haacke, such experiments never became even a minor part of his research agenda. To his lasting credit, Eimer had carried out some regeneration experiments on medusae that paralleled Romanes's similar work, but this had to do with the issue at hand only indirectly.

In certain ways, Eimer played the same role in Germany as did Spencer in England. He described mechanisms for evolution at least a decade before Weismann promoted the notion of a continuity of the germ-plasm. Like Spencer, he assumed the inheritance of acquired characters and was instinctively hostile to utility as the basis for gauging evolution. Three months to the day after Weismann had presented his inaugural address on his new concept of heredity in Freiburg, Eimer delivered an address on his notion of the "Individual" in biology at the Versammlung of scientists and physicians, which that September also met in Freiburg.[22] The contrast between the two addresses—one to a general university audience, and the other at a national scientific forum—was stark. Contrary to the assumed expectations of the different audiences, Weismann's was the more scientifically detailed, and Eimer's the more relaxed and poetic.

Both speakers in Freiburg examined the relationship between individuals and species and referred in passing to the biology of butterflies and bees, among other organisms. Weismann's vision was atomistic and assumed specific material entities were being passed from one generation to the next in the proposed continuity of a theoretical germ-plasm. He explained the rise of this hereditary transmission in terms of a division of labor of the simplest of organisms, the protozoa. His scientific frame of reference was anatomical and embryological. As for understanding the evolution of species, Weismann moved from the collective to the individual where the physical variations to be selected must first appear. With the exception of a few minor enigmatic cases, Weismann had become explicitly hostile to any explanation of evolution that invoked an inheritance of acquired characters. As we have seen, it was also by this time when Weismann introduced his audience to the notion of panmixia as an explanation for degeneration of parts. He still referred to Haeckel's "overgrowth" as a natural way to understand heredity between generations, but in what soon became a denial of that concept, he insisted that variations must be traced to the germ-plasm within the germ. As we have previously suggested, these claims became the stock in trade during the subsequent development of Weismann's germ-plasm theory.

Eimer presented to the collected scientists and physicians, assembled for his lecture in a general session of the 1883 Naturforscher Versammlung, a far more philosophical and literary talk. What he had to say would be echoed in his more elaborate texts that appeared in the future. The continuity in nature, he insisted, must not be confined to the germ or hypothetical germ-plasm but reflected the unity of the organism, or species, or nature as a whole. Where Weismann sharpened his understanding by descending from the whole organism to the nucleus of the individual, Eimer's strategy was to move roughly in the opposite direction. The inheritance of acquired characters for Eimer was not solely an issue of the individual but implied the biological connection between individuals and species. Where natural selection required the chance appearances of variations, Eimer sought general laws of growth to explain the regularity of the same phenomena. In Eimer's view, the sciences of anatomy and embryology were limited in scope because their approach tore apart and dissected the wholeness of nature. Eimer moreover was convinced that natural selection denied both the morality and idealism inherent in humans. He rejected the concept of the continuity of germ-plasm not because Weismann had inadequately demonstrated its existence, but because, to borrow an expression Weismann had used against Spencer, Eimer was fixated on the "all-sufficiency" of the inheritance of acquired characters. His was not a reactionary interpretation of phenomena but a vague and ultimately misguided attempt to attribute the apparent chaos of variations, heredity, and phylogeny to both a mechanistic and a holistic explanation of existence.[23]

Hardly a philosopher with the credentials of Spencer, Eimer nevertheless spoke with the authority of a productive university zoologist. His published texts, with correspondingly lengthy titles, appeared as *Die Entstehung der Arten* (1888–1901) in three volumes and *Die Artbildung und Verwandschaft bei den Schmetterlingen* in two parts with accompanying atlases (1889–1895), both elaborated and hammered away at the themes introduced in his Freiburg address.[24] When one looks through Eimer's *Entstehung der Arten,* it becomes clear that this was a work designed with multiple goals in mind. It presented an expanded version of his theory of growth in eight full chapters, or "Abschnitte," but it is equally clear—even in the title itself—that Eimer assumed as a given the inheritance of acquired characters, which did not need to be proved but needed only to be explained. Instead of a systematic and single-minded examination of a personal theory of evolution, Eimer singled out for particularly critical examination Weismann's theory of continuity, his belief that death was an acquired adaptation, his early experiments on the influence of temperature on seasonal dimorphism in butterfly species, and his explanation of degeneration by means of panmixia.

In retrospect Eimer's *Entstehung der Arten* appears to be a hodgepodge of counterclaims appended to his overall effort to explain evolution through the effects of growth and the inheritance of acquired characters. To cement the connection between this volume and his earlier lecture in Freiburg, Eimer appended a reprint of the latter to his conclusion of the former. Notwithstanding this reminder, the confrontation and antithesis between the approaches to evolution of these two South German professors could not have been clearer.

A brief discussion of Eimer's elaborate, repetitive, and surely exhausting (not exhaustive) discussions of growth is worthwhile, given my general historical thesis that soon after becoming involved in the details of nuclear cytology, Weismann broke away from the traditional model of overgrowth as an explanation of heredity. In contrast, as the full title of *Entstehung der Arten* implied, Eimer did not employ growth simply as an omnipresent physiological process but envisioned "laws of organic growth." His studies of wall lizards and many other organisms, particularly swallowtails, provided an explanation of organic variations that seemed indifferent to the utility implied by the mechanism of natural selection. His theory of evolution instead focused on the lawful production and accumulation of these acquired variations rather than the chance appearance of useful variations and their selection. In brief,

Eimer's theory of evolution offered a complete package for explaining the two major issues concerning evolutionary biologists: variations and heredity.

The finale to his work came when Eimer drew upon recent studies of regeneration, which in his mind must trace precisely the original development of any lost part. In turn, regeneration lent unquestionable proof of his growth laws. "Thus, as it seems to me, as with individual development so also with recrescence," Eimer emphasized, "complete proof is provided of the truth of my theory of the organic growth of the living world."[25] In key places throughout his text, Eimer turned to extensive analyses of Weismann's neo-Darwinian explanations of contemporary issues. Weismann's analyses of pertinent biological phenomena became the fodder for Eimer's assault. It was Eimer's way of avowing an eventual collision between himself and his neighbor on the east side of the Schwarzwald. Eimer's brief allusion to "mein Freund Ziegler," suggests that a more general tension existed between Tübingen and Freiburg.[26]

The disagreement, which may have seemed a local one in 1883, became a general one in 1888, and developed into a confrontation by 1895. By this time, Weismann had responded to German and English critics of his germ-plasm theory by explaining in depth how he imagined variations might arise through germinal selection. It was a purely hypothetical plan, and as with most of the germ-plasm theory, it consisted of an extension of the details of his mechanical model that could and would soon be proved woefully wrong. The confrontation between him and Eimer remained on the level of disputing primary claims about the nature of heredity, rather than specific demonstrations for or against key elements in their respective theories. Not once did Eimer consider Weismann's use of nuclear cytology.

Weismann on Eimer

One side of this complex relationship is partially documented in Weismann's letters. When asked in 1886 by his old friend Anton de Bary, who had moved from Freiburg to Strasbourg, for recommendations for the chair in zoology vacated by the death of Eduard Oskar Schmidt (1823–1886), Weismann included on his list Eimer, by then serving as full professor of zoology and comparative anatomy in Tübingen. Weismann wrote a straightforward and quite laudatory recommendation. He stressed that Eimer had assembled "nice" material that would well serve contemporary studies on evolution and development, but he added the caveat that Eimer had "theorized" about it too much and not clearly. Weismann concluded by describing Eimer as a "lively" lecturer, an "energetic character," and that "he is a thoroughly respectable character, also good tempered, but intense and irritable; one can however easily get along with him." Weismann sent a third letter to de Bary on February 2 in response to de Bary's immediate reply to his. De Bary appeared to have asked his friend about his own possible interest in becoming a candidate for the Strasbourg position, but he also must have had some questions about Eimer. After declining to be considered himself, Weismann explained that he had not been in touch with Eimer about the Strasbourg position, but that the latter had besieged him with "letters and telegrams" for support. This time Weismann was more blunt in his appraisal but with the stipulation that his letter would be for de Bary's eyes only. He closed his long paragraph with a tempered condemnation of Eimer's science:

> If he only had, at least with respect to scientific matters, a less unclear mind! But for him the wholly justified preference in our day is the working up of facts, or as one generally says: to the speculation over the mere dregs, for at the same time he runs afoul of his own thoughts and thereby he will even confuse the observation, because he is strongly inclined to see what he wants to see. This [serves] as a full appraisal of him. All of this, however, does not preclude that he would not be in the position to lead an Institute energetically. To the contrary he would certainly have the stuff and his personality guarantees it. Above all as far as being a representative [of the university] he is the right one.[27]

A year later, Weismann wrote a long, telling letter to Eimer.[28] Indeed, it might have been the only letter he wrote to his colleague across the hills. It was apparently a response to a letter containing criticism of Weismann's early ideas about "continuity." Weismann denied having suggested that "dispositions" of acquired characters could play a role in influencing the germ-plasm. More interesting is that Weismann did not accept Eimer's claim that such acquisitions influence the germ-plasm through nutrition. The letter, however, held a more exacting message for his critic. Weismann made it clear that he did not wish to discuss their scientific differences through correspondence. He argued it would be much better for each of them to publish his ideas and let others know about them and be in a position to adjudicate the issues. "So publish what you will, I will certainly then find the opportunity to answer."[29] And publish Eimer did.

Weismann rarely mentions Eimer in letters to other colleagues. He was reluctant to examine in detail his views of Eimer's science for Gustav Fischer, who was considering the publication of the latter's *Entstehung der Arten.* Weismann found Eimer's factual observations "interesting" even though he himself did not agree with the conclusions.[30]

The Spuler-Eimer Exchange

In his 1890 correspondence with Poulton, who in England was preparing the first volume of the *Essays,* Weismann commented on Eimer's *Entstehung der Arten,* which by then was being translated into English. "His book, as far as I know," Weismann ventured, "is an unusual and confused concoction, that, as I believe, can make no claim for scientific attention. For this reason I have up to now not replied to a single syllable of it and also hope in the future not to be compelled to."[31] As we will see, he kept his word even at the artificially or purposefully engineered confrontation at the international meetings in Leiden.

The operative word in Weismann's letter is "replied." We can, however, go further. As the leader of a thriving institute in zoology that graduated ten doctoral students and accommodated two habilitations between 1888 and 1895, Weismann had the opportunity to guide students into an area of research that challenged Eimer's interpretation of evolution. During this period, he attracted an ideal candidate for the task. As a youth Arnold Spuler was intent on becoming a physician. At the same time, when in the "Secunda" tier of students of the gymnasium in Karlsruhe, he was encouraged by a local architect to collect butterflies, who made his own collection available to the young student for study. Spuler later described himself in his curriculum vitae as a youthful "Schmetterlingsammler" (butterfly collector).[32] Matriculating in Freiburg during the winter semester of 1888–1889, Spuler concentrated in zoology with minors in botany and geology. During that time he also served in Wiedersheim's Anatomical Institute as a voluntary assistant, completed his military obligations, and prepared for state medical examinations. He was a busy and talented student with clear goals in mind. His doctoral research was completed under the supervision of Weismann, assisted by Ziegler and vom Rath. For some unclear but perhaps relevant reason, Gruber recommended Spuler's dissertation to the faculty in October 1890, but the dissertation did not appear in a journal until 1892.[33]

Nevertheless, Gruber's Gutachten quickly got to the heart of the dissertation in a somewhat awkward way by explaining that the form and ontogeny of the venation of the wings of some butterflies displayed the same initial patterns as the most primitive phylogenetic form of the same general type. As Gruber elaborated, Spuler's dissertation was traditional but contained refined embryology, presumably of the development of the wing veins. At the same time, Spuler was reassembling some of his dissertation material into a professional paper with the objective of refuting Eimer's understanding of the evolution based on the stripes of the wings of swallowtails, the Papilionidae.[34] In both his dissertation and the ensuing paper, Spuler included the standard student statement of gratitude about his supervisor "Herrn Geh. Rath Weismann," and thanked both Ziegler and vom Rath for assistance with its composition and technical matters.[35]

Spuler wasted no time in issuing a paper criticizing Eimer's work, which recently had appeared as the first Theil of *Artbildung und Verwandtschaft* (1889). In a lengthy aside of three pages of small print, he chastised Eimer for ignoring related species of butterflies other than those in the genus *Papilio* and for failing to examine the origin of the eleven-striped Papilonidae, which formed the showpiece of Eimer's theory of orthogenesis. Spuler pointed out inconsistencies in the fine details of Eimer's presentation, and he further insulted Eimer by claiming that his theories had found little approval among others. Finally, he ventured that "Eimer's elaborations are composed so unclearly, that to give a precise refutation in brief would scarcely be possible." The rest of Spuler's paper consisted of his own construction of the phylogeny of the family of Papilionidae, a large family of many genera, including the *Papilio* dealt with in Eimer's book. His strategy was to trace as many forms as he could back to an "Urform" using the morphology and embryology of the venation in the wings, including the underwing patterns. Spuler's examination confirmed a branching phylogeny, based in principle on venation and included not only the swallowtails but the parnassians (subfamily Parnassiinae) and the sulphurs and whites (family Pieridae). The wing stripes, which Eimer had identified as the product of individual *Papilio* species pushing in a common developmental direction, turned out for Spuler to be the result of a branching, differentiated phylogeny evolving throughout the entire family. Although Spuler did not push the point, his was the stuff of a Darwinian evolutionary process.[36]

Eimer, of course, could not let this detailed attack from Freiburg pass in silence. There can be little doubt that he was convinced of Weismann's presence behind the scenes. In referring to Spuler, Eimer complained that he had been presented with the most severe reproach that can be made by a scientist, and that he felt compelled to respond.[37] Eimer's response consisted of detailing thirteen minor inaccuracies in Spuler's paper. He rejected with minute countering details Spuler's claim that he, Eimer, had falsely identified specific wing stripes in relationship to the veins of the wings and to the stripes on the thorax. Finally, he attacked Spuler's conclusion that the entire family of swallowtails (Papilionidae) could be traced to the urform *Thais*. Eimer did not confine himself to critiquing a third party in the form of Spuler; he devoted the last four pages of his counter criticism to the "Weismann school" specifically and to Darwinism in general. He made clear that he was hostile to the contemporary zoology and botany that had become so thoroughly focused on flaying their subjects and "picking them to pieces." In doing so, Darwin and the neo-Darwinians had not shown how traits evolved; rather, they relied simply on traits that were already useful, and on so-called continuity of the germ-plasm, and the accidents of sexual mixings. By way of contrast, Eimer had, through the examples of his swallowtails, shown in detail that external influences and constitutional causes—the inheritance of acquired characters—were far more important in evolution.

Eimer on the Inheritance of Acquired Characters

When Eimer responded in 1895 at the International Congress of Zoology in Leiden three days after Weismann had delivered his lecture *On Germinal Selection,* he had been a full professor of zoology at Tübingen for twenty years, and he had written the first of three proposed volumes on the origin of species (1888–1901) and two volumes (the second in press at the time) with atlases on species formation in butterflies.[38] He was well known as an opponent of neo-Darwinians and particularly as a critic of Weismann. His address, separated by three days from Weismann's talk, must have been anticipated. The translator of both Weismann's and Eimer's talks, Thomas J. McCormack, must have relished the connection.

As Weismann had done at the onset, Eimer addressed the assembled zoologists on both the formation of variations and their relevance to evolution. As with Weismann, he also focused on butterflies in his discussion.[39] Eimer's Versammlung talk fell into two parts. The first was a discussion of the causes of transformation, which found that "orthogenesis is a universal law" (more emphatically in the original: "Die Orthogenesis ist ein allgemeines Gesetz").

It quickly becomes clear to the reader what Eimer meant by "Gesetz": the progressive (or retrogressive) developmental regularities he continued to find in the wing patterns of related butterflies from one species to the next, which were easily demonstrated in his illustrations. The same patterns of stronger and weaker wing bands not only spoke of phylogenetic relationships but made clear that evolution proceeded in definite directions.

As a scientific explanation, this orthogenetic regularity far surpassed the chance variations insisted upon by the Darwinians. Furthermore, Eimer had earlier explained such orthogenetic regularities in terms of four laws of growth. As we saw, these laws of growth were descriptive regularities rather than mechanical explanations. In 1895, Eimer expanded the laws of growth to eleven. All, he concluded, "show that evolution everywhere is a definitely directed [*eine bestimmt gerichtete*] process, and . . . is one which advances steadily forward as if conforming to a definite plan, [*wie nach einem bestimmten Plan,*] in diametrical contradiction to the [Darwinian] assertion that variations 'oscillate in the most diverse directions about a zero-point.'"[40]

The second portion of his Leiden talk concentrated on species formation and factual demonstrations. Repeating what he had laid out earlier, Eimer recognized three different patterns in species as a result of his laws of growth. Because he felt these differences were so fundamental to his argument, he endowed them with his own technical nomenclature. "Genepistasis" denoted a cessation in the orthogenetic developmental pattern among certain members of a species, thus allowing varieties to separate from one another without geographic separation. "Halmatogenesis" specified the rise of new developmental patterns due to changes in external influences. A new species formed in correlation with a changed environment and might geographically separate from the original species, a process that would prevent interbreeding. Finally, "Kyesamechania" indicated a state where physiological or morphological changes in reproductive systems prevented interbreeding. The value of the Greek neologisms aside, there is only a limited sense in which Eimer thought in populational terms. At the same time, he explicitly rejected that natural selection could do anything more than preserve the species formed through his laws of growth and orthogenesis. The rest of his talk demonstrated what Eimer did best—detailing the wings of dozens of butterfly species.

As mentioned earlier, the reprint of Eimer's Leiden address appeared as the second chapter of the second volume of *Entstehung der Arten* (1897). The third chapter consisted of a detailed critique of Weismann's Leiden address on germinal selection. In addition to a vituperative attack on Weismann's neglect of his writings and a veiled accusation that Weismann had borrowed some of his ideas without attribution, he provided a discussion of Weismann's words in twenty-four quotations and an equal number of lengthy, sharply worded replies. His most telling point seemed to be that Weismann had sneaked into his new theory of germinal selection a mechanism for orthogenetic patterns made possible by an orthogenetic selection of germinal variations. To some extent, Eimer was right. Weismann had been pushed by both Eimer and Spencer to consider more seriously the appearance of variations. It was clear, however, that Eimer, for whatever reason, did not have a real understanding of selection on any level and that he refused to follow Weismann into the depths of his microscopy to see both the possibilities as well as the shortcomings of the germ-plasm theory. The second volume of *Entstehung der Arten,* with its included reprint of the Leiden address, was Eimer's last publication before his untimely death.[41] With the publication of this volume, the bitter controversy between Weismann and Eimer came to an end.

Bowler Account and Conclusion

In an excellent article, Peter Bowler analyzed the confrontation between Weismann and Eimer.[42] He found that those believing in orthogenesis in general and Eimer in particular clearly identified directional patterns in evolution in terms of internal directive processes—hence the term "Orthogenesis," coined by Haecke and popularized by Eimer. For some, such as Nägeli, a directional phylogeny might be due to a nebulous vital force. For his part, as we have seen, Eimer resorted to a hypothesized chemo/mechanical developmental process. The crux of such "definitely directed evolution" ("bestimmt gerichtete Entwicklung") lay in its anti-neo-Darwinian thrust—after all, here was an evolutionary pattern that Eimer felt could not result from the natural selection of chance variations. Bowler suggested that Eimer's explanation echoed the idealism of *Naturphilosophie.* In contrast, he found Weismann and other neo-Darwinians strained in their efforts to explain away such apparently undeniably directed patterns in terms of utility and survival.

Weismann, who was being pushed by his many critics, explored the phenomenon more deeply and became convinced that a germinal selection of ids and determinants within the germ-plasm served as a logical alternative for explaining orthogenetic patterns. The outcome of such a process reinforced the generality of the Darwinian process while solving an undeniable biological phenomenon. "Germinal selection," Bowler wrote, "was a theory of internally controlled variation which imposed no fixed pattern on the directions that might arise."[43] What initially appeared to be a strategy on Weismann's part to deny the inheritance of acquired characters as promoted by Spencer and Hertwig developed into an elaborate mechanism to refute Eimer's directed growth.

The fire that flew between Weismann and Eimer, however, reflected more than philosophical differences. The initial sparks may have come from an article criticizing Weismann's germ-plasm theory that Eimer had written for a literary journal, but the continuation resulted from Eimer's words and Weismann's self-imposed silence, which was buttressed by a published attack on Eimer's science by a Freiburg student who had not yet completed his dissertation. "I only fear," Weismann informed his inquiring publisher Gustav Fischer, "that [Eimer's article] will seem unclear and all too subjective."[44]

When Eimer died after an abdominal operation at the end of May 1898, Weismann wrote with sincerity to Hermann Vöchting, a botanist in Tübingen who had just inquired to him about possible replacements for Eimer's Lehrstuhl: "The cruel fate, which Eimer had so suddenly met, has indeed deeply shaken me."[45] Weismann's most elaborate comment on his departed opponent came in a letter to Ishikawa, now back in Tokyo, at the end of June of the same year:

> You will have read that my earlier student and later bitter enemy, Eimer has died in Tübingen. It may well say that he owes to me the entire direction of his work, he also has from me the theme of butterflies. Also his earlier works on the color of lizards, of birds and mammals is based on my work of the '70s on the origin of markings in sphinx caterpillars. Instead of a little gratitude he has had only hatred for me. He was incensed that I had not commented about his confused "theory," and to be sure I was silent only out of regard for him, for if I had, then it would have only been a judgment of his views. I have already avoided in acceding to a polemic with him, on the ground that I well knew that nothing would ever result therefrom. Moreover he argued like a lawyer, who wants to help out a villain, or like a political agitator, to whom the momentary effect of his words is the concern.[46]

Weismann is worth quoting in full here, because it says as much about its author as it does about the deceased, revealing how Weismann saw their complex relationship from beginning to end.

19

Seasonal Dimorphism in Butterflies and Parthenogenesis of Drones

Ernst Mayr included a chapter of forty-five pages on variation and genetics in populations in his *Animal Species and Evolution* of 1963. The emphasis in this comprehensive book on animal evolution was well warranted, and the background of population genetics informs the entire work. Sixty years before, species and evolution had been poorly identified as subjects. Classical genetics, of course, was missing from the mix, and the techniques for exploring the behavior of individual variations in different generations hardly existed. Weismann was not the only one who was interested in the matter, but like most of his contemporaries he focused on the morphology and taxonomy of variations rather than trying to identify the grosser, causal patterns in their appearance.

The strategy of mathematizing the consequences of selecting and breeding individual variations was certainly not in Weismann's world of study. The results could be informative though not conclusive for evolutionary and even developmental questions, but they were not easily generalizable as the simple sets of patterns that would become characteristic of modern genetics. Weismann and his generation were basically naturalists, evolutionists, and embryologists. It took a different set of questions to change the focus on the hybridization of trait variations. What follows are two detailed examinations of Weismann's studies in the older framework of explaining the consequences of generational changes as they might enlighten the evolutionist.

Seasonal Dimorphism in Butterflies

We reviewed the early research on the temperature effects of polymorphism by Dorfmeister and Edwards in Chapter 5. Between 1883 and 1895, Weismann extended this earlier work with several species, including *Chrysophanus phlaeas*.[1] He repeated on a larger scale and with a greater variety of manipulations much of his previous work. In most cases, he experimented with 100 to 200 individuals over a period of two to three years. The individual sets of experiments were not extensive in size by later standards, but they were not insignificant in the number of cases studied, and his conclusions did not depend on single noteworthy specimens or broods. As Weismann had done earlier, he often divided broods into subsets, established controls, and exposed the remaining subsets to varied temperature conditions. He sometimes traced results through an annual cycle of generations; at other times, he concentrated on a particular generation.

The lack of consistency at this point signified a very different set of objectives than the hybridization experiments of classical genetics. He recorded percentage changes, to be sure, and he reported his failures as well as his successes, but there seemed to be no sense on his part that statistical analysis could provide a powerful supplementary tool to his experimental manipulations. Finally, because some of the breeding operations appear to have foreshadowed twentieth-century procedures, it is important to observe that Weismann's investigations primarily concerned evolutionary questions about the relative merits of Darwinian, orthogenetic, and neo-Lamarckian mechanisms. To his credit, Weismann indeed observed the scientific method, but his work must not be confused with the experimental tradition that became associated with the rise of Mendelian genetics.[2]

The details of his new experiments on multiple species, including *Vanessa levana* and *V. prorsa,* need not concern us, for Weismann is very clear: "Above all, I was concerned to test my earlier results by more exhaustive and, where possible, more accurate experiments."[3] He paid closer attention to the exact stages of the life cycle when he manipulated the temperature, and he was more precise about the changes in color. Of particular importance were Weismann's studies of the European small copper or fire butterfly, *Chrysophanus phlaeas*.[4] In 1875, he had alluded to its dimorphic character only in passing.

C. phlaeas is a small copper-colored butterfly that is widely distributed across Europe, North Africa, Asia, Japan, and North America. By the nineteenth century, it was well recognized that the upper surface of the forewings of both winter and summer generations of the northern European form (*C. phlaeas*) is a vivid, copper-red color, and the terminal area of the upper surface of the hindwings possesses a broad band of the same vivid color. In contrast, the forewings of all generations of the southern European form (*C. eleus*) are a dusky gray, and the terminal area band is much narrower. It was natural to ask whether these differences were temperature dependent and hereditary. Without going into the details of his temperature experiments, in which Weismann subjected the larvae and pupae of both forms to a variety of low and high temperature manipulations, we should point out that Weismann found that the appearance of both the northern and southern forms could be forced toward the color of the other. It is equally important to realize that the conversions were rarely complete and that the cold weather form was more resistant to change than its warm weather counterpart.

The asymmetrical outcome was not startling, for Weismann had demonstrated similar reactions in his earlier experiments on *V. levana*. Unlike *V. levana* and *V. prorsa,* however, *C. phlaeas* and *C. eleus* offered a consistent way of correlating artificial temperature responses with geographic distributions. In addition to the easily separated northern and southern races, *phlaeas* and *eleus* also presented a seasonally dimorphic population along the Ligurian and Tyrrhenian coasts. The spring brood in both Sicily and Sardinia possesses a similar vivid copper as the red northern race. In contrast, its summer brood is indistinguishable from the dusky southern race. Again employing his temperature experiments on this local variety, Weismann demonstrated that he could convert the upper surface of the wings of the summer form to the northern color, and that he could with less certainty force the spring brood to take on some of the dusky appearance of the southern race. Seasonal dimorphism and geographic dimorphism appeared to have closely linked reactions to a similar set of temperature ranges.

In 1875, Weismann had been prone to explain both seasonal dimorphism and the results of his temperature experiments in terms of the direct influence of the temperature on wing color and pattern formation. After the development of his germ-plasm theory, he was able to invoke germinal determinants to explain the alternative colors. "If we might assume that both forms of colour were adaptations, and possibly afforded protection to the insects, then it might be explained by the supposition of two kinds of determinants in the germ-plasm.[5] Discrete alternative determinants still allowed for the effects of climate or artificial temperature conditions for bringing about the development of one form or the other. This was, after all, his strategy in explaining the determination of sex. The coastal seasonal dimorphic population of *Chryso-phanus,* because it appeared to be a transition between northern and southern populations, validated the evolutionary sequence.

By 1892, Weismann's standard adjustment for seasonal dimorphism consisted of the application of two closely associated sets of determinants. Furthermore, the standard neo-Darwinian response was to consider each type of determinant as an adaptation to peculiar circumstances. But here was the paradox with the coastal seasonal dimorphic population of *Chrysophanus.* Weismann could not for the world envision any adaptive advantage of *eleus* over *phlaeas* during the summer months in Sicily and Sardinia. Consequently, the *eleus* determinants could only be a direct response to the summer heat. Accepting this outcome, Weismann was thrown into a greater dilemma, for the asymmetric results to his experiments on the monomorphic Neapolitan *eleus* and the monomorphic Freiburg *phlaeas* indicated more than just a direct temperature reaction—they demanded a germinal difference, too. Simply put, the evolution from *eleus* to *phlaeas* indicated a germ-plasm change, yet this shift appeared to be a nonadaptive response to an increase in temperature. Weismann seemed to be edging close to a neo-Lamarckian position of accepting the inheritance of indifferent acquired traits, but he steadfastly refused to jump into this potential quagmire.

Instead, Weismann insisted that under certain circumstances the environment would not only cause a direct impact on the developing soma but might influence the sequestered germ-plasm as well. As a mechanist and a believer in the specificity of anlagen, this made sense. Similar determinants in both somatoplasm and germ-plasm might well react to similar stimuli. Their only difference would be one of degree. Later biologists referred to this double impact as "parallel induction." The subsidiary hypothesis allowed Weismann to maintain his neo-Darwinian purity and still account for the diverse experimental and geographic phenomena associated with *phlaeas* and *eleus.* In my original essay, I had commented that "Weismann's ad hoc strategy, however, came at a cost to his basic distinction of the germ-plasm and soma."[6] On reconsideration, such a parallel induction, however, appears no different than his treatment of the impact of nutrition, which according to his theory could influence both the somatoplasm and germ-plasm.

Weismann had never been deeply concerned about devising mechanisms within his germ-plasm theory that would add new phylogenetic variations to the germinal line. He, of course, emphasized sexual reproduction or amphimixis as the biological mechanism for producing new combinations of already extant determinants, but this important innovation did not address the issue of long-term evolutionary change.[7] The fact that the germ was a physical entity always left him with the option of invoking random molecular change, which, like mutations today, could have provided his theory with a continual supply of molecular novelty. Such a device well suits the age of classical genetics and neo-Darwinism. As Ernst Mayr has pointed out, however, spontaneous change had a different meaning for Weismann.[8] In an age of mechanistic biological determinism, spontaneous change implied a nonphysical or vitalistic change. The whole weight of Weismann's mechanistic science bore against his considering randomness as a legitimate scientific part of an explanation.[9] Instead, he chose the only other route available to him. By 1892, and in greater detail in 1895, seasonal dimorphism obliged Weismann to invoke a parallel induction in the germ-plasm.

The case of *C. phlaeas* and *eleus* pushed Weismann's germ-plasm theory to its limits. The resulting ad hoc hypothesis of parallel induction revealed in a particularly forceful way the operation of a nineteenth-century embryological mindset. The determinants involved were at one and the same time subunits of transmission and the units of development. They responded to the external temperature in a similar way. Their growth and multiplication created the phylogenetic novelty, but in the final analysis this novelty was developmental not simply structural. In 1892, Weismann was explicit about the mechanics of germinal novelty:

We cannot avoid assuming that *the elements of the germ-plasm*—i.e., the biophors and determinants during their almost uninterrupted growth *are subject to continual fluctuations of composition [steten Schwankungen in ihrer Zusammensetzung unterworfen sind], and that these very minute fluctuations, which are at first imperceptible to us, are the primary cause of the greater deviations in the determinants, which we finally observe in the form of individual variations.*[10]

Far from being statistical changes based on submolecular events, these minute fluctuations were the result of the inevitable "inequality of nutrition" that supplied the growth and multiplication of the determinants and biophors. The fact that the inequality in nutrition rarely occurred when the determinants and biophors were still nested as part of the isolated germ-plasm did not disguise the identity of the separate "phenotypic" and "genotypic" variations. The similar developmental requirements of the parallel sets of determinants and biophors confounded the separations between germ-plasm and soma that were the initial hallmark of Weismann's science. By 1892, Weismann had already resorted to a competition and selection on the molecular level, but this was largely a competition in the somatoplasm during development. From this perspective, it does not seem a dramatic revision to develop in a much more refined manner the notion of germinal selection three years later when, as we have seen, he was pressed by Herbert Spencer and George John Romanes to explain the degeneration and disappearance of nonfunctioning traits. Experiments on the influence of temperature on the forms of *C. phlaeas* and *eleus* enhanced the neo-Darwinian nature of Weismann's germ-plasm theory. Parallel work by Frederic Merrifield and Frederick Dixey militated in the same general direction. The same kind of temperature experiments, however, carried out by other contemporary lepidopterists such as Max Standfuss and Emil Fischer were interpreted as support for the inheritance of acquired characters.

Parthenogenetic Reproduction of Drones

In 1892, as professor of entomology and curator of the Zoological Institute at the University of Halle, Otto Taschenberg wrote a lengthy history of parthenogenesis.[11] It was appropriate that he spent his career as an entomologist and specialized in Hymenoptera, especially the Cynipidae, for that order and family of insects played a significant role in understanding the boundaries and complexities of parthenogenesis. Taschenberg's summary conclusion to his history reflected on the past results of two generations of the science. "It is a long way from Dzierzon and Siebold to Hertwig and Weismann!" the professor of entomology proclaimed. He then continued,

> Almost a half a century has passed, since parthenogenesis was claimed and proven, and how it was advanced by theory, so from it new theoretical observations now proceed, in order to penetrate the secret of fertilization and heredity. The best, however, that it has brought us, is the fact that it has freed us from the proscription that fertilization is the necessary precondition for development of the animal egg, freed us also from the opposition of a male and female principle. This same process of parthenogenesis, . . . according to the views of the moment, concerns the adaptation to life's conditions and results above all in fertilization, that is, the mixing of the hereditary substances of two individuals or "amphimixis," where "a meaningful advantage for the preservation of species is stored."[12]

Forty-seven years earlier, Richard Owen had been inspired to coin the expression "parthenogenesis" in the first place.[13] As Owen defined parthenogenesis, it was a "Lucina sine concubitu," or birth without copulation. There is no question that the phenomenon became evident to him during his examination of the alternation of generations and from his conviction that the consequent sequence of separate individuals must be bound together by a vital spermatic force.[14] It was a singular way to introduce a new concept into biology, and Owen's explanation failed to survive in an increasingly mechanistically oriented science. Along with a growing recognition that a wide spectrum of variations and associated processes were associated with eggs, spermatozoa, and asexual buds, "Lucina sine concubitu" became by 1860 a widely documented phenomenon, providing a common challenge to all naturalists who wanted to understand sexual reproduction. Taschenberg dedicated the major portion of his historical account to detailing the accelerating discoveries about parthenogenesis in a variety of lower animals (and to a certain extent plants). His twenty-six-page bibliography alone vividly indicates to us how rapidly the study of parthenogenesis grew after midcentury. Siebold's elaborate study of drone production in bees and other organisms firmly established parthenogenesis as an important phenomenon. To be sure, there remained some holdouts, such as Rudolf von Wagner and Franz von Leydig, both of whom passionately insisted that parthenogenesis contravened the established order of nature by destroying the natural partition of life into two sexes, but this problem was easily dismissed by a younger generation.[15]

The Dzierzon System

The particular work done by Siebold and by Weismann's future mentor Leuckart on bees vindicated a recent claim made in the 1840s that drone bees arose parthenogenetically. The claim became central to Weismann's later studies in the 1890s. The episode, however, is worth following from the beginning. As suggested previously with Siebold and what soon became apparent with Leuckart, both found themselves independently focusing on the ideas of Johannes Dzierzon about the production of the three castes in a honeybee hive. Dzierzon, a Silesian minister and avid beekeeper, maintained on the basis of his extensive knowledge and experience with bees that queen and worker bees were the composite product of fertilized eggs and the food supplied to designated cells during the larval stage of development. In contrast, Dzierzon maintained that drones were the product of unfertilized eggs and a simpler diet.

In brief, they were the products of parthenogenetically produced yet viable eggs.[16]

Baron August von Berlepsch of Seebach in Thuringia was an avid beekeeper and one of Dzierzon's staunchest supporters. His arguments were of the standard natural history variety used by Dzierzon and other contributors to the specialized apiary journals of the day.[17] With the discovery of the micropyle in the egg of the honeybee in 1854, a new approach toward testing the Dzierzon system opened up. Each explored with the microscope to see whether the spermatozoa passed through the small opening through the outer membrane or chorion of eggs and whether it would determine the different castes.[18] The simultaneous discovery provided zoologists with an additional way of testing the Dzierzon system, and both Leuckart and Siebold arranged to visit the large apiary owned by von Berlepsch the following year to examine whether spermatozoa penetrated the chorion through the micropyle of the eggs deposited in the cells of the three castes.

Leuckart, who traveled to Seebach at the end of May, had a frustrating time despite the favorable season and the assistance of von Berlepsch. According to Siebold's later account, largely quoting directly from Leuckart's description in the *Bienenzeitung,* Leuckart failed to retrieve a sufficient number of freshly deposited eggs (needed within fifteen minutes of the actual laying). Only twice out of a hundred worker eggs did he find the penetrating sperm; not once did he find a spermatozoan penetrating in twice that number of future drone eggs. "Sie sehen," Leuckart candidly wrote, "das Resultat ist zweifelhaft." ("The result is questionable.") "Es sollte mich, im Interesse der Wissenschaft, unedlich freuen," he continued, "wenn andere beobachter in dieser hinsicht glücklicher wären als ich es gewesen bin." ("It would make me, in the interest of science, infinitely happy, if other observers would be more successful in this respect than I have been.")[19] In August, Siebold was luckier. Berlepsch could make available to him many more eggs, and Siebold developed a technique of pressing the egg to be examined with a coverslip, thereby forcing the yolk to flow out of the egg. It was a delicate exercise, but it was here he found spermatozoa, sometimes multiple spermatozoa, in fifty-two worker eggs; however, none could be found in the twenty-seven drone eggs he examined. Siebold was both lucky and persistent, and his host and an assistant were eager to provide him with the necessary material and help document Siebold's observations. Theirs was an outstanding example of cooperation between a professional scientist and a deeply involved amateur. Together, they verified the Dzierzon system.[20] Forty-plus years later, Weismann was to relive the experience with somewhat different results.

The Dickel Theory

Between Siebold's work on parthenogenesis and Weismann's involvement, other zoologists had questioned and tested the Dzierzon system. Hermann Landois, professor of zoology at the University of Münster, had in the 1860s, using a specially designed technique, transferred eggs from one kind of cell to another. His results convinced him that the selected food delivered by workers rather than the details of fertilization was the determinative factor.[21] The French comparative anatomist and zoologist Jean Pérez hybridized Italian queen bees with northern European bees. Because the former and latter races possess differently colored abdomens, the Dzierzon system predicted that the drones of the next generation should bear the abdomen color of the queen. Out of the 300 drones he examined, Pérez found very different results, including intermediate forms. His work provided an effective argument against the microscopic demonstrations of Siebold and Leuckart and roused a sympathetic hearing among beekeepers who were unfamiliar with microscopic techniques.[22]

The natural history of beekeeping, hybridization procedures, and the microscopy of fertilization vied with one another to supply a definitive evaluation of the Dzierzon system. Beekeepers, general zoologists, and microscopists used different explanatory language, and each confirmation and challenge was embedded in a unique set of understandings and specially learned manipulations.

It was an experienced beekeeper who in the 1890s again galvanized attention to the possible inadequacies of the Dzierzon system. Ferdinand Dickel (1854–1917) taught school in Darmstadt, was an avid beekeeper, and edited the local *Nördlinger Bienenzeitung.*[23] As reported in a biographical note, Dickel formally presented his challenge at the 1898 meetings of the annual Itinerant Gathering of German and Austro-Hungarian Beekeepers [*Wanderversammlung deutscher und österreichisch-ungarischer Beienenwirte*]. His argument was founded not only upon his own experiments but upon the experience of many of the outstanding apiarists of the day. It caused a sensation!

In the same year, Dickel also published his ideas in a seventy-page pamphlet entitled *Das Prinzip der Geschlechtsbildung.* It is to this document that we owe a detailed account of Dickel's early work and challenge to the Dzierzon system.[24] His attack was two-fold. First, he criticized the earlier investigations of von Berlepsch, Leuckart, and Siebold. He was particularly critical of Siebold's microscopical efforts that were carried out at the end of August on a hive that was forced to maintain an aging queen. As every beekeeper knows, queens under such conditions only lay drone eggs. Siebold's evidence, moreover, was limited to the examination of only twenty-seven drone eggs, and given the circumstances this hardly seemed sufficient.[25] Both criticisms reflected a deeper concern. Nature, Dickel argued, was subject to uniform laws, and when Dzierzon maintained that bisexual organisms could reproduce both sexually and parthenogenetically, he appeared to violate this uniformity. According to Dickel, the Dzierzon system "declares . . . drones are derived from unfertilized eggs and moreover stands an unquestionably well-established law of science on its head when at the same time it asserts: the contribution of the father during the act of reproduction prevents the rise of sons, since among bees the successful mating by the father determines that only female beings can come into existence![26] His is perhaps a clumsy sentence, but Dzierzon's apparent violation of common sense and Dickel's outrage against the insult to his manly pride is all too clear.

Second, Dickel presented a range of observations and a few experiments that militated toward the conclusion that worker bees rather than the *Anlagen* or germinal constitution of the egg determined the sex and caste of bees. He observed that worker bees spent considerable time tending to the cells of developing larvae—this would not be necessary, he declared, if the sex of the larvae was already determined. He noted that worker bees have four pairs of secretion glands, so one might suppose that they served different feeding functions. Moreover, the worker bee was sensitive to the different sizes of cells, which were known to give the cue for the normal feeding routine of the larvae inside. Dickel carried out exchange experiments of the kind Landois had done, both with the transfer of eggs and of larvae, and he came to the conclusion that given appropriate steps, one could force any of the three castes from the worker cells.

What emerged from Dickel's lengthy and discursive broadside was a system whose principles contrasted with Dzierzon's:

1. Dickel urged, with a nod to Weismann's germ-plasm theory, that the normal male or female zygote must be hermaphroditic, for it must contain both male and female *Anlagen*. He elaborated that the drone gametes carried the female *Anlagen* and the queen gametes carried the male *Anlagen*.

2. He argued that in their construction of different kinds of cells and because of their nutritional care of the larvae, the workers were the agents that through their secretions determined which of the two *Anlagen* became activated and which of three castes appeared.

3. It followed that the eggs from unmated and aged queens and worker bees could not be hermaphroditic. When such eggs developed, they necessarily led to abnormal drones regardless of which cells they were placed in and what nutrition the larvae received. There was thus, according to this principle, a contrast between the normal, functioning hermaphroditic drone and the abnormal or so-called false drone, which was unable to contribute in any way to the perpetuation of the hive. Dickel's distinction between normal and "false" drones was based on the logic of the situation and not on morphological distinctions.[27]

There existed a balance of a beekeeper's experience and a complex argument undergirding Dickel's replacement system. There can be no doubt that Dickel was a keen observer and was constantly in touch with the writings of other professional apiarists.

Critical of the academic zoologist encroaching upon his field of expertise, Dickel wrote sarcastically about Siebold's naivete in investigating bees. We also know, however, that Dickel had studied for "a few years" at the Darmstadt Technische Hochschule, where he had been exposed to zoology and chemistry under professors Gottlieb von Koch and Wilhelm Städel, respectively. This experience may have encouraged him to comment in a general manner at the end of his text about the importance of finding a unified theory of reproduction that would connect his ideas about sex determination in bees with that of higher organisms.[28] For this reason, he may have recognized that observations and arguments alone would not suffice and that he needed microscopical verification of his ideas, which he himself could not provide. In the summer of 1897, he contacted Rudolf Leuckart, who responded with encouragement but declined to become involved.[29] Dickel had apparently also contacted Weismann in early July of the same year, and this overture initiated a period of four years of collaborative research and thirteen years of correspondence. Unfortunately, we have only one side of the correspondence; nevertheless, the twenty-four letters by Weismann to Dickel, copied in Weismann's copybooks, reveal how very important this issue became for the Freiburg zoologist.[30]

The Dickel-Weismann Collaboration

At first, Weismann begged off involvement with Dickel's project to demonstrate the erroneous nature of Dzierzon's system, perhaps because of his ailing eyes which could no longer tolerate extended microscopical research.

Nevertheless, he encouraged Dickel to keep him informed and to send him copies of his relevant publications.[31] A year and a half later, Dickel obliged him with copies of his *Bienenzeitung* and the recently published *Prinzip der Geschlechtsbildung*. Weismann replied that he found some of its arguments quite convincing. "I consider the main point as proven," he responded enthusiastically, "it seems to me almost certain that the queen fertilizes the drone eggs. You have thus done science a great service, whether or not many of your further conclusions prove to be correct."[32] It also becomes clear that Weismann had not forgotten the issue during the intervening eighteen months, for he had put a student, probably Wilhelm Paulcke, on the project in the summer of 1898. Because it was late in the season and they could only obtain older eggs, this first study was unsuccessful. The second letter also gave Weismann the opportunity to arrange a collaboration with Dickel to examine freshly laid eggs (less than an hour old) under the microscope.

During the season of 1899, Weismann's student, now clearly identifiable as Paulcke, supplied Dickel with vials of conserving fluid.[33] Dickel placed freshly laid worker and drone eggs into separate, labeled vials and sent them back to Freiburg. Paulcke then used the most up-to-date techniques of serial sectioning and staining to examine the eggs for evidence of sperm penetration through the vitelline membrane, for the male pronucleus within the cytoplasm of the egg, and for aster formation at the time of its union with the egg pronucleus. This kind of evidence would be able to confirm definitively Dickel's conviction that the drone eggs were also fertilized. "The conclusion," Weismann pointed out in his next letter, "can only be made with the microscopical investigation of fresh eggs."[34] At the same time, Weismann had been in correspondence with Hans Reepen, an experienced apiarist and a student at the zoological institute in Jena, who must have read about Weismann's encouragement to Dickel in Dickel's *Bienenzeitung*. It seems that he wrote Weismann a letter of caution. For one thing, he informed Weismann, queens are known to err frequently by laying inappropriate eggs in worker and drone cells. Weismann would have immediately recognized that this common error might distort Dickel's findings.[35] Weismann also claimed to have recognized from the outset that "in fact, the theoretical labyrinths of Herr. Dickel are scarcely worthy of attention." Nevertheless, he intimated to Reepen that "since Siebold, Leuckart and Dzierzon we have proceeded in error. Probably not certainly"![36]

Thus, Weismann was clearly playing a double role. On the one hand, he encouraged Dickel and wished to collaborate with him because he recognized a diligent observer and experienced beekeeper who might well have put his finger on a serious problem with the accepted Dzierzon doctrine. On the other hand, he was not enthralled by Dickel's own system, which in a cumbersome manner merged multiple conjectures into a general system.

The collaboration indeed took place. By May 1899, Weismann reported that Paulcke's microscopical investigations were going well. After further exchanges, encouragements, and urging for ever fresher eggs, Weismann in July reported to Dickel some exciting news: "We are in the clear with the matter. The eggs from drone cells are in fact, as you have already long since concluded from your experiments, fertilized."[37] He explained that Paulcke had found sperm pronuclei on five different occasions and at the same time he had located the polar bodies of the egg. Both Weismann and his first assistant, Valentin Häcker, had verified the preparations; all that was needed was to see the astral and chromosomal figures of sperm and egg pronuclei fusion. Weismann ended his glad tidings with cheer and enthusiasm: "It pleases me especially that this theoretically very important question is now finally decided and I wish you luck moreover in removing this fatal error about the sex-determining nature of the spermatozoa from science."[38]

Any celebration that Dickel might have had, however, was short lived. The next day Weismann wrote again: "Unfortunately this morning I had to retract by telegram my glad tidings of yesterday's letter."[39] A new series of slides had indicated that the Freiburg team had confused the egg pronuclei with sperm pronuclei. No matter which arrived first, the letter of the 10th or the telegram of the 11th, one can well imagine the disappointment Dickel must have felt. He evidently tried to offer some explanation as to why Paulcke could not get consistent results, for by the end of the month Weismann had to explain that it was impossible for the two pronuclei to have formed a zygote before the egg was deposited. Further letters brought more discouraging news. Over fifty different series of observations failed to reveal any male pronuclei in drone eggs, but the control series on worker eggs demonstrated the effectiveness of Paulcke's technique of searching for the male pronucleus. Dickel must have objected on further occasions, but each time Weismann explained how they could not avoid the microscopical implications of Paulcke's work. With regret, Weismann reported on August 20 on the end results of the season's investigations: "This is entirely different, from what we had expected from your numerous and careful experiments."[40]

The outcome of Paulcke's microscopical investigations was important not only to Dickel but to Weismann as well. For a number of reasons, Weismann had felt strongly enough about the uncertainties of the Dzierzon system to commit yet another season and another talented student, the Russian emigre Alexander Petrunkevitch, to the project. He also took pains to lay out in a detailed fashion to Dickel what the issues were. Petrunkevitch would determine whether the drone eggs were fertilized; in fact, "that is possible only through the microsc. investigation." No matter what the answer to that question might be, Petrunkevitch would be unable to respond to the second question—the determination of sex—through the microscope. "What determines the sex here," Weismann penned, "is then *your* affair, and your investigations to date speak strongly for a determination by the worker."[41] Weismann went on to urge Dickel to serve science by continuing to provide the institute with fresh eggs. As Petrunkevitch's work got successfully under way, Weismann even held out the possibility that a microscopical difference in the development of worker-laid drone eggs and of queen-laid drone eggs, discovered by Petrunkevitch, might support Dickel's belief in a distinction between "true" and "false" drones.[42]

Despite the encouragement, Dickel must have been on his guard about a project he could not control nor fully understand. At one point, he tested Petrunkevitch's procedure by reversing the labels on the returned vials. Petrunkevitch quickly detected a problem and executed a hurried trip to Darmstadt to sort things out. Weismann's reaction to the episode is of interest: instead of scolding Dickel for the deception, he reasoned that "I find it very natural that you wanted to convince yourself whether we were truly in the position to unearth sound results and whether in addition this was really the right approach."[43] As events unfolded, Petrunkevitch's microscopical studies built up a decisive refutation of Dickel's claims that drone eggs were fertilized, and in the process he developed a new conserving fluid and found the subjects for both his doctorate and *Habilitationsschrift*.[44]

By mid-July, Weismann promised to publish a brief report on the results of three years of research. He was as good as his word: he wrote the report over his summer vacation and sent it to the *Anatomischer Anzeiger* in early October.[45] It was a straightforward report addressing the core of the Dzierzon theory—that is, whether drone eggs were fertilized. Appreciative lines scattered through the text made clear his debt. He commended Dickel for recognizing the weakness in Siebold and Leuckart's forty-five-year-old demonstrations and for his energetic pursuit of the problem through egg-swapping experiments, and he thanked Dickel for supplying both Paulcke and Petrunkevitch with fresh eggs laid at various stages during the maturation process. He pointed out, however, that among other demonstrations Petrunkevitch had found the sperm aster in 100 percent of sixty-two worker bee eggs at the stage of second polar body formation, but by way of contrast, he had found no evidence of the sperm nucleus or its aster in 272 drone eggs at the same stage.[46] Weismann turned his attention to Dickel's nutrition theory and argued that too little was known about the worker secretion(s) to make any claims about their effects. It was clear, however, that worker bees with their secretions could not counteract the deleterious effects on the hive of aging queens that produce only drone eggs. Finally, Weismann did not dismiss out of hand Dickel's concept of true and false drones. After all, he pointed out, wasps often produced different castes of males as well as females. In brief, Weismann did not debunk Dickel's complex theory; he treated it with respect but insisted that microscopical inspection had once again vindicated Dzierzon's original claim about the parthenogenesis of drones.

The ever-resilient Dickel responded in print with an attack on the trustworthiness of Petrunkevitch's method and new conserving fluid. He cited once again the well-known results of hybridization experiments and elaborated upon his secretion theory. The editor of the *Anatomischer Anzeiger,* Karl Bardeleben, an anatomist himself, permitted Weismann a short rebuttal and Dickel a final rejoinder.[47] The ground had now been tilled on both sides and could no longer bear fruit. Weismann explained to Bardeleben, who had apparently asked him about Dickel, that "he [Dickel] had also written *me* earlier often long letters about sex and reproduction and other problems, which all showed that he has an original and restless, active mind, which is, however, deficient in training in reasoning and above all in a knowledge of biological fundamentals."[48] The general tenor of this comment was to be repeated to other professional correspondents over the next few months.

Dickel, however, would not be satisfied. He responded to the publication of Petrunkevitch's dissertation with another defense, this time in the *Zoologischer Anzeiger.*[49] In 1903, he wrote a systematic account, complete with a short history of the controversy over Dzierzon and a synopsis of the experiments that supported his position. What was really at stake, besides the recital of his experiments, came out in a general comment at the outset: "In my opinion the determination in this theoretically so important a question cannot be achieved through the microscope, but only through experimental investigation." It was appropriate that Dickel by this time had turned to the *Archiv für der gesammten Physiologie* as a venue for his defense.[50] Its editor, the Bonn physiologist Eduard Pflüger, added a supportive note later in the year.[51] The physiologist Albrecht Bethe and the physicist Porphiry Ivanovich Bachmetjew also weighed in on Dickel's side.[52]

Weismann's reaction to this flood of support for Dickel was based on the fundamental principle that had been established in zoology in the 1870s when it was shown that *material* from the spermatozoa joined *material* from the egg in the process of fertilization. This principle underwrote the fundamental importance of nuclear microscopy. Some insight into Weismann's response to what he considered outside kibitzers can be gleaned from a letter he wrote to Buttel-Reepen, who by 1903 had received his doctorate in zoology in Freiburg and had returned to Oldenburg to continue his interests in the biology of bees. After thanking his former student for a reprint, Weismann continued:

> I would, however, prefer to dispense mixing in this apiarian confusion; at least for now. I could only repeat what I have said at the beginning. How do *Bethe* and *Pflüger* come to their non-zoological points of view? Bachm. [Bachmetjew], is a physicist, obviously doesn't know about the process of fertilization, otherwise he would be unable to operate with "half" fertilization.[53]

Weismann's side of the correspondence with Dickel dropped away for a while, but three letters toward the end of Weismann's copybooks terminated the exchange. Dickel apparently continued to send Weismann reprints, manuscripts, and many letters—Weismann wrote to Petrunkevitch in 1907 that he had received twelve long letters from Dickel that year alone. Finally, in 1910 Weismann insisted to Dickel that his personal conclusions had not changed and complained that he could no longer understand the beekeeper's arguments. "I want to suggest to you," he counseled in reply to two further reprints, "that you temporarily let these essays rest. If they contain good ideas and promote science, they will without doubt come into their own."[54] By the last letter in the series, however, Weismann had lost his patience. "I cannot ascribe any scientific value to the scientific fantasies you write me. . . . Please do not write me anymore."[55]

Lessons to Be Learned

Shortly after Weismann and his fellow lepidopterists had explored the effects of temperature changes on trait appearances in the butterfly, *C. phlaeas* and *eleus,* and certainly by the time Weismann last corresponded with Dickel about parallel induction of drone reproduction—in short, by the time he had set his germ-plasm theory to solving complex developmental and evolutionary phenomena—the new science of Mendelian genetics was under way. As we will see in Chapter 20, Weismann was aware of the new developments, but he sincerely believed in the power of his germ-plasm theory, or perhaps another comprehensive theory yet devised, to solve transmission as well as developmental and evolutionary phenomena. He was committed to a comprehensive theory to explain all three areas dominating biology during the second half of the nineteenth century: development, heredity, and evolution, in his opinion, could and should be solved together. The fallacy of such a commitment is apparent, as the synthesis masks the inadequacy of its independent parts.

20

Adapting the Germ-Plasm

1900–1914

Throughout the preceding chapters I have alluded to the need to fashion an intellectual bridge between the zoology Weismann was pursuing and twentieth-century advances. The latter appear to us to make more common sense in un-raveling the complexities of biology, but this is only because most of us have been brought up in the scientific schooling of the second half of the twentieth century and when we take for granted many of the assumptions of modern genetics, induction embryology, population biology, and neo-Darwinism evolution. It is only at the turn of the nineteenth century, pace Darwin, Mendel, and Flemming, and Roux and Driesch, that these modern views were generated and became the assumed foundations of modern biology. There was rarely a sharp divide between the biology of the nineteenth and twentieth centuries and people such as de Vries, Johannsen, and Bateson successfully straddled both. So must we.

Biology at the Turn of the Century

Mendelism

Historians are quick to point out that decade and century designations are tokens of human expedience, not matters of intrinsic historical significance. For the history of biology, however, 1900 is an important benchmark, and the decade following it was a time of demarcation between the evolutionary and embryological concerns of the nineteenth century and the developmental and genetic problems of the twentieth century. The year 1900 was, of course, the precise year when three academic plant breeders "rediscovered" Gregor Mendel's 1866 paper on the hybridization of peas. This work confirmed for later biologists the value of thinking in terms of algorithms when determining the significance of hybridization experiments. It provided new categories of analysis to apply to breeding outcomes: dominant and recessive traits, their segregation in successive generations, and their independent assortment when multiple traits were concerned. These breeding concepts would soon expand to include the terms alleles, homo- and heterozygosity, genetics, epistasis, genotype, and phenotype. Finally, the breeding programs that quickly arose succeeded because they were deemed experimental and statistical; they manipulated traits, both seen and unseen, in the crucible of their breeding pens, just as chemists had manipulated unseen molecules in their test tubes to produce observable and predictable results.

It remains an important question that I will take up in this chapter as to what extent Weismann could accept and accommodate to Mendelian genetics as it overtook biology after 1900.

Experimental Embryology

There was a great deal to 1900 as an important benchmark in the history of biology. Modern experimental embryology, commonly associated with the German term "Entwicklungsmechanik," began in 1884 when the young anatomists Wilhelm Roux (b. 1850), Oscar Hertwig (b. 1849), and Gustav Born (b. 1852) independently initiated experiments on the development of frog and sea urchin eggs. As described in Chapter 16, their work was a response to a paper by the famous Bonn physiologist Eduard Pflüger (b. 1829), who had proposed that experiments from his institute indicated that gravitation rather than internal influences determined the initial orientation of the developing frog. The authoritative experimental methodology of a physiologist threatened to trump the morphological histology of the anatomist on the central theme of the latter's research agenda—an explanation of organic form. Belonging to a later generation than Pflüger, each of the three anatomists must have been sensitive to a changing mood in anatomy and zoology. They certainly were convinced that Pflüger's rather simple conclusions could not be right. Each, however, must have instinctively recognized that fire had to be fought with fire. In 1884, each published reports of counter experiments that excluded the importance of external factors in the early orientation of the egg, its plane of cleavage, and the axis of the future embryo. As a consequence, the frontier of embryology—the ultimate science of organic form—became inevitably experimental.

By 1900, experimental embryology became a recognized specialty; it possessed its own journals,[1] it had clear institutional identification, and could boast certain heroic achievements associated with Roux, Hertwig, Hans Driesch, and T. H. Morgan, among many others.[2] Between its commencement in the mid-1880s and the turn of the century, its preoccupation had been largely with regeneration as a unique phenomenon. The experiments asked, What organisms? With what structures? At what stages does regeneration occur? Was there a substantive difference between organic and crystal regeneration? An equally important question, articulated by Roux, was whether regeneration was dependent or independent of surrounding developmental processes. Employing a different diagnostic criterion, I might identify the hundreds of regeneration experiments performed by scores of investigators as manipulations concerned with the disposition and specification of regenerated structures and manipulations consisting of excisions, incisions, and distortions of the developing organism. As these designations imply, such manipulations, for the most part, initiated a pathological process and were rather crude—Roux's attempt to destroy the nucleus after first cleavage in the frog with a hot needle and Driesch's shaking of the cleaving sea urchin egg, hoping some blastomeres would separate and develop, being classic examples of these initial ambiguous first endeavors.[3]

The large majority of these early experiments appeared designed to explore the pressing morphological issues of the 1880s and 1890s. They posed questions concerned with the relationship between the nucleus and cytoplasm, the specificity of the germ layers, the anatomy and phylogeny of the regeneration processes, the detection of relevant preconditions, such as stage, age, and size of the organism, and the comparative validity of mosaic and regulative developmental models, as well as the age-old question of whether development exhibited characteristics of preformation or epigenesis. Not only embryologists but cytologists, botanists, protozoologists, and medical pathologists were all involved in these explorations and attentive to these broad issues.[4]

As we move into the twentieth century, the centrality of these traditional morphological issues appears to recede in importance. New questions, generally associated with younger researchers and more directly with the fine mechanics of normal development, emerged. The new perspective helped fashion a new problematic concerned with position effects and tissue interactions—

which later became identified with terms such as "induction," "double-assurance," and by 1924 with Hans Spemann's "organizer." Spemann himself and Ross Harrison were exemplary in devising new experimental techniques and cutting through the thicket of descriptive regeneration questions to the significant causal questions about normal development.

The new century also appeared to focus more clearly on the cause and effect of single factors, often identified with very specific chemical and physical interferences. The Hertwigs might have led the way in 1887 with an important discussion about the influence of specific chemical agents on fertilization and cleavage, but they failed to pursue the matter.[5] In 1900, Curt Herbst discovered that calcium-free seawater separated the blastomeres of sea urchins, and he insisted on understanding the "formative stimulus" (*formative Reize*) involved. Immediately after the turn of the century, Charles Manning Child used narcotics to identify a parallel between a metabolic and an axial gradient in planaria. Although the data were fraught with problems of interpretation, his general approach signaled better than most the possibilities of searching for single causal, formative factors. The number of experiments on single factors jumped sixfold between 1902 and 1910, by which time Dietrich Barfurth in his annual review of regeneration research referred to them as "the Lieblingsgebiet [favorite subject] of today's experimental research."

We have already seen that the first phase of experimental embryology, with its emphasis on regeneration, necessarily shaped Weismann's mechanical concept of the chromosome. How he might adapt to the second and more functionally aware phase as it began playing a commanding role at the turn of the century remains an important question about Weismann and his students.

Endocrinology

There were still other trends in biology that emerged in importance around the turn of the century. The interpretation of evolution itself grew more varied and elaborate with each passing year. Few professional German biologists disbelieved in an evolutionary process at all, including the newly appointed professor of zoology at Erlangen, Albert Fleischmann.[6] Probably most had serious misgivings about the efficacy of the Darwinian mechanism of natural selection and cast about for other causes. I will return to the state of evolutionary thought in Chapter 21.

An area of physiology that began playing a role in morphology in the first decade of the twentieth century was endocrinology. Developed first in France,

Figure 20.1. Research colleagues in the summer of 1900 assembled in the garden of the institute.
Klaus Sander, ed. *August Weismann (1834–1914) und die theoretische Biologie des 19. Jahrhunderts. Urkunden, Berichte und Analysen.* Freiburger Universitätsblätter, vols. 87, 88. Freiburg: Rombach Verlag, 1985. Helmut Risler, "August Weismanns Leben und Wirken nach Dokumenten aus seinem Nachlass," p. 41.

particularly under the promotion of the French-American physiologist Charles Eduard Brown-Sequard, the first interest in endocrinology was medically oriented and promised eternal youth and cures for certain diseases by the administration of extracts from ductless glands. By the turn of the century, English-trained physiologists, particularly those associated with Edward Schäfer, a German emigré working at University College in London, began a program of analyzing the physiological "mode and site" of secretions. Adrenalin and then "secretin" from the mucous lining of the duodenum were among the first secretions whose origins and place of action were identified. One of the codiscoverers of secretin, Ernest Starling, in 1905 coined the term "hormone" to apply to these physiologically important secretions of ductless glands and in the same year spoke of an unidentified secretion of the thyroid that influenced growth and hence an organism's ultimate form.[7]

It is not surprising that Weismann remained silent about hormones and their potential contribution to development and more specifically to the determination of sex. Nevertheless, we have seen that he, Albert Kölliker, and Marie von Chauvin had performed forced metamorphosis in axolotl in the 1870s, and Weismann had developed a Darwinian explanation of the process. In 1913, J. F. Gudernatsch, an American physician working at the University of Prague, demonstrated the forced effects of thyroid extracts on the metamorphosis of tadpoles.[8] It is also relevant to point out that in the year of Weismann's death the American Edward Calvin Kendall isolated thyroxin as the growth hormone.[9] Too late to alter Weismann's biological understanding, these discoveries would influence zoologists of the next decade, such as Richard Goldschmidt and Julian S. Huxley, who were to exploit endocrinology for their own genetic and evolutionary purposes.[10]

More Nuclear Cytology

For Weismann's germ-plasm theory, the most important biological change to take place with the new century concerned the advancing details and interpretation of cytology. This field had always been close to the core of Weismann's research and theoretical productions. It had focused his initial energies when he had turned away from a career in medicine and taken up human histology; it had provided him with tools of observation as he became engrossed with insect development and with the gamete production in daphnids and hydrozoa, and it had become central to his germ-plasm theory, descriptions of reduction division, and the research program of his advanced students at his institute after 1880. It had been his involvement in cytology that had separated his career from that of the other great German evolutionary biologist of the nineteenth century, Ernst Haeckel.

We have seen the enormous advances that had been made in cytology between 1875 and 1890. Improvements included the microscope, the introduction of new stains and new ways to prepare specimens, and the increased effectiveness of the serial microtome. These had ushered in a new understanding of cell division, of fertilization, and of the nucleus and its structures. Weismann and the staff at his institute immediately took advantage of and contributed to such advances, as we saw in Chapter 19; they continued to pursue important research in nuclear cytology through the end of the nineteenth century. But technical and conceptual advances have a way of overtaking even the most robust of scientific programs. Two of the most contentious cytological issues that remained unresolved—even in their grossest outline—were the process of gamete maturation and the dynamics of reduction division. These are processes that were inherently more complex than the "copulation" of "pronuclei" with fertilization or mitotic division. Together they would soon be identified as "meiosis."[11] We have seen that Weismann and Hertwig had done battle in the 1890s over this tail end of the life cycle. I should add that many other talented microscopists had entered the same quest to detail the fine events of gamete production. Most important among them was Theodor Boveri. It is necessary to understand Boveri's work, for more than the work of any other cytologist, it built on and then altered what Weismann had previously done.

Theodor Boveri and the New Cytology

Unfortunately, Boveri has played a minor role in our interpretation of twentieth-century biology. Textbooks pay cursory attention to the depth of his investigations, and some of the very best professional books in the history of German and American biology—books I admire greatly—ignore or slight his contributions. Older appreciations by Fritz Baltzer, Jane Oppenheimer, and earlier still by Edmund B. Wilson and Hans Spemann leave no doubt about Boveri's importance, but their accounts need to be supplemented.

Coming to Terms with Boveri's Work

When he died in 1915 at the age of fifty-three, Boveri left posterity a literary corpus of fifty-six papers and monographs. This was not an extraordinary number for a major professor at a German university of the day, but it represented exceptionally solid research and had earned him a reputation as one of the empire's most resourceful cytologists. Measures of his contemporary reputation can be gathered from the calls he received to fill other positions. When Weismann retired in 1912, Boveri was offered the Lehrstuhl in zoology at Freiburg. Shortly thereafter, in 1913, the Kaiser Wilhelm Institute for Biology was established, and Boveri was invited to become its first director and the leader of the division of heredity theory and zoology. He politely declined the Freiburg offer after short consideration, and after half a year of serious deliberation, he turned down the second offer as well.[12]

During his professional career, Boveri had been closely affiliated with only three institutions. The first was the university in Munich, where he completed his doctoral work in Karl von Kupffer's anatomical institute and served as first assistant in Richard Hertwig's zoological institute. The second was in Würzburg, where Boveri lived from 1893 to his untimely death and where he served as professor of zoology and comparative anatomy. The third was the Zoological Station in Naples, where on eight separate occasions he pursued his research.[13]

Boveri's published legacy, however, is not easy to plumb. The Anglo-American reader will find that very few of Boveri's works have been translated into English. For those with a command of German, his cytological papers are difficult to sort through with their concentration on only two groups of organisms, the nematode, *Ascaris megalocephala* (*univalens* and *bivalens*), and a few species of sea urchins.[14] Furthermore, the titles of Boveri's papers provide few signposts as to where to begin and how to work through the trends in Boveri's thought.

Boveri's six monographic *Cell Studies* (*Zellenstudien*) would appear to provide a natural point of entry.[15] However, the scholar quickly realizes that some of these classical works are expanded and restated conclusions rather than points of departure. Several general papers, such as a review of fertilization (1892), a formal address on the organism as a historical being, and a lengthy statement about malignant tumors (1914),[16] provide guides of a sort, but the historian can only track Boveri's developing thoughts by splicing all of these monographs together with smaller, generally antecedent papers that appeared in local journals and are somewhat chaotically scattered among the larger benchmarks. One soon discovers that Boveri repeatedly returned to certain general themes throughout his career, such as "fertilization," "merogony," "polyspermi," "chromatic elements," the "relationship of the nucleus and cytoplasm," and "early cleavage." He seemed unable to let go of a theme even as he moved on. Nevertheless, it is possible to detect a major shift in his work, which was particularly relevant to the advance of early twentieth-century biology and to the fate of Weismann's program in particular. It is worth placing this shift within the context of Boveri's commitment to cytology, for it would be appropriate to consider Boveri as the natural heir to the spirit, if not the details, of Weismann's science.[17]

For this purpose Boveri's cytology needs to be understood on at least three levels: first, for its exacting details, second, for the substantive issues that motivated his early microscopy, and third, for Boveri's methods and ontology. I first examine these levels in Boveri's work before 1900. I will then present Boveri's important work on polyspermi in the sea urchin of 1902 to 1907. Finally, I will draw some general conclusions about Boveri and the transformation of biology, which speaks very directly to Weismann's career during the last decade of his life.

The Cytological Details

From the time of his Habilitationsschrift in 1887, it is clear that Boveri was a meticulous practitioner of cytology. He commented candidly, though never petulantly, on the failures in the techniques and illustrations of his contemporaries. Thus, as he embarked on his career with an examination of the formation of polar bodies in *Ascaris,* he challenged some of the most gifted senior microscopists of the day who had already written important monographs on the subject. He found the illustrations of Anton Schneider and Moritz Nussbaum unclear in key details and often based on pathological specimens. He noted that Jean-Baptiste Carnoy and Eduard van Beneden had failed to distinguish between the univalent and bivalent forms and so remained powerless to reconcile their conflicting observations. Furthermore, he noted also that van Beneden had overlooked the fact that an alcohol and weak acid solution failed to kill *Ascaris* eggs immediately, so their development and the subsequent observations thereof remained problematic.[18]

Although Boveri pointed out these and many other deficiencies on the part of contemporaries, he does not appear to have been an enfant terrible among cytology's rising stars. It is equally easy to find him extolling the accomplishments of the very same researchers he had criticized a few lines earlier. Thus, he did not refrain from commending both van Beneden and Carnoy, whose works "are furnished with a detail and a wealth of illustrations, like few other works in the cytology literature."[19] A year later, as Boveri published his second *Cell Studies,* which focused on fertilization and cleavage, he again lavished praise on van Beneden's study of 1883, and "happily admitted how much stimulation and learning he had drawn directly from the same [fine] qualities," of research he had just listed.[20] There can be no question that van Beneden's demonstration that the sperm and egg nucleus in *Ascaris* failed to fuse before first cleavage significantly guided Boveri's subsequent research. In turn, Boveri's work reflected the same exactness he expected of others. His studies of fertilization, polar body formation, and merogonic hybridization—and later on the random distribution of chromosomes at first cleavage—displayed a penchant on his part to record every detail in appropriate order that he possibly could. He was particularly observant about varying the focus of view so as to be able to create a three-dimensional image of the object. These traits spoke openly not only to his style as a microscopist but his belief that cytology was a science about function as well as form. Serial details provided a chronological sequence, which in turn suggested a process. There was no other way, Boveri believed, that one could come to understand the full mechanics of karyokinesis, gamete maturation, or fertilization.

The Substantive Issues

Observations, no matter how expertly made, have little consequence without a focus, and Boveri pursued a number of them in the form of general themes. From 1886 until the turn of the century, a period that included his first four *Cell Studies,* Boveri concentrated on the events of fertilization. He called, in fact, for a "Befruchtungslehre."

Befruchtungslehre included activation—stimulation, or in modern terms fertilization—of the egg, which had always been an important subject in biology since the time of the Hippocratic writers and Aristotle. In modern times, the 1875 demonstrations by Oscar Hertwig that the "pronuclei" of male and female gametes fused initiated our modern understanding of sexual reproduction. Hertwig's discovery revealed that a union of parental materials constituted the essence of fertilization. By the late 1880s, after the general pattern of karyokinesis had been recognized and cellular structures had been identified in far greater detail, Hertwig's observations no longer sufficed as a comprehensive doctrine of fertilization. Moreover, his conclusions about the composition of the hereditary substance were heatedly disputed. It was necessary to unravel the fine details of the process and establish the function of polar body expulsion; to understand better the relationship between the nucleus and cytoplasm; to figure out the origin and action of the achromatic elements of karyokinesis, such as the astral body and individual spindle fibers; to distinguish and track the sperm and egg centrosomes; to find consistent patterns in the condensation, movements, and numbers of chromosomes in a range of organisms; and finally, to determine whether those threads of chromatin—that is, the chromosomes—were secondary or primary elements of the cell. Boveri reflected this expansion of issues in his review on fertilization in 1891: "Briefly said: the report on fertilization will provide a summary of the entire developmental cycle of sex cells from the fertilized egg to the mature gametes (eggs and spermatozoa) and will record all the specific events and aspects, which are consequently demonstrable."[21]

Boveri was not alone in his struggle to understand fertilization on a more basic level. Beneden, Carl Rabl, Oscar and Richard Hertwig, Valentin Häcker, and others were pushing into the same terra incognita. Moreover, Boveri at first appeared more reluctant than most to extend his discussions beyond the testing and retesting of his observations. His focused results, which would eventually congeal into far-reaching conclusions, at first seemed secondary to the meticulously organized information gathered through his microscope. Above all, during the last fifteen years of the century, Boveri was most concerned with the multiple facets of fertilization. First, he sought to understand the function of the polar bodies and the action of the spindle. Second, he intensely investigated the role of the centrosomes in early cleavage. Third, and most important, he analyzed from several angles the nature of the chromosomes. The modern reader must bear in mind that the cytological details worked out at the end of the nineteenth century quickly attained the status of elementary information for biologists within a generation, yet they required intensive research and meticulous presentation in order to be generalized and generally accepted. Let us briefly follow each of these three areas that captured Boveri's attention.

1. THE FUNCTION OF THE POLAR BODIES. When Boveri started investigating the production of polar bodies, Oscar Hertwig and M. A. Giard had already described them as rudimentary cells and the product of an indirect cell division.[22] Weismann and van Beneden had championed two major competing theories concerning their function. As we have seen in Chapter 14, Weismann claimed that the first polar body represented the expulsion of ovogenetic germ-plasm and the second provided the organism with a mechanism for eliminating one-half of the hereditary material before fertilization, at which time that material must be doubled. This short-lived theory reflected his materialistic commitment and his phylogenetic construction of the chromosome. He later revised his ideas by considering both maturation divisions as part of the process of reduction, with the production of the second polar body as the act of reducing from what would soon be called by the Mendelians the "diploid" to "haploid" state.[23] In contrast, van Beneden, among others, had argued that the expulsion of the polar body (he did not consistently consider two or three) was a way for the organism to eliminate male elements before the egg's "rejuvenation" by a new male element from the sperm. His commitment was more to an atomistic understanding of the nuclear chromatin and to a hypothesized consequence of their union—"rejuvenation." The term itself was often used in the 1880s, but like the doctor's soporific drug in Molière's *Le Malade Imaginaire,* it was a hypothesized cause identified and named after a qualitative effect. What was needed

was the display of a precise and continuous series of illustrations that showed how the chromosomes were distributed during polar body expulsion.

This Boveri provided by distinguishing the two forms of *Ascaris* and illustrating with an unrivaled exactitude an unbroken sequence of the karyokinetic events. He concluded that the process consisted of two cycles of normal division and a random assortment of chromosomes that were uninterrupted by a resting stage of the nucleus. Although it was not the focus of his first monograph, Boveri lent credence to Weismann's earlier suggestion about reduction division by showing that polar bodies and the mature egg each contained half the number of chromosomes of the immature egg. Boveri's subsequent conclusion, however, differed from Weismann's in that he argued that the reduction must have occurred before the two maturation divisions. How and when this happened, Boveri could not say, but it was characteristic of his research style not to let suppositions jump too far ahead of his demonstrations.[24]

2. THE ACTION OF THE SPINDLE AND ROLE OF THE CENTRO-SOME. Boveri devoted much of his early cytological research to examining the achromatic elements involved in the karyokinetic process of polar body expulsion. In his second *Cell Studies,* he developed, uncharacteristically, an idiosyncratic theory of the "archoplasma," which detailed a halo around the penetrating spermatozoon and described it as the center of the achromatic material used by the centrosomes of the spermatozoon to form the spindles of the egg's two maturation divisions.[25] He soon discarded the theory, but his concern about the origins and function of the spindle fibers represented part of his exploration for a mechanism of chromosome assortment. Instead of a casual segregation, he found a carefully orchestrated double-halving of the egg's chromatic elements and a precisely regulated mechanism guided by the sperm centrosomes that allowed the halves of each dividing chromosome to become attached to achromatic fibrils and drawn to opposite poles. The randomness in the segregation process was the result of a very precise binary mechanism in the egg and polar bodies and for the cleaving blastomeres to share equally in chromosome numbers. As early as 1887, Boveri was alert to abnormal cases of dispermic eggs that labored under three or four centers of attraction. The binary segregation mechanism still functioned, but the extra centers of attraction disrupted the normal equational distribution of chromosomes—a finding that Boveri would later exploit in his famous multipolar research.

3. THE NATURE OF THE CHROMOSOMES. Jonathan Harwood has expertly detailed the German (and American) communities of geneticists who clashed in the 1920s over whether, in addition to the chromosomes, there existed hereditary elements in the cytoplasm. It was a confrontation between a reigning "Kernmonopol" (monopoly of the nucleus) and the revisionist champions of cytoplasmic inheritance.[26] At the turn of the century, hints of this controversy were beginning to take shape, but the issues remained burdened by nineteenth-century terminology and preoccupations with the antagonism between preformation and epigenesis. Even the most ardent supporters of a chromosomal theory of heredity, such as Weismann and his followers, could not be construed in the terms of the Kernmonopol of the 1920s. As we have seen, the outcome of determinant expression was conjectured as an interplay of physiological as well as chromosomal forces. The same can be said more emphatically for Boveri. In fact, more than anyone, he changed the terms of the debate.[27]

From the outset of his work, Boveri concentrated on the chromosomes for a number of reasons. He was deeply concerned about their mechanical distribution as he sought to understand fertilization. He also became interested by their diminution in size in *Ascaris* in all but a few blastomere lineages during early cleavage and later development. His description of this diminution phenomenon became a classic example of Boveri's constant consideration of the influence of the cytoplasm as well as the nucleus in development.[28] By the following year, the same year, incidentally, in which Waldeyer coined the term "chromosome," Boveri signaled his interest in the constitution of these "chromatic elements"—he often called them "Schleifen" and "Stäbschen."

Chromosomes, however, are transitory structures under the light microscope. They disappear at resting stage and reappear in preparation for karyokinesis only after a process of condensation from a network of chromatin and achromatic threads. Although van Beneden had demonstrated the independence of parental nuclei during first cleavage, his conclusion did not automatically transfer to the separate integrity of the parental chromosomes. Serving as a rejoinder to this transitory behavior, both Boveri and Carl Rabl independently noted that after the chromosomes had disappeared with the completion of the telophase, they often reappeared in daughter cells with the same size and in the same configurations. They therefore felt justified in attributing a certain persistence and individuality to their existence. By 1890 and his third *Cell Studies,* Boveri was also keenly aware of how important the constancy of chromosomal number is in both the fertilization and reduction processes. He recognized that their number is identical in both parents and believed that the tetrad formation seen before first maturation division was a product of two divided chromosomes rather than four independent ones.[29] These regularities also suggested their persistence and individuality. Nevertheless, it is equally clear that during these early years Boveri conceived of chromosomes in the same fundamental manner as Weismann before him. Without invoking the trappings of the latter's germ-plasm theory or the details of his explanation of reduction division, Boveri believed at this time that each chromosome contained the anlagen of all somatic traits (though not their variations) and felt that their persistence and independence were structural rather than functional attributes.[30] Boveri had clearly adopted Weismann's fundamental distinction between germinal matter and the soma.

During these years, Boveri also performed merogonic hybrid experiments on sea urchins, which convincingly but not definitively argued for the importance of the nucleus of the sperm vis-à-vis the protoplasm of the egg in determination.[31] While Boveri's research provided strong arguments for the persistence of the chromosomes and their importance in heredity, he tempered his findings by also investigating the role of the cytoplasm in early development. As mentioned previously, he attributed the diminution of *Ascaris* chromosomes to a cytoplasmic gradient, and in 1901 he performed "elegant" experiments to demonstrate the importance of a horizontal stratification of the cytoplasm of *Paracentrotus* zygotes in blastomere determination after third cleavage.[32]

Methodological Concerns

During the last fifteen years of the century, Boveri was explicit about certain methodological advances and commitments in cytology. To a not unimportant degree, he shared similar commitments with Weismann.

Boveri's deft exploitation of the most recent tools and techniques of cytology extended his observations far beyond what had been possible in the mid-1880s. Most nuclear cytologists took advantage of these advances.[33] In addition Boveri was ultrasensitive to the unique characteristics of the two organisms he repeatedly investigated. *Ascaris,* which had been studied earlier by Anton Schneider and van Beneden, presented an ideal organism in which to investigate fertilization. Its eggs were small relative to its chromosomes, which were large and few in number. The parental chromosomes, as van Beneden had pointed out, remained widely separated during early cleavage and maturation of the egg, and the nuclear envelope disappeared early in karyokinesis. Furthermore, the achromatic structures were easy to follow from one division to the next. "All of these circumstances," Boveri explained, "make the egg of *Ascaris megalocephala* a research object that as yet has no equal."[34] The various species of sea urchins, which Boveri examined when he worked in Naples, were readily available, and the fertilization was easy to control. Because of their transparency, the eggs could be easily observed and followed through the fertilization and cleavage processes. Boveri followed the Hertwig brothers in using it for investigating merogony, polyspermi, and other experimentally induced and naturally occurring abnormal conditions.[35]

During these years, Boveri relied largely on the cogency of comparative anatomy for his general conclusions. In fact, Baltzer has pointed out that throughout his life Boveri continued to publish fine monographs in this field, particularly on *Amphioxus*.[36] As he became involved with cell studies and the process of fertilization, he carried the comparative method to the microscopic level. For example, van Beneden's demonstration of the nonfusion of parental nuclei, mentioned earlier, gained for Boveri the "value of a general law" when the same circumstances were shown in many other organisms.[37] Two years later, he again combined the results of his own investigations with those of other cytologists on eleven different species to conclude, through the comparative method, what soon became known as the law of the constancy in the number, size, and form of chromosomes of males and females of the same species.[38] The comparative method was also applied in working out the details of events in a single species. As with descriptive embryology on the histological level, Boveri preserved hundreds of specimens on the cellular level and from each took dozens of sections, inferring from them in the aggregate a sequential, three-dimensional picture of change.

From the beginning of his professional career, Boveri anticipated an important role for experimentation in cytology. Experimental embryology in general (Boveri rarely used the term Entwicklungsmechanik) and the work of the Hertwig brothers of 1887 in particular taught him the possibilities of manipulating the fertilization process. Nuclear cytology moved more slowly than general embryology into the experimental mode. It was simply harder to devise meaningful manipulations at this level of organization, and it was only with the microtechniques developed by Boveri's student, Hans Spemann, that the cleaving egg was handled with precision. Nevertheless, Boveri was constantly alert to the value of naturally occurring abnormal cases among the multitude of specimens he studied.[39] It was for this reason that merogonic and polyspermy phenomena were of such interest to him from the outset. They were nature's experiments writ for the benefit of discerning investigators.

Driving all his research from the very beginning was Boveri's commitment to the physical details of visualized events, to the *mechanics* of structural movements and changes that could be seen through the light microscope. This made karyokinesis such an interesting process. Boveri insisted on illustrating the full sequences of chromosomal and achromatic activity in polar body production and first cleavage. He pursued what he called the "division mechanics" ("Teilungsmechanik") of these events and was convinced that they could not at this moment of scientific understanding be reduced to chemical and physical processes. "Even the last parts of the cell, which we can demonstrate as definite elementary forms," Boveri insisted, "are still organized structures, which as whole entities manifesting life, ridicule every explanation through chemico-physical forces."[40] Thus, Boveri's sought-after mechanisms should not be confused with contemporary reductionism. Cells, chromosomes, and centrosomes were complex, organized individuals that had to be understood on their own terms.

As a consequence of these methodological considerations, Boveri was emboldened to think of his work in nuclear cytology as a pursuit in a new science. It was the science of "Befruchtungslehre," which was neither Entwicklungsmechanik nor *Entwicklungsgeschichte;* it emulated physiology, chemistry, and physics by being experimental, but it could not be reduced to them. It was with this in mind that he persistently pursued the "constitution of the nucleus" by investigating "the history of the chromatic elements." Even during this period, Boveri recognized that there was another related process besides "Befruchtung" that required clarification, which was "Vererbung" or heredity, "the building up of a cell complex of definite form that reflected the parents."[41]

Polyspermi Experiments

What I have described so far takes us to the end of the century, but Boveri's conclusions, particularly with respect to heredity, were changing. The decisive line of research that demonstrated the material connection between fertilization at the beginning of the life cycle and the transmission of parental traits at the end of the cycle may be found in Boveri's work on polyspermi in sea urchins.[42]

Polyspermi, the penetration of multiple spermatozoa into a single egg, is an abnormal but not rare condition of fertilized sea urchin eggs, which was first described by Emil Selenka in 1874. It quickly became a subject of study for Oscar Hertwig and Hermann Fol. A decade later, both Hertwigs began investigating the physical and chemical conditions that induced multiple fertilization and thereby demonstrated the possibility of producing the condition on demand.[43] Hans Driesch experimented with the process in the 1890s, forcing Boveri to think more critically about his belief in the anlagen equivalency of chromosomes. When in 1900 Curt Herbst developed a technique for disassociating the early blastomeres with calcium-free seawater, Boveri became more interested in the dispermic cases as a way to understand both the origin of the multiple centrosomes and the fundamental nature of the chromosomes themselves. A morphological condition could be used as an experimental opportunity. With his visit to Naples during the winter of 1901–1902, Boveri's focus shifted fully to the chromosomes, and he performed a set of experiments that led to his classic paper on "multipolar mitosis." We should notice particularly the extension of his title: "On Multipolar Mitosis *as a Means of Analysis of the Cell Nucleus.*"

Briefly described, his experiments took advantage of the fact that in the dispermic condition there exists a third again as many chromosomes. Furthermore, there are three or four rather than two poles of attraction, which eventuate in three or four blastomeres at first cleavage. The key to the experiment for Boveri was recognizing that the distribution of chromosomes to the cleavage blastomeres would be random and that a guarantee no longer existed that a bipolar karyokinetic mechanism would distribute to any of the blastomeres one of each homologous chromosome. It was a matter of statistical analysis to work out the probabilities of the random distribution of the chromosomes after the egg cleaved into the first three (or four) blastomeres. After cleavage occurred, Boveri would disassociate the blastomeres from one another, isolate each blastomere, and follow their development to the larval stage. The outcome was close to the predicted number of normal and abnormal larvae and to the anticipated range in severity of the abnormalities. The results could only be interpreted in terms of the individualized contributions of each chromosome to the developmental process. A full complement of half the homolog pairs alone remained sufficient to produce normal development. (It is worth noting in passing that Boveri's statistical analysis came within a year of the recognition among biologists of the statistical power of Mendel's paper—but more on Mendel in a moment.)

Boveri published his paper in 1902, presented a rectorial address on the subject at the end of May 1903, developed in detail his arguments and their consequences in an important monograph in 1904, and presented the definitive analysis of his achievement in 1907 in the form of the sixth and last of his *Cell Studies*.[44] Collectively, these papers demonstrate how Boveri had moved away from an analysis of fertilization to the study of heredity. In other words, he had moved from a "Befruchtungslehre" to, in his terms, an *"Embryonalanalyse* der Chromosomen"—to an embryological analysis of the chromosomes. What flowed from this reorientation in general and from his polyspermi experiments in particular were some remarkable conclusions that were to change the face of biology.

Experimental Results

Above all, Boveri effectively redefined the chromosome in a way that provided the basic paradigm for the entire twentieth century. In 1887, by his own admission, he had believed in the Weismannian doctrine of the essential equivalence of all chromosomes of the same organism. He may have changed his mind about this in the 1890s, but the polyspermi experiments demonstrated beyond doubt how untenable Weismann's view had become.[45] They provided a demonstration that the chromosomes were independent ("Selbständig") in an expanded meaning of that term so that their individuality, that is, their developmental impact, was now seen as an expression of their different genetic values. It followed that a full complement of all the individual types of chromosomes was necessary for normal development.

Part and parcel with his redefinition came a more subtle understanding of tetrad formation and reduction division. Up to 1901, Boveri had recognized the cytological reality, though not the origin, of the former and had deferred an explanation of the latter.[46] Now, drawing upon the work of American cytologists Clarence McClung, Walter Sutton, and T. H. Montgomery, he realized that before the first maturation division the two homolog chromosomes synapsed and then each split longitudinally to form four "chromatids," as McClung called them. The four chromatids, making up each tetrad, were then randomly distributed during the following two uninterrupted karyokinetic divisions. Sutton, more than any other investigator, provided Boveri with the morphological demonstration that supplemented his functional experiments:

How excellently these morphological results complement my physiological ones, goes without saying [*liegt auf der Hand*]. After these facts I believe we may consider this as certain, at least in many nuclei, that a differentiation of the chromosomes to distinctive forms takes place, so that in their totality they only represent in all cells the befitting [*zukommenden*], and for them necessary, traits.[47]

The explication of tetrad formation fitted in with Boveri's new definition of the chromosomes and essentially ended the protracted and often contentious debates about the timing and mechanism of the reduction process. This result was immediately reflected in the general volume (*"Allgemeiner Theil"*) of Eugene Korschelt and Karl Heider's monumental text on comparative embryology. These two authors had catalogued in an elaborate fashion the debates and theories of reduction division, but in the last two paragraphs of their 730-page volume they managed to mention Boveri's recently published paper. They ended their volume with a tribute to his achievement: "With justice Boveri points out in his contribution that facts now take the place of suppositions. Hence forth the way is also indicated, in which we may hope for the further advance of hereditary theory."[48]

Redefinition of the chromosome and an understanding of tetrad formation were both outstanding revelations, but there was much more to the polyspermi work. Boveri included in his first paper a not-so-cryptic footnote that read, in part, "these and related problems, as well as the relevance of this to the results of botanists in studies of hybrids and their descendants, will be discussed separately."[49] His remarks preceded by a few months a similar but more explicit passage by Walter Sutton, which concluded his study of chromosomes in the lubber grasshopper, *Brachystola magna*. In the pattern of pairing and separation of paternal and maternal chromosomes, Sutton saw what "may constitute the physical basis of the Mendelian law of heredity." He further remarked, echoing Boveri, "To this subject I hope soon to return in another place."[50] Boveri and Sutton appeared to have exchanged letters about their respective ideas, and in April 1903 Sutton was good to his published word and wrote a detailed paper on the connection between Mendelism and reduction division.[51] Boveri followed with a four-page discussion of his own on chromosomes and Mendelism in an address at a plenary session of the German Zoological Society in May of the same year; then he followed this with a somewhat more detailed account that became part of an important monograph on the constitution of chromosomes. The work was completed in July 1903.[52]

The two authors differed in perspective, however. Sutton presented a morphological argument derived from his study of the size, form, and shapes of chromosomes. Boveri, although dependent in part on Sutton's work, concentrated on the functional analysis he had pursued in his polyspermi and other cytological studies. Today, depending on our national orientation, we continue to refer to the Boveri-Sutton or the Sutton-Boveri law, but this minor parochialism should not overshadow the great importance of bringing the mechanics of nuclear cytology into line with the onset of Mendelian research.

Boveri also addressed biology's contemporary debates over preformation and epigenesis from his new perspective. Because he had followed the separated blastomeres to the larval stage, he was in a position to posit with precision what abnormalities were due to the absence of chromosomes. (His merogony experiments had never given him that certainty.) Moreover, he argued that the chromosomes played a formative role in development only with and after the appearance of the mesenchyme cells prior to gastrulation. The cytoplasm appeared to guide development up to that point. Such demonstrations, it seems to me, put an end to the preformation or epigenesis question and set up a different antithesis concerned with the function of the cytoplasm and nucleus at specific stages and with respect to specific qualities. The echinid egg for Boveri, borrowing an early expression from Driesch, was nothing less than "a harmonic equipotential system," but only through the first two cleavages; the situation changed thereafter. By the turn of the century, Driesch had moved beyond this inherently mechanistic model and resorted to the "autonomy of living processes" affirmed in the teleological principle of the entelechy.[53] By way of contrast, Boveri had demonstrated that the fine structure of the cytoplasm and the complexity of the chromosomal system could together steer a way through the thicket of uncertainty toward a mechanistic explanation of differentiation.[54]

Furthermore, the polyspermi experiments effectively ended another contentious debate. During the previous decade, Boveri had been critical of the concept of qualitatively unequal nuclear division.[55] His redefinition of the chromosome and his sensitivity to the possible interaction of chromosomes and the cytoplasm empowered him to reject the Roux-Weismann theory. "It appears to me," he judicially explained in his characteristically understated manner, "that the quite peculiar interaction of the cytoplasm with its simple structure and differential division and the nucleus with its complex structure and manifold total multiplication may still achieve what Weismann and Roux attempted to explain with the help of differential nuclear division."[56] Roux quickly assented; Weismann did so only in part at the end of his life.[57]

Finally, Boveri's work shed light onto the perplexing problems of mosaic hybrids and parthenogenesis. Both phenomena had become explosive topics by the end of the century. Boveri could now explain them in terms of his new chromosomes. The mosaic hybrids, such as found occasionally in drone bees and by Boveri in *Echinus,* now could be understood in terms of asymmetric fertilization of the egg at the two- or four-celled stage or even in later development. "In this way, the most diverse mixtures of male and female characters could result, as has been actually observed."[58] Boveri delayed speaking of parthenogenesis until his address to the zoological society in 1903, but here again, regardless of whether one asked about merogonic, or artificial, and obligate parthenogenesis, it was his recognition that once the developmental process was initiated, all that was needed for complete development of an egg was one normal component of chromosomes. In cases of facultative parthenogenesis, Boveri argued that the second polar body might be reabsorbed and that its nucleus might fertilize the egg. Boveri's achievements could not be ignored. The chromosome, as he interpreted it, became the foundation for a new physical basis of heredity. Tetrad formation and two maturation divisions offered in combination a readily acceptable mechanical picture of the reduction event so important to Weismann.

Cytology and Mendelism

Boveri, more than any other, reshaped our understanding of these processes just at the time when they could be effectively integrated into the new wave of Mendelian research. While advancing new concepts for the new century, Boveri at the same time helped diminish the importance of the older, nineteenth-century issues. The endless debates pitting preformation and epigenesis and the specific claims by Weismann and Roux about qualitative nuclear division could be eliminated from serious scientific concern. Driesch's claim to have experimentally demonstrated the entelechy was refuted by even more precise experiments. Even the nineteenth-century preoccupation with the life cycle of individual forms and phenomena such as parthenogenesis, seasonal dimorphism, merogony, and mosaic hybrids now found a new measure for their understanding. It would be an understatement to claim that Boveri's conclusions about the cytoplasm and chromosomes brought biology into the twentieth century.

Weismann and the Weismannians Respond

Prologue: Weismann's Personal Trajectory

The decade between the rediscovery of Mendel's work and the establishment of classical genetics is to some extent forgotten years in the history of biology. It is not that historians and biologists are unfamiliar with the basic out-line of the period—far from it. But the decade, saddled between two major episodes, continues to provide an awkward disjunction between our accounts of the early Mendelians and biometricians on the one hand, and the onset of classical genetics typified by the Morgan school on the other. What is missing is a convincing account of how biologists traveled, or failed to travel, from the particulate view of heredity common at the turn of the century to the marriage of hybridization experiments and chromosomal cytology by 1912. To put it another way, this period comprised years when hybridizers sought to fathom the limits and implications of Mendel's generalizations. They wit-nessed an intellectual trip from a view of continuous, embryologically deter-mined traits to a view of discontinuous, chromosomally transmitted ones. They found in sex determination a phenomenon that seemed amenable to resolu-tion through the new cytology, and as I have indicated above, Boveri's cyto-logical work, which came at the beginning of this period, formed a necessary step along the road. It was also a period during which American biologists began to contribute in substantial numbers to the European-led discussions of heredity, development, and evolution.

Weismann turned sixty-six within a month after the turn of the new cen-tury. He would live for another fourteen years. During most of this period he continued to be an active though senior participant in certain aspects of the biological dialogue. His institute thrived as a research center for general zo-ology, and he continued to teach and direct research projects until his retire-ment in 1912. He had the good fortune to be assisted by Johann Mayer, who served as custodian of the institute at the beginning of the decade to long after Weismann's retirement.[59] At the same time, the institute attracted some of its most talented students. After Häcker and Petrunkevitch left Freiburg, Waldemar Schleip and then Alfred Kühn became assistants in zoology. Both would later establish international reputations.

In addition, Weismann attracted other students who would become pro-fessional scientists in their own rights, among them the ornithologist Emilie Snethlage, the neurophysiologist Viktor Bauer, the systematist and experi-mental zoologist Johannes Strohl, the apiarist Hugo Berthold von Buttel-Reepen, the ichthyologist Reinhard Demoll, and the theoretical physicist Richard Becker. Four other students came from Russia and Great Britain to earn a doctorate under Weismann's supervision, and Georg von Guaita, who earned his degree in 1901, continued to serve as a voluntary assistant in the institute until 1910. On the other hand, August Gruber, by his own admission and quoted earlier, dropped out of productive research in zoology. Although he continued to offer courses until his retirement in 1913, he devoted his creative efforts to Freiburg's civic theater, orchestra, and city museum.

Personal affairs, some rewarding, others taxing, dogged Weismann into the new century. In the summer of 1895, at the age of sixty, with expressions of relief that he would not have to spend his later years alone, Weismann married for a second time. His new wife, Wilhelmina Jesse, was the daughter of a Dutch minister, and she held out the prospect of a loving and compatible mate. "In any case," Weismann informed his lifelong friend Otto Eiser, "I have again found a true friend for life which is for me a great good fortune, whether ill may otherwise come of it.—Also for my Julius it is a blessing, as it has already been shown."[60] Companionship in old age, a mature feminine voice to read to him in the evenings, trips to Switzerland, a refreshing vacation in Greece and Constantinople, and the promise of a mate who would manage the household and extend motherly comfort to his only son, however, were overcome by tragedy.

The newly wedded couple had yet to reach Freiburg after their marriage in the Netherlands when "Mina" became ill. Weismann confided to his friend that "all the beautiful hopes, which I have set on this new relationship, appeared to me highly questionable a[nd] I was not in the position, other than to see very gloomily into the future. Today or actually yesterday the situation turns out now to be better and I have once again seen too darkly."[61] The unspecified symptoms—recurring aches in head, back, and arms in response even to the most minor exertions—turned out to be a chronic problem antedating their marriage, eventually diagnosed by Weismann's colleague Alfred Hegar, one of the founding fathers of modern gynecology, as an infection of the left ovary. Hegar successfully removed the offending organ, which turned out to be enlarged but without a tumor. This seemed to resolve her illness for a while, but by the following year Mina needed a second operation. Whether the operations and associated illnesses led to the next complication is uncertain, but Mina became addicted to morphine.

For a while she and Weismann tried to manage the problem together. Mina assured him on one occasion that she had overcome her addiction. At another time, Weismann participated in a regimen for reducing Mina's dependency. The pull of addiction, however, proved too strong; even as Weismann gradually reduced the medically prescribed dosage for her from one to zero grams per day, Mina secretly found other sources to supplement her cravings. Weismann appears to have been slow at recognizing the signs of relapse and deception, which made it seem all the more traumatic and personally humiliating when he eventually learned the truth. Mina had been spending her housekeeping allowance ("Nadelgeld"), short-changing the faithful family cook with the household's food money, and deceiving doctors and apothecaries in Freiburg into filling prescriptions for more morphine. It was a doctor friend who eventually tipped Weismann off. The only resolution was a legal separation, which included a financial settlement and the placement of Mina in a medical establishment in Illenau.[62]

We only have Weismann's side of the story on the chain of events that led to Mina's illnesses and addiction, but from his perspective his second marriage turned out to be calamitous, and it must have had an impact on his life and his concentration. In 1897, he expressed his disappointment and frustrations with both Mina's health and his own severely limited eyesight in a letter to his son-in-law Will Parker:

> So I will still pass away the time this summer, and should be with this retirement also satisfied and fortunate, if my wife could read to me. That this is not possible and will always fail is a great stroke to my calculation, which leaves me crippled and out of sorts, content to shape my own life. For my work, however, no longer satisfies me, for the eyes are too weak, and I certainly need whatever spiritual sustenance possible. I had the lovely thought to go with Mina to the Alps for a vacation—but without a book in my hand that is also dreary and empty.[63]

If his life had become empty before his second marriage was dissolved, other personal events crowded into Weismann's already despondent existence. The husband of his eldest daughter Therese died suddenly in March 1901. Ernst Schepp had been a government counselor and member of the Landrat in Neustadt and Münster. Weismann and Julius traveled to Bonn for the burial and perhaps to begin the process of family healing. Then, in July of the next year, Julius married Anna Hecker, a singer eight years his senior. Because Julius was only twenty-one at the time, Weismann had reason to feel the marriage was precipitous. He did not attend the wedding in Munich, although whether his absence was an expression of disapproval or for other reasons is not clear. The Wiedersheims, "Uncle" Robert and "Aunt" Tilde, did attend the wedding, as did Therese, who reported back to her father her account of the event. Although Weismann soon accepted the union as a fait accompli, it must have been an additional strain in those years that were already full of anguish. The one happy note amid all his personal setbacks occurred when Therese and her five children moved to Freiburg and took up residence in Weismann's ample house on Stadtstrasse. Therese filled the role of house manager and reading companion where Mina had failed; thereafter, Weismann took great pleasure in watching his grandchildren grow up under his roof.[64]

The Dawn of a New Biology: Overview of Weismann's Research, 1900–1914

Despite the psychological ups and downs in his life, Weismann was able to pursue certain well-proscribed professional undertakings. As we already saw in Chapter 19, he formally concluded his report on the parthenogenesis of drones in 1900.[65] As the new decade advanced, he responded in a carefully crafted paper to Richard Semon's theory of the mneme,[66] wrote a number of introductions to the works of younger scientists,[67] and completed a small paper on regeneration in the salamander *Triton,* which combined experiments he had done in the early 1890s on the regeneration of internal organs with a series of experiments performed in his institute in 1901 by a short-term student Egon Breinig.[68] In addition, he prepared for presentation and publication two papers for the Darwinian celebrations of 1909. The major thrust of all of these items concerned Weismann's life-long preoccupations. These, however, were all small projects compared with Weismann's aspiration to write a textbook explaining evolution in terms of his germ-plasm theory. I will also examine this publication in detail in Chapter 21, but here it is worth inspecting several chapters to observe Weismann's reaction to the new developments in heredity: the rediscovery of Mendel, Boveri's new work on chromosomes, and the Boveri-Sutton law concerning the correspondence between Mendel's law of segregation and reduction division.

The *Vorträge* (*Lectures*) had their origins in Weismann's popular university lectures on "The Theory of Descent." The two-volume text first appeared in 1902 and reappeared in two later, slightly altered editions of 1904 and 1913. The English zoologist, J. Arthur Thomson, and his wife, Margaret Robertson Thomson, brought out an English translation of the second edition the same year that the German had appeared. The work represented the capstone of Weismann's professional career; more than any other text of the day, it served as a synthesis of contemporary heredity and evolution theory. We see with hindsight that the effort was dated even before it appeared, but that makes it no less interesting for the historian. In the context of this chapter, it is valuable to compare the three editions, for in them we can find small but significant changes that tell us much about Weismann's science in the last dozen years of his life. When these changes are intercalated with scraps of comments found in

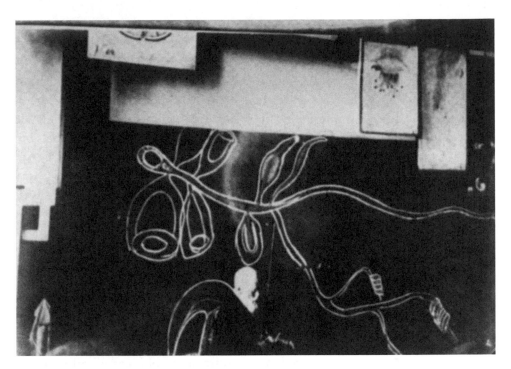

Figure 20.2. Weismann giving a lecture at his institute in 1909. A sketch of a hydromedusa is shown on the blackboard.

Klaus Sander, ed. *August Weismann (1834–1914) und die theoretische Biologie des 19. Jahrhunderts. Urkunden, Berichte und Analysen.* Freiburger Universitätsblätter, vols. 87, 88. Freiburg: Rombach Verlag, 1985. Helmut Risler, "August Weismanns Leben und Wirken nach Dokumenten aus seinem Nachlass," p. 37.

Weismann's correspondence, an interesting picture begins to emerge. At the same time, it is also worth examining the reactions to the new advances of some of Weismann's former students who assisted and advised him during these years. These devoted colleagues were generally faithful to the Weismannian program and so provide a window onto what Weismann himself may have been thinking at the times when his voice remained silent. Overall, Weismann's reflections are surely personal and unique, but the general pattern they represent is just as surely valuable for understanding his reactions to the new developments of his students, other Weismannians, and other biologists in general.

Mendel's Rules and Weismann's First Response

In addition to presenting a broad-ranging text on many biological subjects, Weismann found room to include in the 1902 edition of the *Vorträge* two brief paragraphs on the rediscovery of Mendel's "rules." His words were not the enthusiastic response of a revolutionary. In fact, Weismann pointed out that even one of the codiscoverers of Mendel's paper, Carl Correns, felt that Mendel's rules did not apply in general. Nevertheless, Weismann did not treat the subject as a naysayer and indicated that he looked forward to more information about them from botanists. Zoologists, he pointed out, had a more difficult time of it. He himself had made attempts hybridizing animals but without success.[69] He further reminded his readers that back in 1892 he had suggested that reduction division resulted in a random distribution of ids, and that this now seemed to be confirmed by countless botanical experiments that confuted the Mendelian regularities that had just become known. Weismann's remarks were not fully informed, but considering the vast amount of material he was endeavoring to pull together in his two volumes, they were hardly a sign of a rejection of recent developments.

By October 1904 when the revised edition of his *Vorträge* appeared, Weismann had had only a brief opportunity to alter his text.[70] Again, the same cautious picture appears as he commented on additional contradictory research. On the one hand, William Castle and Allen's study of hybridization of gray and white mice indicated to him a "confirmation" of Mendel's rules. On the other hand, he saw in Bateson and Saunders's first *Report to the Evolution Committee* a statement of limitations to their general application.[71] "Further research," Weismann insisted, "must provide further information; meanwhile we can only say that even today—and to be sure also with reference to hybrids— repeated deviations from the Mendelian rules of separation have been established."[72]

The second edition also made clear for the first time that Weismann had continued to pursue modest hybridization studies of his own. The immediate stimulus appears to have been a letter from the Grönigen tax official C. L. W. Noorduyn (Noorduijn), who out of personal interest had carried out cross-breeding experiments between brown and yellow-gold domesticated races of canaries. The crosses resulted in green-checkered offspring, but after five years of back-crossing the successive generations with the brown race, he had achieved a fully green-tinged race that was similar to the wild canaries from Teneriffe.[73] It was a standard nineteenth-century endeavor, which followed in the footsteps of Carl Friedrich von Gaertner and Darwin. Weismann attempted to repeat the experiments but without much success. He reported Noorduyn's work in the context of discussing reversion to ancestral types.[74]

To us, Noorduyn's reversion and Mendel's segregation experiments lie conceptual worlds apart. In Weismann's *Vorträge,* they were physically and suitably separated by a single page. "Suitably" because the canaries were easily adapted into a germ-plasm theory that explained evolution and reversion in terms of the random nature of reduction division, amphimixis, and the effects of natural selection. Noorduyn and Weismann were interested in such reversions for what they had to say about the evolution process. The latter found the examples of Mendel's peas and Correns's beans interesting but marginally relevant statistical regularities in this more general process—"and so we must avoid the working in of this new material into our theory until a more meaningful broader basis of facts from the botanists is provided."[75] Such a comment reemphasizes Weismann's initial ambivalence toward and profound misunderstanding of Mendel and the new research of the Mendelians.

When in 1906 Hugo Iltis contacted Weismann about contributing to a statue commemorating Mendel, it was in keeping with this opinion for Weismann to reply with reasoned reluctance:

> To be sure Gregor Mendel performed a great and lasting service for science. He lead us a bit deeper into the secrets of heredity, and if the facts, which he uncovered, have a little less far and general validity, as one initially from many sides believed, they lose however nothing of their theoretical significance; that will be shown, the more one works further.
>
> Shouldn't one, however, not wait for a monument, until the new ideas, to which he had given the impulse, are worked through? Until one also surveys the entire meaning of his work? I believe: yes![76]

By the end of the year, and apparently only after prompting from Tschermak, Weismann agreed to support Iltis's proposed monument.[77]

In addition to his early published and private comments on Mendel, it should be mentioned that Weismann made it clear in the second edition of his *Vorträge* that T. H. Montgomery, Valentin Häcker, and Theodor Boveri, among others, were independently and sometimes contentiously reinvestigating the finer events of reduction division.[78]

Weismann's Supporters Respond

Another way to measure the influence that the germ-plasm theory could exert on the reception of new developments in heredity is to examine the comments by three of Weismann's former students: Valentin Häcker, Alexander Petrunkevitch, and H. Ernst Ziegler. By 1905, none of them had yet attained an established position, but all would in their own way distinguish themselves in zoology.

VALENTIN HAECKER.   As we have seen, Häcker had worked as Weismann's first assistant between 1892 and 1900, at which time he was invited to assume the chair of zoology at the Technische Hochschule in Stuttgart. In 1909, he received the call to Halle, where he occupied the Lehrstuhl of zoology until his death in 1927. Häcker was a general zoologist of the nineteenth-century mold. While in Freiburg and Stuttgart, he intensively investigated gametogenesis and germinal continuity in copepods, but when he moved to Halle he plunged into the microscopical study of the origin of hereditary traits. He also published on bird migration and vocalization, freshwater plankton, and animal psychology. Here I need to examine Häcker's early attempt to adjudicate between Weismannism and the new advances in heredity theory. His assessment was part of a *Festschrift* for Weismann's seventieth birthday, so there can be no question that Weismann saw it, but it is less clear that he studied it in detail.[79] If he had, he would have found a survey and evaluation of the Mendelian and chromosomal studies, which together were clearly in the forefront by 1904. The monograph was entitled "Hybridization and Gamete Formation" and featured three sections.[80]

The first consisted of a short survey of hybridization results, starting with an explication of Mendel's rules of segregation and independent assortment. Häcker then proceeded to present the various types of exceptions to these rules. There were the cases of compound allelomorphism, regression to the ancestors, constant hybrid races, impure hybrid matings, and de Vries's *Mutationslehre*. Häcker may have drawn many of these examples from Bateson and Saunders's *First Report* of 1902, but the point I am making is that the Weismannian school was absorbing and evaluating Mendelian genetics as it appeared.

Häcker's second section was devoted to recent studies on chromosome reduction during gametogenesis. There were those, Häcker claimed, such as Boveri, who argued that reduction occurred at prophase before the first maturation division; those such as Korschelt and Heider who claimed that reduction came during the first maturation division; and those such as Sutton and himself who argued that reduction occurred at the second maturation division. Reciting this spread of opinion, best supplied in Korschelt and Heider, was to emphasize the tentative nature of any conclusion for the moment. At the same time, Häcker had completely misunderstood the thrust of Montgomery's 1903 paper by focusing exclusively on the moment of reduction.

In the third section, Häcker attempted to establish a relationship between the hybridization and chromosomal reduction phenomena. He was well aware of the Boveri-Sutton hypothesis of 1902; of Cannon's hypothesis, later more carefully presented by Montgomery, that described a conjugation of homolog chromosomal pairs; of the autonomy of paternal and maternal chromosomes, a subject he had contributed to in his theory of gonomery; and of the contrast between Weismann's theory of the chromosome as a unique package of ancestral ids and Boveri's claim that chromosomes were unique genetic packages that come in paternal and maternal pairings. He was prepared to accept Sutton's proposal that at the metaphases in gametogenesis the different maternal and paternal chromosomes separated randomly into the daughter cells, but his awareness of each of these propositions did not translate into an unqualified endorsement for Boveri's view of the functional dissimilarity of the chromosomal pairs. Häcker could only lay out the variety of outcomes during the maturation process and point out which ones were in accord with the Mendelian results. None of the events mentioned here had persuaded him that the germ-plasm theory was now obsolete.[81]

ALEXANDER PETRUNKEVITCH.   Petrunkevitch had left the institute in Freiburg in late 1903 to assume a lecturing position at Harvard University. He had made the move reluctantly, but he had just married an American student, Wanda Hartshorn, in Freiburg and had come to the conclusion that employment opportunities would be better for him in the United States. Thus, he missed Weismann's seventieth birthday celebrations but took the opportunity to report in Weismann's *Festschrift* on the work he had done at the Zoological Station of the Berlin Aquarium in Rovigno during the months of March and April 1902. In brief, he revisited the exciting phenomena of artificial parthenogenesis opened up by Jacques Loeb and Thomas H. Morgan, and placed their results in the context of his own work on the natural parthenogenesis of drones and Boveri's experiments with merogonic and polyspermic reproduction in the sea urchin. The most important common theme running through his discussion was how this contemporary work reflected on the nature of the chromosomes. There can be no question that Petrunkevitch was aware of several alternative views: (1) that of Yves Delage, which held that the chromosomes were reconstituted at each prophase and therefore could have neither continuity nor individuality; (2) that which held that the chromosomes possessed continuity but were not individually distinct because they were the product of chemicophysical forces of the egg; (3)

that of Weismann, which maintained that the chromosomes possessed continuity and individuality in the sense that they were "essentially the same but qualitatively different"; and (4) that of Boveri, which claimed that the chromosomes manifested continuity and functional individuality in the sense that each of the chromosomes in the gametes possessed the anlagen for different traits. It was the contrast between these latter two views that occupied Petrunkevitch's attention.

According to him, Boveri's polyspermic work was persuasive as far as it went.[82] However, he felt it failed to account for the random nature of reduction division, and it certainly did not provide a source for chromosomal innovation, which was essential to a neo-Darwinian view of evolution. Others had used a similar argument against Weismann until he began emphasizing germinal selection. After discussing in detail his expanding work on parthenogenesis in many organisms, Petrunkevitch concluded that there was an essential difference between natural and artificial parthenogenesis—the eggs of the former alone would develop to normal adulthood because reabsorption of the second polar body allowed the fertilized egg to reestablish a diploid number of chromosomes. The quantity of the chromosomes, each of which possessed "de potentia die Anlage [sic] für die ganzen Organismus" ("potentially the basis for the whole organism"), not the functional qualitative difference of the chromosomes, was at the heart of his argument.[83] He felt that Boveri's polyspermi experiments, in fact, supported his position. His understanding of the chromosomes appears more akin to Weismann's idants.

H. ERNST ZIEGLER.   Whether Häcker and Petrunkevitch were being deferential to Weismann's germ-plasm theory because their presentations were written for Weismann's *Festschrift* is hard to say. Such inhibitions, however, would not have so directly influenced H. Ernst Ziegler as he wrote a small textbook on heredity the following year. Ziegler had completed a dissertation on the development of the salmon egg in Weismann's institute in 1882 and a Habilitationsschrift on the development of a freshwater mussel at Strassburg in 1884.[84] He returned to Freiburg, where he received the *venia legendi,* and through much of the 1890s he served the zoological institute as an assistant and a nonregular außerordentlicher Professor. In 1897, he became Ritter-Professor of Phylogenetic Zoology in Jena, where he worked and taught in Ernst Haeckel's zoological institute. He was a zoologist with broad interests, and he became deeply involved in heredity theory as it related to human evolution. Weismann was his strong supporter, and repeatedly intervened to advance his career.

Ziegler's *Vererbungslehre* consisted of a lecture presented to the twenty-second Congress for Internal Medicine at Wiesbaden in 1905 and five additional essays, which ranged in subject from an analysis of various concepts of chromosomes and a review of Mendel's laws to an assessment of current experiments on the inheritance of acquired characters and a comparative examination of de Vries's and Weismann's theories of heredity.[85] It is the first two subjects that concern us here.

Ziegler identified tetrad formation as a critical event for the understanding of gametogenesis, and he understood enough of the process to recognize that the separation of the maternal and paternal chromosomes from one another at either first or second maturation division designated the moment of true reduction. Moreover, in a footnote he cited Boveri's major statement of 1904 as the modern explanation of how the maternal and paternal chromosomes came together to initiate tetrad formation.[86] Ziegler, however, appeared not yet to have fully grasped what this might have implied for the qualitative differences of the chromosomes themselves. When in his Wiesbaden lecture he illustrated an example of reduction division and fertilization of two parents with eight chromosomes in the diploid state, he lumped together all grandpaternal (and all grandmaternal) chromosomes as qualitatively similar and so arrived at a total of twenty-five rather than thirty-two genotypic differences.[87] Might this confusion have reflected his unconscious adherence to Weismann's concept rather than the Boveri-Sutton concept of the chromosome?

If Ziegler's discussion in April 1905 showed a lingering Weismannism, the chapter on chromosome theory, obviously written thereafter, revealed that he had swung farther into Boveri's camp. Mention of Boveri's 1904 monograph, instead of appearing as a footnote, now formed an important part of his discussion, and his acceptance of qualitative individuality seems clear: "The doctrine of the individuality of the chromosome may be well founded at this time and will scarcely experience a contradiction."[88] At the same time, Ziegler disavowed the gonomery theory of Häcker, his former Freiburg colleague. He put his finger exactly on the contrast between Häcker's and Boveri's work. Häcker, of course, would object to Ziegler's conclusion because he also assumed a qualitative difference between the chromosomes, but his was a Weismannian rather than a Boverian concept of individuality.

Nevertheless, Ziegler showed some of the same reluctance to generalize from Mendel's laws as had many of his contemporaries. He recognized cases of multiple forms in the F1 generation, of deviations from the 3:1 phenotypic ratio in the F2 generation, of imperfect segregation in the F2, and the sudden reversion to distant ancestral forms. Bateson and Saunders's *Report to the*

Evolution Committee, Weldon's critiques of Mendelism, and the mixed results reported by Castle and Tschermak helped fortify his ambivalence. Despite these ambiguities, the regularities of Mendel's results needed to be explained, and it was at this juncture that Ziegler revealed how far he had still to travel. He rejected the Boveri-Sutton account of associating a given factor (Ziegler tends to speak of traits or Eigenschaften) with a particular chromosome for the simple reason that most chromosomes appeared to have the same size and form and so must have the same function. Given this opinion, Ziegler appears not yet to have followed the evidence for the different sizes and forms of chromosomes presented so authoritatively by Sutton. Furthermore, there must be associated factors that could not segregate in the recognized Mendelian fashion.[89] Instead, Ziegler speculated that the Mendelian genotypic ratio of 1:2:1 in the F2 generation could be explained in terms of the preponderance of either dominant or recessive factors or their near equality in number. The whole discussion seemed to be a reversion to an embryological rather than a transmission explanation for the manifestation of a given trait.[90]

As brief as these accounts are of three lengthy and intricate monographs, it should be clear that at least three of Weismann's former students and strong supporters were participating in the discussions about the relationship between Mendel's rules and the chromosomal events in maturation division. This would have been natural, for the Weismannian program had always been concerned with the consequences of reduction division. Three features of their 1904–1905 discussions seem to stand out. (1) There was uncertainty about the general application of Mendel's rules. (2) There was a reluctance to endorse Boveri's over Weismann's very different concepts of the qualitative individuality of the chromosomes. (3) There remained confusion about the formation of the tetrads and the subsequent segregation of maternally and paternally derived chromosomes during maturation divisions. Might it be that Weismann and his former students had a harder time than others in integrating these three elements with a new genetics because they had already been absorbed for a decade in the details of an elaborate chromosomal theory of heredity, development, and evolution?

Weismann Revisits the Issues

As his former students became increasingly involved in accessing the general significance of Mendel's laws and the implications of recent discoveries about the chromosomal events in gametogenesis, Weismann, who was seventy-four years old in January 1904, appeared to take a backseat in the conversation. He wrestled a bit with the ratio of parental chromosomes in the production of drones of the F2 generation.[91] His son-in-law, William Parker, kept him informed about the struggle between the Mendelians and biometricians in England.[92] He also thanked Correns for some unspecified reprints and a copy of his "recht wertvoll" edition of Mendel's letters to Nägeli.[93] In a letter to Arnold Lang, he indicated his familiarity with that researcher's hybridization of land snails, which revealed Mendelian patterns in some experiments.[94] Finally, in mid-1910 he indicated to Wilhelm Schallmayer that "the Mendelian law shows us in a valuable way, what can happen, but it does not clarify for us, what must happen and upon which it depends, that in a definite case occurs a separation of the Anlagen or a mixture of them, and so forth."[95] This final comment reflected Weismann's lifelong desire to have phenomena explained in a necessary manner, which for him meant in terms of a mechanism. None of his comments, however, suggested that Weismann was intently tracking the new developments in hybridization research, let alone following the implications of Boveri's polyspermi experiments or the work of American cytologists on tetrad formation. It is not unlikely, however, that he understood more than his informal and brief correspondence let on.

It is more important to recognize that between 1904 and 1912 Weismann was concerned, to the extent that his eyes and his general health allowed, with promoting his neo-Darwinian agenda. His publications were not numerous but for the most part reflected this orientation. He wrote a lengthy, critical review of Richard Semon's mneme theory and contributed a number of substantial papers on Darwin and natural selection. Until his death in 1914, his personal research was focused on understanding mimicry in butterflies, particularly of the genus *Elymnias*.

In 1912, however, he was forced to confront a decade of new developments as he prepared a third edition of his *Vorträge*. He must have clearly understood the need for assistance, for the Weismann *Nachlass* contains a primer on Mendelian ratios, reduction division, and their relationship written in longhand by his assistant Waldemar Schleip.[96] The line of argument and two of the illustrations reappear in the appropriate chapters of the *Vorträge*. It is to the third edition we must turn to appreciate the extent of Weismann's intellectual growth.

When the third edition appeared in 1913, Weismann was in his seventy-ninth year, and his eyes had gotten increasingly worse. That he managed to revise the text at all speaks to the single-minded work ethic that he maintained throughout his life. Small revisions were made throughout both volumes, but the substantial changes came in two places. The first consisted of an entirely new chapter entitled "The Phenomena of Inheritance in the Narrow Sense" ("Vererbung im engeren Sinne"), which contained a description of Mendel's laws (Gesetze) and an explanation for them in chromosomal terms. The second was a substantial modification of what had been chapter 22, but now appeared as chapter 23, "Share of the Parents in the Building Up of the Offspring—Reversion."[97] It is worth dipping into both chapters to gauge Weismann's response to over a decade of biological research.

Chapter 22 contained much of the material provided by Schleip the year before, but it would be a mistake to describe it as simply a repetition of Schleip's primer. Weismann began with a review of the hybridization experiments of Arnold Lang on *Helix* and Dorothea Charlotte Edith Marryat on *Mirabilis jalapa*.[98] The first provided an animal example of Mendelian dominance and recessiveness and their segregation ratios; the second served as a botanical example of what appeared as incomplete segregation in the heterozygote forms of the F1 and F2 generations. Next turning to dihybrids, Weismann used Mendel's example of round/wrinkled and yellow/green character pairs. It was a solid though not extensive presentation. What seemed important for Weismann, particularly in the case of the dihybrids, was that Mendel's laws reflected a regularity in the transmission process: "Dennoch waltet hier nicht Zufall und Willkür, sondern das Gesetz." ("Moreover accident and arbitrariness do not rule here but law.")[99] In no way, however, should this endorsement be seen as a capitulation of his germ-plasm theory. Instead, Weismann's statement that "Die Mendelsche Lehre ist eine Bestätigung der Grundlagen der Keimplasmatheorie" ("Mendel's law is an affirmation of the foundation of the germ-plasm theory") stood as the underlying theme of this and his next chapter.[100] Weismann comfortably used some of the Mendelian language of the day, such as mono- and dihybrids, homo- and heterozygotes, and recessive and dominant characters, but in addition he vigorously argued that the anlagen for traits must be self-replicating molecules, which he continued to denote as "determinants."

Ever since 1892, Weismann had considered the determinants to be a subset of the "Iden," that is of the ancestral plasm, and these in turn were a subset of the "Idanten," the chromosomes. It was these chromosomes that had always given his discussion an empirical base and, in my mind, rendered his germ-plasm theory a legitimate though limited scientific model. At the outset, however, he had had no grounds for distinguishing the form and size of different chromosomes, and he had assumed that they were qualitatively different only in the sense that they or their subsets, the ids, were derived from different ancestral lines and so might contain variants of or occasionally even different anlagen. By 1904, he recognized that the situation had changed. He had become aware through the work of Sutton and Montgomery that many chromosomes could be morphologically distinguished. Boveri's dispermic experiments on sea urchins provided a strong argument for their physiological differences, and sex-related chromosomal differences were by then being identified in profusion.[101]

As Weismann (and his students) put these assorted types of evidence together, he (and they) recognized that they provided an argument for the qualitative difference in the chromosomes in Boveri's sense that they were the bearers of different types of anlagen rather than representatives of different ancestors. In this new sense, the chromosomes provided a powerful explanation especially for Mendel's dihybrid experiments. In essence, Weismann had found the response to his 1910 statement to Schallmeyer, provided earlier. This is shown more clearly in a new diagram he had drawn for illustration of polar body production in *Ascaris bivalens*. Evident for all to see is tetrad formation, reduction division, and the segregation of two chromatids to the egg and the second polar body. That this was a random segregation is made equally evident two pages later with a schematic illustration of segregation in dihybrids. A mechanical and a symbolic model were brought together to show the correlation between chromosomes and simple Mendelian hybridization ratios.

Weismann assimilated both lines of research into his germ-plasm theory by now conceiving of the chromosomes as partial or "Teilide." Such a concept, particularly in higher animals and plants, helped him explain the morphologically and physiologically qualitative differences highlighted by recent research.[102] However, as we shall see, they did not resolve hybrid complexities beyond the simple Mendelian cases. The move was a compromise for accommodating the Mendelian and Boveri-Sutton worldview into the material foundations of his germ-plasm theory.

A mislabeled *Vortrag* XIII focused deeper into the mechanics of gametogenesis and trait expression. Thirteen of the seventeen pages of the second edition were substantially rewritten for the third edition. What stands out above all is Weismann's new concern for the details of tetrad formation before the prophase of the first maturation division. According to him, the event helped explain the surface phenomena, "Vom Boden der Keimplasmatheorie."[103] What tetrad formation seemed to make clear to Weismann is the real meaning and

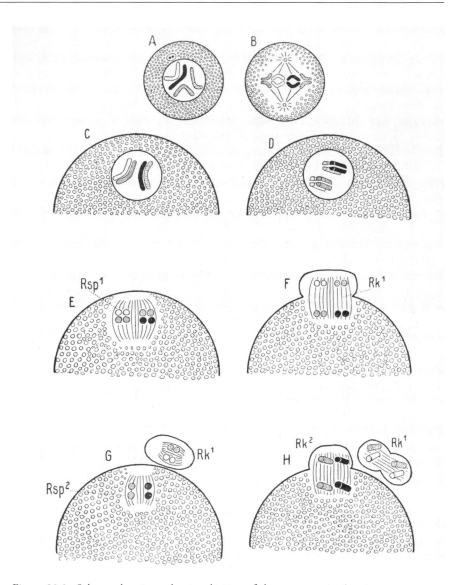

Figure 20.3. Scheme showing reduction division of chromosomes in *Ascaris*.

August Weismann, "Schema der Richtungsteilungen . . . ," *Vorträge* (1913), vol. 2, chapter 22, p. 43. Gustav Fischer, Jena.

significance of homolog chromosomes. It was now clear how a single longitudinal division of this synaptic pair produced the well-known tetrads. It was now clear that the transverse divisions, which he and his school had struggled so long to demonstrate, were simply unusual configurations of the tetrads during subsequent maturation divisions—in fact, by 1913 it became clear to

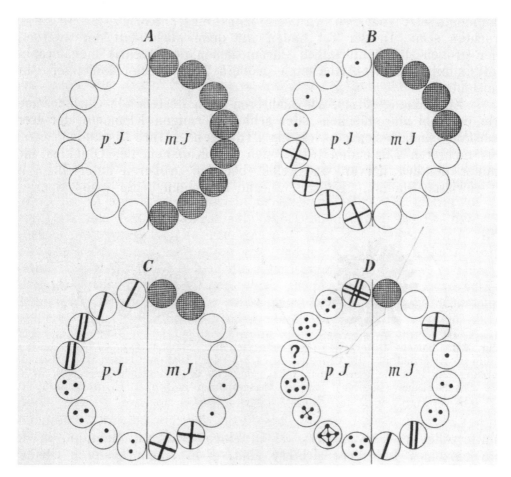

Figure 20.4. The different composition of the id during four successive generations of amphimixes. August Weismann, "Die Keimplasmatheorie . . . ," *Vorträge* (1904), vol. 1, p. 285. Gustav Fischer, Jena.

him that it was irrelevant which of the maturation divisions served as the true reduction division.[104] What became clear and most important for him was that tetrad formation allowed the two maturation divisions to produce a random sorting of the different parental chromatids so that each daughter cell received a mixture of paternal and maternal chromosomes.[105] Thus, gametogenesis produced the geometrically expanding gametic ratios found in dihybrids and polyhybrids. Qualitative differences among the chromosomes, tetrad formation, and "vor allem *Mendel*," that is, the stuff of the new genetics, had for Weismann opened the source for understanding. Gametogenesis moreover

Figure 20.5. Four sketches of *Elymnias* species attributed to Schilling taken from Herta Schepp. The upper and undersides are shown of each butterfly.

Klaus Sander, ed. *August Weismann (1834–1914) und die theoretische Biologie des 19. Jahrhunderts. Urkunden, Berichte und Analysen.* Freiburger Universitätsblätter, vols. 87, 88. Freiburg: Rombach Verlag, 1985. Helmut Risler, "August Weismanns Leben und Wirken nach Dokumenten aus seinem Nachlass," p. 38.

now could be seen as a more precise and powerful mechanism for producing the variety needed for the Darwinian process.[106] Illustrating in a diagrammatic fashion his overall concept of the passage of germ cells from site of origin to the gonophores where they became mature and able to contribute sexually to the next generation, Weismann included two comprehensive overviews of the entire process of germ-cell production and passage in hydroids and medusids. Drawn by Petrunkevitch and printed in the last few pages of volume 1 of the *Vorträge*, they show the phylogenetic passage of subsequent germ cells from the ectoderm of the gastric cavity to the gonophore where they mature. (See Plates 7 and 8.)

Determinant expression, to play with a modern phrase, was a more complex matter for Weismann's changing model. Where the number of traits did not exceed the number of chromosome pairs, Mendelian patterns might pre-

Figure 20.6. Photograph of August Weismann probably taken in 1903, by Dr. Georg von Guaita.

Klaus Sander, ed. *August Weismann (1834–1914) und die theoretische Biologie des 19. Jahrhunderts. Urkunden, Berichte und Analysen.* Freiburger Universitätsblätter, vols. 87, 88. Freiburg: Rombach Verlag, 1985. Reproduced by Gloria Robinson, "August Weismann Heredity Theory," p. 84.

vail. Because this would only rarely be the case with highly complex plants and animals with their hundreds or thousands of traits, Weismann fell back on the mechanisms of determinant competition and germinal selection to produce the overwhelming number of non-Mendelian patterns.[107] It was a justified compromise fashioned to meet the experimental precision of the new genetics while at the same time it remained faithful to the more global expectations of his germ-plasm theory and the nineteenth century.

General Observations

I have often contended that Weismann should be viewed as a transitional figure between the contrasting biology of two centuries. This chapter further promotes that view. Transitional figures from Tycho Brahe to Max Planck have always played an important role in the history of science. They force us to explore how the mind might move with the evidence from one—dare I say paradigm?—to another. Though I do not wish to place Weismann in the same celestial hierarchy as either Brahe or Planck, his historical role appears to have been important and of the same nature. Because he had already developed his own substantial picture of the biological world, he did not readily jump on new bandwagons. Using the language of sociology, I might say he already had in his germ-plasm theory an enormous investment to deploy and protect. On the other hand, he was receptive from the outset and awaited further developments, for he was a scientific realist who accepted reported facts as they became increasingly well established. The accruing evidence of Mendelian hybridization experiments, of Boveri and Sutton's association of meiosis with those Mendelian ratios, of Sutton and Montgomery's description of tetrad formation, and Boveri's polyspermi experiments could not be and were not ignored.

Weismann may be faulted for not throwing out wholesale his germ-plasm theory rather than resorting to the ad hoc adjustment of partial ids, and his strategy certainly lent credence to the general complaint that he was too speculative. These normative judgments, however, were often made in retrospect and isolated from the historical context. They ignore the handicaps under which he labored and even more so overlooked the fact that his germ-plasm theory still answered many more questions about the relationship of heredity and evolution than did the first dozen years of Mendelian genetics. In brief, I might say that for Weismann and the Weismannians Mendel's results seemed interesting but not central to the process of heredity. Their importance increased, however, as the meaning of the chromosomes changed and the cytological understanding of reduction division deepened.

21

A Mechanical Model for Evolution

It is not unusual in Germany to celebrate special anniversaries for outstanding professors, those who have contributed both to science and to the teaching of students of all descriptions. Jacob Henle's fiftieth anniversary was celebrated in 1882 with a *Festschrift* to which Weismann contributed.[1] In a similar fashion, his grateful students organized a *Festschrift* for Weismann's sixtieth birthday in 1894 under the auspices of the *Berichte* of the Naturforschenden Gesellschaft zu Freiburg i. Br.[2] It was an in-house publication; the twelve dedicators were also the twelve authors of papers, and it certainly helped their cause that August Gruber, the editor of volume 8, was also a member of the society. The short papers were thoroughly professional, each devoted to episodes and details of its author's personal research. The status of the contributors ranged from full professors to privatdozents, all of whom had studied with Weismann for a semester or more. It was a modest but meaningful expression of gratitude, but it paled in comparison to the outpouring of good cheer and scholarship that accompanied the next celebration.

Evolution Theory, 1880–1910

During the last thirty years of Weismann's life, evolution theory was undergoing a complex revision. It struggled on a seesaw between the factors of natural selection and internal bodily changes due to environmental, physiological, and selective processes, all of which were pushed forward for recognition and were at times claimed to be dominant within the broad biological community of zoologists, botanists, natural historians, ecologists, paleontologists, and educated speculators. We have witnessed some of the resulting turmoil as it confronted Weismann and his determination to push for the ultimate but not exclusive explanation of evolution in terms of random variations, the value of any given variation under specific circumstances, the continued query about and misunderstanding of appearance and disappearance of uncertain bodily changes, and the complexities of the survival and disappearance of variations in any and all lineages of change. Weismann labored for and contributed to this turmoil within his germ-plasm theory. His goal was to simplify and stabilize the entire process from fertilization through reproduction at the other end of an individual's life as well as the prospect of change from one generation to the next primarily but not exclusively through selection.

Others, too, labored with these myriad processes, mostly without a clearer statement of cause and effect. De Vries's mutation theory vied, often successfully, to be the ultimate picture of the entire process. Nägeli produced a highly theoretical grand theory on the basis of his detailed studies of plant variations of a given type. We have seen others, such as Virchow, Haeckel, Spencer, Haase, O. Hertwig, and Romanes, combine with processes of development, reproduction, micromerism, and Darwinian and cellular selection, phenomena that were believed ultimately to explain the complexities of evolution.

All synthetic efforts to project the ultimate picture of evolution failed until the details of transmission and development were better understood or parceled aside. This included Weismann's germ-plasm theory, but at least its restraints, fashioned in large measure by the mechanics of contemporary cytology, offered one of the most provocative analyses of biology's efforts to paint the full picture of evolution. Weismann's theory failed as well, but in the process it helped focus attention on rigid cytological details and the realities of natural history and reproduction in a way that many appreciated and took into account. Ironically, as Weismann turned seventy the key to his efforts was at last beginning to take shape, but in a form he did not fully comprehend: modern genetics.

Weismann's Seventieth Birthday Celebration

By the time Weismann's seventieth birthday arrived on January 17, 1904, biology had changed dramatically in a way that rendered his germ-plasm theory obsolete but not his focus on the mechanics of chromosomes and his brand of neo-Darwinism. Some of these changes were inevitable because of the recent rediscovery of Mendelian genetics; others came along with the new understanding of the chromosomes that had been fashioned, as we have seen, by Boveri and several American cytologists.

At the time of his seventieth birthday, Weismann must have been deeply involved in improving the second edition of his published *Vorträge*. It augured well, for the need of a new edition suggested that demand had been strong during the two years since the first edition had appeared. It also signified that these university lectures were ripe for translating into English. Success and additional attention to the *Vorträge* may have shielded Weismann from the full implications of post-1900 cytology for his germ-plasm theory. The proceedings of his birthday celebration included some of the most important stars in German biology, and appropriately not a hint can be found in them that Weismann's contributions to current biology were being questioned.

The 1904 celebrations in Freiburg and in print took several forms. On the presentation morning, Weismann's university colleagues and city officials assembled in the hall of Weismann's house, which was nestled below the vineyards that covered the Schlossberg at the time. According to the published account, "Children and children's children, close and distant relatives and friends surround the individual to be celebrated. In the plant bedecked hallway assembled numerous university delegates, professors and students, state officials, and the scientific societies of Freiburg. A concealed bust of Weismann stood in a wreath of palms and blossoming plants, its covering to be dropped at the end of the festivities."[3]

The carefully orchestrated series of official presentations that morning must have lasted about two hours. First, the dean of the philosophical faculty spoke, then the prorector of the university, then the dean of the medical faculty, followed by a student from the philosophical faculty. They spoke directly to Weismann but in a most general way. They mentioned in passing his studies on the rise of life and the development of individuals and species, his understanding of the duration of life, and his examination of heredity. They recognized his involvement in the rights of the university and the defense of freedom of academic studies (*Wissenschaft*), and brought up his broad interests as a German professor who followed national art, music, and literature as well as science. The older speakers noted and the student speakers emphasized how important Weismann's influence had been on the faculty and students in Freiburg.

After the first four laudatory addresses, Weismann himself delivered a general appeal to the students. For him, the most important lesson in the modern university was the necessity of an ethic promoting uninhibited research—to pursue research not for practical gain but for knowledge. This is what he referred to as the "Idealismus der Wissenschaft." To the modern ear, Weismann's metaphors may seem excessively militaristic at times: he depicted an ideal of common striving to lead humanity to an "ever higher level of culture," in which teachers and students become "an intellectual army" (*Geistesarmee*). Together, faculty and students were internally bound together in a common task, and together they would produce a bounty of new knowledge. This, he continued, was not due simply to unrequited striving but to the indispensable assumption that freedom of thought is inherent in the university system.

The Oberbürgermeister of Freiburg and two of his prominent councilors thanked Weismann for working so successfully on the deepest of problems in science and at the same time remaining in Freiburg despite the many "enticing and honorable" attempts to lure him to other universities. The somewhat dry congratulations from the city officials nevertheless prompted Weismann to mention his first visit to Freiburg and his choice of the university for his habilitation three years later. Bertholdstrasse, the main thoroughfare at that time, contained only one row of houses from which one could gain a broad view toward the vineyards of Loretto and Günterstal. When he joined the university in 1863, Weismann particularly remembered enjoying the harvest of the university's vineyards. Forty years later, Freiburg had grown extensively. The vineyards on the Schlossberg had become strewn with dwellings past the walls over which one had to climb to reach the mountain's summit. The "old university" had been converted to a new city hall anachronistically bedecked with gothic ornaments. Additional bridges had been thrown over the Dreisam, additional schools built, the city tower refurbished, a new theater and music hall constructed, and a possible art museum was hoped for by many. A country university town had grown into a culturally diverse city center during Weismann's tenure. Weismann's own understanding of music was also well known and appreciated.

After short laudatory praises from the Freiburger Naturforschenden Gesellschaft and the Deutschen Zoologischen Gesellschaft, Heinrich Ernst Ziegler, by then a professor of zoology at Jena, performed the formal unveiling of the life-sized marble bust fashioned the previous fall by the Frankfurt sculptor Joseph Kowarzik. The committee in charge of eliciting support for the bust reported that 185 academics, fifty-nine of them from Freiburg and thirty-two beyond Germany, had contributed to funding the work of art.[4] Weismann had sat for the artist for fourteen days in the Städelsches Kunstinstitute where his mother had studied, and where he had learned to paint. The association of this current moment of unveiling with his childhood drawing lessons in the very same institute prompted Weismann to reminisce about the importance of his parents for his appreciation of art and music throughout his life. It was appropriate that Weismann closed the morning festivities with a toast to Kowarzik.

A mid-day "Festessen" followed in the main hall of the Europaischen Hoff. Approximately eighty local colleagues, out-of-town zoologists, Julius Weismann, and Kowarzik partook of the two-hour meal. Inevitably this was followed by two additional "table talks." The first was given by a "Hofrat" and faculty member, who again expressed the community's gratitude that Weismann, despite tempting efforts by competing universities, had remained in Freiburg for forty years. Even the advantages of the University of Munich and the international arts of the Bavarian capital had failed to persuade him to leave Freiburg. The second table talk came from Valentin Häcker, by then a professor in Stuttgart, who rendered a synopsis of Weismann's scientific achievements. The scene was set for Weismann's closing address.

The occasion gave Weismann the opportunity for an introductory account of his childhood and an expansion on his interests in natural history and his university studies. He also described his first years in Freiburg and the onset of his eye illness. The talk may well have served as an early version of the considerably longer but uncompleted "Vita Propia" that Weismann left in manuscript form in 1913.[5] Besides the biographical details, Weismann addressed the criticisms that had recently been levied against the theoretical dimensions of his work. "As much as I was inclined to pursue general questions," he reminded his audience, "I knew full well that a certain, yes the largest possible and broadest and at the same time the single most penetrating knowledge must form the foundations of all natural thinking and speculation." The address closed with a sincere tribute to his many students who had provided him with informative and technical support, and he ended with another tribute to the German university system that had supported "the knights of the intellectual order."[6]

The celebrations were not complete without the traditional *Festschrift.* In the form of a supplementary volume to the *Zoologische Jahrbücher,* founded and edited by Johann Wilhelm Spengel, professor of zoology at Giessen, this highly respected zoological journal presented twenty-one professional papers, each averaging over thirty pages in length, to fill a special volume of 750 pages and thirty-three plates.[7] It was a stupendous celebratory achievement, boasting not only papers by Weismann's current and former university colleagues but also contributions that ranged from experimental zoologists such as Theodor

Figure 21.1. August Weismann and his son, Julius, in 1909.

Boveri and Hans Spemann to expert entomologists such as August Forel and Erich Wasmann.[8] Although none of the papers seem in retrospect to have been major contributions, the entire collection measured the esteem in which the zoological community held Weismann.

Darwinism Pursued

Richard Semon (1859–1918)

At the outset of Chapter 4 of this biography appears a thumbnail comparison of the careers of Weismann and his nearly exact contemporary Ernst Haeckel. Haeckel's name also appears throughout this text. Despite the declining frequency of letters between the two, despite common references to their respective careers, and despite mutual support of similar candidates for relevant biological positions, the interaction between the two friends after the first few years appears to have engaged little of general scientific importance. Both were heralded as early and leading proponents of Darwinism, yet the content and style of their respective publications diverged from one another. Weismann was pursuing his science to the cellular and nuclear level of heredity and drawing up a general model of evolution based on an elaborately articulated particulate concept of the organism; Haeckel was expanding his view of the organism to interlock with his mechanistic view of the cosmos—his "monism," as he referred to it. The interaction between these two German evolutionists concerning their respective accomplishments was slight and occasionally antithetical.

It is fortunate for the history of the period that one of Haeckel's former students and a leader in psychology developed an elaborate theory of evolution of certain aspects of the mind that was professionally challenging and yet largely incompatible with Weismann's germ-plasm theory. Among the basic principles this student used to explain the rise and operation of memory throughout the animal world entailed the need to include the inheritance of acquired characters. The theory of the mneme by Richard Semon was strongly supported by Haeckel but denounced by Weismann. Semon provides a formal but contrived closure to the Weismann-Haeckel relationship.

Born and raised in Berlin, Semon became a student in Jena in 1879.[9] In retrospect, his move to Jena for his university studies makes sense, for Semon had already familiarized himself with the works of both Darwin and Haeckel while at the Kaiser Wilhelm Gymnasium. It must have been with a sense of mission that he joined the zoological institute at the height of Haeckel's career. Daniel Schacter provides a balanced view of Semon's experience with Haeckel. According to this modern author, after Semon became interested in the laws of embryology, he turned to marine organisms and traveled to Australia, all particular activities that reflected Haeckel's research interests and research experiences. Semon may have eschewed Haeckel's social Darwinism, his militarism, and his pension for mysticism, but according to Schacter, he shared Haeckel's desire to find a unity in all of nature. "One of the peculiar ironies that we will find in Semon's work," Schacter points out, "is that the same Monistic orientation partially contributed to his scientific downfall— owing to his views on the mechanisms of heredity—and to his strikingly original contributions to the theory of memory."[10] These activities were certainly what led him into a confrontation with Weismann.

Semon's masterwork was published in the summer of 1904. Its original title, *Die Mneme, als erhaltendes Prinzip im Wechsel des organishen Geschehens*, promised much more than the briefer title of the English translation (*The Mneme*), which appeared after Semon's death in 1918.[11] According to Semon's introduction to both German and English versions, his basic psychological approach to evolution was anticipated by Ewald Hering in 1870 and supplemented—perhaps independently—by August Forel, Haeckel, F. Laycock, and several others. Semon's unique terminology and the extensive development of his ideas, however, resulted in a body of literature that directly impinged on the basic premises of Weismann's germ-plasm theory. Out of respect for him but with the feeling that he had nevertheless distorted Darwinian theory, Semon sent Weismann a copy of *Die Mneme*.[12] Although Weismann generally respected Semon's articulation of his position, he felt it important to respond publicly to the aspect of Semon's comprehensive scientific conclusions that ultimately necessitated invoking the inheritance of acquired characters.[13]

The theoretical focus of Semon's theory of evolution was derived from his theory of the "Mneme," which in turn rested on an understanding of the physiological impact of the psychological processes of ontological and evolutionary change. Its starting point, in other words, resided in a completely different domain of the biological sciences from the chromosomal foundations of Weismann's heredity theory. Semon's argument, really a theoretical sketch, focused on the passage by inheritance of the behavior of one generation to the offspring of the next and subsequent generations. It was a straightforward but not simple argument embellished by a specially fashioned vocabulary. Behaviors and patterns of growth, he argued, induced long-term effects of the physiological system. The consequent stimuli, he continued, form "engrams" or physiological effects on the living material. Such effects may be latent at the outset, but they maintained a "memory" of the original engrams and their associations. Collectively these engrams formed a reservoir of potential reactions that collected in the organic body as a mneme, or memory factor. This remained inherent to the organism, was transmitted to the germ cells, and then ultimately was passed on to the next generation.

Semon's details were not morphological but rather nominally implied the imprint of both physiological and behavioral reactions. Semon ignored questions about what structures were involved and about where and how they were provided or measured. In short, he simply imagined the living cells of the organism responding as the consequence of organic reactions. What Semon focused on was the mneme and its memory capacity. This capacity developed with and ensured the consistency in ontogeny, became transmitted to descendants and their descendants, and evolved into the innate patterns and behaviors of specific phylogenetic lineages. To repeat, the engrams and mneme consisted of the physiological memory of past experiences and consequently determined future events. The lack of morphological specificity and Semon's hypothesized extension of basic patterns of experimentally induced changes were strained. They supposedly explained well-known historical behaviors such as those revealed in Marie von Chauvin's earlier demonstrations of the contrasting oviparous and viviparous reproductive behavior in axolotls. By hypothesis, they established the phylogenetic passage of latent but potentially excitable mnemes to later generations.[14] Above all, they endorsed the argument for the inheritance of acquired characters.[15]

Weismann was quick to point out the most important difference between Semon's mneme theory and his own germ-plasm theory. Where the former comprised the functionally developed and inherited reactive properties that became embedded in the cells of the living soma, his ids and determinants were physical, particulate factors that made up much of the germ-plasm. In the case of the germ-plasm theory, hereditary variations of the individual were largely due to nutritional differences (e.g., germinal selection) in the germ-plasm.[16]

After reviewing many instances of similarly induced patterns of plant growth and animal behaviors, of the parallel effects of altitude on tree growth, of the periods of latency and activity in the development of spiders and insects, of the growling of dogs and the automatic bathing of birds, Weismann concluded that in every case these apparent innate activities might have easily arisen through variations and natural selection. He dismissed Semon's mneme theory with the casual conclusion that his "transmitted changes in the germinal substance come primarily not from the body, and besides the germinal substance does not exist of Engramms."[17] He readily admitted, however, that Semon's analogy between memory and embryonic and evolutionary development was a position worth continued exploration.[18]

Darwin Centennial (1909)

After his entanglement with Semon, Weismann turned back to justifying his strong commitment to natural selection as the operative mechanism in evolution. The opportunity to ratify and publicize its important role in evolution came twice in 1909, first at the centennial of Darwin's birth and second at the fiftieth anniversary of the publication of *The Origin of Species*. Weismann commemorated the first in a presentation in Freiburg University's large Paulussaal exactly 100 years after Darwin's birthday of February 12, 1809. A thousand auditors gathered to hear one of the city's most celebrated faculty members.[19] The second appeared as an invited contribution in a volume of nearly 600 pages crammed with twenty-nine articles and edited by A. C. Seward, a professor of botany at Cambridge.[20] The verbal presentation in Freiburg and the more formal discussion in the Cambridge volume soon appeared in the opposite languages. Together, shy of two, they formed the ultimate contributions in both German and English to Weismann's bibliography of 136 items. The two publications that followed entailed a short item on the resting position of a hawkmoth and the third and final edition of Weismann's *Vorträge*.[21] In 1909, forty-one years after Weismann had presented his inaugural address on Darwin ("Antrittsrede as a plansmässige Extraordinarius") as a regular professor in the great auditorium at Freiburg University, he appeared in the same hall for another address on Darwin.

A great deal of progress in the biological sciences had occurred since his first address. Weismann had lived through and participated in a deeper delving into the development of many organisms from lower invertebrates to humans, in clarifying morphological questions in terms of cells, and in discovering different modes of reproduction. Previously new branches of biology had amassed an ever-increasing number of facts between the Napoleonic wars and 1859, but for Weismann this forty-five-year period failed to bring about a deeper understanding of biological causes. Only with Darwin (and Wallace), Weismann insisted, did a new era of synthesis begin, through straightforward nineteenth-century natural history and microscopic biology. Weismann had the opportunity to present the life of Darwin as something significantly different. His sources on Darwin's accomplishments—the *Autobiography, The Origin of Species,* and many of Darwin's other works, including the two volumes on barnacles and many of his later books—followed a standard account that historians today would recognize. It was also a personalized account with a message for the audience. *The Origin of Species,* Weismann insisted, "is simple and straightforward, never sensational in style, but advancing quietly and concretely from one position to another, each supported by a mass of carefully sifted facts." Was Weismann simply interpreting Darwin's practice or was he perhaps reflecting on his own sixty years of work in the vineyards of biology? Again on Darwin, he added, "Every possible objection is duly considered, and the decision is never anticipated, but all the arguments on both sides are carefully and impartially discussed in a manner that is apt to seem to the impatient reader almost too conscientious and cautious."[22]

Weismann did not ask the questions a historian might have. He implied them instead by turning to his own first experience in biology at Schloss Schaumburg. "I myself was at the time in the stage of metamorphosis from a physician to a zoologist, and as far as philosophical views of nature were concerned I was a blank sheet of paper, a tabula rasa." He had read Darwin's book "at a single sitting," and it turned him into an evolutionist and perhaps into a devotee of Darwin's methods. Weismann's talk continued with references to many developments after 1859. He mentioned Fritz Müller's *For Darwin* (1864), Haeckels's *Generelle Morphologie* (1866), his own talk of 1868 on Darwin, and then he turned back to a discussion of Darwin's post-*Origin* works. He appropriately resisted talking about the development of heredity theory, the microscopic study of the nucleus, and the roiling controversies over the inheritance of acquired characters—all reflecting Darwin's legacy, but not the Englishman's own science. Weismann avoided his own practice of science completely, but that became the subject of his next essay.

In his Cambridge essay, entitled "The Selection Theory," Weismann not only addressed Darwin's major biological contribution but created the opportunity to include his own use of selection as a formative process on the germinal level. As with the other contributions to the volume, Weismann's discussion was tailored for the general reader. In this context, Weismann saw his personal invention of germinal selection as an addition to what Darwin had done. Personal selection, original in both the works of Darwin and Wallace, Weismann explained, produced the struggle for existence between variations of the same species, but the selection theory meant only "on an average, the best equipped which survive, in the sense of living long enough to reproduce." Weismann made it clear to the general reader that this selection process produced nonteleological results that statistically kept pace with the ever changing environmental conditions. Elsewhere on the same opening pages he wrote, *"On an average,"* with emphasis, to stress that there was nothing predetermined about individual attributes. "Since the conditions of life cannot be determined by the animal itself, adaptations must be called forth by the conditions." Considering the intended audience, there was nothing novel about the presentation, and Weismann's was a simple and modest message, which years later remains noncontroversial in most educated circles.[23]

The bulk of the essay was essential background and verbal illustrations. Weismann found both the "Lamarckian principle" of the inheritance of acquired characters and the current mutation theory of evolution associated with de Vries to be wanting in comparison with the selection theory. (The editor of the volume balanced out Weismann's account with an essay by de Vries on the subject of "Variation.") Weismann recognized his failure and the failure of others to explain whether at their onset particular variations have selective value. *"We must assume so, but we cannot prove it in any case."*[24] He goes on to imply the statistical nature of this conclusion by asserting that the solution must be found in recognizing that "small variations in different directions present themselves in every species."[25]

Weismann then turned to the contentious issues of coadaptation and degeneration, which a decade earlier had shaped his arguments against Spencer and Romanes. By the time the general reader, who may well have been unfamiliar with the earlier debates, had finished the first half of Weismann's essay, he would have recognized that Weismann's germinal variations and their selection had evolved into the linchpin to solve the unresolvable phenomena on the somatic level. "Germinal selection," Weismann insisted, "supplies the stones out of which personal selection builds her temples and palaces: adaptations."[26] It was a reductive argument of the type Weismann was all too practiced in making during his university lectures. It was the process of moving from the familiar and observable soma to a mechanical explanation found in the selective processes on the theoretical germinal level. In an intentional way, Darwin and Wallace, admired by Weismann above all biologists, had been undergirded by Weismann's germ-plasm theory.

Weismann, however, would not stop here. He devoted the second half of his Cambridge essay to returning to what may have been his original task.

Here he discussed both sexual and natural selection as recognized in living nature. Phenomena such as decorative coloration in the males of many species, sympathetic coloration in butterflies, concealment through camouflage in design and color, the mimicry so expertly promoted by his friend E. B. Poulton—all such phenomena and by implication many more may be understood as adaptations to conditions brought on by natural or sexual selection. It is interesting that Weismann spent much of the second half of his talk providing detailed examples of butterfly patterns, many of which likely came from his own extensive collection. It is worth noting that Weismann's last, but unfortunately uncompleted, study focused on butterfly mimicry in the East Asian genus *Elymnias*.[27] It is also telling that out of the twenty-nine lectures assembled by Seward for this Cambridge volume only Weismann's essay included two of the five illustrations in the book, and the second of these was a dazzling color plate of twelve examples of mimicry in butterflies taken from his *Vorträge*.

Weismann's *Vorträge* on *The Evolution Theory*

At the time of his seventieth birthday celebrations Weismann was deeply involved in editing the second edition of his *Vorträge*.[28] One can imagine, given his thirty-six lectures during the nine weeks of the winter semester, that his course likely had converted to four lectures per week. The lecture structure is more evident in all three of the German editions because each Vortrag starts out with the salutation "Meine Herren!" It was a work that more than any other reveals the breadth of his knowledge of biology as a whole, which is to be expected in a general course given by the professor of zoology at a German university. This breadth is worth emphasizing; over a hundred years later, we tend to overlook the nuts and bolts of a university's course of lectures as we search for the novel orientation in formal statements. With its 918 pages and two color plates, the first edition was the only textbook in either German or English to provide a full picture of the current issues in evolution biology at the turn of the century. That it was also structured to integrate the cellular level of biology with the processes of species formation lent the work a unique perspective. In the eleven years encompassed by its three editions, Weismann took his readers from the natural history of antiquity through the introduction of evolutionary thinking by Goethe, Lamarck, and Chambers. Darwin, not surprisingly, played a central role in the early historical lectures, not only as Weismann's most important predecessor but as an accomplished scientist in the fullest sense of Weismann's understanding of the term. Darwin's voyage, his work on barnacles, and the rest of his published research provided Weismann with the tools for his success: "Never before had a theory of evolution been so thoroughly prepared for, and it is undoubtedly to this that it owed a great part of its success; not to this alone, however, but still more, if not mainly to the fact that it presented a principle of interpretation that had never before been thought of, but whose importance was apparent as soon as attention was called to it—the principle of selection."[29]

After four lectures on the fundamentals of Darwin's theory and the vivid contemporary supporting biological phenomena of coloration in animals and mimicry (lectures 2–5), Weismann left no doubt about the point of his introductory lectures: "the philosophical significance of natural selection lies in the fact, that it shows us how to explain the origin of useful, well adapted structures purely by mechanical forces and without having to fall back on a *directive* force."[30] Above all others, this was undoubtedly an allusion to the evolutionary explanations of Nägeli. He felt it important to add that Darwin not only employed natural selection as the central mechanism for evolutionary change but that he had included the effects of correlation, the direct influence of altered living conditions, and the use and disuse of structures as subsidiary mechanical influences, which Darwin also had willingly folded in with the general effects of natural selection.[31] Protective adaptations in plants, the manifestation of carnivorous plants, the instincts of animals, and symbiosis were common, widely occurring, and by the beginning of the twentieth century well-investigated biological phenomena that served as the foci of Weismann's further supporting lectures (see Plate 9).

In his lecture on the "Origins of Flowers" (lecture 10), Weismann purposefully provided one of the best understood phenomena supporting the operation of natural selection. From the research of Chrisian Konrad Sprengel at the end of the eighteenth century to Hermann Müller's current studies, Weismann presented material that could easily be interpreted as showing the selective interactions in the symbiotic evolution of insects and plants. He examined Nägeli's claim that plants possessed an internal formative power that responded to the visits of pollinating insects. Once again, Weismann's repudiation demonstrated the chasm separating the two evolutionists. Nägeli's objection to natural selection "overlooks the facts that a species of plant and of butterfly consists not of one individual but of thousands or millions, and that these are not absolutely uniform, but in fact heterogeneous. It is precisely in this that the struggle for existence consists—that the individuals of every species differ from one another, and that some are better, others less well consti-tuted."[32] It would have taken an inattentive student listening to such claims in the institute's lecture hall not to have followed such an argument.

Weismann did not stop with a litany of compelling morphological and be-havioral examples but moved on to Darwin's mechanism of sexual selection. He examined this in a long lecture that endorsed with some exceptions the Darwin-Wallace use. He recognized the differences between the two accounts of these English naturalists: where Darwin had focused on female choice, Wal-lace had urged sexual stimuli and natural selection. Weismann demonstrated considerable knowledge about the subject, and referred extensively to the world of butterflies and the experiences of investigators in the tropics. He was not dogmatic about the many examples but insisted that "all that the theory re-quires is that the selective and eliminative processes do, *on the average,* secure their ends, and in the same way the theory of sexual selection does not need the assumption that every female is in a position to exercise a scrupulous choice from among a troop of males, but only that, *on an average,* the males more agreeable to the females are selected."[33] It was his repeated emphasis that se-lection worked "on the average" that brought Weismann's concept of natural selection into the twentieth century far more than his quasi-statistical ver-sion of a particulate theory of heredity.

Lectures 12 to 19 reveal a transition from the natural history that domi-nated the first third of the course to lectures 25 to 36, which comprised the last half of the course and focused on the process of evolution in the context of Weismann's germ-plasm theory. These transitional lectures moved from selection among cells and tissues, to reproduction both in unicellular and multicellular organisms, to differentiation of germ cells, to multiple pro-cesses of fertilization in both plants and animals, to a detailed presentation and an *a priori* justification of his germ-plasm theory. One of many issues that Weismann presented in these transitional lectures included an argu-ment that personal selection rather than cellular selection drove the evolu-tionary process. The selection of one individual over another—that is, per-sonal selection—may indeed depend on the nature and thriving of an individual's cells, but Weismann insisted that there existed no known mech-anism to pass the somatic results of use and disuse or other supposed cellular changes to the germinal line of the individual. It was a generalized version of his oft-repeated arguments against the inheritance of acquired characters. "How can purely local changes, not based in the germ," he had challenged his audience in an early lecture, "but called forth by the chances of living conditions, be transmitted to descendants?"[34] The argument was not only one more denial of what had become associated with neo-Lamarckian mechanisms of evolution, but directly led Weismann to an examination of reproduction in unicellular and multicellular organisms. Consequently, the first half the course concluded with three detailed lectures on his own germ-plasm theory.

Two lectures on the functions and phyletic occurrence of regeneration started off the second volume. As we have seen, the subject had initially played an important role in the early presentation of his germ-plasm theory. From this well-rehearsed subject, Weismann turned to current and compelling issues of the new century. He titled the next lecture "Share of the Parents in the Building Up of the Offspring" ("Anteil der Eltern am Aubau des Kindes"). It was an important topic when he wrote it. It is even more so for us, given our understanding of the science that had just been radically altered by the rediscovery of Mendel's work in 1900 by de Vries, Correns, and Tschermak. Weismann was well aware of these discoveries. In 1902, he responded to them in a single paragraph and added more in the 1904 edition, but in both editions his focus was on how his germ-plasm theory could explain the transmission of traits.

For the modern biologist and historian, the study of heredity had become a transformative undertaking. The statistics of hybridization revealed a regularity that soon called for additional documentation and a physical explanation. Weismann, however, was caught between years of work on his germ-plasm theory and the so-called Mendelian revolution. In the first edition of his *Lectures,* he mentions Mendel in a single paragraph, but in the second edition he described the "Mendelsche Regel" and, as mentioned earlier, showed an awareness of its basics.[35] In the third edition of 1913, he split lecture 22 into two separate presentations and discussed Mendel's achievements in greater depth. However, for Weismann the transmission of parental traits to their offspring had not been a subject for statistical examination. He had, it is true, made an effort to understand reproduction in the terms of the numbers of chromosomes involved in a given species and the advantage of their apparent doubling followed by a halving of the chromosomes during two maturation divisions.[36] It was the 4–8–4–2 process that he had discussed in the previous decade when he worked out the pattern of polar body production during gamete maturation. This was an argument above all to increase variations by means of reduction division as the egg prepared for fertilization. "It requires that each individual should be a *peculiar complex of hereditary characters;* that is that all the fertilized germ-cells of a pair should possess an individually well-marked character."[37] His was an argument supported by the recent work of Häcker

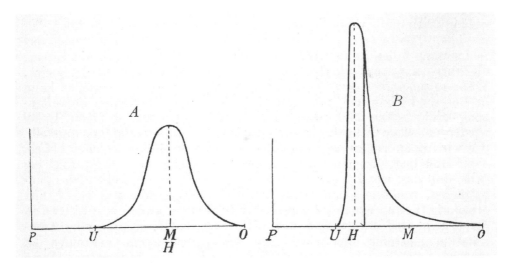

Figure 21.2. Curves according to Ammon. Weismann drew upon his friend Otto Ammon to provide a graphic illustration of symmetrical and asymmetrical variation curves to show the possibilities of variations due to amphimixis.

Vortrag 28, Bd II, 1902, p. 232; lecture 28, p. 232. Also see letter from Weismann to Otto Ammon, 28 May 1896 in Churchill and Risler, *August Weismann: Ausgewahlte Briefe und Dokumente*, Universitatsbibliothek Freiburg i. Br. 1999, 1:266–267.

for the integrity of the mingling and uniqueness of individual chromosomes.[38] Ultimately his was an argument for natural selection, not a statistical examination of hybrid ratios. It entailed a conviction that future research into chromosome distributions would show "it is not chance alone which presides over the re-arrangement of the chromosomes during the reducing divisions; *affinities* play a part also; there are stronger or weaker attractions between the chromosomes, which aid in determining their relative position to one another."[39]

The final thirteen lectures of his course on evolution theory dealt with processes and solutions that Weismann had addressed earlier in his controversies and research. The neo-Lamarckian issue of whether functional modifications were transmitted needed to be addressed once Weismann had laid down the gauntlet with the cytological details of his germ-plasm theory. His biological discussions are elaborate. They include examinations of Brown-Séquard's experiments inducing epilepsy in guinea pigs, the development of the germ theory of disease, the rise of instincts, coadaptations, degeneration, and much more on the relationship of different insect castes to function and nutrition. Lectures 23 and 24 must have been seen as important lessons for the medical students who flooded his course during its final fifteen years. The two lectures on "Germinal Selection" (i.e., lectures 25 and 26) may at first seem excessive until one realizes that they concern the complicated but universal question of the degeneration and disappearance of nonfunctioning parts and behavior. The lectures focus in greater detail and with broader applications on the arguments that Weismann had with Spencer and Romanes in the 1890s. These applications allowed Weismann to move from variations on the somatic level to the operation of the germ-plasm on the germinal level.

By the turn of the century, germinal selection had become for Weismann a theoretical model for molecular processes validating his germ-plasm theory. These lectures also gave him a chance to speculate about the origins of such a universal biological process and the effects on such patterns brought about by inbreeding. His last six lectures allowed Weismann the opportunity to discuss various aspects of species formation such as geographical isolation, the Sarasin cousins' work on species rings in snails in the Moluccas, the nature of physiological selection introduced by Romanes, the importance of isolation and mutual sterility between incipient species, the extinction of species, and the history of and recent experiments on spontaneous generation.

Weismann, with the support of Gustav Fischer Verlag, closed his *Vorträge* with two dazzling plates showing mimicry between different genera and species of butterflies. Combined the plates showed imagoes from Africa, central North America, South America, and Ceylon. Also included are colored illustrations of the caterpillars and pupae of two different mimetic species, *Limnenitis archippus* and the familiar *Danais erippus* (monarch butterfly). These plates emphasized that the adult similarities are a consequence of protection, rather than the result of a closely related phylogeny. In the first color plate (see Plate 10), Weismann emphasized that the adults of one species might mimic or could be confused with the adult of an "immune" species. Thus, the female of the African *Papilio merope* (Plate 10, number 2) differed considerably from its male counterpart (Plate 10, number 1), but could be easily confused by predators of the immune species *Danais chrysippus* (Plate 10, number 3). The same lesson is told in the second plate (see Plate 11), with the added demonstration that biologists could establish an entire "mimicry ring" of adults of different families and genera (Plate 11, numbers 12–15). With such resemblances between differently derived but communal adults, Weismann concluded that natural selection rather than the environment or internal forces induced such parallel results. Mimicry in butterflies was a strong argument for Weismann's entire objective of promoting natural selection as the primary force in the evolutionary process.

The two volumes of *Vortäge* provided not only historical reflections but a veritable trove of information about contemporary biological advances suitable for students interested in evolution and for many of those ultimately bound for a degree in medicine. They offered a mechanistic, though dated, explanation for the origin and evolution of the world of life. It is hard to know how Weismann might have further altered his lectures in response to new developments in heredity and evolution. He was seventy-nine years old when he completed the third edition. He no longer directed the Zoological Institute or the scientific research of its students, and his attention had become distracted by the onset of World War I. All these circumstances certainly diverted his attention from the many relevant new advances in biology.

Retirement and the Final Years

Elymnias, 1912–1914

In addition to detailing in the three editions of his *Vorträge* the operation of heredity and evolution of many specific cases, it was characteristic of Weismann to pursue simultaneously in a research mode the specific details of a specific case of the evolution of clearly identifiable natural morphological characters. He had done this with his work with diptera, cladocerans, and hydromedusae. The last of such exacting investigations was the mimetic phenomenon exhibited by *Elymnias,* an East Asian butterfly.[40] After examining a nearly completed manuscript which Weismann had been preparing for publication at the time of his death, Risler points out that Weismann was interested in mimicry because it offered a prime example of natural selection. In Weismann's words, "There can be no mention that the cause of the change of species lies with the direct influence of external conditions, this leaves only the one possibility that the advantage that the mimicker has is because of its similarity to the immune species. That the mimicker is protected and that on the average and in the course of time this always produces victory."[41] Mimicry for Weismann became a prime example of the action of natural selection on the forms of life.

Figure 21.3. August Weismann with butterflies in case in 1912.
Klaus Sander, ed. *August Weismann (1834–1914) und die theoretische Biologie des 19. Jahrhunderts. Urkunden, Berichte und Analysen.* Freiburger Universitätsblätter, vols. 87, 88. Freiburg: Rombach Verlag, 1985. From Conner Sorensen, "William Henry Edwards, August Weismann and Polymorphism in Butterflies," p. 165.

Retirement

After serving both in the medical and philosophical faculties at Freiburg University since 1863, Weismann wrote the Kultus und Unterrichts Ministerium on November 23, 1911, that he would retire from active duty after the current winter semester. From a modern perspective, there need be no apologies for retirement at the age of seventy-nine. He had been in active service at the university for nearly fifty years, he had led biologists in exploring the cytological details of the heredity process, he had uncompromisingly championed Darwinian natural selection in evolution, and had managed an outstanding research program at his institute. Weismann had presided over his course on Descendenztheorie during the winter semesters since the 1880s and shared the institute's teaching assignments with August Gruber from the early 1880s.

In writing to Petrunkevitch in September 1911, Weismann reported, "The institute is rightfully brave given the institute's thirty concentrators and with its lecture on zoology which attracted over 300 auditors."[42] He then modified his optimistic note: "But I will not have them much longer, for although I speak freely without any written notes, it strains my eyes and it appears to be the moment to unburden the load onto younger shoulders."[43] Two months later, Weismann must have made official his intent to retire at the end of the winter semester. The Kultusminister Franz Böhm expressed its regret and set forth on the process of finding Weismann's replacement.[44]

By mid-December, with the help of the Freiburg faculty, the Minister of the Naturforschende und Mathmatische Fakultät in Karlsruhe recommended as Weismann's replacement Theodor Boveri, Eugen Korschelt, and Valentin Häcker, in that order. Boveri was already recognized as the rising star of German cytology; the other two, well established in their respective universities of Wurzburg and Halle, had been former assistants to Weismann. After the first two declined, the faculty opened up the search again, recommending Franz Doflein, a professor of zoology and specialist in protozoology at the University of Munich, Richard Hesse, a specialist in the morphology of vertebrates at the provincial college in Berlin, and Hans Spemann, an experimental embryologist who was a professor of zoology in Rostock. Doflein, the first choice of the faculty on their second list, accepted and served as professor in Weismann's former institute between 1912 and 1918. Doflein's successor, who clearly continued the Weismann tradition of teaching and outstanding research, would be Hans Spemann, who would have a long and distinguished career at Freiburg's Zoological Institute.

After his retirement, Weismann was granted by Doflein the opportunity to work at the institute in a room of his own, but in the last two years his few letters indicated that his eyesight had worsened to a degree that inhibited sustained microscopical research. Despite the beginning of Germany's war with France and Great Britain, Weismann made the effort to visit with his stepson William Newton Parker in Bern.

The Final Days

Risler was fond of telling the family's story of Weismann's last day. His account follows Gaupp's description and is substantiated in part in Weismann's *Tagesbuch*. During the fall of 1914, Weismann became ill and then bedridden. His son Julius was inclined to please him by playing on the piano the classics and some of Julius's own compositions. On November 5 he did so again and Weismann simply passed away. Julius wrote in the *Tagesbuch* on the date that his father had died.

Epilogue

Over eighty years, from 1834 to 1914, to fathom the mechanics of living forms passing within and between generations. Over eighty years to dedicate a life thoroughly to the pursuit of some of biology's most complex processes, which in nineteenth-century terms were still described as heredity, development, and evolution. Over eighty years to grasp what appeared to be the essentials of these three processes, to do justice to the chemical and physical mechanisms of understanding them with the science of the time, and to use the scientific themes and vocabulary of the time to convince colleagues that the proposed explanations were legitimate. August Weismann's career did not consist of a single event or publication but rather comprised seventy years of observations and explorations of living things—a challenge that posed more specific problems than he could answer, and left as many opponents as it found supporters in the field of biology as he pursued so many important aspects of life, cut a path through a wilderness of ideas and concepts, and fashioned results without coming to specific explanations.

So were those eighty years in the life of a single scientist worth it? For those of us who hold an overwhelming commitment to a mechanistic explanation of life and maintain an ever-increasing commitment to evolution as an undeniable global process brought about by selection on both the physiological and population levels, this might seem close to implausible. It may have seemed foolhardy on Weismann's part to have constrained his explanations to some kind of selection within nature at large, within the physiology of individuals and competing gametes. When so many of his earnest interlocutors manufactured other physical mechanisms of organic change on both individual and population levels, why did Weismann not explore those options? From the

Figure E.1. August Weismann with daughter-in-law Anna and grandson Carl August, 1906.

Klaus Sander, ed. *August Weismann (1834–1914) und die theoretische Biologie des 19. Jahrhunderts. Urkunden, Berichte und Analysen*. Freiburger Universitätsblätter, vols. 87, 88. Freiburg: Rombach Verlag, 1985. Helmut Risler, "August Weismanns Leben und Wirken nach Dokumenten aus seinem Nachlass," p. 41.

very beginning of Weismann's interest in biology, selection appeared to him as the ultimate biological process of change. Why this was the case for Weismann and was not for other contemporary biologists requires a lot more individual study. It may have derived from his first association with Henle, Friedrich Wöhler, and Jacob Henle in Göttingen or Rudolf Leuckart in Giessen, or his first reading of Darwin's *Origin of Species* when he was still pursuing dipteran development in Schaumberg—it is hard to tell from the records currently available. Whatever prompted him in those early years as an advanced student, there was no vacillation on his part once it was clear to him how this Darwinian mechanism could work and even might work on many levels.

On a more specific level, Weismann's early career reveals a lot more of his early involvement than simply the invention or manufacture of certain theories. It was emblematic that much of his early success arose from his preparation of the factual groundwork that helped produce so many models of explanation. Initially, as a chemist, he was not embroiled in biological theories of change. Time and again, we find that he called for repetition in his studies, for the testing of his results on many organisms and different species. His accomplishments were not in uncovering and attempting to explain single phenomena but rather in the establishment of order, of demanding repetition or lack thereof in events under similar or contrasting circumstances, whether in the biology of life or the inorganic chemistry he pursued in Rostock.

Throughout his career, we find Weismann emphasizing the importance of studying multiple organisms, for his science promoted more a natural history of events than a specific discovery of any single process. Weismann's early study of seasonal and sexual dimorphism in the annual cycle of certain species of butterflies revealed far more than a collector's production and cataloguing of variants might have. His study of the migration of the cladocerans in the Bodensee revealed a general pattern of contrasting diurnal and nocturnal migrations—hardly the production of an invertebrate inventory of an ice age lake. It was the cyclical reproduction of many organisms that intrigued him. His was a study of the patterns of variants over a seasonal or annual cycle of *Prosa* and whether that pattern might reveal more about variations than the environmental impact on the taxonomy of related butterflies.

His study of phyletic series in sphinx moth caterpillars or the pattern of seasonal sequences added more to his exploration of a mechanical view of nature than it did to the nature of caterpillar markings—though a detailed examination of the latter promoted the analysis of the former. The study of Cladocera in the Bodensee gave him an enduring sense of population changes in these small creatures. His elaborate, detailed, six-year investigation of hydromedusae evolved into generalizations about the migration of germ cells in the evolutionary process. Here, Weismann combined his mastery of histology with a sense of the life history of specific organisms and on a focus on the process of reproduction.

His interest in the very substance of hereditary transmission led Weismann to follow his predecessors who had wrestled with the nature of and the physical involvement of the material elements of transmission. His intense interest in the cytology of germ cell reproduction through the 1880s added factually to current nuclear cytology. His interest in reproduction was not focused on the facts of cytology themselves; instead, he found in the study of polar body production a key in the form of numbers: the number of chromosomes before and after maturation divisions. This would become the key to understanding sexual change, which in turn became the key to understanding evolutionary change on all organisms above and beyond simple environmental influences.

Actually anticipating Nägeli, Weismann emphasized both the intergenerational and intragenerational continuity of hereditary material. His focus on both aspects of continuity led him directly to examine the transmission of adaptations and evolution. Earlier in his career, he had hinted at the need for micromerism—a reliance on particles to help explain the mechanics of change at the lowest molecular level and transmission of that change to higher organic structures. The contrast between these two "continuities," the molecular and the anatomical, would continue to confound discussion of evolution well into the beginning of the twentieth century. The continued arguments over the inheritance of acquired characters, the confusion between what was innate and what was induced at any and every level of the organism—the molecules or structures from nucleus to soma and gametes, or from the chromosomes to larger body parts—all blended together in a heatedly debated morass of controversies during the last two decades of the century. Weismann's contribution was to articulate the controversies, to sketch them in a lineal sequence as an artist might do, breaking new and clearer boundaries for discussion. Weismann's germ-plasm theory with all its weaknesses served this function and ironically clarified what needed to come next—the unambiguous facts for a new century and a new biology.

After sketching out an elaborate molecular model that would bind together all these factual and aspired dimensions, Weismann spent too much of his time in the 1890s justifying the completeness of his germ-plasm theory as an explanation for heredity and Darwinian evolution. Is this bound to be the outcome of believing too intensely in a single elaborate, well-researched but hardly tested model for heredity and evolution? The irony, of course, would come at the turn of the century with a new, far less complete model based on the spectacular studies of Theodor Boveri (an admirer of Weismann) and the rediscovery of "Mendel's laws" by others who knew Weismann's work well.

Appendix

Works by August Weismann

Weismann, August. 1857. "Ueber die Bildung der Hippursäure beim Menschen." *Zeitschrift für rationelle Medizin* 2:331–343.

Weismann, August. 1857. *Ueber den Ursprung der Hippursäure im Harn der Pflanzenfresser.* Göttingen: Dieterichschen Univ.-Buchdruckerei.

Weismann, August. 1857. "De acidi hippurici in corpore humano generatione." Diseratio inauguralis, Göttingen, Francofurti.

Weismann, August. 1858. "Untersuchungen über den Salzgehalt der Ostsee." *Archiv für Landeskunde in den Grossherzogthümern Mecklenburg* 8:19–304.

Weismann, August. 1858. "Analysen des Ostseewassers." *Archiv für Landeskunde Grossherzogthümern Mecklenburg* 8:437–444.

Weismann, August. 1859. "Ueber Nervenneubildung in einem Neurom." *Zeitschrift für rationelle Medizin* 7:209–218.

Weismann, August. 1861. "Ueber das Wachsen der quergestreiften Muskeln nach Beobachtungen am Frosch." *Zeitschrift für rationelle Medizin* 10:263–284.

Weismann, August. 1861. "Ueber den feineren Bau des menschlichen Nabelstranges." *Zeitschrift für rationelle* 11:140–166.

Weismann, August. 1861. "Ueber die Verbindung der Muskelfasern mit ihren Ansatzpunkten." *Zeitschrift für rationelle Medizin* 12:126–144.

Weismann, August. 1861. "Ueber die Neubildung quergestreifter Muskelfasern. Eine Erwiderung an Herrn Prof. Budge." *Zeitschrift für rationelle Medizin* 12:354–359.

Weismann, August. 1861. "Ueber die Musculatur des Herzens beim Menschen und in der Thierreihe." *Archiv für Anatomie, Physiologie und wissenschaftliche Medicin*, 41–63.

Weismann, August. 1862. "Ueber die zwei Typen contractilen Gewebes und ihre Verteilung in die grossen Gruppen des Thierreichs, sowie über die histologische

Bedeutung ihrer Formelemente." *Zeitschrift für rationelle Medizin* 15:60–103, 279–282.

Weismann, August. 1862. "Nachtrag zu der Abhandlung: Ueber die zwei Typen contractilen Gewebes und ihre Vertheilung in die grossen Gruppen des Thierreichs, sowie über die histologische Bedeutung ihrer Formelemente." *Zeitschrift für rationelle Medizin* 15:279–282.

Weismann, August. 1862. "Ueber die Entstehung des vollendeten Insekts in der Larve und Puppe. Ein Beitrag zur metamorphose der Insekten." *Abhandlungen der Senckenbergischen Naturforschenden Gesellschaft* 4:227–260.

Weismann, August. 1863. "Die Entwickelung der Dipteren im Ei, nach Beobachtungen an Chironomus spec., *Musca vomitoria* und *Pulex Canis.*" *Zeitschrift für wissenschaftliche Zoologie* 13:107–220.

Weismann, August. 1864. "Die nachembryonale Entwicklung der Musciden nach Beobachtungen an *Musca vomitoria* und *Sarcophaga carnaria.*" *Zeitschrift für wissenschaftliche Zoologie* 14:187–336.

Weismann, August. 1864. "Zur Embryologie der Insecten." *Archiv für Anatomie, Physiologie und wissenschaftliche Medicin,* 265–277.

Weismann, August. 1865. "Zur Histologie der Muskeln." *Zeitschrift für rationelle Medizin* 22:26–45.

Weismann, August. 1866. "Die Metamorphose der Corethra Plumicornis. Ein weiterer Beitrag zur Entwicklungsgeschichte der Insecten." *Zeitschrift für wissenschaftliche Zoologie* 16:45–127.

Weismann, August. 1868. *Über die Berechtigung der Darwin'schen Theorie. Ein akademischer Vortrag gehalten am 8. Juli 1868 in der Aula der Universität zu Freiburg im Breisgau.* Leipzig: Wilhelm Engelmann.

Weismann, August. 1872. *Ueber den Einfluss der Isolirung auf die Artbildung.* Leipzig: Wilhelm Engelmann.

Weismann, August. 1873. "Bericht über die Weiterentwicklung der Descedenztheorie im Jahre 1872." *Archiv für Anthropologie* 6:119–145.

Weismann, August. 1873. "Zur Abwehr." *Beilage zur Allgemeinen Zeitung,* 2 Dezember, 5131.

Weismann, August. 1874. "Über Bau und Lebenserscheinungen von *Leptodora hyalina* Lillgeborg." *Zeitschrift für wissenschaftliche Zoologie* 24:348–418.

Weismann, August. 1874. "Ueber den Saison-Dimorphismus der Schmetterlinge." [Studien zur Descendenztheorie. Theil I.] *Annali del Museo Civico di Storia Naturele di Genova* 6:209–302.

Weismann, August. 1875. "Über die Umwandlung des mexikanischen Axolotl in ein Amblystoma." [Studien zur Descendenztheorie. Theil II. Über die letzten Ursachen der Transmutationen.] *Zeitschrift für wissenschaftliche Zoologie* 25 (Supplementband):297–334.

Weismann, August. 1875–1876. *Studien zur Descendenztheorie.* [*Theil I. Über den Saisondimorphismus der Schmetterlinge; Theil II. Über die letzten Ursachen der Transmutationen. 1. Die Entstehung der Zeichnung bei den Schmetterlingsraupen, 2. Über den phyletischer Parallelismus bei metamorphischen Arten, 3. Über die Umwandlung des mexikanischen Axolotl in ein Amblystoma, 4. Über die mechanische Auffassung der Natur.*] 2 vols. Leipzig: Wilhelm Engelmann.

Weismann, August. 1876. "Das Thierleben im Bodensee." *Schriften des Vereins für Geschichte des Bodensee's und seiner Umgebung* Heft 7:132–161.

Weismann, August. 1876. "Zur Naturgeschichte der Daphniden, Theil I. Über die Bildung von Wintereiern bei *Leptodora hyalina.*" *Zeitschrift für wissenschaftliche Zoologie* 27:51–112.

Weismann, August. 1876–1879. *Beiträge zur Naturgeschichte der Daphnoiden.* 7 Abhandlungen vols. Leipzig: Wilhelm Engelmann.

Gruber, August, and August Weismann. 1877. "Ueber einige neue oder unvollkommen gekannte Daphniden." *Abhandlungen der Naturhistorischen Gesellschaft zu Freiburg* 7:50–116.

Weismann, August. 1877. "Zur Naturgeschichte der Daphniden, Theil IV. Über den Einfluss der Begattung auf die Erzeugung vom Wintereiern." *Zeitschrift für wissenschaftliche Zoologie* 28:212–254.

Weismann, August. 1877. "Ueber die Fortpflanzung der Daphnoiden." *Amtlicher Bericht über die 50. Versammlung Deutscher Naturforscher und Ärzte in München vom 17.-22. September 1877*:178.

Weismann, August. 1877. "Diskussionsbemerkung zu dem Vortrag von A. Forel: Über den Ursprung der verschiedenen Faunen unserer Süsswasserseen." *Amtlicher Bericht über die 50. Versammlung Deutscher Naturforscher und Ärzte in München vom 17.-22. September 1877*:172.

Weismann, August. 1877. *On the Change of the Mexican Axolotl to an Amblystoma.* Smithsonian Report, 349–375. Translated by Henry M. Douglass. Washington, D.C., U.S. Government Printing Office.

Weismann, August. 1877. "Die Zur Naturgeschichte der Daphniden, Theil II. Eibildung bei den Daphnoiden." *Zeitschrift für wissenschaftliche Zoologie* 28:95–175.

Weismann, August. 1877. "Zur Naturgeschichte der Daphniden, Theil III. Die Abhängigkeit der Embryonal-Entwicklung vom Fruchtwasser der Mutter." *Zeitschrift für wissenschaftliche Zoologie* 28:176–211.

Weismann, August. 1878. "Zur Naturgeschichte der Daphniden, Theil V. Ueber die Schmuckfarben der Daphnoiden." *Zeitschrift für wissenschaftliche Zoologie* 30 Suppl.:123–165.

Weismann, August. 1878. "Rechtfertigung." *Zeitschrift für wissenschaftliche Zoologie* 30:194–202.

Weismann, August. 1878. "Über Duftschuppen." *Zoologischer Anzeiger* 1:98–99.

Weismann, August, and Robert Wiedersheim. 1878. "Aus dem zoologischen und anatomischen Institut der Universität Freiburg i. Br." *Zoologischer Anzeiger* 1:6–7.

Weismann, August. 1878. *Ueber das Wandern der Vögel.* Berlin: Carl Habel.

Weismann, August. 1879. "Die Bedeutung der Zoologie für das Studium der Medicin." *Beilage zur Allgemeine Zeitung,* 5 April, 1401–1403.

Weismann, August. 1879. "Zur Naturgeschichte der Daphniden, Theil VI. Samen und Begattung der Dapnoiden." *Zeitschrift für wissenschaftliche Zoologie* [1880] 33:55–110.

Weismann, August. 1879. "Zur Naturgeschichte der Daphniden, Theil VII. Die Entstehung der cyclischen Fortpflanzung bei den Daphnoiden." *Zeitschrift für wissenschaftliche Zoologie.* [1880] 33:111–270.

Weismann, August. 1880. "Parthenogenese bei den Ostracoden." *Zoologischer Anzeiger* 3:82–84.

Weismann, August. 1880. "Zur Frage nach dem Ursprung der Geschlechtszellen bei den Hydroiden." *Zoologischer Anzeiger* 3:226–233.

Weismann, August. 1880. "Über den Ursprung der Geschlectszellen bei den Hydroiden. II." *Zoologischer Anzeiger* 3:367–370.

Weismann, August. 1881. "Beobachtungen an Hydroid-Polypen. I. Pulsieren des Körperschlauchs." *Zoologischer Anzeiger* 4:61–63.

Weismann, August. 1881. "Beobachtungen an Hydroid-Polypen. II. Selbständige Bewegungen des Ectoderms." *Zoologischer Anzeiger* 4:63–64.

Weismann, August. 1881. "Beobachtungen an Hydroid-Polypen. III. Die Enstehung der Eizellen in der Gatung *Eudendrium.*" *Zoologischer Anzeiger* 4:111–114.

Weismann, August. 1881. "Über eigenthümliche Organe bei *Eudendrium racemosum* Cav." *Mittheilungen aus der Zoologischen Station zu Neapel* 3 [Jbd. 1882]:1–14.

Weismann, August. 1881. "Über die Dauer des Lebens." In *Tageblatt der 54. Versammlung deutscher Naturforscher und Aerzte in Salzburg vom 18.-24. September 1881.* Salzburg.

Weismann, August [Wiesmann, Auguste]. 1881. "L'origine des cellules sexuelles des hydroides." *Annales des Sciences Naturelles, B: Zoologie et Pathologie Animales* 11:1–37.

Weismann, August. 1882. *Studies in the Theory of Descent.* Translated by Raphael Meldola. 2 vols. London: Sampson Low, Marston, Searle, & Rivington.

Weismann, August. 1882. "On Some Peculiar Organs of *Eudendrium ramosum.*" *Annals and Magazine of Natural History,* ser. 5, 9:201–202.

Weismann, August. 1882. *Über die Dauer des Lebens.* Jena: Gustav Fischer.

Weismann, August. 1882. "Beiträge zur Kenntniss der ersten Entwicklungsvorgänge im Insektenei." In *Festgabe Jacob Henle zum 4. April 1882 dargebracht von seinen Schülern.* Bonn: M. Cohen, 80–111.

Weismann, August. 1882. "Bemerkungen zu Professor Bütschli's Gedanken über Leben und Tod." *Zoologischer Anzeiger* 5:377–380.

Weismann, August. 1883. *Die Entstehung der Sexualzellen bei den Hydromedusen. Zugleich ein Beitrag zur Kenntnis des Baues und der Lebenserscheinungen dieser Gruppe.* (Vol. Mit Atlas.) Jena: Gustav Fischer.

Weismann, August. 1883. [Einige Worte des Willkommens]. In *Amtlicher Bericht über die 56. Versammlung deutscher Naturforscher und Aezte welche zu Freiburg im Breisgau vom 18.–22. September 1883 tagte.* Freiburg i. Br.

Weismann, August, ed. 1883. [*Festrede gehalten bei der Enthüllung des Okendenkmals in Offenburg am 29. Juli 1883*], *Volksfreund.* Offenburg: Adolf Beck.

Weismann, August. 1883. *Ueber die Vererbung* [*Vortrag. gehalten bei der Feier der Uebergabe des Prorectorates in der Aula der Universität Freiburg am 21. Juni 1883*]. Jena: Gustav Fischer.

Weismann, August. 1883. *Ueber die Ewigkeit des Lebens, Programm wodurch zur Feier des Geburtstages seiner koeniglichen hoheit unseres durchlauchtigsten Grossherzogs Friedrich im Namen des academischen Senates die angehoerigen der Albert-Ludwigs-Universitaet einladet der gegenwaertige Prorector Dr. August Weismann.* Freiburg i. Br.: H. M. Poppen & Sohn.

Weismann, August. 1884. *Ueber Leben und Tod.* Jena: Gustav Fischer.

Weismann, August. 1884. "Die Entstehung der Sexualzellen bei den Hydromedusen." *Biologisches Zentralblatt* 4:12–31.

Weismann, August. 1885. "Zur Frage nach der Unsterblichkeit der Einzelligen." *Biologisches Zentralblatt* 4:650–665, 677–691.

Weismann, August. 1885. *Die Continuität des Keimplasmas als Grundlage einer Theorie der Vererbung.* Jena: Gustav Fischer.

Weismann, August. 1885. "Ueber die Bedeutung der geschlechtlichen Fortpflanzung für die Selektionstheorie." In *Tageblatt der 58. Versammlung deutscher Naturforscher und Aerzte in Strassburg 18.–23. September 1885,* edited by J. Stilling. Strassburg: G. Fischbach.

Weismann, August. 1885. [Bemerkungen zu dem Vortrag von Virchow: "Über Acclimatisation."] In *Tageblatt der Versammlung deutscher Naturforscher und Aerzte in Strassburg 18.-23. September 1885,* edited by J. Stilling. Strassburg.

Weismann, August. 1886. "Zur Annahme einer Continuität des Keimplasma's." *Berichte der Naturforschenden Gesellschaft zu Freiburg i. Br.* 1:89–99.

Weismann, August. 1886. *Die Bedeutung der Sexuellen Fortpflanzung für die Selections-Theorie.* Jena: Gustav Fischer.

Weismann, August. 1886. "Über den Rückschritt in der Natur." *Berichte der naturforschenden Gesellschaft zu Freiburg i. Br.* 2:1–30.

Weismann, August. 1886. "Zur Geschichte der Vererbungstheorie." *Zoologischer Anzeiger* 9:344–350.

Weismann, August. 1886. "Zur Frage nach der Vererbung erworbener Eigenschaften." *Biologisches Zentralblatt* 6:33–48.

Weismann, August. 1886. "Richtungskörper bei parthenogenetischen Eiern." *Zoologischer Anzeiger* 9:570–573.

Weismann, August. 1887. *Über die Zahl der Richtungskörper und über ihre Bedeutung.* Jena: Gustav Fischer.

Weismann, August, and Chiyomatsu Ischikawa. 1887. "Ueber die Bildung der Richtungskörper bei thierischen Eiern." *Berichte der naturforschenden Gesellschaft zu Freiburg i. Br.* 3:1–44.

Weismann, August. 1887. "On the Significance of the Polar Globules." *Nature* 36:607–609.

Weismann, August, and Chiyomatsu Ischikawa. 1888. "Weitere Untersuchungen zum Zahlengesetz der Richtungskörper." *Zoologische Jahrbücher, Abt. Anatomie* 3:575–610.

Weismann, August. 1888. "Das Zahlengesetz der Richtungskörper und seine Entdeckung." *Morphologisches Jahrbuch* 14:490–506.

Weismann, August, and Chiyomatsu Ischikawa. 1888. "Ueber die Befruchtungserscheinungen bei den Dauereiern von Daphniden." *Biologisches Zentralblatt* 8:430–436.

Weismann, August. 1888. "Ueber Zahl und Bedeutung der Richtungskörper." *Naturforscher* 31:56–57.

Weismann, August. 1889. "Über die Hypothese einer Vererbung von Verletzungen." In *Tageblatt der 61. Verwammlung deutscher Naturforscher und Ärzte in Köln vom 18.-23. September 1888. Wissenschaftlicher Theil Angeb.: Vorträge der Allgemeinen Sitzungen.* Köln: Albert Ahn.

Weismann, August, and Chiyomatsu Ischikawa. 1889. "Ueber partielle Befruchtung." *Berichte der Naturforschenden Gesellschaft zu Freiburg i. Br.* 4:51–51.

Weismann, August, and Chiyomatsu Ischikawa. 1889. "Nachtrag zu der Notiz über 'partielle Befruchtung.'" *Berichte der Naturforschenden Gesellschaft zu Freiburg i. Br.* 4:55–58.

Weismann, August, and Chiyomatsu Ischikawa. 1889. "Ueber die Paracopulation im Daphnidenei, sowie über Reifung und Befruchtung desselben." *Zoologische Jahrbücher, Abt. Anatomie* 4:155–196.

Weismann, August. 1889. "Gedanken über Musik bei Thieren und beim Menschen." *Deutsche Rundschau,* 1. Oktober 1889, 50–79.

Weismann, August. 1889. "Retrogression in Animal and Vegetable Life." *Open Court* 3:1801–1804, 1827–1831, 1840, 1855–1857.

Weismann, August. 1889. *Essays upon Heredity and Kindred Biological Problems.* Translated by A. E. Shipley and Selmar Schönland. Edited by Edward B. Poulton. 1st ed. Oxford: Clarendon Press.

Weismann, August. 1890. "Prof. Weismann's Theory of Heredity." *Nature* 41:317–323, 373.

Weismann, August. 1890. "Bemerkungen zu Ischikawa's Umkehrungs-Versuchen an Hydra." *Archiv für mikroskopische Anatomie* 36:627–638.

Weismann, August. 1890. "Bemerkungen zu einigen Tagesproblemen." *Biologisches Zentralblatt* 10:1–12, 33–44.

Weismann, August. 1891. *Amphimixis oder: Die Vermischung der Individuen.* Jena: Gustav Fischer.

Weismann, August. 1891–1892. *Essays upon Heredity and Kindred Biological Problems.* 2nd ed. 2 vols. Translated by Lilian J. Gould, Antoine Lüroth, and Fräulein Diestel. Edited by Edward B. Poulton. Oxford: Clarendon Press.

Weismann, August. 1892. *Aufsätze über Vererbung und verwandte biologische Fragen.* Jena: Gustav Fischer.

Weismann, August. 1892. *Das Keimplasma: Eine Theorie der Vererbung.* Jena: Gustav Fischer.

Weismann, August. 1892. "Vorwort in Albert Lang, Über die Knospung bei Hydra und einige Hydropolypen." In *Zeitschrift für wissenschaftliche Zoologie,* vol. 54, pp. 365–385.

Weismann, August. 1892. *Essais sur l'hérédité et la sélection naturelle.* Translated by Henry de Varigny. Paris: C. Reinwald.

Weismann, August. 1893. *The Germ-Plasm: A Theory of Heredity.* Translated by William Newton Parker and Harriet Rönnfeldt. Edited by H. Ellis. Contemporary Science Series. London: Walter Scott.

Weismann, August. 1893. "Historisches zur Lehre von der Continuität des Keimplasma's." *Berichte der Naturforschenden Gesellschaft zu Freiburg i. Br.* 7:36–37.

Weismann, August. 1893. "The All-Sufficiency of Natural Selection: A Reply to Herbert Spencer." *Contemporary Review* 64:309–338, 596–610.

Weismann, August. 1893. *Die Allmacht der Naturzuchtung. Eine Erwiderung an Herbert Spencer.* Jena: Gustav Fischer.

Weismann, August. 1894. *The Effects of External Influences upon Development.* Translated by Gregg Wilson. Romanes Lecture. London: Henry Frowde.

Weismann, August. 1894. *Äussere Einflüsse als Entwicklungsreize.* Jena: Gustav Fischer.

Weismann, August. 1895. "Vorwort." In *Soziale Evolution.* Edited by B. Kidd. Jena: Gustav Fischer.

Weismann, August. 1895. "Neue Versuche zum Saison-Dimorphismus der Schmetterlinge." *Zoologische Jahrbücher. Abteilung für Systematik, Geographie und Biologie der Tiere* 7:611–684.

Weismann, August. 1895. "Wie Sehen die Insekten?" *Deutsche Rundschau* 83:434–452.

Weismann, August. 1895. "Heredity Once More." *Contemporary Review* 68:420–456.

Weismann, August. 1896. *Über Germinal-Selection. Eine Quelle bestimmt gerichteter Variation.* Jena: Gustav Fischer.

Weismann, August. 1896. "New Experiments on the Seasonal Dimorphism of *Lepi-doptera*." *The Entomologist* 29:29–39, 74–80, 103–113, 153–157, 173–208, 240–252.

Weismann, August. 1896. "Germinal Selection." *The Monist* 6:250–293.

Weismann, August. 1896. *On Germinal Selection as a Source of Definite Variation*. 2nd ed. Translated by T. J. McCormack. Chicago: Open Court.

Weismann, August. 1899. "Thatsachen und Auslegungen in Bezug auf Regeneration." *Anatomischer Anzeiger* 15:445–474.

Weismann, August. 1899. "Regeneration: Facts and Interpretations." *Natural Science* 14:305–328.

Weismann, August. 1900. "Ueber die Parthenogenese der Bienen." *Anatomischer Anzeiger* 18:492–499.

Weismann, August. 1900. "Der Gesang der Vögel." *Beilage zur Allgemeinen Zeitung* 25 September, 1–5.

Weismann, August. 1901. "Zu verstehendem Aufsatz auf Herrn Dickel." [Meine Ansicht über die Freiburger Untersuchungsergebnisse von Bieneneiern.] *Anatomischer Anzeiger* 19:108–110.

Weismann, August. 1901. "Vorwort." In *Experimentelle entomologische Studien*. Edited by P. I. Bachmetjew. Leipzig: Wilhelm Engelmann.

Weismann, August. 1902. *Vorträge über Descendenztheorie gehalten an der Universität zu Freiburg im Breisgau*. 2 vols. Jena: Gustav Fischer.

Weismann, August. 1903. "Versuche über Regeneration bei Tritonen." *Anatomischer Anzeiger* 22:425–431.

Weismann, August. 1904. *Vorträge über Descendenztheorie gehalten an der Universität zu Freiburg im Breisgau*. 2nd ed. 2 vols. Jena: Gustav Fischer.

Weismann, August. 1904. *The Evolution Theory*. 2 vols. Translated by J. Arthur Thomson and Margaret Thomson. London: Edward Arnold.

Weismann, August. 1906. "Semons 'Mneme' und die 'Vererbung erworbener Eigenschaften.'" *Archiv für Rassen- und Gesellschafts-Biologie* 3:1–27.

Weismann, August. 1908. "Eine hydrobiologische Einleitung." *Internationale Revue der gesamten Hydrobiologie und Hydrographie* 1:1–9.

Weismann, August. 1909. "The Selection Theory." In *Darwin and Modern Science*. Edited by A. C. Seward. Cambridge: Cambridge University Press.

Weismann, August. 1909. *Die Selektionstheorie. Eine Untersuchung*. Jena: Gustav Fischer.

Weismann, August. 1909. *Charles Darwin und sein Lebenswerk. Festrede gehalten zu Freiburg i. Br. am 12 Februar 1909*. Jena: Gustav Fischer.

Weismann, August. 1909. "Über die Trutzstellung des Abendpfauenauges." *Naturwissenschaftliche Wochenschrift* N. F. 8:721–726.

Weismann, August. 1913. *Vorträge über Descendenztheorie gehalten an der Universität zu Freiburg im Breisgau*. 3rd ed. 2 vols. Jena: Gustav Fischer.

Notes

Preface

1. This work, originally published in Russian in 1971, provides an outline of Weismann's career. Leonid I. Blacher, *The Problem of the Inheritance of Acquired Characters: A History of a Priori and Empirical Methods Used to Find a Solution,* trans. Noel Hess, ed. Frederick B. Churchill (Washington, D.C.: Smithsonian Institution Libraries/National Science Foundation, 1982).

2. See Frederick B. Churchill, "Life before Model Systems: General Zoology at August Weismann's Institute," *American Zoologist* 37 (1997): 260–268.

1. Growing Up and Early Research

1. *Encyclopaedia Britannica,* 9th ed., s.v. "Frankfurt-on-the-Main."

2. All three of these have all been transcribed in Frederick B. Churchill and Helmut Risler, *August Weismann, Ausgewählte Briefe und Dokumente = Selected Letters and Documents,* 2 vols. (Freiburg: Universitätsbibliothek Freiburg, 1999), 2:529, 531–536, 537–606. All translations are by F. B. C.

3. Wend von Kalnein, ed., *Die Düsseldorfer Malerschule* (Mainz: Philipp von Zabern, 1979).

4. Friedrich Justin Bertuch, *Naturgeschichtliche Belustigungen, od. Abbild. Naturhist. Gegenstände* (Weimar: Verlage des Industrie-Comptoirs, 1811).

5. Churchill and Risler, *August Weismann,* 2:547.

6. Ibid., 2:547–548.

7. Ibid., 2:548–549.

8. Ibid., 2:549.

9. Otto Felsberg, "Die Nachkommenschaft des Valentin Weismann aus Weierburg (Niederösterreich) und des Ehrenreich Weismann 1641–1717, württembergischer

Generalsuperintendent & evangelischer Abt des Klosters Maulbronn/Wüttemberg," *Beiträge zur Geschichte der Familie Weismann* 4 (1936), 38–44.

10. Churchill and Risler, *August Weismann,* vol. 1.

11. Ibid., 2:552.

12. Ibid., 2:544.

13. Ibid., 1:212–13.

14. Hans-Heinz Eulner, *Die Entwicklung der medizinishen Spezialfächer an den Universitäten des deutschen Sprachgebietes* (Stuttgart: Ferdinand Enke, 1970), 45.

15. Ibid., 334, 497.

16. Churchill and Risler, *August Weismann,* 2:553.

17. Johannes Müller, *Handbuch der Physiologie des Menschen für Vorlesungen,* 2 vols. (Koblenz: Verlag von J. Hölscher, 1837–1840).

18. Churchill and Risler, *August Weismann,* 2:554–555.

19. August Weismann, *De acidi hippurici in copore humano generatione: Dissertatio inauugralis Göttingen* (Frankfurt am Main: Typis Broennerianis, 1857).

20. August Weismann, "Untersuchungen über den Salzgehalt der Ostsee," *Archive für Landeskunde Grossherzogthümern Mecklenburg* 8 (1858): 289–304; "Analysen des Ostseewasser," *Archive für Landeskunde Grossherzogthümern Mecklenburg* 8 (1858): 437–444.

21. Weismann wrote, "Geduld u. apothkerhafte Genauigkeit beim experimentieren." See Churchill and Risler, *August Weismann,* 2:558.

22. Ibid., 2:558–560.

2. The Age of Development

1. For more detailed accounts of their earliest careers and references to many associated secondary sources, see Timothy Lenoir, *The Strategy of Life: Teleology and Mechanics in Nineteenth-Century German Biology* (Dordrecht: D. Reidel, 1982), 65–71; and Frederick B. Churchill, "The Rise of Classical Descriptive Embryology," in *A Conceptual History of Modern Embryology,* ed. Scott Gilbert (New York: Plenum Press, 1991), 2–3.

2. *De Ovi Mammalium et Hominis Genesi,* trans. Charles O'Malley, in Karl Ernst von Baer and Charles Donald O'Malley, "On the Genesis of the Ovum of Mammals and of Man," *Isis* 47 (1956): 117–157.

3. Karl Ernst von Baer, *Über Entwickelungsgeschichte der Thiere: Beobachtung und Reflexion* (Königsberg: Gebrüdern Borntrager, 1828 and 1837).

4. Jane M. Oppenheimer, "Problems, Concepts and Their History," in *Analysis of Development,* B. H. Willier, Paul Weiss, and Victor Hamburger (Philadelphia: Saunders, 1960), 1–24.

5. Baer, *Über Entwickelungsgeschichte,* 1828, 51.

6. Churchill, "Rise of Classical Descriptive Embryology," 10–12.

7. Baer, *Über Entwickelungsgeschichte,* 1828, 150.

8. Ibid., 157.

9. See Karl Ernst von Baer, "Fragments Relating to Philosophical Zoology: Selected from the Works of K. E. von Baer," in *Scientific Memoirs, Selected from the Transactions of Foreign Academies of Science, and from Foreign Journals of Natural History,* ed. A. Henfrey and T. H. Huxley (London: Taylor and Francis, 1853), vol. 1, pt. II, 186–238.

10. Baer, *Über Entwickelungsgeschichte,* 1828), 225.

11. Baer, "Fragments Relating," 237–238.

12. Boris Eugen'evic Raikov, *Karl Ernst von Baer, 1792–1876. Sein Leben und sein Werk* (Leipzig: Johann Ambrosius Barth, 1968).

13. Karl Ernst von Baer, *Vorlesungen über Anthroplogie, für den Selbstunterricht,* Erster Theil (Königsberg: Gebrüder Borntrager, 1824). See the facsimile reprint in *Carl Ernst von Baer on the Study of Man and Nature,* ed. William Coleman (New York: Arno Press, 1981).

14. Raikov, *Karl Ernst von Baer,* 44–64.

15. Martin Heinrich Rathke, *Untersuchungen über die Bildung und Entwickelung des Flusskrebses* (Leipzig: Leopold Voss, 1829).

16. Christian Pander, *Beiträge zur Entwickelungsgeschichte des Hünchens im Eye* (Würzburg: H. L. Brönner, 1817), 12, all translations by F. B. C.

17. Jane Oppenheimer, "The Non-Specificity of the Germ-Layers," *Quarterly Review of Biology* 15 (1940): 1–27. Reprinted in *Essays in the History of Embryology and Biology,* ed. Jane Oppenheimer (Cambridge, Mass.: MIT Press, 1967), 256–294.

18. Transcribed by Klaus Sander for Weismann's copybooks; see Frederick B. Churchill and Helmut Risler, *August Weismann, Ausgewählte Briefe und Dokumente = Selected Letters and Documents,* 2 vols. (Freiburg: Universitätsbibliothek Freiburg, 1999), 2:745–747.

19. Henry Harris, *The Birth of the Cell* (New Haven, Conn.: Yale University Press, 1999).

20. Churchill and Risler, *August Weismann,* 2:554.

21. John Farley, *Gametes and Spores: Ideas about Sexual Reproduction, 1750–1914* (Baltimore: Johns Hopkins University Press, 1982). See also Frederick B. Churchill, "Sex and the Single Organism: Biological Theories of Sexuality in Mid-Nineteenth Century," *Studies in History of Biology* 3 (1979): 139–177.

22. Farley, *Gametes and Spores,* 43–47.

23. Ibid., 48–54.

24. Chamisso's term for the phenomenon was *"Generationswechsel."*

25. Johannes Japetus Smith Steenstrup, *On the Alternation of Generations: or, the Propagation and Development of Animals through Alternate Generations: A Peculiar Form of Fostering the Young in the Lower Classes of Animals.,* trans. George Busk, from German version by C. H. Lorenzen (London: Ray Society, 1845). For discussions

about Steenstrup and his unique ideas, see Churchill, "Sex and the Single Organism," 143–145; Farley, *Gametes and Spores,* chapter 4.

26. Rudolf Leuckart, "Zeugung," in *Handwörterbuch der Physiologie mit Rucksicht auf physiologische Pathologie,* 4 tomes in 5 vols., ed. Rudolf Wagner (Braunschweig: F. Vieweg u. Sohn, 1849–53), 4:707–1000.

27. This conclusion is emphasized in Churchill, "Sex and the Single Organism," 157–171. For Leuckart's influence, see Frederick B. Churchill, "From Heredity Theory to Vererbung: The Transmission Problem, 1850–1915," *Isis* 78 (1987): 339–342.

28. I have used the English translation, which appeared the following year. Albert Kölliker, *Manual of Human Histology,* 2 vols., ed. and trans. George Busk and Thomas Huxley (London: Sydenham Society, 1853–1854), 6–7. See also Albert Kölliker, *Erinnerungen aus meinem Leben* (Leipzig: Wilhelm Engelmann, 1899), 188–189.

29. Churchill and Risler, *August Weismann,* 2:560.

30. Ibid., 2:560–561.

31. August Weismann, "Ueber das Wachsen der quergestreiften Muskeln nach Beobachtungen am Frosch," *Zeitschrift für rationelle Medizin* 10 (1861): 263–284; and "Ueber die zwei Typen contractilen Gewebes und ihre Vertheilung in die grossen Gruppen des Thierreichs, sowie über die histologische Bedeutung ihrer Formelemente," *Zeitschrift für rationelle Medizin* 15 (1862): 279–282.

3. Becoming an Embryologist

1. Quotation from Loetze provided and translated by Timothy Lenoir, *The Strategy of Life: Teleology and Mechanics in Nineteenth-Century German Biology* (Dordrecht: D. Reidel, 1982), 171.

2. Ibid., 172.

3. Quoted in Klaus Wunderlich, *Rudolf Leuckart. Weg und Werk* (Jena: Gustav Fischer, 1978), 19.

4. Ibid., 21–23, a capsule overview of Leuckart's publications during his Giessen period.

5. Rudolf Leuckart to C. Claus, 14 February 1861, quoted in Wunderlich, *Rudolf Leuckart,* 138fn47.

6. Frederick B. Churchill and Helmut Risler, *August Weismann, Ausgewählte Briefe und Dokumente = Selected Letters and Documents,* 2 vols. (Freiburg: Universitätsbibliothek Freiburg, 1999), 2:566.

7. "Freih Von Anders, erz . . . Kammurversteher (Schaumburg) an Dr. med. Etc. August Weismann, 19th Dezember 1861," unpublished manuscript, December 19, 1861, in my possession. This manuscript contains five conditions for employment at Schaumburg.

8. Churchill and Risler, *August Weismann,* 2:566.

9. Ibid., 2:565–567.

10. In his brief "Autobiographie von August Weismann aus dem Jahre 1896," Weismann described his stay at Schloss Schaumburg as lasting for two years (i.e., "1860–1862"). This miscalculation appeared in the 1903 English version. See Churchill and Risler, *August Weismann,* 2:531.

11. Churchill and Risler, *August Weismann,* 2:568.

12. Ibid., 2:569.

13. Ibid., 2:572–575.

14. Ibid., 2:745.

15. August Weismann, "Über die Entstehung des vollendeten Insekts in der Larve und Puppe: Ein Beitrag zur Metamorphose der Insekten," *Abhandlungen* der *Senckenbergischen Naturforschenden Gesellschaft* 4 (1862–1863), 227–260; *Habilitationsschrift. zu Freiburg. Weismann* (Frankfurt am Mein: M. Brönner, 1863), "Die Entwicklung der Dipteren im Ei, nach Beobachtungen an Chironomus spec., *Musca vomitoria* und *Pulex Canis* [Erster Theil]," *Zeitschrift für wissenschaftliche Zoologie* 13 (1863): 107–220, "Die nachembryonale Entwicklung der Musciden nach Beobachtungen an *Musca vomitoria* und *Sarcophaga carnaria* [Zweiter Theil]," *Zeitschrift für wissenschaftliche Zoologie* 14 (1864): 187–336. The latter two articles appeared soon after as August Weismann, *Die Entwicklung der Dipteren: Ein Beitrag zur Entwicklungsgeschichte der Insecten* (Leipzig: Wilhelm Engelmann, 1864).

16. Francis Maitland Balfour, *A Treatise on Comparative Embryology,* 2 vols. (London: Macmillan, 1880–1881). The German edition appeared almost immediately after the English: Francis Maitland Balfour, *Handbuch der vergleichenden Embryologie,* trans. B. Vetter (Jena: Gustav Fischer, 1880–1881).

17. Balfour, *A Treatise,* 1:1–6; Francis Maitland Balfour, "Address to the Department of Anatomy and Physiology of the British Association, 1880," reprinted in *The Works of Francis Maitland Balfour,* ed. M. Foster and Adam Sedgwick (London: Macmillan, 1885).

18. Eugen Korschelt and Karl Heider, *Lehrbuch der vergleichenden Entwicklungsgeschichte der wirbellosen Thiere* (Jena: Gustav Fischer, 1890–1902). As with Balfour's text, this work was translated into the other language. Eugen Korschelt and Karl Heider, *Text-Book of the Embryology of Invertebrates,* part 1, trans. Edward L. Mark and W. McM. Woodworth; parts 2–4, trans. H. J. Campbell (London: Swan Sonnenschein, 1895). The *Allgemeiner Theil* of 1902, however, was not translated. The quotation is taken from the introduction to volume 1 of the English version.

19. For historical reflections on the tradition, see Jane M. Oppenheimer, "The Non-Specificity of the Germ-Layers" (1940), reprinted in her *Essays in the History of Embryology and Biology* (Cambridge, Mass.: MIT Press, 1967), 256–294, and Frederick B. Churchill, "Living with the Biogenetic Law," in *From Embryology to Evo-Devo: A History of Developmental Evolution,* ed. Manfred D. Laubichler and Jane Maienschein (Cambridge Mass.: MIT Press, 2007), 37–81.

20. Korschelt and Heider, *Text-Book,* 167, 263.

21. Ibid., 352–354.

22. Ibid., 269.

23. Klaus Sander, "August Weismanns Untersuchungen zur Insektenentwicklung 1862–1882," in *August Weismann (1834–1914) und die theoretische Biologie des 19. Jahrhunderts. Urkunden, Berichte und Analysen,* ed. Klaus Sander (Freiburg: Rombach Verlag, 1985), 43–52.

24. Weismann, *Die Entwicklung der Dipteren,* 3–4.

25. Weismann, "Die nachembryonale Entwicklung der Musciden," 227.

26. Rudolf Leuckart, *Zur Kenntniss d. Generationswecksels u. d. Parthenogenesis bei der Insecten* (Frankfurt: Verlag von Meidinger Sohn, 1858); and *Die Fortpflanzung und Entwickelung der Pupiparen nach Beobachtungen an Melophagus ovinus* (Halle: H.W. Schmidt, 1858).

27. Albert Kölliker, *De prim insectorum general* (Turici: Meyeri et Zelleri, 1842).

28. Weismann, *Die Entwicklung der Dipteren,* 90.

29. Weismann, "Die Entwicklung der Dipteren," 204–205.

30. Ibid., 206–207.

31. Sander, "August Weismanns," 47.

32. Weismann, "Die Entwicklung der Dipteren," 208.

33. Sander, "August Weimanns," 46.

34. Weismann, "Die Entwicklung der Dipteren," 211–212.

35. Vitus Graber, *Die Insecten* (München: R. Oldenbourg 1877); John Lubbock, *Origin and Metamorphoses of Insects* (London, Macmillan, 1874).

36. Weismann, "Die nachembryonale Entwicklung der Musciden," 187.

37. Weismann, "Über die Entstehung"; Weismann, "Die nachembryonale Entwicklung der Musciden."

38. For example, Jan Swammerdam, *Biblia Naturae* (1758).

39. Weismann, "Über die Entstehung." Weismann pointed out that this was his Habilitationsschrift (227–228).

40. Louis Agassiz, "The Classification of Insects from Embryological Data," *Smithsonian Contribution to Knowledge* 2 (1851): 1–28.

41. Ibid., 14, 19.

42. Weismann, "Über die Entstehung," 227–228. Weismann provides a lengthy quotation from Burmeister's *Handbuch* documenting his description of early adult parts.

43. Klaus Sander gives three extended quotations from Weismann's two accounts ("August Weismanns," 49).

44. See Weismann, "Die Entwicklung der Dipteren," 157–160. The term *Anlage(n)* presents a problem for translation. Weismann here used it in the sense of a cellular precursor, which would be different from a "rudiment" or miniature in situ.

45. Weismann, "Die nachembryonale Entwicklung der Musciden," 323.

46. Ibid., 325–326.

47. Ibid., 326–327.

48. Quotations in August Weismann, "Zur Embryologie der Insecten," *Archiv für Anatomie, Physiologie, und wissenschaftliche Medicin* (1864): 268–269.

4. Studies in Descent, Part I

1. Frederick B. Churchill and Helmut Risler, *August Weismann, Ausgewählte Briefe und Dokumente* = *Selected Letters and Documents*, 2 vols. (Freiburg: Universitätsbibliothek Freiburg, 1999), 2:568.

2. von Gabriele Blod, Wolfgang Hug, Manfred Lallinger, and Haartmut Zuche, "Unruh im 'Pfaffenstadtchen,' Reaktion, 'Neue Ära' und Kulturkampf (1850–1870)," in *Geischichte der Stadt Freiburg im Breisgau*, vol. 3, ed. Heiko Haumann und Hans Schadek (Stuttgart: Conrad Theiss Verlag, 1992).

3. "Nicht dass ich dort äusserlich viel zu erreichen gehofft hätte, aber ruhige Arbeitszeit schien mir fur mich jetzt das wichtigste, u. musste ich dort finden können." Churchill and Risler, *August Weismann*, 2:573.

4. Churchill and Risler, *August Weismann*, 2:573–574.

5. August Weismann, "Über des vollendeten Inseckts Laarve und Puppe: Ein Beitrag zur Metamorphose der Insekten," *Abhandlungen* der *Senckenbergischen Naturforschenden Gesellschaft* 4 (1862–1863): 227–260.

6. Churchill and Risler, *August Weismann*, 2:577.

7. Weismann to Ernst Haeckel, 13 February 1871, in Georg Uschmann and Bernhard Hassenstein, "Der Briefwechsel zwischen Ernst Haeckel und August Weismann," in *Kleine Festgabe aus Anlass der hundertjährigen Wiederkehr der Gründung des Zoologischen Institutes der Friedrich-Schiller-Universität Jena im Jahre 1865,* ed. M. Gersch (Jena: Friedrich-Schiller-Universität, 1965), 27.

8. Churchill and Risler, *August Weismann*, 2:538–540.

9. Ibid., 2:552.

10. Personal communication.

11. Weismann to Haeckel, 15 October 1865, in Uschmann and Hassenstein, "Der Briefwechsel."

12. The majority of the letters are Weismann's to Haeckel. These were preserved in the "Villa Medusa" when Haeckel's residence was deeded to the Carl-Zeiss-Stiftung with the proviso that Haeckel could live out his life there. In 1920, it was officially opened as a museum and as a repository of Haeckel's archives. Haeckel's letters to Weismann came to light only after World War II when in 1964 a granddaughter of Weismann, Hertha (neé Schepp) von Dechend, sold them to Freiburg University. These appear to be the remnants of a "trunk full" of incoming letters, most of which were probably destroyed during the immediate postwar period. See the introduction in Uschmann and Hassenstein, "Der Briefwechsel."

13. Uschmann and Hassenstein, "Der Briefwechsel," Weismann to Haeckel, 15 October 1865.

14. Ibid., 12.

15. Ibid.

16. Ibid., 14.

17. Ibid., Haeckel to Weismann, 4 June 1866; Haeckel to Weismann, 28 June 1867.

18. Ibid., Weismann to Haeckel, 21 May 1867.

19. Ibid., Weismann to Haeckel, 12 July 1866; Weismann to Haeckel, 13 February 1871.

20. Ibid., Haeckel to Weismann, 3 May 1871.

21. Ibid.

22. Ibid., Weismann to Haeckel, 13 February 1871; Haeckel to Weismann, 3 May 1871; Weismann to Haeckel, 31 May 1871; Haeckel to Weismann, 2 January 1877.

23. Ernst Haeckel, *Die Kalkschwämme: Eine Monographie,* 3 vols. (Berlin: Georg Reimer, 1872), 483–484; Ernst Haeckel, "On the Calcispongiae, Their Position in the Animal Kingdom, and Their Relation to the Theory of Descendance," *Annals and Magazine of Natural History* 11, series 4 (1873): 241–262, 421–430; quotation from p. 430.

24. See Jane Oppenheimer, "The Non-Specificity of the Germ-Layers," *Quarterly Review of Biology* 15 (1940): 1–27, reprinted in Jane Oppenheimer, *Essays in the History of Embryology and Biology* (Cambridge, Mass.: MIT Press, 1967), 256–294; Frederick B. Churchill, "Living with the Biogenetic Law: A Reappraisal," in *From Embryology to Evo-Devo: A History of Developmental Evolution,* ed. Manfred D. Laubichler and Jane Maienschein (Cambridge Mass.: MIT Press, 2007), 37–81.

25. Haeckel, *Die Kalkschwämme,* 1:483.

26. Uschmann and Hassenstein, "Der Briefwechsel," Weismann to Haeckel, 3 January 1873.

27. Ibid., Weismann to Haeckel, 3 January 1873.

28. Ibid., Haeckel to Weismann, 12 December 1873.

29. Ibid., 36, Weismann to Haeckel, 27 January 1874.

30. Ibid., Haeckel to Weismann, 6 February 1874. See also August Weismann, "John Lubbock's *Monograph of the Collembola and Thysanura* [1873] and *On the Origin of Metamorphosis of Insects* [1874]," *Jenaer Literaturzeitung* 18 (1874): 262–265; John Lubbock, *Monograph of the Collembola and Thysanura* (London: Ray Society, 1873), and *On the Origin and Metamorphosis of Insects* (London: Macmillan, 1874).

31. Georg Uschmann, ed., *Ernst Haeckel: Forscher, Künstler, Mensche. Briefe* (Leipzig: Urania Verlag, 1961), 142–151.

32. Ernst Haeckel, "Die Gastrula und die Eifurchung der Thiere," *Jenaische Zeitung für Naturwissenschaft* 9 (1875): 499–501.

33. Uschmann and Hassenstein, "Der Briefwechsel," 37–38, Weismann to Haeckel, 27 December 1875.

34. Eugen Korschelt and Karl Heider, *Text-Book of the Embryology of Invertebrates,* Part 3, trans., H. J. Campbell (London: Swan Sonnenschein, 1895), 289–290.

35. These two terms, coined in *Anthropogenie* in 1874, were fully explicated in Haeckel, "Die Gastrula."

36. Uschmann and Hassenstein, "Der Briefwechsel," 37–38, Weismann to Haeckel, 27 December 1875.

37. Ibid., Haeckel to Weismann, 26 October 1877.

38. Ernst Haeckel, *Der Monismus als Band zwischen Religion und Wissenschaft* (Bonn: E. Strauss, 1892).

39. Ernst Haeckel, "Zur Phylogenie der australischen Fauna. Systematische Einleitung," in *Zoologische Forschungsreisen in Australien und dem Maylayischen Archipel,* band. 1, part 1: *Ceratodus,* I–XXIV, ed. Richard Semon (Jena: Gustav Fischer, 1893–1913).

40. See Ernst Haeckel, *Generelle Morphologie der Organismen* (Berlin: Georg Reimer, 1866), 2:170–190, in which he contrasts in their many forms both "conservative" and "progressive Vererbung."

41. Ernst Haeckel, *Natürliche Schöpfungsgeschichte* (Berlin: Georg Reimer, 1889), xxii, 149, and *Anthropogenie; oder, Entwichlungsgeschichte des menschen* (Leipzig: Engelmann, 1891), 836.

42. Uschmann and Hassenstein, "Der Briefwechsel," 54, Haeckel to Weismann, 15 January 1894.

43. Ibid., 56, Weismann to Haeckel, 19 January 1894.

44. Ibid., 63, Weismann to Haeckel, 8 February 1914.

45. Churchill and Risler, *August Weismann,* 2:711–721.

46. August Weismann, *Über die Berechtigung der Darwin'schen Theorie* (Leipzig: Wilhelm Engelmann, 1868).

47. August Weismann, *Über den Einfluss der Isolirung auf die Artbildung* (Leibzig: Wilhelm Engelmann, 1872).

48. August Weismann, *Studien zur Descendenztheorie,* 2 vols. (Leipzig: Wilhelm Engelmann, 1875–1876). This important series of essays was soon translated into English by Raphael Meldola, with a brief "Prefatory Notice" by the elderly Charles Darwin and soon appeared in translation as *Studies in the Theory of Descent* (London: Sampson Low, Marston, Searle, & Rivington, 1882).

49. Weismann, *Über die Berechtigung der Darwin'schen Theorie.*

50. Ingrid Schumacher, "Die Entwicklungstheorie des Heidelberger Paläontologen und Zoologen Heinrich Georg Bronn (1800–1862)" (PhD diss., Ruprecht-Karl-Universitat, Heidelberg, 1975), chapter 7; Sander Gliboff, *H. G. Bronn, Ernst*

Haeckel, and the Origins of German Darwinism: A Study in Translation and Transfor-mation (Cambridge, Mass.: MIT Press, 2008).

51. For more extensive contemporary bibliographies of Darwinian literature, see Georg Seidlitz, *Die Darwin'sche Theorie: Elf Vorlesungen über die Entstehung der Thiere und Planzen durch Naturzüchtung* (Dropat: C. Mattiesen, 1871); J. W. Spengel, *Die Darwinische Theorie, Verzeichniss der über dieselbe in Deutschland, England, Amerika, Frankreich, Italien, Holland, Belgien und den Skandinavischen Reichen erschienen Schriften und Aufsätze* (Berlin: Wiegandt und Hempel, 1872).

52. Weismann, *Über die Berechtigung der Darwin'schen Theorie*, 14–15.

53. Ibid., 29.

54. Ibid., 24.

55. Uschmann and Hassenstein, "Der Briefwechsel," 14, Weismann to Haeckel, 4 December 1865.

56. Moritz Wagner, *Die Darwin'sche Theorie und das Migrationsgesetz der Organismen* (Leipzig: Duncker & Humblot, 1868).

57. Frank J. Sulloway, "Geographic Isolation in Darwin's Thinking: The Vicissitudes of a Critical Idea," *Studies in History and Philosophy of Biology and Biomedical Sciences* 3 (1979): 23–65.

58. Wagner, *Die Darwin'sche Theorie*, 37.

59. Ibid., 4.

60. Francis Darwin, ed., *Life and Letters of Charles Darwin* (New York: D. Appleton, 1893), 336–337.

61. Weismann delivered his inaugural address on 8 July 1868; he signed the preface of the printed version on August 2. The addendum responding to Wagner would have been added just before returning the proofs to the publisher.

62. Moritz Wagner, "Ueber den Einfluss der geographischen Isolirung und Colonienbildung auf die morphologischen Veränderungen der Organismen," *Sitzungsberichte der königl. Bayerischen Akademie der Wissenschaften* 10 (1870): 154–174.

63. Ibid., 155.

64. Ibid., 166–171.

65. Weismann, *Über den Einfluss*.

66. F. Darwin, *Life and Letters*, 334–336, Darwin to Weismann, 5 April 1872.

67. Uschmann and Hassenstein, "Der Briefwechsel," Weismann to Haeackel, 13 February 1871, 31 May 1871.

68. Franz Hilgendorf, "Über Planorbis multiformis im Steinheimer Süsswasskalk," *Akademie der Wissenschaften zu Berlin Montasberichte* (1866): 274–503; Wolf-Ernst Reif, "The Search for a Macroevolutionary Theory in German Paleontology," *Journal of the History of Biology* 19 (1986): 79–130; Wolf-Ernst Reif, "Endemic Evolution of *Gyraulus Kleini* in the Steinheim Basin (Planorbid snails,

Miocene, Southern Germany)," in *Sedimentary and Evolutionary Cycles*, ed. Ulf Bayer and Adolf Seilacher (Berlin: Springer-Verlag, 1985). I follow closely my account of Hilgendorf in Churchill and Risler, *August Weismann*, 2:777–781.

69. Weismann, *Über den Einfluss*, 19.

70. Ibid., 28.

71. Ibid., 77–84.

72. Ibid., 48–51.

73. August Weismann, "Bericht über die Weiterentwicklung der Descedenztheorie im Jahre 1872," *Archiv für Anthropologie* 6 (1873): 139.

74. Ibid.

75. August Weismann, "Zur Abwehr," *Beilage zur Allgemeinen Zeitung*, Nr. 336, S. 5131, 2 (December 1873).

76. Alfred R. Wallace, "On the Phenomena of Variation and Geographic Distribution as Illustrated by the Papilionidae of the Maylayan Region, or Swallow-tailed Butterflies, as Illustrative of the Theory of Natural Selection," *Transactions of the Linnaean Society* 25 (1864): 1–71. A later, somewhat altered version of the text of this paper appeared in Alfred R. Wallace, ed., *Contributions to the Theory of Natural Selection* (London: Macmillan, 1870). A facsimile edition of this volume was issued by the AMS Press of New York. Both 1870/1973 versions of "the Maylayan Papilionidae" included a reprint of the 1864 original text but with tables, plates, and references to plates omitted. Some of the scientific names had also been changed by Wallace owing to the data from Cajetan and Rudolf Felder of the *Novara* Expedition, in "Lepidoptera," in *Reise der österreichischen Fregatte Novara um die Erde in den Jahren 1857, 1858, 1859 unter den Befehlen des Commodore B. von Wüllerstorf-Urbair*, 3 vols. (Wien: Kaiserlich-königliche Hof- und Staatsdruckerei, 1864–1875).

77. Both quotations from Wallace, *Contributions to the Theory of Natural Selection*, 131–132.

78. Ibid., 142–185.

79. Ibid., 143–165.

80. Uschmann and Hassenstein, "Der Briefwechsel," Weismann to Haeckel, 13 February 1871; Haeckel to Weismann, 3 May 1871. See also Alfred R. Wallace, "Mimicry and Other Protective Resemblances among Animals," *Westminster Review*, n.s., 32, no. 1 (1867): 1–43.

81. William Charles Kimler, "One Hundred Years of Mimicry: History of an Evolutionary Exemplar" (PhD diss., Cornell University, 1983); Muriel L. Blaisdell, *Darwinism and Its Data, The Adaptive Coloration of Animals* (New York: Garland, 1992).

82. Weismann, *Über den Einfluss*, 58–59.

83. August Weismann, *Studien zur Descendenztheorie*, 2 vols. (Leipzig: Wilhelm Engelmann, 1875–1876), 1; August Weismann, *Studies in the Theory of Descent*, 2

vols., trans. Raphael Meldola (London: Sampson Low, Marston, Searle & Rivington, 1882), 1–2.

84. Weismann, *Über den Einfluss,* 19–21.

85. Weismann, *Studien zur Descendenztheorie,* 1.

86. Georg Dorfmeister, "Ueber Arten und Varietäten der Schmetterlinge," *Mitteilungen des Naturwissenschaftlichen Vereines für Steiermark* 2 (1864): 95–98; Georg Dorfmeister, "Ueber die Einwirkung verschiedener, während der Entwicklungsperioden angewendeter Wärmegrade auf die Färbung und Zeichnung der Schmetterlinge," *Mitteilungen des Naturwissenschaftlichen Vereines für Steiermark* 2 (1864): 99–108.

87. W. Conner Sorensen, *Brethren of the Net: American Entomology, 1840–1880* (Tuscaloosa: University of Alabama Press, 1995), chapter 10. See also W. Conner Sorensen, "William Henry Edwards, August Weismann and Polymorphism in Butterflies," in *August Weismann (1834–1914) und die theoretische Biologie des 19. Jahrhunderts. Urkunden, Berichte und Analysen,* ed. Klaus Sander (Freiburg: Rombach Verlag, 1985).

88. William Edwards, *Memoirs of Col. William Edwards Formerly of Stockbridge and Northampton, Mass., Later of Hunter, Greene Co. N.Y. and of Brooklyn, N.Y., Written by Himself, in his 76th Year, 1847, with Notes and Additions by His Son, William W. Edwards, and by his Grandson, William Henry Edwards* ([Washington, D.C.: W. F. Roberts, 1897]).

89. William H. Edwards, *A Voyage up the River Amazon, Including a Residence at Pará* (New York: D. Appleton, 1847), 254–556. See also Sorensen, *Brethren of the Net.*

90. William H. Edwards, *The Butterflies of North America,* 3 vols. (Boston: Houghton Mifflin, 1874, 1884, 1897).

91. Sorensen, *Brethren of the Net,* 216.

92. William Henry Edwards, "Rearing Butterflies from the Egg," *Canadian Entomologist* 3 (1871): 70–73. At the time, Weismann asked for samples of *Grapta album* and some other butterflies.

93. Only two complete letters from Weismann to Edwards (the second from 1895) appear to be extant, but the publication of extracts from three other letters and their contents suggests that the correspondence was two-sided and beneficial to both the American businessman turned lepidopterist and the German zoologist. A decade later, the English chemist and lepidopterist Raphael Meldola revisited and updated Edwards's breeding experiments of the 1870s and included comments and data in the form of footnotes and an appendix to his translation of Weismann's *Studies in the Theory of Descent,* pp. 33, 126–160.

94. The scientific nomenclature of the Ajax/Zebra complex has also changed considerably. Alexander B. Klots designates the early spring form "marcellus," the later spring form "telamonides," and the summer form "lecontei." E. Shull uses the

same nomenclature, but his plates are confusing. Today, the species goes under the binomial name *Eurytides marcellus*. To be consistent, I will maintain the names of the 1870s.

95. W. H. Edwards, "An Abstract of Dr. Aug. Weismann's paper on 'The Seasonal Dimorphism of Butterflies,'" *Canadian Entomologist* 7 (1875): 228–240.

96. Weismann, *Studien zur Descendenztheorie*, 5.

97. Weismann to Edwards on Tharos, extract printed in Edwards, "History of Phycoiodes Tharos, A Polymorphic Butterfly," *Canadian Entomologist* 9 (1877), 1–10, 51–58. The same remark is reproduced by Meldola in his translation of Weismann, *Studies in the Theory of Descent*, Appendix II, pp. 140–148.

5. Studies in Descent, Part II

1. August Weismann, *Vorträge über Descendenztheorie gehalten an der Universität zu Freiburg im Breisgau*, 2 vols. (Jena: Gustav Fischer, 1902); "Vorwort," in Ernst Gaupp, *August Weismann: Sein Leben und sein Werk* (Jena: Gustav Fischer, 1917), 14.

2. August Weismann, *Studien zur Descendenztheorie*, 2 vols. (Leipzig: Wilhelm Engelmann, 1875–1876). These were soon published in English as *Studies in the Theory of Descent*, trans. Raphael Meldola, 2 vols. (London: Sampson Low, Marston, Searle & Rivington, 1882).

3. Weismann, *Studien zur Descendenztheorie*, 62.

4. Ibid., 105.

5. Ibid., 57.

6. Ibid., 30–33, and Weismann, *Studies in the Theory of Descent*, 40–42.

7. Rudolf Leuckart, *Die Fortpflanzung und Entwickelung der Pupiparen nach Beobachtungen an Melophagus ovinus* (Halle: H. W. Schmidt, 1858), 111; discussion in Frederick B. Churchill, "Sex and the Single Organism: Biological Theories of Sexuality in Mid-Nineteenth Century," *Studies in the History of Biology* (1979): 139–177.

8. Rudolf Leuckart, "Zeugung," in *Handwörterbuch der physiologie: Mit rücksicht auf physiologische pathologie*, 4 tomes in 5 vols., ed. Rudolf Wagner (Braunschweig: Vieweg und Sohn, 1853), 4:707–1000; Rudolf Leuckart, *Zur Kenntniss d. Generationswecksels u. d. Parthenogenesis bei der Insecten* (Frankfurt: Verlag von Meidinger Sohn, 1858); John Farley, *Gametes and Spores: Ideas about Sexual Reproduction, 1750–1914* (Baltimore: Johns Hopkins University Press, 1982), particularly chapter 3; Otto Taschenberg, "Rudolf Leuckart: Eine biographische Skizze," *Leopoldina—Amtliches Organ der kaiserlichen Leopoldino-Carolinischen Deutschen Akademie der Naturforscher* 35 (1899), 62–66, 82–94, 102–112, which points out that Leuckart used "Heterogonie" in his work on *A. nigrovenosa* (1865). "Metagenesis" was introduced into the literature by Richard Owen, *On Parthenogenesis; or, The*

Successive Production of Procreating Individuals from a Single Ovum (London: John von Voorst, 1849), 22, to distinguish the apparent alternation of generations of the *Pluteus* larvae into young echinoderms from the process of "metamorphosis." His example turned out to be just the opposite, but the thrust of the word, implying a radical alternation of one sexual form into another, continued to be useful. See also Churchill, *Sex and the Single Organism*.

9. For the basics, see Erich Hintzsche's sketch of Kölliker's career and work in the *Dictionary of Scientific Biography*, vol. 7, ed. Charles Gillispie (New York: Charles Scribner's Sons, 1970–1980), 437–440; Albert von Kölliker, *Erinnerungen aus Meinem Leben* (Leipzig: Wilhelm Engelmann, 1899). As an impressive interpretation of the generational dynamics of German anatomy and zoology, I highly recommend Lynn K. Nyhart, *Biology Takes Form: Animal Morphology and the German Universities, 1800–1900* (Chicago: Chicago University Press, 1995). Nyhart includes an excellent account of Kölliker's theory of heterogenesis.

10. See Nyhart's translation and discussion of this prospectus, in Nyhart, *Biology Takes Form,* 94–96.

11. Albert Kölliker, "Ueber die Darwin'sche Schöpfungstheorie," *Zeitschrift für wissenschaftliche Zoologie* 14 (1864): 174–186.

12. Ibid., 182.

13. Frederick B. Churchill, "August Weismann and a Break from Tradition," *Journal of the History of Biology* 1 (1968): 91–112.

14. Ernst Haeckel, *Generelle Morphologie der Organismen,* 2 vols. (Berlin: Georg Reimer, 1866), esp. 2:91–109.

15. Ibid., 2:107. He did not use either term in his *Zur Entwicklungsgeschichte der Siphonophoren* (1869).

16. Haeckel simply discusses the alternation of generations (metagenesis and sexual reproduction) without elaborating upon the great variety of permutations. See Ernst Haeckel, *Naturliche Schöpfungsgeschichte,* 2nd ed. (Berlin: Georg Reimer, 1870); it can be found in H. Ziegler and E. Bresslau, *Zoologisches Wörterbuch* (Jena: Fischer, 1927) but is simply defined as "Generationsfolge."

17. Nyhart, *Biology Takes Form,* esp. 179–193; Carl Claus, *Die Typenlehre und E. Haeckel's Gastraea-Theorie* (Vienna: Manz, 1874); Carl Grobben, "Carl Claus," *Arbeiten aus den zoologischen Instituten der Universität Wien und der zoologischen Station in Triest,* 11 (1899): 1–12.

18. Carl Claus, *Grundzüge der Zoologie: Zum Gebrauche a Universitäten und Höheren Lehranstalten sowie zum Selbststudium,* 2nd ed. (Marburg und Leipzig: N. G. Elwert'sche Universitäts-Buchhandlung, 1872). His basic divisions of *Urzeugung, Theilung, Sprossung, Keimbildung,* and *geschlechtliche Fortpflanzung* followed Leuckart's "Zeugung."

19. Ibid., 35.

20. Nyhart, *Biology Takes Form,* 187. Quoting a letter from Weismann to Claus, written 20 October 1874, Nyhart suggests that Weismann was uneasy with Haeckel's most recent publications.

21. Weismann, *Studies in the Theory of Descent,* 81–85.

22. Richard Owen, *On Parthenogenesis,* 76.

23. Weismann, *Studies in the Theory of Descent,* 85–86.

24. Ibid., 89, quoting Claus, *Grundzüge der Zoologie.*

25. Weismann, *Studies in the Theory of Descent,* 92–93.

26. Ibid., 98–99, quotation from p. 96. Weismann recognized, one could imagine, that evolutionary scenarios in which metagenesis might evolve a parthenogenetic form of reproduction as well as a complete contraction in metamorphic stages, such as with the summer generation of *Leptodora hyalina,* made the role of parthenogenesis clearer.

27. Overall, Claus was critical of Weismann's strategy, for it appeared to assign the insect *Cecidomyia* to regressive metagenesis and tunicates to heterogenesis. It is unclear, however, whether Claus intended a phylogenetic or ontogenetic objection. At the time, tunicates were considered a class of mollusks with affinities to Amphioxus and vertebrates (Claus, *Grundzüge der Zoologie* [1872], 690–695). Weismann generally avoided broad phylogenetic claims but often took generic and familial relationships into account, which will become clearer in Chapter 6.

28. Weismann, *Studies in the Theory of Descent,* 102.

29. Ibid., 100–107.

30. Ibid., 112.

31. Ibid., 115–116.

32. Haeckel to Weismann, 31 December 1875, Georg Uschmann and Bernhard Hassenstein, "Der Briefwechsel zwischen Ernst Haeckel und August Weismann," in M. Gersch, ed., *Kleine Festgabe aus Anlass der hundertjährigen Wiederkehr der Gründung des Zoologischen Institutes der Friedrich-Schiller-Universität Jena im Jahre 1865 durch Ernst Haeckel* (Jena: Friedrich-Schiller-Universität, 1965), 39.

33. Weismann, *Studies in the Theory of Descent,* 611.

34. Thomas Horsfield and Frederick Moore, *A Catalogue of Lepidopterous Insects in the Museum of the East India Company* (London: W. H. Allen, 1857).

35. Samuel Hubbard Scudder, the New England lepidopterist and curator of entomology at the Museum of Comparative Zoology, provided an exception to this generalization. See A. G. Mayor, "Samuel Hubbard Scudder, 1837–1911," *Memoirs of the National Academy of Sciences* 17 (1924): 74–104.

36. See Weismann, *Studies in the Theory of Descent,* 187, 209.

37. Given this chronology, it appears that Stephen Jay Gould, *Ontogeny and Phylogeny* (Cambridge, Mass.: Harvard University Press, 1977), and Frederick B. Churchill and Helmut Risler, *August Weismann, Ausgewählte Briefe und Dokumente = Selected*

Letters and Documents, 2 vols. (Freiburg: Universitätsbibliothek Freiburg, 1999) earlier have overemphasized the influence of Haeckel's biogenetic law on Weismann's caterpillar studies.

38. Weismann, *Studies in the Theory of Descent,* 168.

39. Ibid., 309.

40. Ibid., 273–275.

41. Ibid., 177.

42. For quotations and other differences between the two genera, see ibid., 227–229.

43. Ibid., 361–362.

44. For a discussion of the relevance of Weismann's caterpillar studies to his species concept, see Churchill and Risler, *August Weismann,* 2:782–784.

45. Weismann, *Studies in the Theory of Descent,* 435–439.

46. Ibid., 440.

47. Ibid., 471–480.

48. Ibid., 461–462 (emphasis mine); see also 479–480.

49. Ibid., 492–493. In this passage, Weismann is not quoting himself. The publication of his observations first appeared in his habilitationsschrift published in 1863–1864.

50. Weismann recognized that Fritz Müller first pointed out this counterintuitive phylogeny, a conclusion later supported by others. Meldola provided a long passage by Balfour, who emphasized in his textbook of 1880 "that the characters of the majority of existing larva forms of insects have owed their origin to secondary adaptations" (Weismann, *Studies in the Theory of Descent,* 493–494).

51. Ibid., 501.

52. Robert Wiedersheim, "Zur Anatomie des Abmblysoma weismanni," *Zeitschrift für wissenschaftliche Zoologie* 32 (1879): 216–236.

53. Weismann, *Studies in the Theory of Descent.*

54. John B. Armstrong and George M. Malacinski, *Developmental Biology of the Axolotl* (New York: Oxford University Press, 1989), see chapter 1 by Hobart M. Smith, "Discovery of the Axolotl and Its Early History in Biological Research," 3–12, and chapter 2 by Ronald A. Brandon, "Natural History of the Axototl and Its Relationship to other Ambystomiatid Salamanders," 13–21.

55. Kölliker, *Erinnerungen aus Meinem Leben,* 30.

56. August Weismann, *Ueber den Einfluss der Isolirung auf die Artbildung* (Leipzig: Wilhelm Engelmann, 1872), 32–34.

57. Weismann, *Studies in the Theory of Descent,* 571; August Weismann, "On the Change of the Mexican Axolotl to an Amblystoma," *Annual Report of the Regents of the Smithsonian Institution for the Year 1877* (Washington, D.C.: Government Printing Office, 1878): 349–375.

58. The circumstances that led to Kölliker's gift to Weismann in 1872 are unclear. Weismann mentioned the event in a way that implies that Kölliker personally presented him with the specimens, but no mention is made of where or how. See Weismann, *Studies in the Theory of Descent,* 556. Although Kölliker mentioned axolotl on several occasions, he did not describe experimenting with them or passing them on to Weismann or anyone else. Although he was the assistant to Kölliker between 1872–1876 and had met Weismann and would in 1873 become his brother-in-law, the role Robert Wiedersheim may have played in securing specimens for Weismann is unclear. In 1876, he became Weismann's colleague in Freiburg, first as prosector of anatomy and in 1878 as außerordentlich Professor. Wiedersheim mentioned that Kölliker visited Lindenhof in the summer of 1873, where both he and Weismann were working on daphnids, but that would have been a year after Kölliker's gift. See Albert Kölliker, *Lebenserinnerungen* (Tübingen: J. C. B. Mohr, 1919), 57–58, 63.

59. Weismann, *Studies in the Theory of Descent,* 557.

60. Weismann enumerates these and other differences in *Studies in the Theory of Descent,* 575–577, 623–624.

61. Blanchard reported the exception in 1876 and two more in 1879. Nevertheless, Weismann held his ground in a lengthy, quite acerbic footnote, which he altered considerably for the English edition of 1882: "Nach Herrn Blanchard ist also nun Alles wieder in Ordnung und der unbequeme Fall vom Axolotl ist glücklich aus der Welt geschafft!" See Weismann, *Studies in the Theory of Descent,* 594–595, and M. Blanchard, "Reproduction in Amblysoma," *Journal of Natural History* 17 (1876): 414–415.

62. Weismann, *Studies in the Theory of Descent,* 595.

63. See also Weismann's postscript, ibid., 613–622.

64. Ibid., 612.

65. At times Weismann appears to use these two senses indiscriminately, but see ibid., 597–598, in which he explained the two cases of reversion as being in opposite directions.

66. Ibid., 571–572.

67. Ibid., 589.

68. Eduard von Hartmann, *Eine Kritische Darstellung der organishen Entwicke-lungstheorie* (Berlin: Carl Dunker's, 1875).

69. Ibid., chapter 5. Hartmann actually discusses first the struggle for existence and correlation of parts, but Weismann takes these up after the other two. It is not clear that von Hartmann made a distinction between variability (the capacity to vary) and variations (the differences between parent and offspring). At the time, there was little consideration of a statistical spread of variations or of a "genotypic" variation in contrast to a somatic variation.

70. Weismann, *Studies in the Theory of Descent,* 659.

71. Ibid., 646–653.

72. Ibid., 663.

73. Ibid., 666–670.

74. Ibid., 677.

75. Ibid., 681.

76. Franz Eilhard Schmidt, *Über die Cuninen-Knospenähren im Magen von Geryonien* (Graz: Naturwissenschaftlicher Verein für Steiermark, 1875).

77. Weismann, *Studies in the Theory of Descent,* 718.

6. *Daphnia* Research and the Ecology of Lakes

1. I am following the classification found in Alfred Kaestner, *Invertebrate Zoology: Crustacea,* trans. Herbert W. Levi (New York: Interscience Publishers, 1970). Robert Barnes, *Invertebrate Zoology* (Philadelphia: Saunders, 1986) considers Cladocera a suborder.

2. W. T. Calman, *A Treatise on Zoology,* part 7, *Appendiculata, Third fascicle Crustacea,* ed. Edwin Ray Lankester (London: Adam and Charles Black, 1909), 29–31.

3. For Hensen's work with surveying marine plankton and fish, see John Lussenhop, "Victor Hensen and the Development of Sampling Methods in Ecology," *Journal of the History of Biology* 7 (1974): 319–337. It is probable, in fact, that Hensen himself coined the term *plankton.*

4. August Weismann, *Das Thierleben im Bodensee* (Lindau: Johann Thomas Stettner, 1877), 55. Weismann's lecture is rich with other descriptions of *Leptodora* feeding habits, its swimming behavior, its anatomy, its role as predator and prey in the economy of the lake, and the cycle of organic material in the world beyond.

5. August Weismann, "Über Bau und lenbenerscheinungen von *Leptodora hyalina* Lilljeborg," *Zeitschrift für wissenschaftliche Zoologie* 24 (1874): 349–418; Robert Wiedersheim, *Lebenserinnerungen* (Tübinben: J. C. B Mohr, 1919), 57–58, 62–63.

6. Weismann, "Über Bau."

7. "The resting eggs may even survive passage through a bird's digestive tract." Alfred Kaestner, *Invertebrate Zoology: Crustacea,* 2nd ed., trans. Herbert W. Levi and Lorna R. Levi (New York: Interscience, 1970), 107.

8. August Weismann, *Beiträge zur Naturgeschichte der Daphnoiden,* 7 vols. (Leipzig: Wilhelm Engelmann, 1876–1879). See also Francis M. Balfour, *Handbuch der vergeichenden Embryologie,* vol. 1, trans. B. Vetter (Jena: Gustav Fischer, 1880–1881), 48. Balfour points out that the formation of winter eggs in *Leptodora* maintains a close analogy ("innige Analogie") with insect egg development.

9. Weismann, *Beitrag,* 1:5–21.

10. Weismann, *Beitrag,* 4:218–222.

11. For Weismann on Hertwig's work, see Weismann to Haeckel, 27 December 1875, in Georg Uschmann and Bernhard Hassenstein, "Der Briefwechsel zwischen Ernst Haeckel und August Weismann," in *Kleine Festgabe aus Anlass der hundertjährigen Wiederkehr der Gründung des Zoologischen Institutes der Friedrich-Schiller-Universität Jena im Jahre 1865,* ed. M. Gersch (Jena: Friedrich-Schiller-Universität, 1965), 7–68. Weismann explained that he purposely withdrew from publication in 1875 that which turned out to be *Beitrag* IV in 1877 because he wanted to confirm in other species his discoveries of the reabsorption during winter egg development in *Leptodora,* as described in *Beitrag* I. For details on the publication sequence, see Weismann, *Beitrag,* 2:93–95.

12. Weismann, *Beitrag,* 2:173.

13. Ibid., 3:209–211.

14. August Weismann, "Zur Naturgeschichte der Daphniden, Theil V. Ueber die Schmuckfarben der Daphnoiden," *Zeitschrift für wissenschaftliche Zoologie* 30, supplement (1878): 123–165.

15. Weismann, *Beitrag,* 5:163.

16. Ibid., 5:142. Weismann does not identify what he thought Darwin's two laws of heredity were, but he may well have meant Darwin's own version of homochronic inheritance and the theory of pangenesis.

17. Ibid.

18. Weismann, *Beitrag,* 6:108.

19. See Weismann to Marion, 18 April 1879, in Frederick B. Churchill and Helmut Risler, *August Weismann, Ausgewählte Briefe und Dokumente = Selected Letters and Documents,* 2 vols. (Freiburg: Universitätsbibliothek Freiburg, 1999), 1:44–45, and William Bateson's later account in *Materials for the Study of Variation Treated with Especial Regard to Discontinuity in the Origin of Species* (London: Macmillan, 1894): 96–101.

20. Weismann, *Beitrag,* 7:123.

21. These enumerations appear ibid., 7:343–393.

22. Ibid., 7:393–407.

23. Ibid., 7:243.

24. Ibid., 7:256–260.

25. These letters now reside in the Staatsbibliothek zu Berlin, Handschriftenabteilung. Sig. Darmst. Lc 1889 (23): Weismann, August; Bl. 51r–60v. The fifth letter was written ten years later and was perfunctory in nature.

26. Weismann to Claus, 10 December 1874, Staatsbibliothek zu Berlin, Lc 1889: Weismann, August, Berlin.

27. Weismann, "Über Bau."

28. Lynn Nyhart, *Biology Takes Form: Animal Morphology and the German Universities, 1800–1900* (Chicago: University of Chicago Press, 1995), 187. Haeckel's *Anthropogenie* (1874) contained a popular version of the gastraea theory. Carl Claus,

Die Typenlehre und E. Haeckel's sog. Gastraea-Theorie (Wien: G. J. Manz, 1874). See also Uschmann and Hassenstein, "Der Briefwechsel," Weismann to Haeckel, 27 January 1874; Weismann to Haeckel, 27 December 1875; and Haeckel to Weismann, 31 December 1875.

29. Weismann to Claus, 10 December 1874, Staatsbibliothek zu Berlin, Lc 1889: Weismann, August.

30. Ibid., Weismann to Claus, 3 February 1875.

31. Ibid., Weismann to Claus, 8 January 1876.

32. According to Weismann, he received this advanced communication on "11 Nov. 1876," but he may not have understood its full implications until after his cordial letter to Claus on November 14. The full paper did not appear until later in the Academy's *Denkschriften*. Carl Claus, "Zur Kenntniss des Baues und der Organisation der Polyphemiden," *Denkschriften der Kaiserlichen Akademie der Wissenschaften* (Wein) 37 (1877): 137–160.

33. According to Weismann *Beitrag* IV had been completed in the fall of 1875, and *Beiträge* II and III were completed in the winter of 1875–1876, but then *Beitrag* II had been rewritten to accommodate his findings that the reabsorption phenomenon found in *Leptodora* was not unique but common to many cladocerans.

34. Weismann, *Beitrag*, 2:93–95.

35. Carl Claus, "Zur Berichtigung und Abwehr," *Zeitschrift für wissenschaftliche Zoologie* 28 (1877): 471–474; August Weismann, "Rechtfertigung," *Zeitschrift für wissenschaftliche Zoologie* 30 (1877): 194–202.

36. "Hätte Claus sich etwas mehr Zeit gegönnt, so würde er ohne Zweifel diesen 'fundamentalen' Irrthum vermieden haben." Weismann, "Rechtfertigung," 199.

37. Claus, "Zur Kenntniss des Baues," 137, 154.

38. Ibid., 154–156.

39. Emil du-Bois Reymond, "Darwin versus Galiani," in *Der Leipzig-Sitzung der Akademie de Wissenschaften*, am 6. July 1876; reprinted in his *Reden von Emil Du Bois-Reymond*, 2nd ed., 2 vols. (Leipzig: Von Veit, 1912), 1:540–566. Claus had earlier been highly critical of Haeckel's gastraea theory for failing to produce a common origin of all metazoans. C. Claus, *Die Typenlehre und E. Haeckel's sog. Gastraea-Theorie* (Wien: G. J. Manz, 1874).

7. From Germ Layers to the Germ-Plasm

1. Weismann to Haeckel, 27 January 1874, in Georg Uschmann and Bernhard Hassenstein, "Der Briefwechsel zwischen Ernst Haeckel und August Weismann," in *Kleine Festgabe aus Anlass der hundertjährigen Wiederkehr der Gründung des Zoologischen Institutes der Friedrich-Schiller-Universität Jena im Jahre 1865*, ed. M. Gersch (Jena: Friedrich-Schiller-Universität, 1965), 35–36.

2. Ibid.

3. Edward Forbes, *A Monograph of the British Naked-eyed Medusae* (London: Ray Society, 1848), 1.

4. Haeckel to Anna Sethe, 27 August 1859, in *Ernst Haeckel Biographie in Briefen,* ed. Georg Uschmann (Gütersloh: Prisma, 1984).

5. Robert J. Richards, *The Romantic Conception of Life and the Philosophy in the Age of Goethe* (Chicago: University of Chicago Press, 2002).

6. Alfred Goldsborough Mayer, *The Hydromedusae* (Washington, D.C.: Carnegie Institution, 1910), 2.

7. John Graham Dalyell, *The Powers of the Creator Displayed in the Creation: or, Observations on Life amidst the Various Forms of the Humbler Tribes of Animated Nature: With Practical Comments and Illustrations* (London: J. Van Voorst, 1851–1858).

8. Mary P. Winsor, *Reading the Shape of Nature: Comparative Zoology at the Agassiz Museum* (Chicago: University of Chicago Press, 1991), chapter 2.

9. Mayer, *Medusae of the World,* 2.

10. Mary P. Winsor, *Starfish, Jellyfish, and the Order of Life, Issues in Nineteenth-Century Science* (New Haven, Conn.: Yale University Press, 1976).

11. Leuckart coined the term "Coelenterates"; Haeckel still called them Zoophytes, and P. J. van Beneden called them polypes, Acalephes, and Spongiaire syn. with Coelenterates.

12. E. Ray Lankester, ed., *A Treatise on Zoology, Part II. The Porifera and Coelentera* (London: Adam & Charles Black, 1900). The chapter on "The Hydromedusae" was written by G. Herbert Fowler.

13. E. Bresslau and H. E. Ziegler, eds., *Zoologisches Wörterbuch* (Jena: Gustav Fischer, 1927), 644.

14. E. Korschelt and K. Heider, *Text-Book of the Embryology of Invertebrates,* trans. Edward Laurens Mark and Martin Fountain Woodward (London: S. Sonnenschein, 1895), 45–47.

15. Libbie Hyman, *The Invertebrates: Protozoa through Ctenophora,* vol. 1 (New York: McGraw-Hill, 1940), 370.

16. I am following here W. J. Rees, ed., *The Cnidaria and their Evolution.* Symposia of the Zoological Society of London, No. 16 (London: Academic Press, 1966); and Hyman, *Invertebrates.*

17. William K. Brooks, "The Life-History of the Hydromedusae: A Discussion of the Origin of the Medusae and the Significance of Metagenesis," *Memoirs of the Boston Society of Natural History* 3 (1886): 411, 425. See also Keith R. Benson, "American Morphology in the Late Nineteenth Century: The Biology Department at Johns Hopkins University," *Journal of the History of Biology* 18 (1985): 163–205.

18. See Frederick B. Churchill, "Weismann's Continuity of the Germ-Plasm in Historical Perspective," in *August Weismann (1834–1914) und die theroretische Biologie des 19. Jahrhunderts: Urkunde, Berichte und Analysen,* ed. K. Sander, *Freiburger Universitätsblätter* 87/88 (1985): 109–111.

19. See Eduard van Beneden, "De la distinction originelle du testicule et de l'ovaire, Caractère sexuel des deux feuillets primordiaux de l'embryon, Hermaphrodisme morphologique de toute individualité animale, Essai d'une theéorie de la fecundation," *Bulletin de l'Académie royale de Belgique. Classe des sciences,* 2nd ser., 37 (1874) 535–536; Thomas Henry Huxley, "On the Anatomy and the Affinities of the Family of the Medusae," *Philosophical Transactions of the Royal Society of London* 139 (1849): 413–434; and Thomas Henry Huxley, *The Oceanic Hydrozoa: A Description of the Calycophoridæ and Physophoridæ Observed during the Voyage of H.M.S. "Rattlesnake," in the Years 1846–1850* (London: Ray Society, 1858), 16.

20. Beneden, "De la distinction originelle," 541.

21. Ibid., 591.

22. J Ciamician, "Zur Frage über die Entstehung der Geschlechtsstoffe bei den Hydroiden," *Zeitschrift für wissenschaftliche Zoologie* 30 (1878): 507.

23. See Nicolaus Kleinenberg, "Über die Entstehung der Eier bei Eudendrium," *Zeitschrift für wissenschaftliche Zoologie* 35 (1881): 326–332.

24. Francis M. Balfour, *A Treatise on Comparative Embryology Treatise,* vol. 2 (London: Macmillan, 1881), 611.

25. Weismann to Dohrn, 23 March 1872, in Frederick B. Churchill and Helmut Risler, *August Weismann, Ausgewählte Briefe und Dokumente = Selected Letters and Documents,* 2 vols. (Freiburg: Universitätsbibliothek Freiburg, 1999), 1:35–36; Dohrn to Haeckel, quoted in *Emil du-Bois Reymond (1818–1896) Anton Dohrn (1840–1901) Briefwechsel,* ed. Christiane Groeben, Klaus Hierholzen, and Ernst Florey (Berlin: Springer-Verlag, 1985), 143; Georg Uschmann and Bernhard Hassenstein, "Der Briefwechsel," Weismann to Haeckel, 27 December 1869. Dohrn commented to du Bois-Reymond on Weismann's nervousness and distrustfulness and concluded that Weismann would make, "as I believe, a really uncomfortable colleague." Dohrn nevertheless praised Weismann as "undoubtedly intellectually the most significant, since Siebold." Quoted in Groeben et al., *Emil du-Bois Reymond,* 145.

26. Churchill and Risler, 1:36–37, Weismann to Anton Dohrn, 24 October 1875.

27. Uschmann and Hassenstein, "Der Briefwechsel," 40, Haeckel to Weismann, 2 January 1877.

28. Churchill and Risler, *August Weismann,* 1:39, Weismann to Marion, 9 May 1878.

29. Ibid., vol. 1, Weismann to Marion, 9 May 1878; 18 May 1878; 10 June 1878; 3 July 1878; 17 September 1878; 31 January 1879; 18 April 1879. Weismann signed off on the seventh and final daphnid contribution on 1 August 1879.

30. He recommended a 0.05% chromic acid solution as a killing agent, and suggested the specimen be dehydrated in several changes of absolute alcohol after twenty-four hours of washing. Osmium acid might be substituted for the chromic acid.

31. Ibid., Weismann to Dohrn, 8 May 1880; Weismann to Marion, 10 April 1880; see also the appendix, no. 47–49.

32. Julien Fraipont, "Origine des Organes sexuels chez les Campanularides," *Zoologischer Anzeiger* 51 (1880): 135–138.

33. August Weismann, "Über den Ursprung der Geschlechtszellen bei den Hydroiden. II," *Zoologischer Anzeiger* 3 (1880): 367–370 (submitted 14 June).

34. See his "Autobiographie aus dem Jahre 1896," in Churchill and Risler, *August Weismann,* 2:533.

35. August Weismann, "Über die Dauer des Lebens," *Tageblatt der 54. Versammlung deutscher Naturforscher und Aerzte in Salzburg vom 18.–24. September 1881* (Salzburg); expanded in 1882 as August Weismann, *Über die Dauer des Lebens* (Jena: Gustav Fischer, 1882).

36. Churchill and Risler, *August Weismann,* 2:533. See his autobiography (1896) in the same volume.

37. Churchill and Risler, *August Weismann,* 1:54–55, Weismann to Marion, 26 February 1881.

38. Alexander Goette, "Ein neuer Hydroid-polyp mit einer neuen Art der Fortpflanzung," *Zoologischer Anzeiger* (1880), 352–358, esp. 353–354.

39. August Weismann, *Die Entstehung der Sexualzellen bei den Hydromedusen. Zugleich ein Beitrag zur Kenntnis des Baues und der Lebenserscheinungen dieser Gruppe,* vol. mit atlas (Jena: Gustav Fischer, 1883), 98–112.

40. Auguste Wiesmann, "Observations sur l'orgine des cellules sexuelles des hydroides," *Annales des Sciences naturelles, B. Zoologie et Pathologie Aniomales* 11 (1881), 36.

41. Weismann, *Die Entstehung,* 109.

42. Wiesmann, "Observations sur l'orgine," 36, "Explication des Fig. 1." There is a confusion with the published form of the "Supplement" because the designated numbers of the figures in the "Explication des Figures. Planche 10" are different from the designated numbers within "Planche" 10 itself.

43. August Weismann, "Beobachtungen an Hydroid-Polypen. I. Pulsieren des Körperschlauchs," *Zoologischer Anzeiger* 4 (1881): 61–63, "Beobachtungen an Hydroid-Polypen. II. Selbständige Bewegungen des Ectoderms," *Zoologischer Anzeiger* 4 (1881): 63–64, "Beobachtungen an Hydroid-Polypen. III. Die Enstehung der Eizellen in der Gatung," *Eudendrium. Zoologischer Anzeiger* 4 (1881), 111–114.

44. August Weismann, "Über eigenthümliche Organe bei Eudendrium racemosum Cav.," *Mittheilungen aus der Zoologischen Station zu Neapel* 3 (1881): 1–14; Churchill and Risler, *August Weismann,* vol. 1, Weismann to Dohrn, 27 April 1881.

45. Christiane Groeben, "August Weismann und Neapel," in *August Weismann (1834–1914) und die theoretische Biologie des 19. Jahrhunderts. Urkunden, Berichte und Analysen,* ed. Klaus Sander (Freiburg: Rombach Verlag, 1985), 141–156.

46. The correspondence might have been attended to at other times of the day and at his rented living quarters beyond the station.

47. Groeben, "August Weismann und Neapel," 152.

48. Carl Claus, *Untersuchungen über die Organisation und Entwicklung der Medusen* (Prag/Leipig: F. Tempsky/G. Freytag, 1883).

49. Nicholaus Kleinenberg, *Hydra. Eine Anatomisch-Entwicklungsgeschichtliche Untersuchung* (Leipzig: Wilhelm Engelmann, 1872).

50. William Keith Brooks, "The Life-History of the Hydromedusae: A Discussion of the Origin of the Medusae and the Significance of Metagenesis," *Memoirs of the Boston Society of Natural History* 3 (1886): 359–430.

51. Churchill and Risler, 1:61, August Weismann to Anton Dohrn, 11 December 1882.

52. The following section follows Frederick B. Churchill, "Weismann, Hydromedusae, and the Biogenetic Imperative: A Reconsideration," in *A History of Embryology,* ed. T. J. Horder, J. A. Witkowski, and C. C. Wylie (Cambridge: Cambridge University Press, 1986), 15–24.

53. August Weismann, "Zur Frage nach dem Ursprung der Geschlechtszellen bei den Hydroiden," *Zoologischer Anzeiger* 3 (1880): 226–233; "Beobachtungen an Hydroid-Polypen," *Zoologischer Anzeiger* 4 (1881): 111–114.

54. August Weismann, "Zur Frage nach dem Ursprung der Geschlechtszellen bei den Hydroiden," *Zoologischer Anzeiger* 3 (1880): 226–233; "Beobachtungen an Hydroid-Polypen," *Zoologischer Anzeiger* 4 (1881): 111–114; "Zur Frage nach dem Ursprung der Geschlechtszellen bei den Hydroiden," *Zool. Anzeiger* 3 (1880): 226–233; "Beobachtungen an Hydroid-Polypen," *Zool. Anzeiger* 4 (1881): 111–114.

55. The taxonomic terms of lower invertebrates was in considerable flux at mid-century. Agassiz considered Acalephs, stinging animals, as a class within the Radiata. Our Anthozoa and *Hydra* were for him simply lumped together as polyps without a medusa stage at all.

56. For example, see Louis Agassiz's *Essay on Classification* (1857) and his *Contributions to the Natural History of America, Second Monograph* (Boston: Little, Brown, 1860–1861). Agassiz's example is discussed in *Contributions*, vol. 3, 113–119: "Considering the mode of reproduction of the Acalephs in general, the highest Hydroids would, of course, be those in which the medusoid elements prevail, and the lowest, those in which the hydroid elements are most prominent."

57. George James Allman, *A Monograph of the Gymnoblastic or Tubularian Hydroids,* vol. 1 (London: Ray Society, 1871), 43–45.

58. Carl Claus, *Grundzüge der Zoologie zum Gebrauche an Universitäten und Höheren Lehranstalten sowie zum Selbststudium* (Marburg: N. G. Elwert, 1876), 220–221.

59. Gottlieb von Koch, "Vorlaufige Mittheilungen über Colenteraten," *Jenaische Zeitschrift für Medizin und Naturwissenschaften* 7 (1873): 464–470, 512–515; Otto Hamann, "Der Organismus der Hydroidpolypen," *Jenaische Zeitschrift für Medizin und Naturwissenschaften* 15 (1882): 473–544; Oscar and Richard Hertwig, *Der Organism us der Medusen und seine Stellung zur Keimblättertheorie* (Jena: Gustav

Fischer, 1878), 64–66. For von Koch's and Hamann's general relationship with Haeckel, see Georg Uschmann, *Geschichte der Zoologie und der zoologiscehen Anstalten in Jena 1779–1919* (Jena: Gustav Fischer, 1959), 87–89, 117–120.

60. Carl Gegenbaur, *Grundzüge der vergleichenden Anatomie,* 2nd ed. (Leipzig: W. Engelmann, 1878): 94–95, 121.

61. Weismann, *Die Entstehung,* 254–266. For a turn-of-the-century appraisal, see Eugen Korschelt and Karl Heiderk, *Text-Book of the Embryology of Invertebrates* (London: Swan Sonnenschein, 1895), part 1, 47–48; for a later view, see Hyman, *Invertebrates,* 427–431.

62. Weismann, *Die Entstehung,* 258–262.

63. Ibid., 249–253.

64. Ibid., 275.

65. Weismann discusses at length how his findings support the Hertwigs' conclusion that hydroids are Ektocarpen. Ibid., 284–293.

66. Ibid., 232–233, 264.

67. Ibid., 288–289. I discuss this in Frederick B. Churchill, "Weismann's Continuity of the Germ-Plasm in Historical Perspective," in *August Weismann (1844–1914) und die theoretische Biologie des 19. Jahrhunderts. Urkunden, Berichte und Analysen,* ed. K. Sander (Freiburg: Rombach, 1985), 24. The translation with some modifications is from N. J. Berrill and C. K. Liu, "Germplasm, Weismann, and Hydrozoa," *Quarterly Review of Biology* 23 (1948): 124–132, quotation from p. 127.

68. Weismann, *Die Entstehung,* 236; Churchill, "Weismann's Continuity," 109–110.

69. Weismann, *Die Entstehung.* This is best seen in plate III, fig. 7.

70. Churchill, "Weismann's Continuity," 112.

71. Weismann *Die Entstehung,* 236–237.

72. "Das wäre nicht nur unbequem, sondern auch unnütz und sinnlos." Ibid., 293.

73. Ibid., 295.

74. August Weismann, *Ueber die Vererbung* [*Vortrag. gehalten bei der Feier der Uebergabe des Prorectorates in der Aula der Universität Freiburg am 21. Juni 1883*] (Jena: Gustav Fischer, 1883); and *Ueber die Ewigkeit des Lebens, Programm wodurch zur Feier des Geburtstages seiner koeniglichen hoheit unseres durchlauchtigsten Grossherzogs Friedrich im Namen des academischen Senates die angehoerigen der Albert-Ludwigs-Universitaet einladet der gegenwaertige Prorector Dr. August Weismann* (Freiburg: H. M. Poppen & Sohn, 1883).

75. August Weismann, "Die Entstehung der Sexualzellen bei den Hydromedusen," *Biologisches Zentralblatt* 4 (1884): 13–31.

76. Clemens Hartlaub, *Beobachtungen über die Entstehung der Sexualzellen bei Obelia* (Leipzig: Wilhelm Engelmann, 1884).

77. Johannes Thallwitz, "Ueber die Entwicklung der männlichen Keimzellen bei den Hydroideen," *Jenaische Zeitschrift für Naturwissenschaft,* neue Folge 11, 18

(1885): 385–444. For further information on Hartlaub and Thallwitz, see Churchill and Risler, *August Weismann,* vol. 2, section 3.2.

78. Weismann's recommendations to the university were supportive but in both these cases lukewarm. See Churchill and Risler, *August Weismann,* 2:639, 681.

79. Weismann, "Die Entstehung der Sexualzellen bei den Hydromedusen," *Biologisches Zentralblatt* 4 (1884): 13–31.

80. Henry Nottidge Moseley, "Dr. August Weismann on the Importance of Sexual Reproduction for the Theory of Selection," *Nature* 34 (1886): 629–632.

81. For a general historical reflection on Weismann's *Hydromedusen* and today's historians, see Churchill, "Weismann, Hydromedusae," 7–33, esp. 24–33.

8. From Egg to Heredity

1. August and Mary had six children: Therese (1868), Hedwig (1870), Elise (1871), Berta (1873), Meta (1877), and Julius (1879). See "Die Nachkommenschaft des Valentin Weismann aus Weierburg (Niederösterreich) und des Ehrenreich Weismann," *Beiträge zur Geschichte der Familie Weismann* (1936), 37.

2. Weismann signed the Preface to his *Hydromedusen* on "10 Mai 1883."

3. Frederick B. Churchill and Helmut Risler, *August Weismann, Ausgewählte Briefe und Dokumente = Selected Letters and Documents,* 2 vols. (Freiburg: Universitätsbibliothek Freiburg, 1999), 2:533–534.

4. August Weismann, *Ueber das Wandern der Vögel* (Berlin: Carl Habel, 1878).

5. August Weismann, "Über die Dauer des Lebens," *Tageblatt der 54. Versammlung deutscher Naturforscher und Aerzte in Salzburg vom 18.–24 September 1881* (1881), 98–114; August Weismann, *Über die Dauer des Lebens* (Jena: Gustav Fischer, 1882); English translation in August Weismann, *Essays upon Heredity and Kindred Biological Problems,* 2 vols., trans., A. E. Shipley and Selmar Schönland, ed., Edward B. Poulton (Oxford: Clarendon Press, 1889).

6. August Weismann, *Über die Ewigkeit des Lebens* (Freiburg: H. M. Poppen & Sohn, 1883), *Ueber Leben und Tod* (Jena: Gustav Fischer, 1884).

7. August Weismann, *Festrede gehalten bei der Enthüllung des Okendenkmals in Offenburg am 29. Juli 1883, Volksfreund* (Offenburg: Adolf Beck, 1883).

8. August Weismann, "Einige Worte des Willkommens," *Amtlicher Bericht über die 56. Versammlung deutscher Naturforscher und Aezte welche zu Freiburg im Breisgau vom 18.–22. September 1883 tagte* (Freiburg: Universitäts-Buch und Kuntshandlung, 1883).

9. August Weismann, *Studies in the Theory of Descent,* 2 vols., trans. Raphael Meldola (London: Sampson Low, Marston, Searle, & Rivington, 1882). For Meldola's career, see Frederick B. Churchill, "Meldola, Raphael," in *Dictionary of Scientific Biography,* Supplement, vol. 2, Frederic Lawrence Holmes (New York: Charles Scribner's Sons, 1990), 616–620.

10. Churchill and Risler, *August Weismann,* 2:534.

11. E. Ray Lankester, *On Comparative Longevity in Man and the Lower Animals* (London: Macmillan, 1870).

12. In the English translation of his Salzburg address, the editor added a footnote referencing Lankester's book. Weismann, *Essays upon Heredity*, 1:66.

13. Lankester, *On Comparative Longevity*, 3.

14. Ibid., 50.

15. Ibid, 15.

16. The original version as read in 1881 appeared as Weismann, "Über die Dauer des Lebens." The expanded version of the following year bore the same title and, when published as a separate piece, ran to ninety-four pages including a large appendix (Weismann, *Über die Dauer des Lebens*). See also August Weismann, *Aufsätze über Vererbung und verwandte biologische Fragen* (Jena: Gustav Fischer, 1892).

17. Weismann, *Essays upon Heredity*, 1:20.

18. Ibid., 25.

19. E. B. Poulton, the editor of the English edition, printed a paragraph written by Alfred Russel Wallace between 1865 and 1870, in which the latter had also made a utility argument for "Old Age, Decay, and Death." Ibid., 23–24.

20. Ibid., 28.

21. Ibid.

22. Ibid., 29, 31.

23. Ibid., 34–35. In Weismann's original this passage reads: "Doch nicht der Besitz der vollen Wahrheit, sondern das Forschen nach ihr ist unser Theil, befriedigt, erfüllt unser Leben, ja beseligt."

24. Karl Grobben in *Almanach der Kaiserliche Akademie der Wissenschaften in Wien* 72 (1923): 171–174; Georg Uschmann, "Goette, Alexander Wilhelm," *Neue deutsche Biographie* 6 (1983): 579; Hans Querner, "Goette, Alexander Wilhelm," in *Dictionary of Scientific Biography*, vol. 5, ed. Charles Gillispie (New York: Charles Scribner's Sons, 1972), 446–447. There is very little published information on Goette. The latter two secondary entries are largely dependent upon the first. None of them is clear about whether Goette earned his venia legendi with the traditional Habilitationsschrift at this newly founded German university. Standard bibliographies are no more enlightening.

25. Alexander Goette, *Die Entwickelungsgeschichte der Unke (Bombinator igneus) als Grundlage einer vergleichenden Morphologie der Wirbelthiere* (Leipzig: Leopold Voss, 1875).

26. Carl Gegenbaur, "Einige Bemerkungen zu Götte's 'Entwickelungsgeschite der Unke als Grundlage einer vergleichenden Morphologieder Wirbelthiere,'" *Morphologisches Jahrbuch* 1 (1876): 299–345.

27. For a penetrating discussion of Gegenbaur's study of limbs, see Lynn K. Nyhart, *Biology Takes Form, Animal Morphology and the German Universities, 1800–1900* (Chicago: Chicago University Press, 1995), 251–262.

28. See Churchill and Risler, *August Weismann,* 1:67–72, Weismann to Anton Bary, 22, 27, and 29 January 1886.

29. Alexander Goette, *Über den Ursprung des Todes* (Hamburg: Leopold Voss, 1883).

30. Querner, "Goette."

31. H. E. Ziegler, *Zoologisches Wörterbuch* (Jena: G. Fischer, 1927).

32. How pleased he would have been to learn that, in addition, "the Volvocida display all gradations of sexual reproduction from isogamy to highly developed heterogamy." Robert D. Barnes, *Invertebrate Zoology* (Philadelphia: W. B. Saunders, 1968), 16.

33. This aspect of Goette's argument was briefly summarized in L. Will, "Goette, Ueber den Ursprung des Todes," *Biologisches Centralblatt* 3 (1883–1884): 734–735.

34. Brandt habilitated in Königsberg at Carl Chun's institute in 1885, and succeeded Karl Möbius as professor of zoology in Kiel in 1888. From this position, he became one of Germany's leading marine zoologists. Johannes Krey, "Carl Brandt," *Neue Deutsche Biographie* 2 (1955). See Carl Brandt, "Über Actinosphaerium Eichornii" (PhD diss., Halle, 1877).

35. Reinhard Mocek, *Die Werdende Form: Ene Geschichte der Kausalen Morphologie* (Marburg an der Lahn: Basilisken-Presse, 1998), 170–171.

36. Ernst Haeckel, *Evolution of Man* (New York: Appleton, 1879), 178–183, 443.

37. By the end of the 1870s, the understanding of egg maturation had changed due largely to the work of Eduard van Beneden and Hermann Fol, who described the formation of polar bodies after the dissolution of the germinal vesicle. See Francis Balfour, *Treatise on Comparative Embryology,* vol. 1 (London: Macmillan & Son), chapters 1 and 2.

38. Goette, *Die Entwickelungsgeschichte,* 65, 842. Quoted by Weismann, *Über Leben und Tod: Eine Biologische Untersuchung* (Leipzig: Gustav Fischer, 1884), reprinted in Weismann, *Aufsätze,* 145–146. This rough translation of a complex sentence is taken from Weismann, *Essays upon Heredity,* 1:126.

39. For letters concerning Weismann, see Churchill and Risler, *August Weismann,* 2:726–729, S. Lösser to August Weismann, 13 December, 1880, and J. Lotz (Ministerium der Justiz, des Kultus und Unterrichts, an den Senat der Universität Freiburg) to August Weismann, 9 May 1883.

40. In reworking the talk for a biological readership, Weismann retitled it *Über Leben und Tod. Eine biologische Untersuchung* (Jena: Gustav Fischer, 1884). The new title fails to impart the overtones of that fine celebration in the summer of 1883. The English translation followed this second title, "Life and Death."

41. See "On Life and Death," in Weismann, *Essays upon Heredity,* 1:120.

42. Ibid., 127.

43. Barnes, *Invertebrate Zoology,* 155.

44. Goette, *Über den Ursprung,* 5, quoted in Weismann, *Essays upon Heredity,* 1:137.

45. Weismann, *Essays upon Heredity,* 1:141. He still had to explain how the last component of immortality disappeared from the somatic line.

46. Ibid., 1:149.

47. Ibid., 1:152.

48. Ibid., 1:154–160.

49. Weismann, *Aufsätze,* 79. Here and elsewhere, I refer to the collected German edition of Weismann's *Essays upon Heredity.* See Frederick B. Churchill, "From Heredity Theory to Vererbung: The Transmission Problem," *Isis* 78 (1987): 339–342, for a more complete discussion of the mismatch between the German expressions of "Fortpflanzung" and "Vererbung" and the English terms heredity, inheritance, and transmission. Where the English version freely uses "transmission," Weismann uses "Vererbung" and "Vererbbarkeit." Compare Weismann, *Aufsätze,* 89, with Weismann, *Essays upon Heredity,* 80.

50. Weismann kept the Jacques-Louis David portrait of Lamarck above his writing desk. See the photograph circa 1900 in Churchill and Risler, *August Weismann,* frontispiece of vol. 1.

51. Weismann, *Essays upon Heredity,* 81. Francis Galton, "Hereditary Talent and Character," *Macmillan's Magazine* 12 (1865): 157–166, 318–327; Wilhelm His, *Unsere Körperform und das physiologissche Problem Ihrer Entstehung. Briefe an einem befreundeten Naturforscher* (Leipzig: Verlag von F.C.W. Vogel, 1874), 157–158; Georg Seidlitz, *Die Darwin'sche Theorie. Elf Vorlesungen über die Entstehung der Tiere und Pflanzen durch naturzüchtung* (Dorpat: C. Mattiesen, 1871), 95–102. For more extensive discussions on challenges to the belief in the inheritance of acquired characters, see Leonid I. Blacher, *The Problem of the Inheritance of Acquired Characters: A History of a Priori and Empirical Methods Used to Find a Solution,* ed. Frederick B. Churchill (New Delhi: Amerind Publishing, 1982); Peter Bowler, *Eclipse of Darwinism* (Baltimore: Johns Hopkins University Press, 1983); and Frederick B. Churchill, "Rudolf Virchow and the Pathologist's Criteria for the Inheritance of Acquired Characteristics," *Journal of the History of Medicine and Allied Sciences* 31 (1976): 117–148.

52. M. D. Olmsted, *Charles-Edouard Brown-Séquard, a Nineteenth-Century Neurologist and Endocrinologist* (Baltimore: Johns Hopkins University Press, 1946), 168–195. For a summary tabulation of the hereditary aspects of his experiments, see Charles-Edouard Brown-Séquard, "On the Hereditary Transmission of Effects of Certain Injuries to the Nervous System," *Lancet* 1 (1875): 7–8; Weismann, *Essays upon Heredity,* 83; Blacher, *Problem of the Inheritance,* 185–187; Churchill, "Rudolf Virchow," 118, 145–146.

53. Weismann, *Essays upon Heredity,* 85.

54. Ibid., 105.

55. Ibid., 91.

56. Ibid., 103–106. The English version mistakenly translated "Continuität des Keimprotoplasms" as "continuity of the germ plasm." Strasburger used "Continuität des Keimplasma" in E. Strasburger, *Neue Untersuchungen über den Befruchtungsvorgang bei den Phanerogamen als Grundlage für eine Theorie der Zeugung* (Jena: Gustav Fisher, 1884), 136, with reference to August Weismann, *Ueber die Vererbung* [*Vortrag. gehalten bei der Feier der Uebergabe des Prorectorates in der Aula der Universität Freiburg am 21. Juni 1883*] (Jena: Gustav Fischer, 1883). The German reprint of Weismann's "Vererbung" used "Keim-Plasma."

57. Rasmus Winther, "August Weismann on Germ-Plasm Variation," *Journal of the History of Biology* 34 (2001): 517–555. See also Rasmus Winther, "Darwin on Variation and Heredity," *Journal of the History of Biology* 33 (2000): 425–455.

58. Weismann, *Essays upon Heredity*, 76.

59. Ibid., 75–80, quotation from p. 80.

9. A Perspective on Heredity

1. See Staffan Müller-Wille and Hans-Jorg Rheinberger, *A Cultural History of Heredity* (Chicago: University of Chicago Press, 2012).

2. The published works of Jean Gayon, J. Thirsk Goody, Ludmilla Jordanova, and Carlos Lopez-Beltrán have been particularly important in informing Müller-Wille and Rheinberger's perspective.

3. Karl Ernst von Baer, *Über Entwickelungsgeschichte der Thiere. Beobachtung und Reflexion* (Königsberg: Gebrüdern Borntrager, 1828 and 1837), part 1, 150. Baer returns to a similar description of the connection of generation and growth in part 2, 4.

4. Scheiden's "Beiträge zur Phytogenesis" first appeared in Müller's *Archiv für Anatomie und Physiologie und wissenschaftliche Medicin* 5 (1838); the first seven signatures (1–112) of Schwann's *Mikroskopische Untersuchungen über die Übereinstimmung in der Struktur und im Wachstum der Thiere und Pflanzen* appeared as a separate issue in 1838 and then *in toto* as *Mikroskopische Untersuchungen über die Uebereinstimmung in der Struktur und dem Wachstum der Thiere und Pflanzen* (Berlin: Sanders, 1839). I refer to the Sydenham Society version of the latter, *Microscopical Researches into the Accordance in the Structure and Growth of Animals and Plants,* trans. Henry Smith (London: Sydenham Society, 1847).

5. Schwann, *Microscopical Researches,* 10–35.

6. Ibid., 36–160.

7. Ibid., 168–169.

8. Emil du Bois-Reymond, *Reden von Du Bois-Reymond,* 2nd ed., 2 vols. (Leipzig: Von Veit, 1912), 1:135–317; Gottfried Koller, *Johannes Müller: Das Leben des Biologen 1801–1858* (Stuttgart: Wissenschaflische Verlagsgesellschaft, 1958), 38–50;

Johannes Steudel, "Müller, Johannes Peter," in *Dictionary of Scientific Biography,* ed. Charles Gillispie (New York: Charles Scribner's Sons, 1973), 7:567–574; Nicholas J. Wade, "Introduction," *Müller's Elements of Physiology* (Bristol: Thoemmes Press, 2003).

9. Johannes Müller, *Handbuch der Physiologie des Menschen für Vorlesungen,* 2 vols. (Koblenz: Verlag von J. Hölscher, 1837–1840).

10. Timothy Lenoir, *The Strategy of Life: Teleology and Mechanics in Nineteenth-Century German Biology* (Dordrecht: D. Reidel, 1982).

11. Schwann, *Mikroskopische Untersuchungen.* For a discussion of the importance of embryology for the cell theory, see F. Duschesneau, *Genèse de la théorie cellulaire* (Montreal: Bellarmin, 1987).

12. As noted above, Schwann separately published the first seven signatures of his *Mikroskopische Untersuchungen* in 1838. Müller also refers to this edition in his *Handbuch.* Furthermore, Schwann published an outline of his cell theory in the form of three letters to F. H. Weber in Ludwig Friedrich and Robert Froriep, *Neue Notizen aus dem Gebiete der Natur and Heilkinde* 5 (1838), 33–36, 225–229, and 6 (1838): 21–23; Johannes Müller, *Ueber den feineren Bau und Formen der Krankhaften Geschwülste* (Berlin: G. Reimer, 1838). This work was soon translated into English as *On the Nature and Structural Characteristics of Cancer, and of Those Morbid Growths Which May Be Confounded with It,* trans. Charles West (London: Sherwood, Gilbert and Piper, 1840). On Schwann and Müller, see Frederick B. Churchill, "Rudolf Virchow and the Pathologist's Criteria for the Inheritance of Acquired Characteristics," *Journal of the History of Medicine and Allied Sciences* 31 (1976): 121–126.

13. Müller, *Handbuch der Physiologie,* 2:613.

14. Ibid.

15. Ibid., 1:21.

16. Ibid., 2:617.

17. Du Bois-Reymond, "Gedächnisrede auf Johannes Müller" (8 July 1858), re-printed in du Bois-Reymond, *Reden,* 1:135–317. Koller follows closely du Bois-Reymond's discussion.

18. Koller provides an undated letter from the seventy-one-year-old Alexander von Humboldt to Müller, which contains glowing but unspecific remarks about Müller's "Monographie" on Zeugung. Koller, *Johannes Müller,* 133. Humboldt's age and the fact he had been a strong supporter of Müller for many years puts his re-marks in perspective.

19. Du Bois-Reymond, *Reden,* 1:200. See Karl Reichert in his *Entwicklungsleben im Wirbelthier-Reich* (Berlin: August Hirschwald, 1840).

20. Reichert, *Entwicklungsleben,* 1–4; Lenoir, *Strategy of Life,* 219–224. Lenoir also carries the story of the teleo-mechanists and the cell theory to Virchow's *Cellularpathologie* (1858).

21. The two other students associated with this biophysically oriented cohort were du Bois-Reymond and Ernst von Brücke.

22. Robert Remak, *Untersuchungen über die Entwickelung der Wirbelthiere* (Berlin: G. Reimer, 1855), 84–85. See Rudolf Virchow, *Cellular Pathology as Based upon Physiological and Pathological Histology,* trans. Frank Chance (Philadelphia: J. B. Lippincott, 1863). A new facsimile edition (New York: Dover, 1971) contains a valuable introductory essay by L. J. Rather.

23. Ernst Haeckel, *Generelle morphologie der organismen: Allgemeine grundzüge der organischen formen-wissenschaft, mechanisch begründet durch die von Charles Darwin reformirte descendenztheorie* (Berlin: G. Reimer, 1866), 1: Vorwart, xvii.

24. Ibid., Haeckel, *Generelle Morphologie,* xiii–xxiv.

25. See particularly Haeckel's discussion on "Morphologie und Physik," Ibid., 10–12.

26. See Ernst Haeckel, *Anthropogenie; oder, Entwichlungsgeschichte des menschen,* 3rd ed. (Leipzig: Wilhelm Engelmann, 1877), 2:754fn199; "Interoperation of Natural Forces," in the English edition, *The Evolution of Man* (London: C. Kegan Paul, 1879), 2:490fn199.

27. Haeckel, *Anthropogenie,* 737. The translation is taken from Ernst Haeckel, *Evolution of Man,* 2:457. As did Müller before him, Haeckel disavowed both materialism and spiritualism and insisted on a monistic concept of the world (456).

28. Haeckel, *Generelle Morphologie,* 2:16, 171; Frederick B. Churchill, "August Weismann and a Break from Tradition," *Journal of the History of Biology* 1 (1968): 97.

29. See G. Heberer, *Der gerechtfertiqe Haeckel* (Stuttgart: G. Fischer, 1968) for a critique of Haeckel's monism.

30. Johann Wolfgang von Goethe, *Die Wahlverwandtschaften* (Berlin: J. G. Cottaische Buchhandlung, 1809); Ewald Hering, "Über das Gedächtnis als eine allgemeine Funktion der organisierten Materie"(Vortrag gehalten in der feierlichen Sitzung der kaiserlichen Akademie der Wissenschaften in Wien am 30.5. 1870) (Vienna: kaiserlich-königliche Hof- und Staatsdruckerei, 1870), 171–194.

31. Ernst Haeckel, *Die Perigenesis der Plastidule oder die Wellenzeugung der Lebensteilchen* (Berlin: Georg Reimer, 1876), 45.

32. The details of and influences on Haeckel's theory have been carefully described by Gloria Robinson, *A Prelude to Genetics: Theories of a Material Substance of Heredity, Darwin to Weismann* (Lawrence, Kans.: Coronado Press, 1979), chapter 3.

33. Haeckel, *Die Perigenesis,* 47–50.

34. Ibid., 55.

35. Georg Seidlitz, *Die Darwini'sche Theorie. Elf Vorlesungen über die Entstehung der Thiere und Pflanzen durch Naturzüchtung* (Dorpat: C. Mattiesen, 1871), 94.

36. Carl Gegenbaur, "Die Stellung und Bedeutung der Morphologie," in *The Interpretation of Animal Form,* Sources of Science No. 15, trans. by William Coleman (New York: Johnson Reprint, 1967), 50.

37. August Weismann, *Essays upon Heredity and Kindred Biological Problems,* 1st ed., trans. A. E. Shipley and Selmar Schönland, ed. Edward B. Poulton (Oxford: Clarendon Press, 1889), 72. See also Weismann's reassertion of his belief in "overgrowth" in the penultimate paragraph of his address. Because of the state of his eyes, he had the *Generelle Morphologie* read to him when it appeared. Haeckel at minimum sent him a copy of the third edition of the *Natürliche Schöpfungsgeschichte.* There is no indication, however, that Weismann paid much attention to the details of Haeckel's discussion of heredity. See August Weismann to Ernst Haeckel, 21 May, 1867 and Ernst Haeckel to August Weismann, 28 December 1872, in Georg Uschmann and Bernhard Hassenstein, "Der Briefwechsel zwischen Ernst Haeckel und August Weismann," in M. Gersch, ed., *Kleine Festgabe aus Anlass der hundertjährigen Wiederkehr der Gründung des Zoologischen Institutes der Friedrich-Schiller-Universität Jena im Jahre 1865* (Jena: Friedrich-Schiller-Universität, 1965).

38. Gerald Geison, "Darwin and Heredity: The Evolution of His Hypothesis of Pangenesis," *Journal of the History of Medicine and Allied Sciences* 24 (1969): 386.

39. Robert Stauffer, ed., *Charles Darwin's Natural Selection; Being the Second Part of His Big Species Book Written from 1856 to 1858* (Cambridge: Cambridge University Press, 1975), 1–2, 11–12, 34–35, 387.

40. The following account of Darwin's discussion of heredity as presented in *The Variation* is in part drawn from my earlier article: Frederick B. Churchill, "From Heredity Theory to Vererbung. The Transmission Problem, 1850–1915," *Isis* 78 (1987): 343–346. For a broader examination of Darwin's ideas, see particularly M. J. S. Hodge, "Darwin as a Lifelong Generation Theorist," in *The Darwinian Heritage,* ed. David Kohn (Princeton, N.J.: Princeton University Press, 1985); and Janet Browne, *Charles Darwin: Voyaging* (London: Jonathan Cape, 1995).

41. Differing accounts of the origin and motivation of Darwin's pangenesis theory are found in R. C. Olby, *Origins of Mendelism* (London: Constable, 1965), 84–86; Geisen, "Darwin and Heredity," 398–409; Robinson, *Prelude to Genetics,* 3–19; and Hodge, "Darwin as a Lifelong Generation Theorist."

42. R. C. Olby, *Origins of Mendelism,* 2nd ed. (Chicago: University of Chicago Press, 1985), points out that Darwin studied the English translation of Johannes Müllers's *Handbuch* and annotated portions "which directly bear on pangenesis . . . and (on) budding as a process in which superfluous material is separated from the organism" (85).

43. Charles Darwin, *The Variation of Animals and Plants under Domestication*, 2 vols. (London: John Murray, 1868), 2:357.

44. Ibid., 2:372.

45. Roy Macleod, "The X-Club a Social Network of Science in Late-Victorian England," *Notes and Records of the Royal Society of London* 24 (1970): 305–322.

46. The first volume of Spencer's *System der Synthetischen Philosphie* appeared in 1875; the *Principien der Biologie,* consisting of volumes 2 and 3 of the *System,* appeared

in 1876–1877. Spencer has left us with a lengthy *Autobiography,* which is well stocked with selections from his correspondence as well as a fascinating sketch of the origin of his ideas in "Filiation." Less useful is David D. Duncan, *Life and Letters of Herbert Spencer* (New York: D. Appleton, 1908). We still wait for an in-depth study of the sources and implications of Spencer's morphological ideas.

    47. Herbert Spencer, *Principles of Biology,* 2 vols. (New York: D. Appleton, 1884), 1:219–221.

    48. Ibid., 508.

    49. William K. Brooks, *The Law of Heredity: A Study of the Cause of Variation, and the Origin of Living Organisms* (Baltimore: John Murphy, 1883).

    50. The clearest sketch of Brooks is M. V. Edds's entry on Brooks in the *Dictionary of Scientific Biography,* vol. 1, ed. Charles Gillispie (New York: Charles Scribner's Sons, 1970). Other somewhat idiosyncratic sketches include: Edwin Grant Conklin, "Biographical Memoir of William Keith Brooks, 1848–1908," *Biographical Memoirs, National Academy of Sciences* 7 (1910): 25–88; and "William Keith Brooks: A Sketch of His Life by Some of His Former Students," *Journal of Experimental Zoology* 9 (1910): 1–52.

    51. Brooks, *The Law of Heredity,* 16–18.

    52. William K. Brooks, "A Provisional Hypothesis of Pangenesis," *American Naturalist* 12 (1877): 144–147.

10. Carl Nägeli and Inter- and Intragenerational Continuity

    1. Carl Nägeli, *Mechanische Theorie der Abstammungslehre* (Munich: R. Oldenbourg, 1884).

    2. Gerald Geisen, "The Protoplasmic Theory of Life and the Vitalist-Mechanist Debate," *Isis* 60 (1969): 272–292.

    3. Nägeli published many of his early papers in this journal. The most important were translated into English by Arthur Henfrey; see *Reports and Papers on Botany,* 2 vols. (London: Ray Society, 1847–1849).

    4. Robert Olby, "Nägeli, Carl Wilhelm von," *Dictionary of Scientific Biography,* vol. 9, ed. Charles Gillispie (New York: Charles Scribner's Sons, 1974): 600–602.

    5. Brigitte Hoppe, "Die Entwicklung der biologischen Fächer an der Universität München im 19. Jahrhundert unter Berücksichtigung des Unterrichts," in *Die Ludwig-Maximillians-Universität in ihren Fakultäten,* ed. Laetitia Boehm and Johannes Spörl (Berlin: Duncker & Humblot, 1972), 378–389.

    6. "Die beschränkte Befähigung des Ich gestattet uns somit nur eine äusserst fragmentarische Kenntnissnahme des Weltalls," in Carl Nägeli, "Die schrankender naturwissenschaftlichen Erkenntnis," *Amtlicher Bericht der 50 Versammlung Deutscher Naturforscher und Aertz in Münchin* (Munich: Straub, 1877), 570.

7. Ibid., 582. It would be wrong to imply that Nägeli was a neo-Kantian. His former student Simon Schwenderer insisted that Nägeli opposed the belief of innate, a priori categories, which are fundamental to the Kantian tradition.

8. Ibid., 598.

9. Ibid.

10. Olby, "Nägeli," 600.

11. Nägeli and Cramer edited the monograph series *Pflanzenphysiologische Untersuchungen,* which appeared in four *Heften* (1855–1858); the second *Hefte,* actually published fourth, consisted of the work on starch grains: Carl Nägeli, *Die Stärkenkörner. Morphologische, physiologische, chemischphysikalische und systematischbotanische Monographie* (Zürich: Friedrich Schulthess, 1858).

12. J. S. Wilkie, "Nägeli's Work on the Fine Structure of Living Matter—I," *Annals of Science* 16 (1960): 11–41.

13. Ibid., 35.

14. Ibid.

15. J. S. Wilkie, "Nägeli's Work on the Fine Structure of Living Matter—II," *Annals of Science* 16 (1960): 171–207.

16. Hoppe, "Die Entwicklung, 386.

17. Nägeli, *Mechanische Theorie,* 3.

18. Nägeli, *Die Stärkenkörner,* 33; quoted in his *Mechanische Theorie,* 15.

19. For the extent of Nägeli's details on this subject see the very useful chapter on Nägeli in Gloria Robinson, *A Prelude to Genetics: Theories of a Material Substance of Heredity, Darwin to Weismann* (Lawrence, Kans.: Coronado Press, 1979).

20. Nägeli, *Mechanische Theorie,* 66.

21. Ibid., 67–73, quotation from p. 67. See also Robinson, *Prelude to Genetics,* 113–114.

22. Nägeli, *Mechanische Theorie,* 69–82.

23. Ibid., 23. For a discussion of Nägeli's use of the term "anlage," see "Translator's Notes," Carl Nägeli, *A Mechanico-physiological Theory of Organic Evolution,* trans. V. A. Clark (Chicago: Open Court, 1898), 49–51.

24. Nägeli, *Mechanische Theorie,* 193.

25. No wonder Mendel's classic paper of 1865 made so little impact on Nägeli! Mendel accounted for the distribution of his seven characters in peas by means of the distribution of the "elements" during reproduction. Only in the situation of the heterozygotes did Mendel have to invoke the principle of dominance and latency, which tacitly called for an activation principle. This, of course, Mendel did not, nor could not provide.

26. Nägeli, *Mechanische Theorie,* 213–214, 199–201, 211, 201–202.

27. Mendel's paper on Hieracium finds the Augustinian monk turning to broader and more inclusive generalizations.

28. Nägeli, *Mechanische Theorie,* 236–245.

29. Elizabeth B. Gasking, "Why Was Mendel's Work Ignored?" *Journal of the History of Ideas* 20 (1959): 60–84, esp. 73–77.

30. See "Summary," in Nägeli, *Mechanico-physiological,* 1–44. Because of its skeletal nature, it fails to convey to the reader the richness and comprehensiveness of Nägeli's theory.

31. Nägeli, *Mechanische Theorie,* 273–274.

32. Ibid., 275.

33. Ibid.

34. Ibid., 277–288.

35. Ernst Mayr, *The Growth of Biological Thought: Diversity, Evolution, and Inheritance* (Cambridge, Mass.: Belknap Press/Harvard University Press, 1982), 671.

11. The Emergence of Nuclear Cytology

1. Ernst Gaup, *August Weismann: sein Leben und sein Werk* (Jena: Fischer, 1917), 32.

2. Francis Balfour, *A Treatise on Comparative Embryology,* vol. 1 (London: Macmillan, 1880), 120–122, 395–406. Centrolecithal eggs are those with the yolk aggregated in the center.

3. These pole nuclei are the product of first cleavage and should not be confused with the pole cells later.

4. Two gall wasps, *Rhodites rosae* and *Biorhiza aptera,* a cricket, *Gryllotalpa vulgaris,* and an unidentified species of midge, *Chironomus.*

5. Vitus Graber, "Vorläufige Ergebnisse einer grösseren Arbeit über vergleichende Embryologie der Insekten," *Archiv für mikroskopische Anatomie* 15 (1878), 630–640.

6. Klaus Sander, "August Weismanns Untersuchungen zur Insecktenentwicklung 1862–1882," *Freiburger Universitätsblätter* 24 (1885): 43–52.

7. August Weismann, "Beiträge zur Kenntniss der ersten Entwicklungsvorgänge im Insektenei," in *Festgabe Jacob Henle zum 4: April 1882 dargebracht von seinen Schülern* (Bonn: Cohen, 1882), 81.

8. Weismann to Haeckel, 27 January 1874 in Georg Uschmann and Bernhard Hassenstein, "Der Briefwechsel zwischen Ernst Haeckel und August Weismann," in *Kleine Festgabe aus Anlass der hundertjährigen Wiederkehr der Gründung des Zoologischen Institutes der Friedrich-Schiller-Universität Jena im Jahre 1865,* ed. M. Gersch (Jena: Friedrich-Schiller-Universität, 1965), 35–36.

9. Alexander Brandt, *Über das Ei und seine Bildungsstätte. Ein vergleichend-morphologischer Versuch mit zugrundgelegung des Insecteneies* (Leipzig: Wilhelm Engelmann, 1878). At the time, Brandt held a doctorate and served as curator at the Royal Academy of Science in St. Petersburg.

10. Weismann, "Beiträge zur Kenntniss, 106.

11. Ibid., 105.

12. Francis Maitland Balfour, "A Treatise on Comparative Embryology," in *The Works of Francis Maitland Balfour* [Memorial Edition], 4 vols., ed. M. Foster and Adam Sedgwick (London: Macmillan, 1885), 1:88–92. Balfour had reviewed in a clear and judicious manner the status of polar body formation and impregnation two years earlier. See Francis Maitland Balfour, "On the Phenomena Accompanying the Maturation and Impregnation of the Ovum," *Quarterly Journal of Microscopical Science*, 1878, reprinted in *The Works of Francis Maitland Balfour*, 1:521–548.

13. Balfour, "Treatise," 1:90–91.

14. Francis Maitland Balfour, *Handbuch der Vergrleichenden Embryologie*, 2 vols., trans. B. Vetter (Jena: Gustav Fischer, 1880–1882); Wilhelm Waldeyer, "Francis Maitland Balfour, Ein Nachruf," *Archiv für mikroskopische Anatomie* 21 (1882): 828–835.

15. Most recently, see Henry Harris, *The Birth of the Cell* (New Haven, Conn.: Yale University Press, 1999).

16. Although synthetic stains had become available by the 1870s and were used by histologists, most of the advances in nuclear cytology during the decade were made using natural dyes, such as hematoxylin from logwood and carmine extracted from cochineal bugs. George Clark and Frederick H. Kasten, *History of Staining* (Baltimore: Williams and Wilkins, 1983), chapters 5 and 6.

17. Strasburger and Flemming had already presented much of their research in scholarly journals. Flemming does present a short chronological account of the period at the end of his volume.

18. Later in the decade, Walther Flemming chose this year and a publication by Anton Schneider as the opening of the "new findings" on cell division. Walther Flemming, "Beiträge zur Kenntniss der Zelle und ihrer Lebenserscheinungen," *Archiv für mikroskopische Anatomie* 16 (1878): 398, 410–412.

19. For Goodsir (1845), Remak (1852), and Virchow (1855), see Erwin Ackerknecht, *Rudolf Virchow: Doctor, Statesman, Anthropologist* (Madison: University of Wisconsin Press, 1953), 82–84. To what extent is there confusion between "nucleus" and "nucleolus"? Both were common terms since the days of Schwann. It is also possible that some microscopists confounded the centrosomes with the nucleolus.

20. Incidental observations on rays associated during egg cleavage had been made prior to 1873. Flemming, "Beiträge zur Kenntniss," 411.

21. Anton Schneider, "Untersuchungen über Platthelminthen," *Bericht der Oberhessischen Gesellschaft für Natur- und Heilkunde* 14 (1873): 69–140.

22. Hermann Fol, "Die erste Entwickelung des Geryomideneies," *Jenaische Zeitschrift für Medizin und Naturwissenschaft* 7 (1873): 471–492.

23. O. Bütschli, "Beiträge zur Kenntniss der freilebenden Nematoden," *Nova acta Academiae Caesareae Leopoldino-Carolinae Germanicae Naturae Curiosorum* 36, no. 5 (1873): 1–144.

24. Walther Flemming, "Ueber die ersten Entwicklungs-erscheinungen am Ei der Teichmuschel," *Archiv für mikroskopische Anatomie* 10 (1874): 257–292.

25. Gustav Born, "Leopold Auerbach," *Anatomischer Anzeiger* 14 (1897): 257–267; Bruno Kisch, "Forgotten Leaders in Modern Medicine: Valentin, Gruby, Remak, Auerbach," *Transactions of the American Philosophical Society,* n.s., 44 (1954): 297–313.

26. Leopold Auerbach, *Organologische Studien. Zur Characteristik und Lebensgeschichte der Zellkerne* (Breslau: Morgenstern, 1874).

27. According to E. B. Wilson, *The Cell in Development and Inheritance* (New York: Macmillan, 1996), 202, and G. Hamoir, "The Discovery of Meiosis by E. van Beneden, a Breakthrough in the Morphological Phase of Heredity," *International Journal of Developmental Biology* 36 (1992): 9–15, van Beneden coined the term "pronucleus" in 1875. At that time, he distinguished between the "pronucleus périphérique et centrale," that is, the male and the female pronucleus. The terms not only reflect van Beneden's hermaphroditic concept of fertilization but emphasize the distinction between the nuclei of fertilization and of first cleavage.

28. Walther Flemming, *Zellsubstanz, kern und zelltheilung* (Leipzig: F.C.W. Vogel, 1882), 182.

29. Oscar Hertwig, "Beiträge zur Kenntnis der Bildung, Befruchtung und Theilung des thierischen Eies," *Morphologisches Jahrbuch* 1 (1876): 425.

30. Otto Bütschli, *Studien über die ersten Entwicklungsvorgänge der Eizelle, die Zelltheilung und die Conjugation der Infusorien* (Frankfurt: Christian Winter, 1886). This monograph is described in greater detail in Frederick B. Churchill, "The Guts of the Matter: Infusoria from Ehrenberg to Bütschli: 1838–1876," *Journal of the History of Biology* 22 (1989): 206–211. At that time, I focused on Bütschli's effort to reassert Theodor Siebold's 1845 analogy between infusoria and the single cells of metazoans.

31. Natasha X. Jacobs, "From Unit to Unity: Protozoology, Cell Theory, and the New Concept of Life," *Journal of the History of Biology* 22 (1989): 215–242.

32. Flemming, *Zellsubstanz,* 327, 393.

33. See Paul Weindling, *Darwinism and Social Darwinism in Imperial Germany: The Contribution of the Cell Biologist, Oscar Hertwig, 1849–1922* (Stuttgart: G. Fischer, 1991) for additional details of Hertwig's work.

34. Oscar Hertwig, *Beiträge zur Kenntniss der Bildung, Befruchtung und Theilung des thierischen Eies* (Leipzig: Engelmann, 1875), also in *Morphologisches Jahrbuch* 1 (1876): 347–434.

35. Hertwig summarized the nineteenth-century efforts to sort out dynamic from material theories of heredity in *Beiträge zur Kenntnis* (1876), 390–398. He seems to be responding to Haeckel's perigenesis!

36. Weindling, *Darwinism and Social Darwinism,* 80–85. For an undeveloped comment on sex transmission, see Hertwig, *Beiträge zur Kenntniss,* 386fn. In this footnote, Hertwig was responding to van Beneden's recent statements that all developed organisms are hermaphroditic and egg maturation is a process of eliminating the paternal residues in preparation for a new male element arriving upon fertilization. Notwithstanding his interest in intergenerational continuity, Hertwig was not yet concerned about trait transmission. For comments pointing out Hertwig's neglect of heredity at this point in his career, see William Coleman, "Cell, Nucleus, and Inheritance: An Historical Study," *Proceedings of the American Philosophical Society* 109 (1965): 124–158, 348–349; and Frederick B. Churchill, "From Heredity Theory to Vererbung: The Transmission Problem," *Isis* 78 (1987): 337–364.

37. Weindling has Strasburger otherwise; see his *Darwinism and Social Darwinism,* 80–86; Hertwig, *Beiträge zur Kenntniss* (1876), 418–420.

38. Hertwig, *Beiträge zur Kenntniss* (1876), 418.

39. Flemming, *Zellsubstanz,* 392; Edward Mark, "Maturation, Fecundation and Segmentation of *Limax campestris* Binny," *Bulletin of the Museum of Comparative Zoology* 6, no. 12 (1881): 341–342.

40. I have followed the detailed discussion of Strasburger's second edition of *Zellbildung und Zelltheilung* to be found in Mark, "Maturation," 372–385.

41. Richard Hertwig, "Beiträge zu einer einheitlichen Auffassung der verschiedenen Kernformen," *Morphologisches Jahrbuch* 2 (1876): 63–82.

42. Ibid., 79.

43. Heindrich Waldeyer, "Über Karyokinese und ihre Beziehungen zu den Befruchtungsvorgängen," *Archiv für mikroskopische Anatomie und Entwicklungsmechanik* 32 (1888): 9. "Karyokinesis" is a term still used in some quarters today rather than "mitosis."

44. Walther Flemming, "Beobachtungen über die Beschaffenheit des Zellkerns," *Archiv für mikroskopische Anatomie* 13 (1877): 693–717.

45. Boettgher "introduced" the technique in 1869 but apparently did not publicize it. George Clark and Frederick H. Kasten, *History of Staining,* 3rd ed. (Baltimore: Williams and Wilkins, 1983), 80.

46. Flemming, "Beobachtungen," 704–708.

47. Ibid., 708–712.

48. Walther Flemming, "Beiträge zur Kenntniss der Zelle und ihrer Lebenserscheinungen," Theil. 1, *Archiv für mikroskopische Anatomie* 16 (1878): 302–436.

49. Ibid.

50. Flemming placed a question mark at this point; elsewhere, he appeared to exclude consideration of the longitudinal fusion in the reverse metamorphosis. Flemming, "Beiträge zur Kenntniss," 391.

51. Ibid., 410–423.

52. Ibid., 412–414.

53. Walther Flemming, "Beiträge zur Kenntniss der Zelle und ihrer Lebenserscheinungen," Theil. 2, *Archiv für mikroskopische Anatomie* 18 (1880): 151–259.

54. Ibid., 192–193.

55. Eduard Strasburger, *Zellbildung und zelltheilung* (Jena: G. Fischer, 1880), 5–6.

56. Ibid.

57. Strasburger, *Zellbildung*, 294–300.

58. Strasburger appears to use "Zellplatte" when describing the future location of the new cell wall in plants, which traversed the site of the "Kernplatte." The latter was equivalent to Flemming's "Aequatorialplatte."

59. Strasburger, *Zellbildung*, 370–371.

60. Ibid., 322.

61. Ibid., 331.

62. Ibid., 109–117, 171–184.

63. Ibid., 374. The continued differences of Strasburger's and Flemming's views between 1880 and 1882 are spelled out in Flemming, *Zellsubstanz,* 398–399.

64. Flemming, *Zellsubstanz,* 1–9.

65. Flemming, "Beiträge," 409, and *Zellsubstanz,* 266.

66. Flemming, *Zellsubstanz,* 235.

67. E. Strasburger, "Die Controversen der indirecten Kerntheilung," *Archiv für mikroskopische Anatomie* 23 (1884): 17–40. The term "telophase" was introduced by Haidenheim in 1894 to designate the reconstitution of the chromatic material into the daughter nuclei. E. B. Wilson, *The Cell,* 342.

68. Flemming, *Zellsubstanz,* 1–7.

69. Ibid., 70 (emphasis added). It consists of a key ending to the passage, which unfortunately was not included in Ilse Jahn, Rolf Löther, and Konrad Sendglaub, *Geschichte der Biologie: Theorien, Methoden, Institutionen, Kurzbiographien* (Jena: G. Fischer, 1982), 356.

70. Beneden had developed his own conclusions on the basis of *Hydractinia* alone.

71. Carl Rabl, "Édouard van Beneden und der gegenwärtige Stand der wichtigsten von ihm behandelten Probleme," *Archiv für Mikroskopische Anatomie* 88 (1915): 29–30.

72. Among those at the time involved were Nelson, Newport, Meissner, Bischoff, Allen Thomson, Clarapede, and Munk. Édouard van Beneden, "L'Apparaeil sexuel femell de l'Ascaride mégalocéphale," *Archives de Biologie* 4 (1883): 95–142; Hamoir "Discovery of Meiosis," 32–37.

73. *Ascaris* egg has a diameter of 70 ± 10 μm compared with the sea urchin's egg of around 0.1 mm. Hamoir, "Discovery of Meiosis," 36–37.

74. The term "chromosome" was coined by Waldeyer in 1888 and denoted attributes of these chromatic elements that had not been worked out in 1884. It soon became apparent that *Ascaris* as well as the sea urchin eggs occasionally contained supernumerary chromosomes. Wilson, *The Cell,* 147.

75. Édouard van Beneden, "Recherches sur la maturation de l'oeuf et la fécondation. *Ascaris megalocephala,*" *Archives de Biologie* 4 (1884): 265–640.

76. Hamoir, "Discovery of Meiosis," 11. Rabl and Brachet both correctly cite this monograph as appearing in 1884; other accounts, including Hamoir's, consider the volume year of 1883 to be the publication date.

77. Ibid., 43–44.

78. Beneden, "Recherches," 419–429, and plates XIV–XV; Hamoir, "Discovery of Meiosis," 39–40.

79. "L'on ne peut donc pas comparer la génèse du premier globule polaire à une division cellulaire indirecte." Beneden, "Recherches," 447.

80. Ibid., 480–483. I am particularly indebted to Hamoir's explication of this difficult section, but I cannot accept his conclusion that van Bendeden was consequently the discoverer of meiosis.

81. Ibid., 524–530. A year thereafter, Rabl was able to confirm this continuation of the separate paternal and maternal nuclear elements in development. Wilson, *The Cell,* 215.

82. Beneden, "Recherches," 620.

83. Ibid., 621.

84. Wilhelm Roux, *Ueber die Bedeutung der Kerntheilungsfiguren: Eine hypothetische Erörtung* (Leipzig: Wilhelm Engelmann, 1883). Besides his own autobiography, "Wilhelm Roux in Halle a. S.," *Die Medizin der Gegenwart in Selbstdarstellungen,* 2 vols., ed. L. R. Grote (Leipzig: Felix Meiner, 1923), 1:141–206, the most thorough account of Roux's life and the experimental embryological tradition of the last twenty years of the nineteenth century is Reinhard Mocek, *Die Werdende Form: Eine Geschichte der Kausalen Morphologie* (Marburg an der Lahn: Basilisken-Presse, 1998). A contemporary assessment of Roux's research with short contributions by the anatomist Hermann Braus, the surgeon Georg Magnus, the experimental embryologists Hans Spemann, Hans Driesch, and Dietrich Barfurth, and the botanist Ernst Küster may be found in "Wilhelm Roux zur Feier seines siebzigsten Geburtstages," *Die Naturwissenschaften* 8 (1920): 431–459. Also relevant is Frederick B. Churchill, "Roux, Wilhelm," in *Dictionary of Scientific Biography,* vol. 11, ed. Charles Gillispie (New York: Charles Scribner's Sons, 1975), 570–575.

85. His initial work was a prelude to a series of Beiträge on functional adaptation in blood vessels, connective tissue, and the skeleton.

86. Wilhelm Roux, *Ueber die Leistungsfähigkeit der Principien der Deszendenzlehre zur Erklärung der Zweckmässigkeiten des thierischen Organismus* (Breslau: S. Schottlaender, 1880).

87. Wilhelm Roux, *Der Kampf der Theile im Organismus. Ein Beiträge zur vervollständigung der mechanischen Zweckmässigkeitslehre* (Leipzig: Wilhelm Engelmann, 1881).

88. Roux had foreshadowed the notion of a struggle of parts in earlier papers.

89. Darwin to Romanes, 16 April 1881, in *Life and Letters of Charles Darwin,* 2 vol., ed. Francis Darwin (New York: D. Appleton, 1897), 22:419–420. Romanes did review the work in which he claimed that Herbert Spencer had already put forth a similar idea. Georges J. Romanes, "The Struggle of Parts in the Organism," *Nature* 24 (1881): 505–506.

90. August Weismann, *Studies in the Theory of Descent,* 2 vols., trans. Raphael Meldola (London: Sampson Low, Marston, Searle & Rivington, 1882), 1:88.

91. This episode was recorded at the end of Roux's life. "Wilhelm Roux in Halle a. S.", in *Die Medizin der Gegenwart in Selbstdarstellungen,* 2 vols., ed. L. R. Grote (Leipzig: Felix Meiner, 1923), 1:141–206; Churchill, "Roux, Wilhelm," 571.

92. Roux relied on the work of Carl Grobben and Moritz Nussbaum, which indicated that germ cells arose before germ-layer formation. Roux appears not to dispute the practical value of the biogenetic law when applied to anatomically localized attributes. Roux, *Der Kampf der Theile,* 59.

93. Ibid., 61.

94. Quoted by Roux; ibid., 161.

95. Ibid., 34–47.

96. Ibid., 213–216, quotation from p. 214.

97. Wilhelm Roux, "Allgemeine Entwickelungs-geschichte und Zeugung," *Jahresberichte über Fortschritte der Anatomie und Physiologie* 10 (1882): 383–422; see esp. 395–396.

98. Roux, *Ueber die Bedeutung,* 129.

99. These popular German sayings are repeated in Ernst Haeckel, *Generelle morphologie der organismen: Allgemeine grundzüge der organischen formen-wissenschaft, mechanisch begründet durch die von Charles Darwin reformirte descendenztheorie,* 2 vols. (Berlin: G. Reimer, 1866), 2:170–171.

100. There quickly followed a reaction to "the monopoly of the nucleus." See Alfred Barthelmess, *Vererbungs-Wissenschaft* (Orbis Academicus. Problemgeschichten der Wissenschaft in Dokumenten und Darstellungen) (Freiburg/München: K. Alber, 1952), and for a later period, Jan Sapp, *Beyond the Gene: Cytoplasmic Inheritance and the Struggle for Authority in Genetics* (New York: Oxford University Press, 1987).

12. A New Perspective on Heredity

1. Weismann to Ernst Haeckel, 4 November 1877, Georg Uschmann and Bernhard Hassenstein, "Der Briefwechsel zwischen Ernst Haeckel und August Weismann," in M. Gersch, ed., *Kleine Festgabe aus Anlass der hundertjährigen Wiederkehr der Gründung des Zoologischen Institutes der Friedrich-Schiller-Universität Jena im Jahre 1865* (Jena: Friedrich-Schiller-Universität, 1965), 44–45.

2. See Chapter 7 and Frederick B. Churchill, "Weismann, Hydromedusae, and the Biogenetic Imperative: A Reconsideration," printed in Symposium 8 of the British Society for Developmental Biology entitled *A History of Embryology*, ed. T. J. Horder, J. A. Witkowski, and C. C. Wylie (Cambridge: Cambridge University Press, 1985).

3. August Weismann, *Die Continuität des Keimplasmas als Grundlage einer Theorie der Vererbung* (Jena: Gustav Fischer, 1885).

4. August Weismann, "Zur Geschichte der Vererbungstheorie," *Zoologischer Anzeiger* 9 (1886): 350. See also August Weismann, "Zur Annahme einer Continuität des Keimplasma's," *Berichte der Naturforschenden Gesellschaft zu Freiburg i. Br.* 1 (1886): 89–99, August Weismann, "Historisches zur Lehre von der Continuität des Keimplasma's," *Berichte der Naturforschenden Gesellschaft zu Freiburg i. Br.* 7 (1893): 36–37, and August Weismann, *The Germ-Plasm. A Theory of Heredity*, trans. William Newton Parker and Harriet Rönnfeldt, ed. H. Ellis, *The Contemporary Science Series* (London: Walter Scott, 1893), 260–265. Patrick Geddes and J. Arthur Thomson, *The Evolution of Sex* (1889), 2nd ed. (London: Walter Scott, 1901), 99–101. L. Bounoure, *Origine des cellules reproductrices et le problème de la lignée germinale* (Paris: Gauthier-Villars, 1939) contains the most elaborate examination of the older literature.

5. Richard Owen, *On Parthenogenesis; or The Successive Production of Procreating Individuals from a Single Ovum* (London: John Van Voorst, 1849), 72.

6. Frederick B. Churchill, "Sex and the Single Organism," *Studies in the History of Biology* (1979) 145–149; John Farley, *Gametes and Spores. Ideas about Sexual Reproduction 1750–1914* (Baltimore: Johns Hopkins University Press, 1982), 101–104.

7. Francis Galton, "Hereditary Talent and Character," *MacMillan's Magazine* 12 (1865): 157–166, 318–327, quotation from p. 322.

8. Francis Galton, "On Blood-Relationship," *Proceedings of the Royal Society of London* 20 (1871–1872): 394–402.

9. Francis Galton, "A Theory of Heredity," *Contemporary Review* 27 (1875–1876): 80–95, quotations from pp. 81 and 88.

10. Ruth Schwartz Cowan, *Sir Francis Galton and the Study of Heredity in the Nineteenth Century* (New York: Garland, 1969), 184.

11. Gloria Robinson, *A Prelude to Genetics: Theories of a Material Substance of Heredity, Darwin to Weismann* (Lawrence, KS: Coronado Press, 1979), 42.

12. Ibid., 30.

13. Weismann to Galton, 23 February, 1889, in Frederick B. Churchill and Helmut Risler, *August Weismann, Ausgewählte Briefe und Dokumente = Selected Letters and Documents*, 2 vols. (Freiburg: Universitätsbibliothek Freiburg, 1999), 1:130–131. Also in Karl Pearson, *Life and Letters of Francis Galton*, 3 vols. (Cambridge: Cambridge University Press, 1914–30), 3:340–341.

14. Ruth Schwartz Cowan, "Nature and Nurture: The Interplay of Biology and Politics in the Work of Francis Galton," *Studies in the History of Biology* 1 (1977): 133–208.

15. Sachs spelled out his priority claim in 1886 and at that time reproduced the relevant passage. Julius v. Sachs, "Continuität der embryonalen Substanz," *Naturwissenschaftliche Rundschau,* Jahrg. I, no. 5 (1886): 33–34. The English translation of his *Vorlesungen* appeared the following year. *Lectures on the Physiology of Plants* (Oxford: Clarendon Press, 1887), see especially 769–770.

16. August Weismann, "Zur Annahme einer Continuität des Keimplasma's," *Berichte der Naturforschenden Gesellschaft zu Freiburg i. Br.* 1 (1886): 18–22.

17. W. Lubosch, "August Rauber. Sein Leben und seine Werke," *Anatomischer Anzeiger* 58 (1924): 129–172, quotation from p. 136. All biographical information comes from this obituary.

18. Rauber's *Lehrbuch de Anatomie des Menschen* (1892) was the fourth edition of Carl E. E. Hoffman's *Lehrbuch de Anatomie des Menschen* (1877–1887). This, in turn, was a translation and reworking of Jones Quain's *Elements of Anatomy* (1st. ed.; 1824). By 1906–1909, Rauber's *Lehrbuch* was edited by F. Kopsch. It continued to be published as Rauber-Kopsch *Lehrbuch der Anatomie des Menschen* at least into the 1960s.

19. August Rauber, *Formbildung und Formstörung in der Entwicklung von Wirbelthieren* (Leipzig: Wilhelm Engelmann, 1880). This monograph was reverentially dedicated to Haeckel and first appeared in Gegenbaur's *Morphologisches Jahrbuch.*

20. Rauber, *Formbildung,* 45.

21. Ibid., 63.

22. August Rauber, "Personaltheil und Germinaltheil des Individuum," *Zoologischer Anzeiger* 9 (1886): 166–171.

23. Specifically, the *Jahresbericht der Zoologischen Station zu Neapel vom Jahre* (1880).

24. August Weismann, "Zur Geschichte der Vererbungstheorie," *Zoologischer Anzeiger* 9 (1886): 345–348.

25. Johannes Müller habilitated in comparative anatomy and physiology in 1824 and in 1830 rose to the rank of a full professor. Johannes Steudel, "Müller, Johannes Peter," in *Dictionary of Scientific Biography,* vol. 9, ed. Charles Gillispie (New York: Charles Scribner's Sons, 1974), 567–574.

26. Moritz Nussbaum, "Adolph Johannes Hubert Freiherr von la Valette St. George," *Chronik der rheinischen Friedrich-Wilhelm Univerität zu Bonn* (1910): 2–5.

27. Harrison's manuscript sketch, dated March 16, 1920, and simply entitled "Moritz Nussbaum," may be found in the Harrison Papers, Yale University Library. The author wishes to thank Jane Maienschein for pointing out the existence of this biographical sketch.

28. Moritz Nussbaum, "Zur Differenzirung des Geschlechts im Thierreich," *Archiv für mikroskopische* 18 (1880): 1–120.

29. Nussbaum, "Zur Differenzirung," 105.

30. Ibid., 109.

31. Ibid., 112.

32. Ibid., 111–112.

33. Jane Maienschein, "What Determines Sex? A Study of Converging Approaches, 1880–1916," *Isis* 75 (1984): 457–480.

34. Nussbaum, "Zur Differenzirung," 113.

35. Moritz Nussbaum, "Ueber die Veränderungen der Geschlechtsproducte bis zur Eifurchung; ein Beitrag zur Lehre der Vererbung," *Archiv für mikroskopische* 23 (1884): 155–213.

36. Ibid., 186–187. With the frog *Rana fusca* the primordial germ cells maintain not only their embryonic character but retain the same size as some of the late-division blastomeres. This suggested more strongly a direct link with the cleavage stage.

37. Ibid., 183.

38. Ibid., 182.

39. Ibid., 188–192.

40. At least two of his later studies discuss heredity, but he continues to be concerned with the distinction between the separation of germ cells from embryonic differentiation. For example, Nussbaum, *Ueber Vererbung* (Bonn: Max Cohen & Sohn, 1888), and "Beiträge zur Lehre von der Fortpflanzung und Vererbung," *Archiv für mikroskopische* 41 (1893): 119–145.

41. Eduard Strasburger, *Die Controversen der indirecten Kerntheilung* (Bonn: Max Cohen & Sohn, 1884).

42. "Einstweilen würde dies nicht zu beweisen, auch nicht zu widerlegen sein." From Walther Flemming, *Zellsubstanz, kern und zelltheilung* (Leipzig: F. C. W. Vogel, 1882), 235. Quoted by Strasburger, *Die Controversen,* 16.

43. Strasburger, *Die Controversen,* 13.

44. Ibid., 7–58.

45. Eduard Strasburger, *Neue Untersuchungen über den Befruchtungsvorgang bei den Phanerogamen als Grundlage für eine Theorie der Zeugung* (Jena: Gustav Fischer, 1884).

46. Strasburger also investigated the union of male and female derived nuclei in gymnosperms, in which the chain of continuity is less complex and dramatic (ibid., 49–55). For a contemporary description of phanerogam fertilization, see Julius von Sachs, *Lectures on the Physiology of Plants,* trans. H. Marshall Ward (Oxford: Clarendon Press, 1887), from the 5th German edition of 1882, lectures 52–53.

47. Strasburger, *Neue Untersuchungen,* 86. Quoted also in Frederick B. Churchill, "From Heredity Theory to Vererbung. The Transmission Problem, 1850–1915," *Isis* 78 (1987): 347–348.

48. Strasburger, *Neue Untersuchungen,* 77.

49. "Hyaloplasm" was a common term at the time referring to the clear proto-plasmic matrix of the cell and nucleus.

50. Strasburger, *Neue Untersuchungen,* 114.

51. Ibid.

52. Ibid., 133.

53. Ibid., 138–171.

54. Ibid., 103–104.

55. Oscar Hertwig, "Das Problem der Befruchtung und der Isotropie des Eies, eine Theorie der Vererbung," *Jenaische Zeitschrift für Medicin und Naturwissenschaft,* n.s., 18, no. 11 (1884–1885): 276–318. The paper was signed by Hertwig on "An-fang October 1884" and dated November 28, 1884; the reprint is dated "1884", and a summary review, written by "C," appeared in *Biologisches Centralblatt* on "15 Mai 1885." Some of my discussion here follows Churchill, "From Heredity Theory to Vererbung," 348–351.

56. Hertwig, "Das Problem," 276 (emphasis added).

57. Ibid., 280–283. In his earlier studies Hertwig had focused on sea urchins, in which polar body formation occurs prior to or during sperm penetration.

58. Ibid., 283–291.

59. Ibid., 290–302.

60. Ibid., 309–311.

61. Ibid., 301–302 (emphasis added). The lead into the quotation, this paragraph, and the translation follow closely Churchill, "From Heredity Theory to Vererbung," 350–351.

62. Hertwig, "Das Problem," 309–318.

63. The context of this work will be discussed in Chapter 16.

64. Paul Julian Weindling has discussed this episode in detail in *Darwinism and Social Darwinism in Imperial Germany: The Contribution of the Cell Biologist Oscar Hertwig (1849–1922)* (Stuttgart/New York: Gustav Fischer, 1991), 109–115. For the relationship between embryology and heredity, see particularly Reinhard Mocek, *Die Werdende Form. Eine Geschichte der Kausalen Morphologie* (Marburg: Basilisken-Presse, 1998). As Mocek points out, in the pre-Flemming era, Wilhelm His consid-ered heredity (the transmission of form from one generation to another) to be the natural consequence of growth and physical processes (166–169).

65. Albert Kölliker, *Errinerungen aus meinem Lebern* (Leipzig: Wilhelm Engel-mann, 1899), 8–9. See also Erich Hintzsche, "Kölliker, Rudolf Albert von," *Dic-tionary of Scientific Biography,* vol. 7, ed. Charles Gillispie (New York: Charles Scrib-ner's Sons, 1970–1980), 437–440; Erhart Kahle, "Kölliker, Albert Ritter v.," *Neue Deutsche Biographie* 12 (1980): 322–323.

66. Kölliker, *Errinerungen,* 9.

67. Kölliker included in his memoirs eighteen lengthy letters written to his mother about their youthful travels. Ibid., 49–84.

68. Brigette Hoppe, "Nägeli, Carl Wilhelm v.," *Neue Deutsche Biographie* 18 (1996): 702–704; Erhart Kahle, "Kölliker." For a discussion of the term "Wissenschaft" in the first half of the century and these two journals, see Lynn Nyhart, *Biology Takes Form: Animal Morphology and the German Universities, 1800–1900* (Chicago: University of Chicago Press, 1995), 37–47, 94–96. Nyhart unfortunately leaves the impression that the *Zeitschrift für wissenschaftliche Botanik* never appeared, but she is probably referring to an effort on the part of Alexander Braun and Nägeli to resurrect it in 1848. For a discussion of specialist journals during the nineteenth century, see Ilse Jahn, Rolf Löther, and Konrad Senglaub, *Geschichte der Biologie: Theorien, Methoden, Institutionen, Kurzbiographien* (Jena: Gustav Fischer, 1982), 328–329.

69. Albert Köelliker, *Grundriss der Entwickelungsgeschichte des Menschen und der höheren Tiere* (Leipzig: Wilhelm Engelmann, 1884), 18–19. Kölliker often refers to the hermaphroditic nature of the nucleus but generally only in the sense that it contains paternal and maternal idioplasm.

70. In this context, Kölliker here mentioned the plant physiologist Julius Sachs who in 1882 had maintained that only a small portion of the egg cell bore the burden of heredity. Albert Kölliker, "Ueber Vererbung (Idioplasma)," *Sitzungsberichte der physical. Medicin. Gesellschaft.* (Wurzburg), *Jahrgang 1885* (1885): 46–49. See also his strong comments about the superiority of Nägeli's contribution to evolution theory in Kölliker, *Errinerungen*, 323.

71. Kölliker, "Ueber Vererbung," 47–48.

72. Ibid., 49.

73. Albert Kölliker, "Die Bedeutung der Zellenkerne für die Vorgänge der Vererbung," *Zeitschrift für wissenschaftliche Zoologie* 42 (1885): 1–46.

74. Kölliker, "Die Bedeutung," 45.

75. Ibid., 24–37. At the same time he hypothesized that the idioplasm corresponded with the nuclein discovered and named by Fr. Miescher. Albert Kölliker, "Das Karyoplasm und die Vererbung, eine Kritik de Weismann'schen Theorie von der Kontinuität des Keimplasma," *Zeitschrift für wissenschaftlisher Zoologie* 44 (1886): 228–238.

76. Kölliker, "Das Karyoplasm," 10.

77. Kölliker continued to oppose Weismann's Darwinian explanation of variations in almost identical words in his memoirs at the end of his life. Kölliker, *Errinerungen*, 357–358.

78. August Weismann, "Die Continuität des Keimplasmas als Grundlage einer Theorie der Vererbung" (1885), reprinted in his *Aufsätze über Vererbung und verwandte biologische Fragen* (Jena: Gustav Fischer, 1892), 213. This English translation from his *Essays upon Heredity and Kindred Biological Problems,* trans. A. E. Shipley and Selmar Schönland, ed. Edward B. Poulton (Oxford: Clarendon Press, 1889), 181.

79. August Weismann, *Die Continuität des Keimplasmas als Grundlage einer Theorie der Vererbung* (Jena: Gustav Fischer, 1885), 213; Weismann, *Essays,* 181.

80. Weismann, *Essays,* 177. He also pointed out that Balfour at the same time had come to the same conclusion in his *Treatise.* It seems that Kölliker did as well. See the footnote in the English version, 177–178.

81. Weismann, *Aufsätze,* 197; Weismann, *Essays,* 167.

82. After consultation with his mathematics colleague, Jakob Lüroth, Weismann reported the following: "We get the formula $(1/p)^n$, where p represents the possibilities, and n the characters. Thus if n and p are but slightly larger than we assumed above [i.e., 10], the probabilities become so slight as to altogether exclude the hypothesis of a re-transformation of somatic Idioplasm into germ-plasm." Weismann, *Aufsätze,* 239–240; Weismann, *Essays,* 201–202.

83. Weismann, *Aufsätze,* 197–207; Weismann, *Essays,* 168–176.

84. August Weismann, "Die Continuität" and *Essays upon Heredity.* A full four-column explication of Weismann's monograph appeared in the newly founded *Naturwissenschaftliche Rundschau* 1 (1886): 6–7.

85. See particularly Mocek, *Die werdende Form.*

86. August Gruber, "Ueber künstliche Teilung bei Infusorien," *Biologischen Centralblatt* 4 (1885): 717–722; and 5 (1885–1886): 137–141. Also see Weismann, *Essays,* 188, on Gruber and parallel work by Nussbaum.

87. Camillo W. J. Thallwitz, "Über die Entwicklung der männlischen Keimzellen bei den Hydroiden," *Jenaische Zeitschrift für Medicin und Naturwissenschaft* 18 (1884–1885): 385–444; and Clemens Hartlaub, *Beobachtungen über die Entstehung der Sexualzellen bei Obelia* (Leipzig: Wilhelm Engelmann, 1884), or *Zeitschrift für wissenschaftliche Zoologie* 41 (1885): 159–185. For Weismann's appraisal of these dissertations, see Churchill and Risler, *August Weismann,* 2:639, 681.

88. Weismann, *Aufsätze,* 230–231; taken from *Essays,* 195 (emphasis added).

89. Weismann, *Aufsätze,* 228; *Essays,* 193.

90. Weismann, *Aufsätze,* 253–254; *Essays,* 214.

91. Weismann, *Aufsätze,* 265–267; *Essays,* 223–225.

92. Rudolf Leuckart, *Zur Kenntniss des Generationswechsels und der Perthenogenesis bei den Insekten* (Frankfurt a. M.: Meidinger Sohn & Comp., 1858), 110–111. John Farley, in *Gametes and Spores: Ideas about Sexual Reproduction 1750–1914* (Baltimore: Johns Hopkins University Press, 1982), has viewed parthenogenesis and related discoveries in a social context. Churchill in "Sex and the Single Organism" discussed this quotation in its theoretical context (see p. 163).

93. August Weismann, "Zur Naturgeschichte der Daphniden, Theil. VII. Die Entstehung der cyclischen Fortpflanzung bei den Daphnoiden," *Zeitschrift für wissenschaftliche Zoologie* 33 (1880): 111–270.

94. Weismann, *Essays,* 232–237. The taxonomic distinction between the univalent and bivalent forms of *Ascaris* was not yet clear.

95. Ibid., 247.

96. In a note appended on June 22, 1885, to the proofs, Weismann reported he had indeed found polar bodies in production of summer eggs of Daphnidae (ibid., 255).

97. Ibid., 250.

98. See Churchill, "From Heredity Theory to Vererbung."

99. Weismann, *Essays,* 170.

13. Transmission of Adaptations and Evolution

1. August Weismann, *Die Entstehung der Sexualzellen der Sexualzellen bei den Hydromedusen. Zugleich ein Beitrag zur Kenntnis des Baues und der Lebenserscheinungen dieser Gruppe* (Jena: Gustav Fischer, 1883).

2. August Weismann, "Neue Versuche zum Saison-Dimorphismus der Schmetterlinge," *Zoologische Jahrbucher. Abteilung für Systematik* 7 (1895): 611–684.

3. August Weismann, "An Autobiographical Sketch with an Introduction by Professor Josiah Royce," trans. Herbert Ernest Cushman, *The Lamp, a New Series of the Bookbuyer* 26 (1903): 21–28.

4. See particularly Conway Zirkle, "The Early History of the Idea of the Inheritance of Acquired Characters and of Pangenesis," *Transactions of the American Philosophical Society,* n.s., 35, pt. 2 (1946): 89–151; and Leonid I. Blacher, *The Problem of the Inheritance of Acquired Characters. A History of a Priori and Empirical Methods Used to Find a Solution,* trans. Noel Hess, ed. F. B. Churchill (Washington, D.C.: Smithsonian Institution Libraries, 1981).

5. The French noun *hérédité* became established in a medical context in the third decade of the nineteenth century. Carlos López-Beltrán, "The Medical Origins of Heredity," in *Heredity Produced: At the Crossroads of Biology, Politics, and Culture, 1500–1870,* ed. Staffan Müller-Wille and Hans-Jörg Rheinberger (Cambridge, Mass.: MIT Press, 2007). The *Oxford English Dictionary* claims that "inheritance" is derived from the old French "enheriter," first "to make an heir," later "to receive as heir." Tennyson used "inheritance" in an anatomical sense in 1862, and Darwin used it in the 1873 edition of the *Origin.* Herbert Spencer first used "heredity" in his *Principles of Biology* of 1863, and Francis Galton used it as an adjective in his *Hereditary Genius* in 1869.

6. Peter J. Bowler, *The Eclipse of Darwinism: Anti-Darwinian Evolution Theories in the Decades around 1900* (Baltimore: Johns Hopkins University Press, 1983). Involved also were national interests, particularly among those who felt Darwinism was simply an updating of Lamarckism, which brings us back to Weismann. For France in particular, see Bowler's chapter 5.

7. There are many accounts of Lamarck's life and work. For my purposes, I have found the most useful to be Richard W. Burkhardt Jr., *The Spirit of System: Lamarck and Evolutionary Biology* (Cambridge, Mass.: Harvard University Press, 1977).

8. Burkhardt, *Spirit of System,* 42, 69–71.

9. Jean Baptiste Lamarck, *Philosophie Zoologique* (Paris: Dentu, 1809), 1:235. Hugh Elliot translated these laws in 1914 in the following way: "First Law. In every animal which has not passed the limit of its development a more frequent and continuous use of any organ gradually strengthens, develops and enlarges that organ, and gives it a power proportional to the length of time it has been so used; while the permanent disuse of any organ imperceptibly weakens and deteriorates it, and progressively diminishes its functional capacity, until it finally disappears. Second Law. All the acquisitions or losses wrought by nature on individuals, through the influence of the environment in which their race has long been placed, and hence through the influence of the predominant use or permanent disuse of any organ; all these are preserved by reproduction to the new individuals which arise, provided that the acquired modifications are common to both sexes, or at least to the individuals which produce the young." J. B. Lamarck, *Zoological Philosophy: An Exposition with the Regard to the Natural History of Animals* (Reprint; New York: Hafner, 1963), 113.

10. Burkhardt, *Spirit of System,* 151.

11. See George John Romanes, *An Examination of Weismannism* (Chicago: Open Court, 1893), 213.

12. Ernst Haeckel, "Ueber die Naturanschauung von Darwin, Göthe und Lamarck," *Tageblatt der Gesellschaft Deutscher Naturforscher und Ärzte* 55 (1882): 81–91.

13. Ibid., 87.

14. See, among others, Galton, His, and du Bois-Reymond for some contemporary critics of the inheritance of acquired characters.

15. August Weismann, "Ueber die Bedeutung der geschlechtlichen Fortpflanzung für die Selektionstheorie," in *Tageblatt der 58. Versammlung deutscher Naturforscher und Aerzte in Strassburg 18.–23. September 1885,* ed. J. Stilling (Strassburg: G. Fischbach, 1885).

16. Ibid., 48. Translations taken from August Weismann, *Essays upon Heredity and Kindred Biological Problems,* trans. A. E. Shipley and Selmar Schönland, ed. Edward B. Poulton (Oxford: Clarendon Press, 1889), 274–275.

17. August Weismann, *Die Bedeutung der Sexuellen Fortpflanzung für die Selections-Theorie* (Jena: Gustav Fischer, 1886). I have followed closely the changes in the two texts. In the English version of Weismann's *Essays,* the translation follows closely the expanded German edition, *Die Bedeutung.*

18. Weismann, "Ueber die Bedeutung," 51; *Essays,* 283. The original German version recorded incorrectly "1020 verschiedenartige Keimplasmen" for 2^{10}. Weismann did not deal with the messy situations of the interbreeding of close relatives.

19. Weismann used a number of different terms: "erblicher individueller Charaktere," "vererbbarer individueller Unterschiede," or "Merkmale," or "Verschieden-

heiten," or rarely "Anlagen." Weismann, "Ueber die Bedeutung"; *Die Bedeutung,* 325–327.

20. Weismann, "Ueber die Bedeutung," 49; *Die Bedeutung,* 277.

21. Weismann, "Ueber die Bedeutung," 49–50; *Essays,* 279–280.

22. Weismann, "Ueber die Bedeutung," 51; *Essays,* 282.

23. Weismann, "Ueber die Bedeutung," 52; *Essays,* 285.

24. Weismann recognized competing theories about the origin of sexuality. Victor Hensen, van Beneden, and others had offered varying versions of a theory requiring rejuvenation of the lineage, but such theories assumed a difference between male and female contributions and eventually got hung up on organisms with complete parthenogenesis, which "can hardly be brought into accordance with the usual conception of life as based upon physical and mechanical forces." "Ueber die Bedeutung," 55; *Essays,* 290.

25. Virchow's career has been detailed by Erwin Ackerknect, *Rudolf Virchow: Doctor, Statesman, Anthropologist* (Madison: University of Wisconsin Press, 1953).

26. Rudolf Virchow, "Reizung und Reizbarkeit," *Archiv für pathologische Anatomie und Physiologie und für klinische Medizin* 14 (1858): 40. See also Frederick B. Churchill, "Rudolf Virchow and the Pathologist's Criteria for the Inheritance of Acquired Characteristics," *Journal of the History of Medicine and the Allied Sciences* 31 (1976): 117–148.

27. Virchow's *Cellular Pathology as Based upon Physiological and Pathological Histology* was originally translated into English by Frank Chance (1863). A new facsimile edition (New York: Dover, 1971) contains a valuable introduction by Leland J. Rather.

28. I have presented such an analysis in Churchill, "Rudolf Virchow," 128–132. I borrow here some of my text and Virchow's passages to support it.

29. Virchow, *Cellular Pathology,* 88–89.

30. Ibid., 351.

31. Rudolf Virchow, "Ueber Acclimatisation," *Amtlicher Bericht über die Versammlung deutscher Naturforscher und Ärzte* 58 (1885): 540–550; Weismann, "Ueber die Bedeutung."

32. Weismann, "Ueber die Bedeutung," 541.

33. Virchow, "Ueber Acclimatisation," 547.

34. Weismann to Fritz Müller, 3 December 1885, in Frederick B. Churchill and Helmut Risler, *August Weismann, Ausgewählte Briefe und Dokumente = Selected Letters and Documents,* 2 vols. (Freiburg: Universitätsbibliothek Freiburg, 1999), 1:65. For similar comments, see Weismann to F. E. Schulze, 2 March 1886, and 18 March 1886.

35. Rudolf Virchow, "Descendenz und Pathologie," *Virchows Archiv für pathologische Anatomie und Physiologie und für klinische Medizin* 103 (1886): 1–14, 205–215,

413–436. The same paper also appeared later in *Biologisches Zentralblatt* 6 (1886–1887): 97–108, 129–137, 161–178.

36. J. Kollmann, "Aug. Weismann, 'Ueber die Bedeutung der geschlechtlichen Fortpflanzung für die Selektionstheorie' [and] Rudolf Virchow, 'Ueber Akklimatisation,'" *Biologisches Zentralblatt* 5 (1885–1886): 673–679, 705–710. Kollmann was an exact contemporary of Weismann and Haeckel, having been born in 1834. He had a productive career in anatomy, human embryology, and race theory.

37. August Weismann, "Zur Frage nach der Verrebung erworbener Eigenschaften," *Biologisches Zentralblatt* 6 (1886): 34–48, quotation from p. 34.

38. For a sharper comment on Kollmann and similar people, see Churchill and Risler, *August Weismann*, 1:74, Weismann to de Bary, 11 February 1886.

39. "In order to lay in front of our readers all the material and to clear up the grounds for the differences of opinion, which came to light between Virchow and Weismann at the recent Versammlung, [and] before their nature may be reviewed, we need to reproduce here the article of Mr. Virchow in its essential parts and in his own words, for we continue to believe, it lies in the interest of science, to explain the basis of this difference. "Descendenz und Pathologie," *Biologisches Zentralblatt* 6 (1886): 97–108, 129–137, 161–178, quotation from p. 97.

40. Such as Wilhelm Waldeyer, Carl Weigert, and Albert von Kölliker. See Churchill, "Rudolf Virchow," 119–120fn11.

41. Dr. W. Detmer, *Allgemeine Zeitung*, 1886.

42. Weismann had actually been invited by Virchow to present his ideas at the Versammlung the following year in Berlin, but Weismann confided to Franz E. Schulze that the experience in Strassburg was too painful to accept picking up the issue again on Virchow's own turf. Churchill and Risler, *August Weismann*, 1:82–83, Weismann to F. E. Schulze, 18 March 1886.

43. August Weismann, "Über die Hypothese einer Vererbung von Verletzungen," *Tageblatt der 61. Versammlung deutscher Naturforscher und Ärzte in Köln vom 18.–23. September 1888. Wissenschaftlicher Theil Angeb.: Vorträge der Allgemeinen Sitzungen* (Köln: Albert Ahn, 1889), 45–57. August Weismann, *Über die Hypothese einer Vererbung von Verletzunge* (Jena: Gustav Fischer, 1889). This is an expanded version of "Über die Hypothese einer Vererbung." See Weismann, *Essays*, 431–461.

44. Earlier in the year he had published "Vermeintliche botanische Beweise für eine Vererbung erworbener Eigenschaften," *Biologisches Zentralblatt* 8 (1888): 65+ and 97+.

45. O. Zacharias had demonstrated bobtailed cats at the Wiesbaden Versammlung the previous year. This condition is recognized today as due to a dominant gene, which had become prevalent in the cats of the Crown protectorate. When it appears in the homozygous state, the gene is lethal.

46. Weismann, "Über die Hypothese einer Veraerbung," 50–51, and *Über die Hypothese einer Vererbung*. Weismann reported on five offspring generations with a sixth on the way—all kept in appropriately numbered breeding cages or preserved.

The norm for the length of tails this time was reported to be between 10.5 and 12 mm. In a footnote to the reprint in *Aufsätze über Vererbung und verwandte biologische Fragen,* Weismann reported eighteen generations of mice had been raised by 1892—all again showing the same results. The English translation in Weismann, *Essays,* follows *Über die Hypothese einer Veraerbung.*

47. This passage was translated into German from the Latin of Blumenbach's *De generis humani varietate native,* 3rd ed. (Göttingen: Vandenhoek et Ruprecht, 1795). Johannes Brock, "Einige ältere Autoren über die Vererbung erworbener Eigenschaften," *Biologisches Zentralblatt* 8 (1888–1889): 491–499. Quotation is on p. 499. For Brock's varied and short career in natural history, comparative anatomy, and embryology, see E. Ehlers, "Johannes Brock," *Leopoldina* 25 (1889): 118–121, 138–141.

48. Weismann, *Über die Hypothese einer Veraerbung,* 57; translation from *Essays,* 461.

49. Rudolf Virchow, "Über künstliche Verunstaltung des Körpers," *Versammlung deutscher Naturforscher und Ärzte in Köln vom 18.–23. September 1888. Wissenschaftlicher Theil.: Vorträge der Allgemeinen Sitzungen* (Köln: Albert Ahn, 1889), 65–73.

50. Ibid., 72.

51. Churchill and Risler, *August Weismann,* 1:85, Weismann to Otto Eiser, 4 April 1886.

52. Ibid., 1:77–78, Weismann to Carlo Emery, 20 February 1886.

53. Ibid., 1:86–88, Weismann to Jakob van Rees, 30 April 1886.

54. Ibid., 1:92–93, Weismann to Fritz Müller, 29 May 1886.

55. Ibid., 1:84, August Weismann to Gustav Adolf Ernst, 19 March 1886.

56. Ibid., 1:98, Weismann to Henry Nottidge Moseley, 16 July 1886.

57. On Meldola, see Frederick B. Churchill, "Meldola, Raphael," in *Dictionary of Scientific Biography,* Supplement, vol. 18, ed. F. L. Holmes (New York: Charles Scribner's Sons, 1990), 616–620.

58. See Charles Darwin "Prefatory Notice" in August Weismann, *Studies in the Theory of Descent,* trans. Raphael Meldola, 2 vols. (London: Sampson Low, Marston, Searle, & Rivington, 1882), v–vi.

59. Alfred R. Wallace, "Two Darwinian Essays," *Nature* 22 (1880): 141–142; "The Theory of Descent," *Nature* 24 (1881): 457–458; "The Theory of Descent," *Nature* 26 (1882): 52–53.

60. Wallace, "Theory of Descent," 53.

61. H. N. Moseley, "Professor August Weismann on the Sexual Cells of the Hydromedusae," *Nature* 29 (1883): 114–118.

62. H. N. Moseley, "The Continuity of the Germ-Plasma Considered as the Basis of a Theory of Heredity," *Nature* 33 (1885–1886): 154–157; "Dr. August Weismann on the Importance of Sexual Reproduction for the Theory of Selection," *Nature* 34 (1886): 629–632.

63. Arthur E. Shipley, "Death," *Nineteenth Century* 17 (1885): 827–832.

64. Churchill and Risler, *August Weismann,* 1:88–90, Weismann to A. E. Shipley, 14 May 1886. The Cambridge rocking microtome had just come onto the market in 1885. It was designed by Horace Darwin and sold by the Cambridge Scientific Instrument Co. It was so arranged that "the ribbon of sections fell by its own weight onto the bench, obviating need for the elaborate silk bands required by other designs." Brian Bracegirdle, *A History of Microtechnique* (Ithaca, N.Y.: Cornell University, 1978), 250. For a further description of the "Cambridge Rocker" and for Weismann's purchase in 1885 of a Thoma-Mikrotom Modell II, see *Journal of the Royal Microscopical Society* (1885): 549–533, and reflectively Klaus Sander, "Biologie vor unserer Zeit: Das Thoma-Mikrotom und seine Benutzer," *Biologie in unserer Zeit* 12 (1982): 108–112.

65. Churchill and Risler, *August Weismann,* 1:88–90, Weismann to Shipley, 14 May 1886; ibid., 1:98, Weismann to Moseley, 16 July 1886.

66. August Weismann, "On the Signification of the Polar Globules," *Nature* 36 (1887): 607–609.

67. Churchill and Risler, *August Weismann,* 1:115, Weismann to Anton de Bary, 18 October 1887.

68. See also Frederick B. Churchill, "August Weismann: A Developmental Evolutionist," in Churchill and Risler, *August Weismann,* 2:768–770.

69. Alfred Russel Wallace, *The Geographical Distribution of Animals, with a Study of the Relations of Living and Extinct Faunas as Elucidating the Past Changes of the Earth Surface* (London: Macmillan, 1876), and *Darwinism. An Exposition of the Theory of Natural Selection with Some of Its Applications* (London: Macmillan, 1889). Quotation appears in a footnote on p. 443 of the latter.

70. Ross A. Slotten, *The Heretic in Darwin's Court: The Life of Alfred Russel Wallace* (New York: Columbia University Press, 2004). For a good account of Wallace's involvement with spiritualism, see chapters 10–14. Quotation appears on p. 351.

71. An excellent description of his life and scientific career may be found in John E. Lesch, "Romanes, George John," *Dictionary of Scientific* Biography, vol. 11, ed. Charles Gillispie (New York: Charles Scribner's Sons, 1975), 516–520. Romanes's friendship and correspondence with Darwin is presented by Joel S. Schwartz, "George John Romanes's Defense of Darwinism: The Correspondence of Charles Darwin and His Chief Disciple," *Journal of the History of Biology* 28 (1995): 281–316. A more recent evaluation of Romanes's career may be found in Slotten, *Heretic,* chapter 17.

72. George J. Romanes, "Physiological Selection: An Additional Suggestion on the Origin of Species," *Linnaean Society Journal* 19 (1886): 337–411.

73. Ibid., 408.

74. Alfred Russel Wallace, "Romanes versus Darwin. An Episode in the History of the Evolution Theory," *Forthnightly Review* 46 (1886): 300–316; Francis Galton, "The Origin of Varieties," *Nature* 34 (1886): 395–396; Francis Darwin, "Physiological Selection and the Origin of Species," *Nature* 34 (1886): 407.

75. Raphael Meldola, "Physiological Selection and the Origin of Species," *Nature* 34 (1886): 384–385.

76. All letters to the editor by George J. Romanes: "Physiological Selection and the Origin of Species," *Nature* 34 (1886): 407–408; "Physiological Selection and the Origin of Species," *Nature* 34 (1886): 439; "Physiological Selection and the Origin of Species," *Nature* 34 (1886): 545; "The Origin of Species," *Nature* 35 (1886–1887): 124–125; "Mr. Wallace on Physiological Selection," *Nature* 35 (1886–1887): 247–248; "Mr. Wallace on Physiological Selection," 35 (1886–1887): 247–248.

77. G. Herbert Fowler, "Professor A. Weismann's Theory of Polar Bodies," *Nature* 37 (1887): 134–136.

78. P. Chalmers Mitchell, "The Duration of Life," *Nature* 37 (1888): 541–542, "Weismann on Heredity," *Nature* 38 (1888): 156–157.

79. August Weismann, "On the Supposed Botanical Proofs of the Transmission of Acquired Characters," and "The Supposed Transmission of Mutilations." The first of these originally appeared in *Biologisches Centralblatt* 8 (1888): 65–79, 79–109; the second originated as Weismann's third plenary address at the Cologne meeting of the Versammlung of 1888, Weismann, "Über die Hypothese einer Veraerbung."

80. Weismann, *Essays;* August Weismann, *Essais sur l'hérédité et la selection naturelle,* trans. Henry de Varigny (Paris: C. Reinwald, 1891); August Weismann, "Vorwort," in Albert Lang, Über die Knospung bei Hydra und einige Hydropolypen," in *Zeitschrift für wissenschaftliche Zoologie* 54 (1892): 365–385.

81. See from T. G. B. Osborn, "Sydney Howard Vines," revised by D. J. Mabberley, in *Dictionary of National Biography* (Oxford: Oxford University Press, 2004), 881–882, online ed. Oct. 2005; F. O. Bower, "Sydney Howard Vines," *Proceedings of the Linnaean Society of London,* session 133–134 (1921–1922): 173–179; and from J. Reynolds Green, *A History of Botany in the United Kingdom from the Earliest Times to the End of the 19th Century* (New York: E. P. Dutton, 1914).

82. Sydney Howard Vines, "Reminiscences of German Botanical Laboratories in the 'Seventies and 'Eighties of the Last Century," *New Phytologist* 34 (1925): 1–8.

83. S. M. Walters, *The Shaping of Cambridge Botany: A Short History of Whole-Plant Botany in Cambridge from the Time of Ray into the Present Century* (Cambridge: Cambridge University Press, 1981), chapter 6.

84. Sidney Howard Vines, *Lectures on the Physiology of Plants* (Cambridge: Cambridge University Press, 1886).

85. Julius von Sachs, *Vorlesungen über Pflanzen-physiologie* (Leipzig: Wilhelm Engelmann, 1882); Julius von Sachs, *Lectures on the Physiology of Plants,* trans H. Marshall Ward (Oxford: Clarendon Press, 1887). Ward was also a fellow of Christ's College, Cambridge, and he acknowledged Vines's assistance.

86. Sachs, *Lecture,* 107.

87. Vines, *Lectures on the Physiology,* 27.

88. Ibid., 654–655.

89. Ibid., 660.

90. Ibid., 663.

91. Although he appears to have left this impression, it is doubtful that Weismann ever held such a narrow view of the origin of variations. For example, "I am also far from asserting that the germ-plasm—which, as I hold, is transmitted as the basis of heredity from one generation to another—is absolutely unchangeable or totally uninfluenced by forces residing in the organism within which it is transformed into germcells." Weismann, however, was also clear that such germinal changes did not imply adaptive changes, for adaptation was a condition determined by natural selection. See August Weismann, *Die Continuität des Keimplasmas als Grundlage einer Theorie der Vererbung* (Jena: Gustav Fischer, 1885), 172.

92. Vines, *Lectures on the Physiology*, 680–681.

93. Churchill and Risler, *August Weismann*, 1:121, August Weismann to Sydney Howard Vines, 7 May 1888.

94. Sydney H. Vines, "An Examination of Some Points in Prof. Weismann's Theory of Heredity," *Nature* 40 (1889): 621–626.

95. Weismann makes this point in his reply. August Weismann, "Prof. Weismann's Theory of Heredity," *Nature* 41 (1890): 317–323, 373.

96. Ibid., 319.

97. Vines, "An Examination," 626.

14. Polar Body Research

1. Wilson referred to "van Beneden's Law" as the demonstration in *Ascaris* that the female and male pronuclei contain the same number of chromosomes, hence the product of their union had twice as many. Edmund B. Wilson, *The Cell in Development and Heredity* (New York: Macmillan, 1925), 426–431.

2. Wilson refers to cases of independence of the male and female derived chromosomes during cleavage and beyond as the "Theory of Gonomery." Wilson, *The Cell*, 431–434. Boveri's 1902 experiments on multipolar mitosis in the sea urchin will be discussed in Chapter 20.

3. Wilhelm Waldeyer, "Ueber Karyokinese und ihre Beziehungen zu den Befruchtungsvorgängen," *Archiv Für mikroskopische Anatomie* 32 (1888): 1–122. "Chromosomen" and quotation appear on p. 27. See also "Karyokinesis and Its Relation to the Process of Fertilization," trans. W. B. Benham, *Journal of Microscopy and Natural Sciences, London* 30 (1889–1890): 159–281; quotation from p. 181.

4. At the time Waldeyer rarely used Flemming's neologism "mitosis."

5. Waldeyer, "Karyokinesis," 164–209.

6. "Tous les phénomènes caryocynétique sont variables: aucun d'eux ne parâit essentiel." Waldeyer, "Ueber Karyokinese," 47 (translation from Waldeyer).

7. Waldeyer, "Karyokinesis," 201.

8. Waldeyer, "Ueber Karyokinese," 52; translation taken from "Karyokinesis," 207.

9. Waldeyer, "Ueber Karyokinese," 45; translation taken from "Karyokinesis," 199.

10. E. Korschelt and K. Heider, *Lehrbuch der vergleichenden Entwicklungsgeschichte der wirbellosen Tiere* (Jena: G. Fischer, 1902), 553; Wilson, *The Cell,* 398, follows their account.

11. Waldeyer does not list his work.

12. Waldeyer, "Karyokinesis," 225–231. See also Korschelt and Heider, *Lehrbuch,* 539–567.

13. Waldeyer, "Karyokinesis," 229.

14. Weismann to August Gruber, 10 February, 1887, in Frederick B. Churchill and Helmut Risler, *August Weismann, Ausgewählte Briefe und Dokumente = Selected Letters and Documents,* 2 vols. (Freiburg: Universitätsbibliothek Freiburg, 1999), 1:100.

15. Waldeyer, "Karyokinesis," 260–264.

16. Edmund Wilson, *The Cell in Development and Inheritance* (New York: Macmillan, 1896), 129–156. The table is on pp. 154–155.

17. Weismann was by no means alone in his pursuit of the polar bodies. He also did not work alone with the material he examined. In the 1880s and early 1890s as his new institute for zoology took shape, Weismann could count on his brother-in-law August Gruber for administrative assistance and for manning some of the basic courses.

18. Wilson, *The Cell,* 500–503; quotation from p. 500. The essay, more a monograph, under discussion was August Weismann, *Über die Zahl der Richtungskörper und über Ihrer Bedeutung für die Vererbung* (Jena: Gustav Fischer, 1887). This was translated and published in the first volume of his *Essays upon Heredity* as "On the Number of Polar Bodies and Their Significance in Heredity"; August Weismann, *Essays upon Heredity and Kindred Biological Problems,* trans. A. E. Shipley and Selmar Schönland, ed. Edward B. Poulton (Oxford: Clarendon Press, 1889), 343–396.

19. The last paper concerned observations of an accessory nucleus or "Nebenkern," which occasionally appeared during the maturation process of the germinal vesicle. August Weismann and Chiyomatsu Ischikawa, "Ueber die Paracopulation im Daphnidenei, sowie über Reigung und Befruchtung desselben," *Zoologische Jahrbücher* 4 (1889): 155–196.

20. Ishikawa's contribution in the microscopy cannot be overestimated. For the purposes of discussing the theoretical implications of the work, however, I will refer only to Weismann.

21. August Weismann and Chiyomatsu Ischikawa, "Ueber die Bildung der Richtungskörper bei thierischen Eiern," *Berichte der naturforschenden Gesellschaft zu Freiburg i. Br* 3 (1887): 24–25.

22. Chiyomatsu Ischikawa, "Trembley's Umkehrungsversuche an Hydra nach neuen Versuchen erklärt," *Zeitschrift für wissenschaftliche Zoologie* 49 (1889): 433–460. Throughout his career, Ishikawa preferred the English spelling of his name, but Weismann always used the German spelling (Ischikawa). For more on Ishikawa, see R. B. Goldschmidt, *Portraits from Memory* (Seattle: University of Washington Press, 1956), 155–158; J. Bartholomew, *The Formation of Science in Japan: Building a Research Tradition* (New Haven, Conn.: Yale University Press, 1989), 59. For eight letters from Weismann to Ishikawa, see Churchill and Risler, *August Weismann,* vol. 1. Also significant is Weismann's Gutachten on Ishikawa's dissertation (ibid., 2:644).

23. August Weismann, *Über die Zahl der Richtungskörper und über ihre Bedeutung* (Jena: Gustav Fischer, 1887); Weismann and Ischikawa, "Ueber die Bildung"; August Weismann, "On the Significance of Polar Globules," *Nature* 36 (1887): 607–609. Weismann alone or with Ishikawa published six more papers on polar bodies between 1888 and 1889. They are listed in the Appendix.

24. Weismann, *Über die Zahl;* Weismann, *Essays.*

25. Weismann, *Essays,* 371.

26. Ibid., 375.

27. Ibid., 371–381.

28. Ibid., 376, figs. II and III.

29. Ibid., 374.

30. See Theodor Boveri, *Zellen-studien* (Jena: Gustav Fischer, 1887–1907).

31. August Weismann and Chiyomatsu Ischikawa, "Weitere Untersuchungen zum Zahlengesetz der Richtungskörper," *Zoologische Jahrbücher* 3 (1888): 575–610.

32. Ibid., 605.

33. Weismann, *Essays.* I have discussed Weismann's break from that tradition in "August Weismann and a Break from Tradition," *Journal of the History of Biology* 1 (1968), and associated it with his rejection of an epigenetic theory of development. Such a break turns out to be more complicated and works on different levels at different times: development, evolution, cytology, and heredity.

15. Protozoa and Amphimixis

1. Weismann to Lüroth, 12 January 1892, in Frederick B. Churchill and Helmut Risler, *August Weismann, Ausgewählte Briefe und Dokumente=Selected Letters and Documents,* 2 vols. (Freiburg: Universitätsbibliothek Freiburg, 1999), 1:168–169.

2. August Weismann, *Amphimixis oder: Die Vermischung der Individuen* (Jena: Gustav Fischer, 1891); see also in *Aufsätze über Vererbung und verwandte biologische Fragen* (Jena: Gustav Fischer, 1892), 673–826; and "Amphimixis or the Essential Meaning of Conjugation and Sexual Reproduction," in *Essays upon Heredity and Kindred Biological Problems,* trans. Lilian J. Gould, Antoine Lüroth, and Fräulein

Diestel, ed. Edward B. Poulton, 2nd ed., 2 vols. (Oxford: Clarendon Press, 1891–1892), 2:99–222.

3. *Essays,* 2:101. Slightly modified: the first two emphases are mine; the last two are Weismann's.

4. Ibid., 1:101. For his definition of "individualities," see pp. 130–131.

5. Ibid., 2:365.

6. Ibid., 2:117.

7. Ibid., 2:122.

8. As we will see in Chapter 20, the tetrad of "chromatids" was understood only in 1900.

9. Weismann, *Essays,* 2:135.

10. Weismann generally wrote of "Idants" rather than "chromosomes."

11. Weismann, *Essays,* 2:134. See also Churchill and Risler, *August Weismann,* 1:168–169, Weismann to Frau Lüroth, 12 January 1892, for mention of Lüroth's discussion. These were incorporated into Weismann's *Aufsätze.*

12. Weismann, *Amphimixis,* 49, reprinted in *Essays,* 2:99–222.

13. Weismann, *Essays,* 2:139–144; quotation on p. 140.

14. Churchill and Risler, *August Weismann,* 1:143, Weismann to Henking, 6 June 1890. For information on Spuler see ibid., 2:675 and Weismann, *Essays,* 2:140. For Weismann's suggested model of reducing ids and thereby multiplying variations with an arbitrary cleavage through the wreath of a double idant, see pp. 2:144–146.

15. Weismann, *Amphimixis,* reprinted in *Essays,* 2, see figures on pp. 137, 141, and 145.

16. Churchill and Risler, *August Weismann,* 1:169–171, Weismann to Valentin Häcker, 14 January 1892; ibid., 1:334, Weismann to Otto vom Rath, 15 February 1900 (see also 2:693fn). Rath's expert microscopical work was noted by Leopoldina at the time of his premature death in 1900.

17. Weismann, *Essays,* 2:164.

18. Ibid., 166.

19. Calkins went on to list in a footnote nine other types and ended with an "etc.," that all-purpose and evasive ampersand. Gary N. Calkins, *The Biology of the Protozoa* (Philadelphia: Lea & Febiger, 1926), 113.

20. Frederick B. Churchill, "Introduction: Toward the History of Protozoology," *Journal of the History of Biology* 22 (1989): 185–187; Frederick B. Churchill, "The Guts of the Matter. Infusoria from Ehrenberg to Bütschli: 1839–1876," *Journal of the History of Biology* 22 (1989): 189–213; Natasha X. Jacobs, "From Unit to Unity: Protozoology, Cell Theory and the New Concept of Life," *Journal of the History of Biology* 22 (1989): 215–242; Marsha L. Richmond, "Protozoa as Precursors of Metazoa: Germ Cell Theory and Its Critics at the Turn of the Century," *Journal of the History of Biology* 22 (1989): 185–278.

21. Johannes P. Müller, "Einige Beobachtungen an Infusorien," *Monatsberichte der Königlichen Preussische Akademie des Wissenschaften zu Berlin* (1856): 389–393; Édouard-Géard Balbiani, "On the Existence of a Sexual Reproduction in the Infusoria," *Annals and Magazine of Natural History,* 3rd ser., 1 (1858): 435–438, translated from the *Comptes Rendus* of 29 March 1858; Edouard Gerard Balbiani, "Recherches sur les phénomes sexuels des infusoires," *Journal de la Physiologie de l'Homme et des Animaux* 4 (1861): 102–130, 465–521. The terms "macronucleus" and "micronucleus" were later coined by Émile Maupas.

22. H. G. Bronn, *Die Klassen und Ordnungen des Their-reichs* (vol. 1, part 1) *Amorphozoen* (Leipzig und Heidelberg: C. F. Winter'sche Verlagshandlung, 1859); (vol. 2, part 2) *Strahlenthiere (Actinozoa)* (1860); (vol. 3, part 1) *Weichthiere (Malacozoa. Acephala)* (1862), Fortgesetzt von Wilhelm Keferstein; (vol. 3, part 2) *Weichthiere (Malacozoa cephalophora)* (1862–1866).

23. Otto Bütschli, *Dr. H. G. Bronn's Klassen und Ordnungen des Thier-reichs,* vol. 1 (Protozoa, part 1) *Sarkodina und Sporozoa* (Leipzig und Heidelberg: C. F. Winter'sche Verlagshandlung, 1880–1882); (part 2) *Mastigophora,* 1883–1886; (part 3) *Infusoria u. System der Radiolaria* (1887–1889).

24. O. Bütschli (1873), "Beitäge zur Kenntniss der freilebenden Nematoden," *Nova Acta Academiae Caesareae Leopoldino-Carolinae* 36, no. 5 (1873): 1–144. See also Otto Bütschli, *Studien über die ersten Entwicklungsvorgänge der Eizelle, die Zellteilung und die Conjugation der Infursorien* (Frankfurt am Main: Christian Winter, 1876). Bütschli signed off on the Vorwort of his *Studien* in Frankfurt in November 1875.

25. For another reaction, see Haeckel to Weismann, 18 February 1878 in Georg Uschmann and Bernhard Hassenstein, "Der Briefwechsel zwischen Ernst Haeckel und August Weismann," in *Kleine Festgabe aus Anlass der hundertjährigen Wiederkehr der Gründung des Zoologischen Institutes der Friedrich-Schiller-Universität Jena im Jahre 1865,* ed. M. Gersch (Jena: Friedrich-Schiller-Universität, 1965), 48.

26. August Weismann, "Bemerkungen zu Professor Bütschli's Gedanken über Leben und Tod," *Zoologischer Anzeiger* 5 (1882): 377–380; "Zur Frage nach der Vererbung erworbener Eigenschaften," *Biologisches Zentralblatt* 6 (1886): 33–48; Otto Bütschli, "Gedanken über Leben und Tod," *Zoologischer Anzeiger* 5 (1882): 64–67.

27. This discussion of Bütschli's *Studien* is drawn in part from my account in Churchill, "The Guts of the Matter," 206–211. The first quotation is from Bütschli's *Studien,* translated on p. 216. The second quotation is my conclusion, which appears on p. 211. The article also discusses Friedrich von Stein's (1854 and 1859–1876) and Haeckel's (1866) views of protozoa as cells, but in my mind both appear to compromise the status of the nucleus as it was understood by the mid-1870s. For the best discussions of Bütschli and his successors who developed and fought over the analogy between the cell and protozoa, see Natasha X. Jacobs, "Unit to Unity," and Marsha Richmond, "Protozoa as Precursors."

28. See also Calkins, *Biology of Protozoa,* chapter 1.

29. Bütschli, *Infusoria u. System,* 1636–1637.

30. Ibid., 1637. Maupas also summarizes these conclusions.

31. Ibid., 1638.

32. Frederick B. Churchill, "August Weismann Embraces the Protozoa," *Journal of the History of Biology* 43 (2010): 783–789. The standard account of Maupas's career may be found in Edmond Sergent, "Émile Maupas. Prince des Protozoologistes," *Archives de L'Institut Pasteur d'Algérie* 33 (1955): 59–70.

33. See Sergent, "Émile Maupas," and particularly, Jean Théodoridès, "Maupas, Émile François," in *Dictionary of Scientific Biography,* vol. 9, ed. Charles Gillispie (New York: Charles Scribner's Sons, 1974), 185–186.

34. For a summary of a thirty-five-year account of Maupas's work, see Calkins, *Biology of Protozoa,* the chapter on "Vitality."

35. E. Maupas, "Le rajeunissement karyogamique chez les Ciliés," *Archives de zoologie expérimentale* 7 (1889): 149–168, quotation from p. 161.

36. E. Maupas, "Recherches expérimentales sur la multiplication des Infusoires ciliés," *Archives de zoologie expérimentale* 6 (1888): 264. For a further example of provocative language, see pp. 267–268fn2.

37. Maupas, "Le rajeunissement," 477–479.

38. Maupas, "Recherches," 165–277.

39. Maupas, "Le rajeunissement," 487–493; quotations from pp. 487, 493.

40. Ibid., 93–94.

41. Ibid., 495.

42. Churchill and Risler, *August Weismann,* 1:41–42, Weismann to Marion, 3 July 1878. Spermatophoren are packets of sperm. Weismann had married Mary Gruber on June 20, 1867.

43. The twelfth voting member was unable to attend. See Gutachten submitted by Weismann to the Akademischen Senat, 3 February 1880, in Churchill and Risler, *August Weismann,* 2:632. This process, which did not include a Habilitationsschrift, was no longer common and must have reflected in part Weismann's own standing in the senate.

44. The Royal Society Catalogue of Scientific Papers lists forty-nine of his contributions, both notes and substantial papers, published between 1878 and 1900.

45. It could hardly have been just a coincidence that Gruber investigated copepods with Leuckart at the time that Weismann was deeply involved with the same organism at the Gruber estate on the Bodensee or that he switched to Protozoa when that phylum became important for understanding fertilization and heredity issues. For references and further details, see the footnote in Churchill and Risler, *August Weismann,* 2:632.

46. Chiyomatsu Ishikawa, "Vorläufige Mitteliungen über die Konjugationserscheinungen bei den Noctiluceen," *Zoologischer Anzeiger* (1891), and "Studies of

Reproductive Elements. II. *Noctiluca millaris,* Sur.; Its Division and Spore Formation," *Journal of the College of Science, Imperial University, Japan* 6 (1894): 297–334. See Churchill and Risler, *August Weismann,* 1:144, Weismann to Ishikawa, 8 June 1890; 1:221, Weismann to Ishikawa, 12 July 1894. Wilson described this work in *The Cell,* 168–169, but in his bibliography attributed it to "M. Ishikawa."

47. August Weismann, "Ueber die Bedeutung der geschlechtlichen Fortpflanzung für die Selektionstheorie," *Tageblatt der 58. Versammlung deutscher Naturforscher und Aerzte in Strassburg 18.–23. September 1885,* ed. J. Stilling (Strassburg: G. Fischbach, 1885), 75. See Rasmus G. Winther, "Darwin on Variation and Heredity," *Journal of the History of Biology* 33 (2001): 425–455.

48. Significantly, the title of this part was "Amphimixis als Conjugation und Befruchtung."

49. Weismann, *Essays,* 2:177–180. Weismann pretty much confined his response to Maupas's harsh criticisms of him to a footnote where he gave the French microscopist a lesson about the importance of mistakes for the advancement of science (see 177fn).

50. Ibid., 185–186, quotation from p. 186. Both Weismann and Maupas suggested that "male" and "female" micronuclei were arbitrary designations that simply distinguished the micronucleus that transferred and the recipient micronucleus that stayed put.

51. Ibid., 189 (emphasis in the original).

52. Ibid., 190–195.

53. It was appropriate to use Flemming's terms "mitosis" or "karyomitosis" for the phenomenon that a decade later would be distinguished as "meiosis." Much still had to be worked out to arrive at that distinction.

54. August Weismann, "Zur Naturgeschichte der Daphniden. Theil VII. Die Entstehung der cyclischen Fortpflanzung bei den Daphnoiden," *Zeitschrift für wissenschaftliche Zoologie,* 33 (1880): 111–270; and *Die Entstehung der Sexualzellen bei den Hydromedusen. Zugleich ein Beitrag zur Kenntnis des Baues und der Lebenserscheinungen dieser Grupp* (Jena: Gustav Fischer, 1883).

55. Frederick B. Churchill, "Weismann and a Break from Tradition," *Journal of the History of Biology* 1 (1967): 91–112.

56. Weismann, *Amphimixis,* 180.

16. A Model for Heredity

1. Yves Delage, *La structure du protoplasm les theories sur l'hérédité et les grands problèms de biologie générale* (Paris: Librairie C. Reinwald Schleicher Fréres, 1895).

2. Ibid., 412.

3. See Weismann to Delage, 16 November 1893, in Frederick B. Churchill and Helmut Risler, *August Weismann, Ausgewählte Briefe und Dokumente = Selected Let-*

ters and Documents, 2 vols. (Freiburg: Universitätsbibliothek Freiburg, 1999), 1:204–205, and footnote 5 for a contradictory reaction.

4. Yves Delage," La nouvelle théorie de, M. Weissman [*sic*]," *Revue philosophique* 35 (1893): 561–589.

5. Churchill and Risler, *August Weismann,* 1:197, 297, Weismann to Delage, 27 July 1893 and 5 July 1898.

6. E. Th. Brücke, "Ernst W. Brücke (1819–1892)," *Neue Osterreichische Biographie* 5 (1912): 66–73; Erna Lesky, "Brücke, Ernst Wilhelm von," *Dictionary of Scientific Biography* 2 (1970): 530–532. For an introduction to the extensive literature of these four students of Johannes P. Mueller, see Paul F. Cranefield, "The Organic Physics of 1847 and Biophysics of Today," *Journal of the History of Medicine and Allied Sciences* 10 (1957): 407–423; Everett Mendelsohn, "Physical Models and Physiological Concepts: Explanation in Nineteenth-Century Biology," *British Journal for the History of Science* 2 (1965): 201–219.

7. Ernst Brücke, "Die Elementarorganismen," *Sitzberichte der Königliche Akademie der Wissenschaften in Wien—Mathematisch-naturwissenschaftliche Klasse* 44 (1861): 381–406.

8. Ibid., 382.

9. In reference to his life, I have followed closely the short accounts by Ilse Jahn, "Johannes von Hanstein," *Neue Deutsche Biographie* 7 (1965): 640–641; and F. Schmitz, "Johannes v. Hanstein," *Leopoldina* 17 (1881): 75–80.

10. Johannes v. Hanstein, *Das Protoplasma als Träger der planzlichen und thierischen Lebensverrichtungen: Für Laien und Sachgenossen dargestellt* (Heidelberg: Frommel und Pfaff, 1880).

11. Johannes von Hanstein, "Beiträge zur allgemeinen Morphologie der Pflanzen," *Botanische Abhandlungen aus dem Gebiet der Morphologie und Physiologie* 4 (1882): 1–244.

12. Delage, *La structure,* 519–520.

13. Hanstein, *Das Protoplasma,* 132.

14. Ibid., 232–262. See fig. 1, p. 251, illustration of a growing cell. Notice the arrows of flowing protoplasm.

15. Ibid., 261–262. The word "protoplast" becomes confusing. De Vries recognizes von Hanstein's "elementary" aspect but then goes on to use "Protoplasts" as elementary organisms with "individual organs." See his *Intracellulare Pangenesis* (1889), reprinted in Hugo de Vries, *Opera e periodicis collata,* 7 vols. (Utrecht: A. Oosthoek, 1918–1927), 5:1–149, esp. 126–127. Brücke may be to blame here for choosing his title as "Elementarorganism."

16. Hanstein, *Das Protoplasma,* 282.

17. Ilse Jahn, "von Hanstein," 640–664.

18. "Die Protoplasten sind Künstler, Werkzeug und plastischer Stoff zugleich." Hanstein, *Das Protoplasma,* 302–305.

19. Ibid., 308.

20. Ibid., 230.

21. August Weismann, *The Germ-Plasm: A Theory of Heredity,* trans. William Newton Parker and Harriet Rönnfeldt, ed. H. Ellis (London: Walter Scott, 1893), 12–20, quoted from p. 12.

22. Biographical details of de Vries's life and professional career are found in Peter W. van der Pas, "Vries, Hugo de," *Dictionary of Scientific Biography* 14 (1981): 95–104.

23. Hugo de Vries, *De invloed der temperatuur op de levensverschijnselen der planten* ('s-Gravenhage: Martinus Nijhoff, 1870); for the full text of his dissertation, see Hugo de Vries, *Opera e Periodicis Collata,* 7 vols. (Utrecht: A. Oosthoek, 1918–1927), 1:1–85.

24. Hugo de Vries, *Intracellulare Pangenesis* (1889), 5:1–149. See also Hugo de Vries, *Intracellular Pangenesis,* trans. C. Stuart Gager (Chicago: Open Court, 1910). Van der Pas wrote in his biographical sketch of de Vries, "About 1890 he abruptly abandoned the study of plant physiology and devoted himself exclusively to heredity and variation" (Van der Pas, "Vries, Hugo de," 96).

25. De Vries, *Intracellulare Pangenesis,* 5:6–7.

26. De Vries, *Intracellular Pangenesis,* 67 and 126–127.

27. De Vries, *Intracellulare Pangenesis,* 5:67.

28. De Vries, *Intracellular Pangenesis,* 195–199, quotation is from p. 195 (emphasis is de Vries's).

29. Ibid., 209–210.

30. Ibid., 19 (modified).

31. "Panmeristic" was an adjective used at the end of the nineteenth and early part of the twentieth centuries. It has all but vanished from the scientific literature. Derived from *pan* = all, and *meros* = part. J. H. Kenneth, *A Dictionary of Scientific Terms,* 7th ed. (Princeton, N.J.: Princeton University Press, 1960), defines it as follows: "an ultimate protoplasmic structure of independent units."

32. De Vries, *Intracellular Pangenesis,* 126–127.

33. Jacob H. van't Hoff, *Ansichten über die Organische Chemie* (Braunschweig: Friedrich Vieweg, 1881), part 1:26.

34. Quoted in de Vries, *Intracellular Pangenesis,* 38–39.

35. Ibid., 50–61.

36. Ibid., 73–75.

37. De Vries, *Intracellulare Pangenesis,* 5:38–41.

38. De Vries, *Intracellular Pangenesis,* 110–121.

39. Ibid., 81; Julius Sachs, *Vorlesungen über Pflanzen-Physiologie,* 2nd ed. (Leipzig: Wilhelm Engelmann, 1887), chapters 27–28.

40. De Vries, *Intracellular Pangenesis,* 120–121 (italics in the original).

41. Translated by the author. The original text of both drafts may be found in Churchill and Risler, *August Weismann,* 1:132–133.

42. August Weismann, *Das Keimplasma: Eine Theorie der Vererbung* (Jena: Gustav Fischer, 1892), 616; Weismann, *Germ-Plasm,* 468. This is my modification of the translation about particulate theories.

43. Weismann, *Germ-Plasm,* 19.

44. Weismann had been impressed by the 1880 observations by Miescher on the "nuclein" content of the head of salmon spermatozoa and its origin in the nuclei. He also accepted the very recent discussion of the size of the chromosomes and the liberation of nuclein into the cytoplasm by J. R. Rückert. See also Weismann, *Germ-Plasm,* 49–51, 85–91, where he draws upon the *Popular Lectures and Addresses* of William Thomson (1889).

45. Weismann, *Germ-Plasm,* 67.

46. Ibid., 83–84. For a discussion of the size and number of the ids and determinants in the nucleus see pp. 86–99.

47. Ibid., 39–53.

48. Ibid., 81–82.

49. On strong and weak versions of the biogenetic law at the end of the nineteenth century, see Frederick B. Churchill, "Living with the Biogenetic Law: A Reappraisal," in *From Embryology to Evo-Devo: A History of Developmental Evolution,* ed. Manfred Dietrich Laubichler and Jane Maienschein (Cambridge, Mass.: MIT Press, 2007), 37–81.

50. Weismann, *Germ-Plasm,* 106.

51. Reinhard Mocek, *Die Werdende Form: Eine Geschichte der Kausalen Morphologie* (Marburg: Basilisken, 1998), Teil II and III; Frederick B. Churchill, "The Rise of Classical Descriptive Embryology," in *Developmental Biology: A Comprehensive Synthesis,* ed. Scott F. Gilbert (London: Plenum Press, 1991), 1–29; Frederick B. Churchill, "Regeneration, 1885–1901," in *A History of Regeneration Research: Milestones in the Evolution of a Science,* ed. Charles E. Dinsmore (Cambridge: Cambridge University Press, 1991), 111–131.

52. A genus of common but primitive salamanders often used for regeneration experiments during this period. It is a curious matter that very few regeneration experiments were carried out in Weismann's institute.

53. The concept of induction was developed only after the turn of the century.

54. Weismann, *Das Keimplasma,* 124–152.

55. Weismann drew upon a long well-known tradition of experimentation dating back to Lazzaro Spallanzani in the eighteenth century as well as contemporary ones. Weismann, *Das Keimplasma,* 156–168; Weismann, *Germ-Plasm,* 103–126.

56. August Weismann, "Versuche über Regeneration bei Tritonen," *Anatomischer Anzeiger* 22 (1903): 425–431.

57. See particularly section 5, "Die Regeneration an thierischen Embryonen und die Principien der Ontogeny," Weismann, *Das Keimplasma,* 179–192; Weismann, *Germ-Plasm,* 134–144.

58. Dinsmore, *A History of Regeneration,* contains twelve essays on eighteenth-century through early twentieth-century regeneration research. For Lereboullet, see Jane Oppenheimer, "Historical Introduction to the Study of Teleostean Development," *Osiris* 2 (1936): 124–148; for Dareste, see Frederick B. Churchill, "Chabry, Roux, and the Experimental Method in Nineteenth-Century Embryology," in *Foundations of Scientific Method: The Nineteenth Century,* ed. Ronald N. Giere and Richard S. Westfall (Bloomington: Indiana University Press, 1973), 161–205.

59. Paul Fraisse, *Die Regeneration vom Geweben und Organen bei den Wirbelthieren besonders Amphiien und Reptilien* (Cassel: Theodor Fischer, 1885).

60. I have discussed Fraisse's career in somewhat more detail in Churchill, "Regeneration," 113–131.

61. That is, Oscar Hertwig (1849), Wilhelm Roux (1850), Gustav Born (1850), and Laurent Marie Chabry (1855). For a clear account of the following events, see Jane Maienschein, *Transforming Traditions in American Biology, 1880–1915* (Baltimore: Johns Hopkins University Press, 1991), 145–158.

62. See particularly Wilhelm His, *Unsere Körperform und das physiologische Problem ihrer Entstehung: Briefe an einen befreundeten Naturforscher* (Leipzig: F. C. W. Vogel, 1874) and William Thierry Preyer, *Specielle Physiologie des Embryo: Untersuchungen ueber die Lebenserscheinungen vor der Geburt* (Leipzig: Th. Grieben, 1885).

63. *Archiv für gesammte Physiologie des Menschen und Thiere* was founded in 1869.

64. Eduard Pflüger, "Ueber den Einfluss der Schwerkraft auf die Theilung der Zellen," in Pflüger's *Archiv für die gesammte physiologie des Menschen und der Tiere* 32 (1883): 64.

65. In 1882, Gustav Born became the father of Max Born, who later became a leading physicist in the Weimar Republic.

66. Gustav Born, "Beiträge zur Bastardirung zwischen den einhemischen Anurenarten," *Archiv für die gesammte Physiologie des Menschen und der Thiere* 32 (1883): 479–481.

67. Gustav Born, "Ueber den Einfluss der Schwere auf das Froschei," *Archiv für Mikroskopische Anatomie* 22 (1885): 534; Gustav Born, "Ueber den Einfluss der Schwere auf das Froschei," *Jahres-Bericht der Schlesischen Gesellschaft für vaterländlische Cultur* (1884): 75–83. The full results were submitted for publication on November 1, 1884, in "Biologische Untersuchungen. I. Ueber den Einfluss der Schwere auf das Froschei," *Archiv für Mikroskopische Anatomie* 22 (1885): 475–545, quotation from p. 534.

68. Wilhelm Roux, *Über die Zeit der Bestimmung der Hauptrichtungen des Froschembryo: Eine biologische Untersuchung* (Leipzig: Wilhelm Engelmann, 1883); Wilhelm Roux, "Wilhelm Roux in Halle a.S.," in *Die Medizin der Gegenwart in Selbstdarstellungen,* 2 vols., ed. L. R. Grote (Leipzig: Feliz Meiner, 1923), 1:141–206, see p. 157.

69. Wilhelm Roux, "Ueber die Entwickelung der Froscheier bei Aufhebung der richtenden Wirkung der Schwere," *Breslauer ärzlicher Zeitschrift* 22 (1884): 1–16.

Also in Roux, *Gesammelte Abhandlungen über Entwickelungsmechanik der Organismen* (*Beiträge* II, vol. 2) (Leipzig: Wilhelm Engelmann, 1895), 256–276.

70. Roux wrote both "Entwickelungsmechanik" and "Entwicklungsmechanik." In later years, Roux wrote treatise upon treatise explaining his contributions to experimental embryology. These were collected together with many of his descriptive morphological papers in his grand, two-volume *Gesammelte Abhandlungen*.

71. "Kurzer Bericht über die Sitzungen der vereinigten 5. und 9. Sektion für Zoologie und Anatomie der 60. Versammlung deutscher Naturforscher und Ärtze im Wiesbaden," *Anatomischer Anzeiger* 2 (1887): 763–764.

72. Wilhelm Roux, "Beiträge zur Entwickelungsmechanik des Embryo. Nr. V. Ueber die künstliche Hervorbringung halber Embryonen durch Zerstörung einer der beiden ersten Furchungszellen, sowie über die Nachentwicklung (Postgeneration) der fehlenden Körperhälfte," *Archiv für pathologisches Anatomie und Physiologie und für klinische Medizin* 114 (1888): 113–153, 246, 291; quotation from the English translation in *Foundations of Experimental Embryology,* ed. Benjamin H. Willier and Jane M. Oppenheimer (Englewood Cliffs, N.J.: Prentice-Hall, 1964), 31.

73. For discussion of Roux's post-generation, see Reinhard Mocek, *Die werdende Form: Eine Geschichte der Kausalen Morphologie* (Marburg: Basilisken, 1998), 241–244. The English translation did not include Roux's section on post-generation.

74. Churchill and Risler, *August Weismann,* 1:114, Weismann to Roux, 24 July 1887. Weismann's salutation and greetings from Gruber suggest a previous visit to Freiburg on Roux's part. It was physically possible, however, for Weismann to have also seen Roux's demonstration, for he was returning from his trip to the British Association for the Advancement of Science (BAAS) in Manchester, Scotland, London, and the Isle of Wight in early September and arrived in Freiburg on September 22. Wiesbaden would have been an easy layover for him.

75. Wilhelm Roux, "Beiträge zur Entwickelungsmechanik, VII, Ueber Mosaikarbeit und neuere Entwicklungshypothesen," *Gesammelte Abhandlungen über Entwicklungsmechanik der Organismen*, 2:818–871.

76. Laurent Marie Chabry, "Contribution à l'embryologie normale tératologique des ascidies simple," *Journal de l'anatomie et de la physiologie normales et pathologique de l'homme et des animaux* 23 (1887): 167–324.

77. Known in the popular literature as sea squirts or sessile tunicates, the ascidians are considered Urochordates. The decorticated egg of *Ascida aspersa,* the species Chabry worked with, is approximately 2 mm in diameter.

78. That is, $n = a^c$, where n is the number of monsters created, a is the number of blastomere combinations, and c is the number of cells in the embryo when assaulted. The single normal outcome is not included in Chabry's figure.

79. There remains a priority question concerning the time Chabry learned of Roux's earlier "Anstichversuche."

80. For an analytic comparison of Chabry's and Roux's experiments, see Churchill, "Chabry, Roux," 161–205, and for an update and disagreement, see Mocek, *Die werdende,* 420–423.

81. Mocek has analyzed the ambiguities of these two concepts in Roux's and contemporary embryologists' discussions. Mocek, *Die werdende,* 233–253.

82. Weismann, *Germ-Plasm,* 136–137.

83. Oscar Hertwig, "Welchen Einfluss übt die Schewerkraft auf die Theilung der Zellen?" *Jenaische Zeitschrift für Naturwissenschaft* 18 (1884–1885): 175–205.

84. Weismann and Haeckel b. 1834; A. Rauber b. 1841; O. Hertwig b. 1849; Roux, Born, and R. Hertwig b. 1850, Chabry b. 1855. Because Jahn, Maienschein, and Mocek did not include Rauber in their portrait of experimental embryology of the 1880s, I have lifted the following account with some alterations and additions from Frederick B. Churchill, "Wilhelm Roux and a Program for Embryology" (PhD diss., Harvard University, 1966). O. Hertwig (1884–1885) included Rauber's work among his pantheon of Pflüger antagonists.

85. W. Lubosch, "August Rauber, Sein Leben und seine Werke," *Anatomischer Anzeiger* 58 (1924): 129–172.

86. Ibid., 136.

87. Mocek, *Die werdende,* 191–200.

88. August Rauber, *Formbildung und Formstörung in der Entwickelung von Wirbelthieren* (Leipzig: Wilhelm Engelmann, 1880), 31.

89. Ibid., 52–59.

90. Ibid., 43, 61–63, 142–143.

91. August Rauber, "Schwerkraftversuche an Forelleneiern," *Berichte der naturforschenden Gesellschaft zu Leipzig,* 12 February 1884; August Rauber, "Einfluss der Schwerkraft auf die Zelltheilung und das Wachsthum," *Berichte der naturforschenden Gesellschaft zu Leipzig,* 11 November 1884; O. Hertwig, "Welchen Einfluss," 179–180.

92. Mocek, *Die werdende,* 257–332, devotes a chapter of nearly 100 pages to Driesch, his education, research career, and philosophy of biology. See also Hans Driesch, *Lebenserrinerungen* (Munich: Ernst Reinhardt, 1951).

93. Only after an unorthodox but profitable research existence of nearly a dozen years did Driesch settle in Heidelberg and slowly work his way into an academic career—in philosophy.

94. Hans Driesch, "Entwicklungs-mechanische Studien. I. der Werth der beiden ersten Furchungszellen in der Echinodermenentwicklung. Experimentelle Erzeugung von Teil- und Doppelbildungen," *Zeitschrift für wissenschaftliche Zoologie* 53 (1891–1892): 160–178. An extensively abbreviated English translation appears in Benjamin H. Willier and Jane M. Oppenheimer, *Foundations of Embryology* (Englewood Cliffs, N.J.: Prentice-Hall, 1964), 38–50.

95. Driesch, "Entwicklungs-mechanische," 172, as translated in Weismann, *Germ-Plasm*, 137. There is a similar but not identical passage in the summary of Driesch's work.

96. Driesch, "Entwicklungs-mechanische," 172, 178.

97. Weismann, *Germ-Plasm*, 138 (italics added). This sentence in the original is somewhat differently constructed, but it contains the same point that nuclear division is observable. Weismann, *Das Keimplasma*, 185.

98. Weismann, *Germ-Plasm*, 139–144.

99. Churchill and Risler, *August Weismann*, 1:182–183, Weismann to Hans Driesch, 28 December 1892. This letter was in part a paraphrase of a passage in *Das Keimplasma*, 184. The English translation *Germ-Plasm* (p. 137) deviates somewhat from the original German version but not in its general message (see also p. 138).

100. By the third review (in 1893), Barfurth extended the title to "Regeneration und Involution." For an examination of the first ten reviews of this series, see Churchill, "Regeneration."

101. See Victor Hamburger, *The Heritage of Experimental Embryology: Hans Spemann and the Organizer* (Oxford: Oxford University Press, 1988); Mocek, *Die Werdende*.

102. Frederick B. Churchill, "Wilhelm Johannsen and the Genotype Concept," *Journal of the History of Biology* 7 (1974): 5–30.

103. On sex chromosomes, see Scott F. Gilbert, "The Embryological Origins of the Gene Theory," *Journal of the History of Biology* 11 (1978): 307–351; Jane Maienschein, "What Determines Sex? A Study on Converging Approaches, 1880–1916," *Isis* 75 (1984): 456–480.

104. Weismann, *Das Keimplasma*, 315; quotation from Weismann, *Germ-Plasm*, 240.

105. Carl Rabl, "Über Zelltheilung," *Morphologisches Jahrbuch* 10 (1885): 241–330; see particularly pp. 323–325 and fig. 12a, b.

106. Theodor Boveri, "Zellenstudien II. Die Befruchtung und Teilung des Eies von *Ascaris megalocelphala*," *Jenaische Zeitschrift für Naturwissenschaft* 22 (1888): 685–882. For a convenient evaluation of both Rabl and Boveri, see E. B. Wilson, *The Cell in Development and Inheritance* (New York: Macmillan, 1896), 215–219.

107. Valentin Häcker, "Die ibildung bei Cyclops und Canthocamptus," *Zoologische Jahrbücher* 5 (1892): 211–248, esp. pp. 243–244 and figs. 28–29. See also Wilson, *The Cell*, 193–195.

108. Weismann, *Das Keimplasma*, 261.

109. For seven letters to Ammon, see Churchill and Risler, *August Weismann*, vol. 1; Hilkea Lichtsinn, *Otto Ammon und die Sozialanthropologie* (Bern: Lang, 1987). The names of other investigators of somatic variations familiar to the students of pre-Mendelian genetics buttressed his arguments: de Vries, Nägeli, and

Galton, of course; Köllreutter, Gärtner, Prosper Lucas, and the animal breeder Stettegast, and others. Except for his own research on the ostracod *Cypris reptans,* detailed in *Amphimixis,* Weismann focused on breeding literature for examples of contrasting traits, reversions, polymorphisms, and ambiguous cases of heredity.

110. Weismann had briefly mentioned the possibility of germinal competition six years earlier in his essay on the "Significance of Sexual Reproduction." August Weismann, *Essays upon Heredity and Kindred Biological Problems,* 1st ed., trans. A. E. Shipley and Selmar Schönland, ed. Edward B. Poulton (Oxford: Clarendon Press, 1889), 330–332.

111. Weismann, *Das Keimplasma,* 332; *Germ-Plasm,* 254. The English expression "predetermined" rather than Weismann's original "im Voraus bestimmt" becomes the rallying cry for Weismann's opponents.

112. Weismann, *Germ-Plasm,* 256–274, modified quotation from p. 273. "Ontogenetischen Verschiebung der Vererbungs-Resultate." Weismann, *Das Keimplasma,* 357.

113. Weismann, *Das Keimplasma,* 392.

114. Ibid., 392–396, quotation from p. 395; Weismann, *Germ-Plasm,* 514–520, quotation from p. 518 (italics in the original).

115. Weismann, *Germ-Plasm,* 397 and footnotes.

116. Ibid., 413.

117. Weismann also mentioned variant ids being produced by the same process, but it is unclear how their growth and competition would be different from the growth and competition between their homologous determinants.

118. Weismann, *Germ-Plasm,* 420–422, quotation from p. 420.

119. Ibid., 418.

120. Ibid., 420–428, quotation from p. 427.

121. Wilson, *The Cell,* 153–156. Wilson lists forty-five species that confirm what he called a "general law with a very high degree of probability." The list includes studies by Weismann's assistants Eugen Korscheldt, Otto vom Rath, and Valentin Häcker, as well as the first observations by van Beneden and many observations by Boveri and Rückert, among others. *Ascaris* and possibly the gastropod *Arion* were thought to have plurivalent chromosomes. The regularity in the number of chromosomes per species and between gametes and somatic cells appeared lawful. The distribution in terms of chromosome numbers within larger groups appeared irregular.

122. August Weismann, *Über die Dauer des Lebens* (Jena: Gustav Fischer, 1882).

123. Weismann, *Germ-Plasm,* 466 (italics in the original).

17. Controversies and Adjustments

1. Gerhard H. Müller, "Johann Wilhelm Haake (1855–1912), Biologe, Vererbungsforscher und Kritiker Weismanns," *Freiburg Universitätsblätter* 87/88

(1985): 167–181; Hans Stubbe, *Kurze Geschichgte der Genetik bis zur Wiederentdeckung der Vererbungsregeln Gregor Mendels* (Jena: Gustav Fischer, 1963), 187–188; L. I. Blacher, *The Problem of the Inheritance of Acquired Characters: A History of a Priori and Empirical Methods Used to Find a Solution,* ed. Frederick B. Churchill, trans. from Russian by Noel Hess (Washington, D.C.: Smithsonian Institution Libraries, 1982), 94–96.

2. See Georg Uschmann, *Geschichte der Zoologie und der zoologischen Anstalten in Jena 1779–1919* (Jena: Gustav Fischer, 1959), 115–117; Johann Wilhelm Haacke, "Zur Blatologie der Korallen: eine morphologische Studie," *Jenaische Zeitschrift* 13 (1879): 269–320. "Blastologie" would be best described as the study of the blastoderm and its development.

3. Uschmann, *Geschichte der Zoologie* (1959), 117; Georg Uschmann, "Haacke, Johann Wilhelm," in *Neue Deutsche Biographie* 7 (1965): S.367. Haeckel's words, quoted by Uschmann, are "gründlicher und gewissenhafter" and "weniger oberflächlich in der Behandlung schwieriger allgemeiner Probleme."

4. Wilhelm Haacke, "Das Ergebnis aus Weismann's Schrift, 'Ueber die Zahl der Richtungskörper und über ihre Bedeutung für die Vererung,'" *Biologisches Zentralblatt* 8 (1888–1889): 282–287.

5. Ibid., 287.

6. Wilhelm Haacke, *Gestalt und Vererbung: Eine Entwickelungsmechanik der Organismen* (Leipzig: T. O. Weigel Nachfolger, 1893), 9. On the same page Haacke also explained that preformation held that all parts of the future organism were formed in advance, whereas epigenesis denoted that the organism developed gradually from "completely, homogeneous, uniform construction material."

7. Haacke, *Gestalt und Vererbung,* 17.

8. Ibid., 18.

9. Ibid., 23–34, quotation from p. 27.

10. Ibid., 43.

11. At the time, the status of the centrosome was highly controversial.

12. Haacke, *Gestalt und Vererbung,* 64.

13. Ibid., 73–80. It is tempting to translate these concepts to the modern distinction of individual and group selection, but the two categories are not the same.

14. Haacke tends to write about "biophors" rather than "determinants."

15. Haacke, *Gestalt und Vererbung,* 94.

16. Stubbe, *Kurze Geschichgte,* 186–188.

17. Haacke, *Gestalt und Vererbung,* 103.

18. Ibid., 105.

19. See Jacques Roger, *Les Sciences de la vie dans la penseé Française du XVIII Siècle: La generation des animaux de Descartes a l'encyclopédie* (Paris: Armand Colin, 1963); Shirley A. Roe, *Matter, Life, and Generation. Eighteenth-Century Embryology and the Haller-Wolff Debate* (Cambridge: Cambridge University Press, 1981).

20. Haacke, *Gestalt und Vererbung,* 107 (emphasis is Haacke's).

21. Ibid., 125.

22. Ibid., 108.

23. Ibid., 111.

24. Weismann to Gustav Fischer, 9 November 1893 in Frederick B. Churchill and Helmut Risler, *August Weismann, Ausgewählte Briefe und Dokumente=Selected Letters and Documents,* 2 vols. (Freiburg: Universitätsbibliothek Freiburg, 1999), 1:202.

25. Ibid., Weismann to Wagner, 15 November 1893, 3 December 1893; Weismann to Roux, 7 January 1894. Haacke had mentioned the dictation while thanking his stenographers in his foreword.

26. Oscar Hertwig, *Die Zelle und die Gewebe,* Bd. 1 (Jena: Fischer, 1893). The second volume did not appear until 1898, and its emphasis was more on the rise and causes of differentiation. Weismann signed his preface on 19 May 1892; Hertwig signed his in October 1892.

27. Oscar Hertwig, "Vergleich der Ei- und Samenbildung bei Nematoden: Eine Grundlag für cellüläre Streitfragen," *Archiv für mikroskopische Anatomie* 36 (1890): 1–36. The other was Gustav Platner.

28. Oscar Hertwig, *Lehrbuch der Entwicklungsgeschichte des Menschen und der Wirbeltiere* (Jena: Eischer, 1886–88); *Text-Book of the Embryology of Man and Mammals,* trans. Edward L. Mark (London: Swan Sonnenschein, 1892). For details of Hertwig's appointment in Berlin, see Paul Julian Weindling, *Darwinism and Social Darwinism in Imperial Germany: The Contribution of the Biologist Oscar Hertwig (1849–1922)* (Stuttgart: Gustav Fischer, 1991), 205–213.

29. Weindling, *Darwinism,* particularly chapters 3, 5, and 6.

30. At the end of his introduction to his *Zeit- und Streitfragen der Biologie* (Leipzig: G. Fischer, 1894), 15, Hertwig explains that he felt it necessary to embark on a "Polemik."

31. The second volume was published in 1898. I have had access only to the English edition.

32. Hertwig provided a list of seven of his publications between 1879 and 1893, which led him to his epigenetic perspective. Hertwig, *Zeit- und Streitfragen,* 138–140; *Lehrbuch,* 13.

33. The full title of the Heft was *Präeformation oder Epigenese? Grundzüge einer Entwicklungstheorie der Organismen* (Jena: Gustav Fischer, 1894). The English version appeared two years later as Oscar Hertwig, *The Biological Problem of To-day: Preformation or Epigenesis? The Basis of a Theory of Organic Development,* trans. Peter Chalmers Mitchell (London: William Heinemann, 1896). This translation was republished with an introduction by Joseph Anthony Mazzeo by Dabor Science in Oceanside, New York, in 1977.

34. Wilhelm Roux, "Zur Orientirung neben einige Probleme der Embryologen Entwicklung," *Zeitschrift für Biologie* 21 (1885): 411–526. When Roux republished

this paper in his *Gesammelte Abhandlungen über Entwickelungsmechanik der Organismen,* he led off the second volume with the paper's "Einleitung" and placed the rest of the paper as contribution Nr. 18 with the original title. Wilhelm Roux, *Gesammelte Abhandlungen über Entwickelungsmechanik der Organismen* (Leipzig: Wilhelm Englemann, 1895), 2:1–23, 144–255. It was Roux's habit to include the original pagination in the body of the reprint.

35. For an attempt to identify this historical development, see Frederick B. Churchill, "Regeneration, 1885–1901," in *A History of Regeneration Research: Milestones in the Evolution of a Science,* ed. Charles E. Dinsmore (Cambridge: Cambridge University Press, 1991), 113–131.

36. Roux, *Gesammelte Abhandlungen,* 2–5, quotation from p. 5.

37. Ibid., 1–10, quotation from p. 6.

38. Ibid., 21.

39. Ibid., fn, 5–6.

40. Hertwig, *Präeformation,* 12–13, esp. fn4. See also Hertwig, *Biological Problem,* 12–14.

41. Hertwig, *Zeit- und Streitfragen,* 14.

42. Hertwig, *Biological Problem,* 38–72.

43. Ibid., 84–87, quotation from pp. 85 and 87.

44. Ibid., 93.

45. Ibid., 98. Weindling has persuasively argued that Hertwig carried this argument of supracellular correlations to the social level and maintained that the diversity in human society resulted from the interaction and cooperation of individuals.

46. Hertwig, *Präeformation,* 98–100; quotations follow Hertwig, *Biological Problem,* 103–105.

47. Haacke, *Gestalt und Vererbung,* 319. "Für den Begriff des Präformismus kommt es nicht darauf an, dass man im Keim ein mikroskopisches Abbild des fertigen Organismus erblickt, sondern man braucht nur, wie Hertwig es thut, eine vorgebildete Anordnung qualitative ungleicher Idioblasten in der Gesamtanlage anzunehmen, um mit vollen Segeln in den Hafen des Präformismus hineinzusteuern." Quoted in Hertwig, *Präeformation,* 132; trans. with minor differences in Hertwig, *Biological Problem,* 135.

48. Hertwig, *Biological Problem,* 140.

49. August Weismann, *Amphimixis oder: Die Vermischung der Individuen* (Jena: Gustav Fischer, 1891). Notice in this title he described it as a mixing up of individuals!

50. There were ways to mediate between these rival claims both experimentally and microscopically. The discussion here follows my more detailed article: Frederick B. Churchill, "Hertwig, Weismann, and the Meaning of Reduction Division circa 1890," *Isis* 61 (1970): 429–457. See also Weindling, *Darwinism,* chapter 6.

51. Translated and quoted in Churchill, "Hertwig," 441.

52. See Weismann, *Essays,* 1:xii–xv.

53. August Weismann, *The Germ-Plasm: A Theory of Heredity* (London: Walter Scott, 1893). W. Newton Parker was a professor at the University College of South Wales and Monmouthshire at the time. He was assisted by Harriet Rönnfeldt, his colleague Franck Arnold, and George Haswell Parker, a zoologist at Harvard University.

54. Churchill and Risler, *August Weismann,* 1:181–182, Weismann to Walter Scott, 16 December 1892.

55. In this section, I will closely follow and at times quote verbatim from my argument developed in Frederick B. Churchill, "The Weismann-Spencer Controversy over the Inheritance of Acquired Characters," in *Human Implications of Scientific Advance,* Proceedings of the XV International Congress of the History of Science (Edinburgh 10–19 August 1977), ed. E. G. Forbes (Edinburgh: Edinburgh University Press, 1978), 451–468.

56. J. V. Jensen, "The X Club: Fraternity of Victorian Science," *British Journal for the History of Science* 5 (1970): 63–72. See also Roy M. MacLeod, "The X-Club: A Social Network of Science in Late-Victorian England," *Notes and Records of the Royal Society of London* 24 (1970): 305–322; Ruth Barton, "'Huxley, Lubbock, and Half a Dozen Others': Professionals and Gentlemen in the Formation of the X Club, 1851–1864," *Isis* 89, no. 3 (1998): 410–444.

57. Peter B. Medawar, *The Art of the Soluble* (London: Methuen, 1967); Peter J. Bowler, "The Changing Meaning of Evolution," *Journal of the History of Ideas* 36 (1975): 95–114.

58. Herbert Spencer, *First Principles of a New System of Philosophy* (London: Williams and Norgate, 1862; New York: D. Appleton, 1865).

59. This and other quotations from the *Principles of Biology* are taken from a later edition. Herbert Spencer, *The Principles of Biology,* 2 vols. (London: William and Norgate, 1884), 1:81.

60. Ibid., 1:133fn.

61. Ibid., 151–152.

62. Ibid., 219–222. Abstracted in Churchill, "Weismann-Spencer Controversy," 455.

63. Herbert Spencer, "Factors of Organic Evolution," *Nineteenth Century* 19 (1886): 570–589, 749–770.

64. Churchill, "Weismann-Spencer Controversy," 451–468.

65. P. C. Mitchel, "The English Translation of Weismann's 'Essays,'" *Nature* 40 (1889): 618–619.

66. Alfred Russel Wallace, *Darwinism: An Exposition of the Theory of Natural Selection with Some of Its Applications* (London: Macmillan, 1889), 440–444, quotation from p. 443.

67. Spencer, "Factors."

68. This and the next two paragraphs are taken directly from my article on Spencer; see Churchill, "Weismann-Spencer Controversy."

69. Herbert Spencer, "The Inadequacy of Natural Selection," *Contemporary Review* 63 (1893): 152–166, 439–456.

70. August Weismann, "The All-Sufficiency of Natural Selection: A Reply to Herbert Spencer," *Contemporary Review* 64 (1893): 309–338, 596–610; and *Die Allmacht der Naturzüchtung: Eine Erwiderung an Herbert Spencer* (Jena: Gustav Fischer, 1893).

71. Herbert Spencer, "A Rejoinder to Professor Weismann," *Contemporary Review* 64 (1893): 893–912.

72. August Weismann, *The Effect of External Influences upon Development* (the Romanes Lecture 1894) (London: Henry Frowde, 1894); *Äussere Einflüsse als Entwicklungsreize* (Jena: Gustav Fischer, 1894).

73. Ethel Romanes, *Life and Letters of George John Romanes* (London: Longmans, Green, 1895), 347–348.

74. Herbert Spencer, "Weismann Once More," *Contemporary Review* 66 (1894): 592–608.

75. August Weismann, "Heredity Once More," *Contemporary Review* 68 (1895): 420–456, and *Neue Gedanken zur Vererbungsfrage. Eine Antwort an Herbert Spencer* (Jena: Gustav Fischer, 1895).

76. Herbert Spencer, "Heredity Once More," *Contemporary Review* 68 (1895): 608.

77. Although Wilson had worked in the Anatomical Institute, Weismann wrote a strong evaluation of his dissertation for the university senate. He was less pleased with Wilson's too literal translations of his Romanes Lecture and eventually complained to Will Parker about the final versions; Churchill and Risler, *August Weismann*, 1:237–238, Weismann to Parker, 15 May 1895. See also Weismann to Gregg Wilson, 11 March 1895 (1:228–229); Weismann's favorable letter and testimonial for Wilson (2:686); and Weismann to Gregg Wilson, 17 December 1895 (1:258–259).

78. Weismann, "All-Sufficiency of Natural Selection," 318–319.

79. Ibid., 337 (emphasis in original).

80. August Weismann, *Ueber die Vererbung* [*Vortrag. gehalten bei der Feier der Uebergabe des Prorectorates in der Aula der Universität Freiburg am 21. Juni 1883*] (Jena: Gustav Fischer, 1883); German reprint, *Aufsätze über Vererbung und verwandte biologische Fragen* (Jena: Gustav Fischer, 1892), 101–102; and the English translation, *Essays upon Heredity and Kindred Biological Problems,* 2nd ed., 2 vols., trans. A. E. Shipley and Selmar Schönland, ed. Edward B. Poulton (Oxford: Clarendon Press, 1891), 1:90–91.

81. Charles Darwin, *On the Origin of Species* (1859, facsimile reprint; Cambridge, Mass.: Harvard University Press, 1964), 454–455.

82. Charles Darwin, *The Variation of Animals and Plants under Domestication* (1868, facsimile reprint; Brussels: Culture et Civilization, 1969), 2:397–398.

83. Charles Darwin, "On the Males and Complemental Males of Certain Cirrepedes, and on Rudimentary Structures," *Nature* 8 (1873): 431–432, reprinted in

The Collected Papers of Charles Darwin, ed. Paul H. Barrett (Chicago: Chicago University Press, 1977), 2:177–182.

84. Weismann, *Ueber die Vererbung,* quotations from the English translation in Weismann's *Essays.*

85. Ibid., 90–91.

86. The word *panmixia* appears only four times in the first volume of his *Essays.*

87. August Weismann, "Über den Rückschritt in der Natur," *Berichte der Naturforschenden Gesellschaft zu Freiburg i. Br.* 2 (1886): 1–30, reprinted in *Aufsätze,* 547–596; trans. in *Essays,* 2:1–30.

88. August Weismann, "Über die Hypothese einer Vererbung von Verletzungen," in *Tageblatt der 61. Verwammlung deutscher Naturforscher und Ärzte in Köln vom 18.–23. September 1888. Wissenschaftlicher Theil Angeb.: Vorträge der Allgemeinen Sitzungen* (Köln: Albert Ahn, 1889), reprinted in *Aufsätze,* 507–546.

89. Weismann, *Ueber die Vererbung,* German reprint, *Aufsätze,* 101–102; *Essays,* 1:90–91.

90. Weismann, "Über die Hypothese," reprinted in *Aufsätze,* 507–546.

91. George J. Romanes, "Natural Selection and Dysteleology," *Nature* 9 (1874): 361–362. *Dysteleology* is Haeckel's term.

92. George J. Romanes, "Rudimentary Organs," *Nature* 9 (1874): 440–441.

93. George J. Romanes, "Disuse as a Reducing Cause in Species," *Nature* 10 (1874): 164.

94. George J. Romanes, *An Examination of Weismannism* (Chicago: Open Court, 1893).

95. Romanes, "Natural Selection."

96. August Weismann, *The Germ-Plasm: A Theory of Heredity,* trans. William Newton Parker and Harriet Rönnfeldt, ed. H. Ellis (London: Walter Scott, 1893), 83–84, 430–431.

97. August Weismann, *Das Keimplasma: Eine Theorie der Vererbung* (Jena: Gustav Fischer, 1892), 565–566. Translation is taken from the English translation *Germ-Plasm,* 430–431. The emphasis is Weismann's.

98. Ethel Romanes, *Life and Letters,* 323–324.

99. George J. Romanes, "The Spencer-Weismann Controversy," *Contemporary Review* 64 (1893): 50–59; "A Note on Panmixia," *Contemporary Review* 64 (1893): 611–612, quotation from p. 612.

100. For comments on the contrast between Spencer and Darwin, see Gloria Robinson, *A Prelude to Genetics: Theories of a Material Substance of Heredity, Darwin to Weismann* (Lawrence, Kans.: Coronado Press, 1979), 11.

101. George J. Romanes, "Panmixia," *Nature* 49 (1894): 599–600.

102. Churchill and Risler, *August Weismann,* 1:194–195, Weismann to Parker, 28 April 1893.

103. Ibid., 1:195–196, Weismann to Romanes, 25 May 1894 and 12 November 1893.

104. The Romanes Lecture Series continues to this day.

105. August Weismann, *The Effects of External Influences upon Development* (The Romanes Lecture, 1894) (London: Henry Frowde, 1894), and *Äussere Einflüsse als Entwicklunsreize* (Jena: Gustav Fischer, 1894).

106. Churchill and Risler, *August Weismann*, 1:124–125, 162, Weismann to Poulton, 26 October 1888; Weismann to Parker, 1 December 1891.

107. Weismann, *Äussere*, 12–16, quotation from p. 15.

108. Ibid., 35 (emphasis is Weismann's).

109. Hertwig, *Zeit- und Streitfragen*, 131–133, quotation from p. 132. Hertwig's original was a more generalizable, stark claim: "Weismann . . . den wichtigen Unterschied zwischen Anlage und Bedingung ganz übersehen hat" (quotation from p. 70 of Weismann's text).

110. Auguste Forel, *Fourmis de la Suisse. Systématique. Notices anatomique et physiologiques. Architecture. Distribution géographique. Nouvelles experiences et observations de moeurs.* (Geneva: Lyon: H. Georg, 1874).

111. Churchill and Risler, *August Weismann*, vol. 1, Weismann to Forel, 4 July 1893, 16 November 1893, and 6 December 1893; Weismann to Forel, 28 November 1894.

112. See Forel, *Fourmis,* 311–323, for a full account of Forel's experiments.

113. Bickford had been a teacher at the Bryn Mawr Preparatory School and an assistant at the Johns Hopkins Biological Laboratory before studying in Leipzig (1893–1894) and then researching and writing a dissertation in Freiburg (1894–1895). She titled her work "Über die Morphologie und Physiologie der Ovarien der Ameisen-Arbeiterrinnen." For more details on Bickford and her dissertation, see Weismann's "Gutachten," in Churchill and Risler, *August Weismann*, 2:619.

114. Elizabeth Bickford, *Über die Morphologie und Physiologie der Ovarien der Ameisen-Arbeiterinnen* (Jena: G. Fischer, 1895), and in *Zoologische Jahrbücher. Abteilung für Systematik, Ökologie und Geographie der Tiere* 9 (10 July 1895): 23–24.

115. See the entry on Bickford in Churchill and Risler, *August Weismann*, 2:619.

18. The Germ-Plasm and the Diversity of Living Phenomena

1. Frederick B. Churchill, "Life before Model Systems: General Zoology at August Weismann's Institute," *American Zoologist* 37 (1997): 260–268.

2. His standard summer semester lecture course became selected topics in zoology, which eventually was listed simply as "Spezielle Zoologie" to contrast with his "Allgemeine Zoologie (Descendenztheorie)" in the winter. From year to year

slight variations appeared in the catalog of courses, but this general pattern continued until Weismann's retirement.

3. Weismann left heavy colored crayon lines in the margins to flag important sections of the off-prints he read.

4. Ernst Mayr, *Animal Species and Evolution* (Cambridge, Mass.: Belknap Press of Harvard University Press, 1963).

5. Darwin mentioned, but made little of an intermediate level between the two, namely, the cellular layer. Gerald L. Giesen, "Darwin and Heredity: The Evolution of His Hypothesis of Pangenesis," *Journal of the History of Medicine and the Allied Sciences* 24 (1969): 375–411.

6. Gloria Robinson, *A Prelude to Genetics: Theories of a Material Substance of Heredity: Darwin to Weismann* (Lawrence, Kans.: Coronado Press, 1979).

7. August Weismann, "Über Germinal-Selektion," *Congrès international de Zoologie. Compte rendu. (Leiden)* 3 (1895): 35–70; August Weismann, "On Germinal Selection as a Source of Definite Variation," trans. T. J. McCormack, *The Monist* 6 (1896): 250–293.

8. August Weismann, *Über Germinal-Selection: Eine Quelle bestimmt gerichteter Variation* (Jena: Gustav Fischer, 1896); *On Germinal Selection as a Source of Definite Variation,* 2nd English ed., trans. Thomas J. McCormack (Chicago: Open Court, 1896).

9. August Weismann, "Heredity Once More," *Contemporary Review* 68 (1895): 420–456, and *Neue Gedanken zur Vererbungsfrage: Eine Antwort an Herbert Spencer* (Jena: Gustav Fischer, 1895); Weismann to v. Wagner, 30 November 1895, in Frederick B. Churchill and Helmut Risler, *August Weismann, Ausgewählte Briefe und Dokumente=Selected Letters and Documents,* 2 vols. (Freiburg: Universitätsbibliothek Freiburg, 1999), 1:257.

10. "Autobiographie von *August Weismann* aus dem Jahre 1896," in Churchill and Risler, *August Weismann,* 2:535. My translation of Weismann's manuscript differs somewhat from Herbert Ernest Cushman's. The latter eventually appeared as "An Autobiographical Sketch with an Introduction by Professor Josiah Royce," in *The Lamp, a New Series of the Bookbuyer* 26 (1903): 21–28.

11. Weismann was well aware that Newton's famous claim that "Hypotheses non fingo" had only a limited referent.

12. Weismann, *Über Germinal-Selection,* x.

13. August Weismann, *On Germinal Selection,* xi–xii.

14. A focus on the historical relationship between ecology and evolution has become an important specialty in the history of biology. As a start, see the *Journal of the History of Biology* 19 (1986), and for the Weismann period in particular, see William Coleman, "Evolution into Ecology? The Strategy of Warming's Ecological Plant Geography," 181–196; Joel B. Hagen, "Ecologists and Taxonomists: Diver-

gent Traditions in Twentieth-Century Plant Geography," 197–214; and William C. Kimmler, "Advantage, Adaptiveness, and Evolution Ecology," 215–233.

15. Weismann, *On Germinal Selection,* 5–11, quotations from pp. 10 and 8 (pp. 13 and 10 of the German edition). The second is emphasized in the German version.

16. For perceptive accounts of Weismann and Eimer, see Peter J. Bowler, "Theodor Eimer and Orthogenesis: Evolution by 'Definitely Directed Variation,'" *Journal of the History of Medicine and Allied Sciences* 34 (1979): 40–73; Peter J. Bowler, *The Eclipse of Darwinism: Anti-Darwinian Evolution Theories in the Decades around 1900* (Baltimore: Johns Hopkins University Press, 1983), chapter 7.

17. Theodor Eimer, *Die Entstehung der Arten auf Grund von Vererben erworbner Eigenschaften nach den Gesetzen organischen Wachsens. Ein Beitrag zur einheitlichen Auffassung der Lebenwelt,* 3 vols. (Jena and Leipzig: G. Fischer, 1888, 1897, and posthumously 1901), quotation from 1:2.

18. There are no Freiburg University records of Eimer, but see C. B. Klunzinger, "Theodor Eimer. Ein Lebensabriss mit Darstellung der Eimer'schen Lehren nach ihrer Entstehung der Arten," in *Jahreshefte des Vereins für vaterländische Naturkunde in Württemberg* 55 (1899): 1–22; Georg Uschmann, "Eimer," *Neue Deutsche Biographie* 4 (1959): 393–394; Frederick B. Churchill, "Eimer, Theodor Gustav Heinrich," *Dictionary of Scientific Biography,* Supplement 2, vol. 17, ed. Charles Gillispie (New York: Charles Scribner's Sons, 1990), 261–264.

19. Marie von Linden, "Professor Dr. Theodor Eimer," *Biologisches Centralblatt* 18 (1898): 721–725.

20. Churchill and Risler, *August Weismann,* 1:253, Weismann to O. Staudinger, 5 November 1895; ibid., 1:441, Weismann to E. Müller, 15 October 1905. The university records are unfortunately silent about Eimer's brief graduate career in Freiburg. When Eimer had earlier spent the year of 1863–1864 in Freiburg as a general medical student, Weismann had just habilitated and was a privatdozent teaching zoology under the nominal supervision of Prof. Otto Funke.

21. Theodor Eimer, *Zoologische Studien auf Capri. II.* Lacerta muralis coerulea. *Ein Beitrag zur Darwin'schen Lehre* (Leipzig: Wilhelm Englmann, 1874), 42.

22. Theodor Eimer, "Ueber den Begriff des tierischen Individuums" (21 September 1883), *Amtlicher Bericht der Versammlung der deutschen Naturforscher und Ärzte,* 45–53, printed from a stenograph. A condensed version of this address was printed in *Humboldt. Monatschrift für die gesammten Naturwissenschaften* 2 (1883): 437–440. According to the stenograph, Eimer was invited to present this address only a few days earlier. An English translation of the stenograph is included in J. T. Cunningham's translation of the first volume of *Entstehung der Arten, Organic Evolution as the Result of the Inheritance of Acquired Characters According to the Laws of Organic Growth* (London: Macmillan, 1890), 413–435.

23. For a contemporary account of Eimer, see Vernon L. Kellogg, *Darwinism Today* (New York: Henry Holt, 1908), 281–285.

24. The full titles of Eimer's texts: *Die Entstehung der Arten auf Grund von Vererben erworbener Eigenschaften nach den Gesetzen organischen Wachsens. Ein Beitrag zur einheitlichen Auffassung der Lebenwelt,* 3 vols. (Jena: G. Fischer, 1888–1901); and *Die Artbildung und Verwandschaft bei den Schmetterlingen: Eine systematische Darstellung der Abänderungen, Abarten und Arten der Segelfalter-ähnlichen Formen der Gattung Papilio* (Jena: G. Fischer, 1889–1895). The first work, according to its translator, J. T. Cunningham, was quickly rendered into English because of the "uncritical acceptance accorded to Professor Weismann's theories of heredity and variation by many English evolutionists." Theodor Eimer, *Organic Evolution, the Result of the Inheritance of Acquired Characters According to the Laws of Organic Growth,* trans. J. T. Cunningham (London: Macmillan, 1890), vii. The comment is of added interest because it was published before Weismann's *Keimplasma* had been written and before Weismann's collected *Essays* had appeared.

25. Eimer, *Die Entstehung,* 427; *Organic Evolution,* 397; quotation modified from Cunningham's translation.

26. Eimer, *Die Entstehung,* 198–201; *Organic Evolution,* 184–187.

27. Churchill and Risler, *August Weismann,* 1:73–74, Weismann to de Bary, 2 February 1886 (translation mine).

28. Ibid., 1:100–101, Weismann to Eimer, 21 February 1887.

29. Ibid., 1:101.

30. Ibid., 1:107–108, Weismann to Gustav Fischer, 4 May 1887.

31. Ibid., 1:145, Weismann to E. B. Poulton, 14 June 1890 (translation mine).

32. Ibid., 2:675.

33. Arnold Spuler, "Zur Phylogenie und Ontogenie des Flügelgeäders der Schmetterlinge," *Zeitschrift für wissenschaftliche Zoologie* 53 (1892): 597–646. Gruber wrote a few other Gutachten for the dissertations submitted to the faculty by the zoological institute, but at those moments Weismann was out of Freiburg.

34. Arnold Spuler, "Zur Stammesgeschichte der Papilioniden," *Zoologische Jahrbücher. Abteilung für Systematik, Geographie und Biologie der Thiere* 6 (1892): 465–498.

35. Spuler, "Zur Stammesgeschichte," 465 and 495; "Zur Phylogenie," 644. He also thanked Gruber in the dissertation. Spuler identified himself as "Dr." in his dissertation of October 1891, but not in the paper submitted in June 1891.

36. To the modern eye, the pattern of venation appears aerodynamically more enduring and phylogenetically more significant than wing stripes except in cases of mimicry and protective coloration. These latter phenomena were gladly accepted by Weismann as demonstrably useful in the Darwinian sense.

37. Theodor Eimer, "Bemerkungen zu dem Aufsatz von A. Spuler zur Stammesgeschitche der Papilioniden," *Zoologische Jahrbücher. Abteilung für Systematik, Geographie und Biologie der Tiere* 7 (1893–1894): 186–205.

38. Eimer, *Die Entstehung;* Eimer, *Die Artbildung;* Theodor Eimer, "Bemerkungen zu dem Aufsatz von A. Spuler zur Stammesgeschitche der Papilioniden," *Zoologische Jahrbücher. Abteilung für Systematik, Geographie und Biologie der Tiere* 7 (1893–1894): 186–205.

39. A German version of Eimer's address may be found in *Entstehung*, 2:1–49. This version includes an introduction, additions to the ending, and nineteen illustrations of butterflies, none of which are included in the English translation. Otherwise, the two follow each other closely. Eimer used the material assembled in Theil II of his *Artbildung* published the same year as his Leiden address.

40. Eimer, *Entstehung*, 2:29.

41. The third volume of Eimer's *Entstehung der Arten* was edited and published in 1901 by his former admirers C. Fickert (still at Tübingen) and Marie Countess von Linden (by then at Bonn).

42. Bowler, "Theodor Eimer and Orthogenesis"; *Eclipse of Darwinism,* chapter 7.

43. Bowler, "Theodor Eimer and Orthogenesis," 71.

44. Churchill and Risler, *August Weismann,* 1:76, Weismann to Gustav Fischer, 11 February 1886.

45. Ibid., 294–296, Weismann to Vöchting, 7 June 1898.

46. Ibid., 300, Weismann to Ishikawa, 31 July 1898 (translation mine).

19. Seasonal Dimorphism in Butterflies and Parthenogenesis of Drones

1. This section has been taken, with small alterations, from my essay "August Weismann: A Developmental Evolutionist," which appears in Frederick B. Churchill and Helmut Risler, *August Weismann, Ausgewählte Briefe und Dokumente = Selected Letters and Documents,* 2 vols. (Freiburg: Universitätsbibliothek Freiburg, 1999), 2:772–776.

2. Frederick B. Churchill, "Lepidopteran Research and the Rise of Classical Genetics: 1880–1920," in *Proceedings of the Sixteenth International Congress of the History of Science* [Bucharest, Romania, August 26–September 3, 1981] (Bucharest: Academy of the Socialist Republic of Romania, 1981), section A: Scientific, 212 (abstract only).

3. August Weismann, "Neue Versuche zum Saison-Dimorphismus der Schmetterlinge," *Zoologische Jahrbücher, Abteilung, für Systematik, Geographie und Biologie der Tiere* 8 (1896): 611–684; and "New Experiments on the Seasonal Dimorphism of Lepidopera," *Entomologist* 29 (1896): 105–106.

4. Today, this species goes by the genus name *Lycaena.* In many of his works, Weismann referred to the genus as *Polyommatus* as well as *Chrysophanus.*

5. Weismann, "New Experiments."

6. Churchill and Risler, *August Weismann,* 2:775.

7. August Weismann, *Die Bedeutung der sexuellen Fortpflanzung für die Selectionstheorie* (Jena: Gustav Fischer, 1886). This paper was originally presented as an ad-

dress to the Versammlung Deutscher Naturforscher und Ärzte in Straßburg, September 1885.

8. Ernst Mayr, "Weismann and Evolution," *Journal of the History of Biology* 18 (1985): 318–320.

9. August Weismann, "Ueber die mechanische Auffassung der Natur," in *Studien zur Descendenztheorie,* vol. 2 (Leipzig: Wilhelm Engelmann, 1876), fn11.

10. August Weismann, *Das Keimplasma. Eine Theorie der Vererbung* (Jena: Gustav Fischer, 1892), fn41, 547. This quotation is taken with important modifications from the English edition (italics in the original).

11. O. Taschenberg, "Historische Entwickelung der Lehre von der Parthenogenesis," *Abhandlungen der Naturforschenden Gesellschaft zu Halle/herausgegeben von ihrem Vorstande* 17 (1892): 111–197.

12. Ibid., quotation from p. 169 (the final paragraph of the reprint).

13. Richard Owen, *On Parthenogenesis; or, The Successive Production of Procreating Individuals from a Single Ovum* (London: John van Voorst, 1849).

14. Frederick B. Churchill, "Sex and the Single Organism," *Studies in the History of Biology* 3 (1979): 145–152.

15. Taschenberg, "Historische Entwickelung," 114–115; Churchill, "Sex and the Single Organism," 163–165.

16. For a fuller account of what became known as the Dzierzon "system" and an associated bibliography, see Frederick B. Churchill, "August Weismann and Ferdinand Dickel: Testing the Dzierzon System," in *Science, History and Social Activism: A Tribute to Everett Mendelsohn,* ed. Garland Allen and Roy MacLeod (Dordrecht: Kluwer Academic, 2001).

17. A fuller account of his work is provided in Churchill, "Sex and the Single Organism," 152–163.

18. The microplyle of the honeybee egg is only 1/5,000 of a millimeter in diameter and requires a delicate technique on the part of the microscopist. Microplyles had already been known to exist in the eggs of other invertebrates, but in 1854 Leuckart and Georg Meissner independently became the first to find and describe it in the honeybee egg. See Karl Siebold, *Wahre Parthenogenesis bei Schmetterlingen und Biene: Ein Beitrag Zur Fortplfanzungsgeschichte der Thiere* (Leipzig: Wilhelm Engelmann, 1856), 101–102.

19. Ibid., 108–109.

20. In a systematic way, Siebold, followed by others, discovered three types of parthenogenetic reproduction in an assortment of lower organisms: arrenotoky, found with bees, where parthenogenesis produced a male offspring; thelytoky, where parthenogenesis produced a female; and amphitoky, where parthenogenesis produced either male or female offspring. Further details and references may be found in Churchill, "Sex and the Single Organism," fn45–49.

21. Hermann Landois, "Ueber das Gesetz der Entwickelung der Geschlechter bei den Insecten," *Zeitschrift für wissenschaftliche Zoologie* 17 (1867): 375–379.

22. This paragraph is a condensation of my discussion in Churchill, "August Weismann and Ferdinand Dickel," 63–64. See Hans Nachtsheim, "Die parthenogenesis bei der Honigbiene: Ein historischer Ueberblick über den Kampf um die Dzierzonsche Theorie," *Bienen-wirtschaftliches Centralblatt* 49 (1913): 298–328; Hans Buttel-Reepen, *Leben und Wesen der Bienen* (Braunschweig: F. Vieweg & Sohn, 1915), 36–37. A brief biographical account of Pérez may be found in the *Archives biographiques françaises.*

23. For a brief overview of Dickel's life, see Erich Schwärzel, *Durch Sie Wurden Wir: Biographie der Grossmeister und Föderer der Bienenzucht im deutschsprachigen Raum* (Giessen: Die Bienen, 1985), 45–46. I do not intend to enter into the extensive and important literature on amateurs in science. In the German context of this episode, I refer to Dickel as an amateur because he was a well-read and enthusiastic apiarist but did not have a university degree in science or, for that matter, in any other subject. In other words, he was not a "Wissenschaftler" or experienced scholar practiced in "scientific" methodology.

24. Ferdinand Dickel, *Das Prinzip der Geschlechtsbildung bei Tieren geschlechtlicher Fortpflanzung entwickelt auf Grundlage meiner Bienenforschung* (Nördlingen: C. H. Beck, 1898). Later, Dickel would defend and slightly alter his ideas in print.

25. Dickel failed to mention that Siebold also examined fifty-two worker bee eggs and found spermatozoa in 58 percent.

26. Dickel, *Das Prinzip,* 13–20, quoted from pp. 14–15.

27. It is interesting to note that Dickel must have relied on reduction division as a mechanism for segregating the percentage of *anlagen*—and so had Weismann.

28. Erich Schwärzel, *Durch SieWurden Wir-Biographie du Grossmeister und Förderer der Bienenzuchtim deutschsprachigen Raum* (Gieseen: Die Bienen, 1985), 45–46; Dickel, *Das Prinzip,* 66–67.

29. Dickel, *Das Prinzip,* 20. Leuckart died the following February.

30. All twenty-four letters are printed in full in Churchill and Risler, *August Weismann,* vol. 1.

31. Ibid., 1:311, Weismann to Dickel, 9 July 1897.

32. Ibid., 1:306, Weismann to Dickel, 3 December 1898.

33. Wilhelm Paulcke was an experienced student, having worked in Weismann's institute since 1895, having spent a year in Arnold Lang's institute for zoology and comparative anatomy in Zürich, and having worked at the Naples Zoological Station for two months. The work complemented his dissertation "Ueber die Differenzirung der Zellelemente im Ovarium der Bienenkönigin (*Apis mellifica*)," *Zoologische Jahrbücher* 14 (1901): 77–202.

34. Churchill and Risler, *August Weismann,* 1:311, Weismann to Dickel, 18 February 1899.

35. One can infer from the cautionary contents from Weismann's letter to Reepen, 15 January 1899. Reepen soon expanded his name to von Buttel-Reepen and became a student at Weismann's institute. He was an experienced apiarist but for professional reasons did his dissertation on parasitic trematodes. After his promotion, he could speak with authority as both a zoologist and beekeeper. He became one of the noteworthy interpreters of bee biology for his generation. See Karl von Frisch's account in *Neue Deutsche Biographie* 3 (1957), 80, and Schwärzel, *Durch SieWurden,* 35–36.

36. The full sentence reads: "Wenn ich Hrn. Dickel den betreffenden Passus der Anerkennung geschrieben u. ihm durch das Recht, denselben abzudrucken zugestanden habe, geschah es, weil mir seine Versuche, die in der Bienenzeitung veröffentlicht wurden, in der That recht warscheinlich machten, daß wir uns seit Siebold, Leuckart u. Dzierzon in einem Irrtum bewegt haben. Wahrscheinlich nicht gewiß!" Churchill and Risler, *August Weismann,* 1:307, Weismann to Dickel, 15 January 1899.

37. Ibid., 1:322, Weismann to Dickel, 10 July 1899.

38. Ibid.

39. Ibid., 1:323, Weismann to Dickel, 11 July 1899. The mistake may indicate how badly Weismann, Häcker, and Paulcke wanted to refute the Dzierzon system.

40. Ibid., 1:328–329, Weismann to Dickel, 20 August 1899.

41. Ibid., 1:336, Weismann to Dickel, 24 April 1900.

42. Ibid., 1:338, Weismann to Dickel, 20 June 1900.

43. Ibid.

44. Alexander Petrunkewitsch [Petrunkevitch], "Die Richtungskörper und ihr Schicksal im befruchteten und unbefruchteten Bienenei," *Zoologische Jahrbücher* 14 (1901): 573–608; and "Das Schicksal der Richtungskörper im Drohnenei: Ein Beitrag zur Kenntniss der natürlichen Parthenogenese," *Zoologische Jahrbücher* 17 (1903): 481–516. When he married an American and moved to the United States, Petrunkevitch anglicized the spelling of his name. After some short-term positions, including that of an independent researcher, he became an internationally recognized expert on the morphology and taxonomy of arachnids and a distinguished professor of zoology at Yale University. For biographical information, see G. Evelyn Hutchinson, "Alexander Petrunkevitch: An Appreciation of His Scientific Works and a List of His Published Writings," *Transactions of the Connecticut Academy of Arts and Sciences,* 30 (1945): 9–24.

45. August Weismann, "Ueber die Parthenogese der Bienen," *Anatomischer Anzeiger* 18 (1900): 492–499. A slightly altered version with a different title appeared at the same time in *Die Biene* (Giessen), the official journal of the *Wanderversammlung.*

46. Petrunkevitch, "Die Richtungskörper," 581; Weismann, "Ueber die Parthenogese," 495. Petrunkevitch found that it was easier to locate the much larger sperm

aster than the sperm pronucleus in his serial sections. This dictated that Dickel supply him with eggs at the stage of second polar body formation.

47. Ferdinand Dickel, "Meine Ansicht über die Freiburger Untersuchungsergebnisse von Bieneneiern," *Anatomischer Anzeiger* 19 (1901): 104–108; August Weismann, "Erwiderung auf Ferdinand Dickel," *Anatomischer Anzeiger* 19 (1901): 108–110; Dickel, "Thatsachen entscheiden, nicht Ansichten," *Anatomischer Anzeiger* 19 (1901): 110–111.

48. Churchill and Risler, *August Weismann*, 1:347–348, August Weismann to Karl Bardeleben, 13 January 1901.

49. Ferdinand Dickel, "Über Petrunkewitsch's Untersuchungsergebnisse von Bieneneiern," *Anatomischer Anzeiger* 25 (1902): 20–27. Dickel emphasized that Petrunkevitch was not justified in assuming that the absence of the astral rays in drone eggs was equivalent to the absence of the much smaller sperm pronucleus. He further questioned the efficacy of his staining methods and doubted that the queen could open and close the sphincter of her spermatheca in a systematic manner.

50. Ferdinand Dickel, "Die Ursachen der geschlechtlichen Differenzierung im Bienenstaat (Ein Beitrag zur Vererbungsfrage)," *Archiv für die gesammte Physiologie des Menschen und der Tiere* 55 (1903): 66–106, quoted from p. 70.

51. Eduard Pflüger, "Ueber die jungfräuliche Zeugung der Bienen," *Archiv für die gesammte Physiologie des Menschen und der Tiere* 99 (1903): 243–244.

52. Albrecht Bethe, "Dürfen wir den Ameisen und Bienen psychische Qualitäten zu schreiben?" *Archiv für die gesammte Physiologie des Menschen und der Tiere* 70 (1898): 15–100; Porphiry I. Bachmetjew [Bakhmetev], "Ein Versuch, die Frage über die Parthenogese der Drohnen mittels der analytisch-statistichen Methode zu lösen," *Allgemeine Zeitschrift für Entomologie* 8 (1903): 37–44.

53. Churchill and Risler, *August Weismann*, 1:477–478, Weismann to Petrunkevitch, 29 September 1907. There is internal evidence in later correspondence to Dickel that Weismann wrote Dickel at least once in 1904. This letter has not been identified.

54. Ibid., 1:504, Weismann to Dickel, 27 February 1910.

55. Ibid., 1:505, Weismann to Dickel, 18 May 1910.

20. Adapting the Germ-Plasm

1. Barfurth's twenty annual reviews of "Regeneration und Involution" began in 1991; Roux's *Archiv für Entwicklungsmechanik* first appeared in 1895; the American-based *Journal of Experimental Zoology* materialized in 1904.

2. Fifteen years younger than the founders of Entwicklungsmechanik, T. H. Morgan (b. 1866) nevertheless entered into the spirit of the early regeneration experiments. His collaboration with Driesch, his *Development of the Frog's Egg* (New

York: Macmillan, 1897), and his *Regeneration* (New York: Macmillan, 1901) all reflected the same nineteenth-century morphological perspective.

3. Reinhard Mocek, *Die Werdende Form: Eine Geschichte der Kausalen Morphologie* (Marburg: Basilisken-Presse, 1998).

4. See Churchill, "Regeneration, 1885–1901," in *A History of Regeneration Research: Milestones in the Evolution of a Science,* ed. Charles E. Dinsmore (Cambridge: Cambridge University Press, 1991), 113–131.

5. Oscar and Richard Hertwig, *Uber den Befruchtungs- und Teilungsvorgang des tierischen Eies unter dem Einfluss äusserer Agentien* (Jena: G. Fischer, 1887).

6. Albert Fleischmann (1862–1942) studied comparative embryology at Erlangen, was appointed an außerordentlicher Professor in 1896, and was summoned to occupy the Lehrstuhl for Zoology in 1898. In private, Weismann was critical of his career for having been promoted within a single university. See Frederick B. Churchill and Helmut Risler, *August Weismann, Ausgewählte Briefe und Dokumente = Selected Letters and Documents,* 2 vols. (Freiburg: Universitätsbibliothek Freiburg, 1999), 2:242, Weismann to Jacob Rees, 22 July 1895. He then was incensed at Fleischmann's anti-evolutionary books, which appeared after the turn of the century (Weismann to the Editor of the *Deutsche Medizin. Wochenschrift,* 8 November 1903).

7. Merriley Borrell, "Origins of the Hormone Concept: Internal Secretions and Physiological Research, 1889–1905" (PhD diss., Yale University, 1976). See also Cornelius Medvei, *A History of Endocrinology* (Lancaster: MTP Press, 1982).

8. J. F. Gudernatsch, "Feeding Experiments on Tadpoles," *Archiv für Entwicklungsmechanik der Organismen* 53 (1913): 457–483.

9. Borell, "Origins of the Hormone," and Medvei, *A History of Endocrinology.* The expression "mode and site" of a hormone's activity is Borell's and is a convenient way for discriminating the concern for the "mechanics" of modern endocrinology from the less guided search for therapeutics and age-resisting substances.

10. Julian Huxley's *Problems of Relative Growth* (1934) describes his research begun in 1919. See also my introductory essay "On the Road to the *k* Constant: A Historical Introduction" in the recent reprint of Julian Huxley, *Problems of Relative Growth* (Baltimore: Johns Hopkins University Press, 1993).

11. Farmer and Moore coined the neologism "meiosis" in 1905. In Greek, the word literally means "diminution."

12. For purposes of research and administration, the biological institute was separated into five departments. Boveri was invited to become the director of the institute and leader of the first department for "Vererbungslehre und Biologie der Tiere." Boveri was reluctant to pull up roots in Würzburg in part because of his wish to remain in a teaching environment, in part because he wished to remain in his native state of Franconia, and perhaps also because he recognized signs of waning health.

(His Würzburg colleague, the physicist Willi Wien, wrote that health was not a primary concern.)

13. Fritz Baltzer's book *Theodor Boveri: Life and Work of a Great Biologist,* trans. Dorothea Rudnick (Berkeley: University of California Press, 1967) provides an informative biography and provides a general description of his scientific work.

14. Boveri also did a number of comparative anatomical studies on *Amphioxus.*

15. Theodor Boveri, *Zellenstudien* (Jena: G. Fischer, 1887, 1888, 1890, 1901, 1905, and 1907).

16. Theodor Boveri, *Das Problem der Befruchtung* (Jena: G. Fischer, 1902), *Die Organismen als historische Wesen* (Würzburg: Kgl. universitätsdruckerei von H. Stürtz, 1906), and *Zur Frage der Entstehung maligner Tumoren* (Jena: G. Fischer, 1914).

17. Boveri encourages such a view on a number of occasions. He was one of the few non-Freiburg zoologists to write a contribution to Weismann's seventy-year Festschrift. Although overturning many of Weismann's ideas, he wrote respectfully of his contributions. It was appropriate that the Freiburg Senate and Baden Cultus minister invited Boveri to become Weismann's successor—unsuccessfully, as it turned out. The only subject in which Boveri differed in spirit from Weismann was his appeal to Lamarckian processes as part of the evolutionary mechanism, although he wrote seldom on this subject, which was also a contrast with Weismann.

18. Boveri, *Zellenstudien, I,* 5–9; *II,* 13–26.

19. "[M]it einem Detail und einem Reichtum an Abbildungen ausgestattet sind, wie wenig andere Werke der Zellen-Litteratur" (ibid., *I,* 6). Boveri considered van Beneden's work a "Schlusstein" for the study of the distribution of chromosomal elements during polar body formation and perhaps early cleavage, which also contributed to the renewed interest in heredity. Furthermore, Boveri praises van Beneden for revealing the two-sidedness of the spindle and the mechanics of separating out the chromosomal elements via the contraction of the spindle threads, but he considered van Beneden's preparations weak, which accounts for the neglect of his accomplishments by others investigating Teilungsmechanik (ibid., *II,* 4). *Zellenstudien II* also lavishes praise on the new facts and ideas in van Beneden's work and his newer contributions on spindle formation and element segregation.

20. Ibid., *II,* 4.

21. Theodor Boveri, "Befruchtung," *Ergebnisse der Anatomie und Entwicklungsgeschichte* (1891): 391–392.

22. Ibid., 447.

23. It is worth noting in passing that Eduard Strasburger coined the modern terms "diploid" and "haploid" only in 1907, which suggests that a lot had to be resolved about the maturation process in the intervening dozen years.

24. Theodor Boveri, "Ueber die Befruchtung der Eier von *Ascaris megalocephala,*" *Sitzungsberichte der Gesellschaft für Morphologie und Physiologie zu München,* 2

(1887): 76–77. The terms "haploid" and "diploid" denote the concept of pairs of homologous chromosomes.

25. Theodor Boveri, "Die Befruchtung und Teilung des Eies von *Ascaris megalocephala*," *Zellenstudien II*, 65–77.

26. See Jonathan Harwood, *Styles of Scientific Thought: The German Genetics Community, 1900–1933* (Chicago: University of Chicago Press, 1993), chapters 2 and 9.

27. Boveri recognizes this in his 1902 paper on polyspermi.

28. Theodor Boveri, "Uber Differenzierung der Zellkerne während der Furchung des Eies von *Ascaris megalocephala*," *Anatomischer Anzeiger* 2 (1887): 688–693. Because the blastomere lineages with undiminished chromosomes produced the primary germ cell line, it was tempting at first glance to consider their existence as a demonstration of Weismann's theory of continuity of the germ-plasm. Neither Boveri nor Weismann, however, felt that that was the case, for it seemed clear from the outset that a polarity in the cytoplasm was the determinitive factor. In his *Vorträge*, Weismann cites Boveri's work, "which points to the correctness of the conception of the germ-plasm," but by the end of his footnote Weismann backs off again. August Weismann, *Vorträge über Descendenztheorie gehalten an der Universität zu Freiburg im Breisgau*, 2 vols. (Jena: Gustav Fischer, 1902), 1:415.

29. Theodor Boveri, *Zellenstudien III*, 50–53. The "X chromosome" was identified by Henking in 1891, but the prevalence and then importance of an odd-numbered or odd-sized chromosome was not recognized until shortly after 1900. Weismann and his students held the opposite view.

30. Ibid., 55–56.

31. Theodor Boveri, "Ein geschlechtlich erzeugter Organismus ohne mütterliche Eigenschaften," *Sitzungsberichte der Gesellschaft für Morphologie und Physiologie zu München* 5 (1889): 73–80.

32. Boveri, *Zellenstudien IV*. Wilson uses the expression "elegant in execution" for these experiments. See Baltzer, *Theodor Boveri*, for an analysis of these experiments and Boveri's criticism of Hans Driesch's concept of the egg as a harmonious equipotential system.

33. Boveri, "Befruchtung," 394. Note that Weismann encouraged the use of new equipment and Petrunkevitch developed new conservation material.

34. Boveri, *Zellenstudien II*, 7. "Alle diese Umstände," Boveri explained, "machen das Ei von *Ascaris megalocephala* zu einem Untersuchungsobjekt, dem sich bis jetzt kein zweites an die Seite stellen kann."

35. Hertwig and Hertwig, "Uber den Befruchtungs."

36. Baltzer, *Theodor Boveri*, 126–130.

37. Boveri, *Zellenstudien II*, 3. Boveri lists other nematodes, other worms (i.e., *Sagitta*), Coelenterates, echinoderms, mollusks, and tunicates which displayed the same nonfusion phenomenon.

38. Ibid., 59. This and other uses of the comparative method reveal that Boveri was not trying to extrapolate universal laws from single "model organisms," a practice that became common after 1910.

39. Boveri, *Zellenstudien II*, 168.

40. "[V]ielmehr sind noch die letzten Bestandteile der Zelle, die wir als bestimmte Formelemente nachweisen können, abermals organisierte Gebilde, die als Ganzes in ihren Lebensäusserungen jeder Erklärung durch chemisch-physicalische Kräfte spotten" (ibid., *II*, 8). For further comments about the mechanics of division, see pp. 102–120 and particularly his assessment of van Beneden and Neyt in this regard. See also figure 64 for an idealized picture of the forces of cell division.

41. Ibid., *II*, 4–12; see also "Befruchtung," 391–394. The manner in which he expressed this maintained the nineteenth-century understanding of heredity as a process that ran from one cell to the next, as well as from one personal generation to the next.

42. Over the years, Boveri was to work on a number of species of sea urchins. In 1902, he mentions *Echinus microtuberculatus,* which contains eighteen chromosomes in the diploid state, and another species of which the number was less certain. Boveri used thirty-six chromosomes for his calculations.

43. Boveri, "Befruchtung," 398–99, 425–427; Emil Selenka, *Zoologische Studien, I. Befruchtung des Eies von* Toxopneustes variegatus (Leipzig: Englemann, 1878), 94; Hertwig and Hertwig, *Uber den Befruchtungs.*

44. Theodor Boveri, "Uber die Konstitution der chromatischen Kernsubstanz," *Verhandlungen der Deutschen Zoologischen Gesellschaft* 13 (1903):10–33; Theodor Boveri, "Zellstudien VI: Die Entwicklung dispermer Seeigeleier. Ein Beitrag zur Befruchtungslehre und zur Theorie des Kernes," *Jenaische Zeitschrift für Naturwissenschaft* 43 (1907): 1–292.

45. Boveri, "Uber die Konstitution," 20–21.

46. It is possible that he envisioned a degeneration of half of the chromosomes early during the oocyte and spermatocyte preparation for maturation division. Boveri, *Zellenstudien III,* 63.

47. Boveri, "Uber die Konstitution," 26.

48. Edmund B. Wilson, *The Cell in Development and Inheritance* (New York: Macmillan, 1900). I had access to a 1911 printing that reflected all the ambiguities in 1900. "Mit Recht weist Boveri in seiner Mittheilung darauf hin, dass nun an Stelle von Vermuthungen Thatsachen getreten sind. Hier mit ist auch der Weg angedeutet, auf dem ein weiterer Fortschritt der Vererbungslehre zu erhoffen ist." E. Korschelt and K. Heider, *Lehrbuch der vergleichenden Entwicklungsgeschichte. Allgemeiner Theil* (Jena: Gustav Fischer, 1902), 728–730, quotation from p. 730. Korschelt and Heider immediately saw the implications of Boveri's redefinition of the chromosome for a different understanding of reduction division.

49. This passage is taken from the Willier and Oppenheimer translation of Boveri's "Uber mehrpolige Mitosen," in *Foundations of Experimental Embryology,* ed. Benjamin H. Willier and Jane M. Oppenheimer (Englewood Hills, N.J.: Prentice-Hall, 1964).

50. Walter S. Sutton, "On the Morphology of the Chromosome Group in Brachystola Magna," *Biological Bulletin* 4 (1902): 24–39, quotation from p. 39. Sutton referenced Boveri's "Mehrpolige Mitosen" paper at the outset and claimed, "The appearance of Boveri's recent remarkable paper . . . has prompted me to make a [this] preliminary communication" (24).

51. Walter S. Sutton, "The Chromosomes in Heredity," *Biological Bulletin* 4 (1903): 231–251.

52. Boveri, "Uber die Konstitution." The congress of the Deutschen Zoologischen Gesellschaft met in Würzburg. The associated monograph had a similar title: *Ergebnisse über die Konstitution der chromatischen Substanz des Zellkerns* (Jena: Gustav Fischer, 1904). Despite its publication date, Boveri's preface was signed in July 1903.

53. Theodor Boveri, "Ueber mehrpolige Mitosen als Mittel zur Analyse des Zellkerns," *Verhandlungen der Physikalischen-medizinischen Gesellschaft zu Würzburg,* n.s., 35 (1902): 67–90.

54. Boveri had also criticized Driesch in 1901 in the context of his work on cytoplasmic stratification.

55. Boveri had sought to demonstrate the qualitatively similar distribution of chromosomes during karyokinesis in 1888. *Zellenstudien II,* 182–188.

56. Boveri, "Ueber mehrpolige," 92–93.

57. Wilhelm Roux, "Ueber die Ursachen der Bestimmung der Hauptichtungen des Embryo im Froschei," *Anatomische Anzeiger* 23 (1903): 141–142. Weismann did so at the end of his life only indirectly when he speculated about the existence of "Teilide" as well as "Vollide" in response to Mendel's and Boveri's results. *Vortäge über Descendenztheorie,* 2:48–50.

58. Boveri, "Ueber mehrpolige" (trans. Willier and Oppenheimer), 94.

59. Johann Mayer served not only Weismann but his successors Franz Doflein, Hans Spemann, and Otto Mangold.

60. Churchill and Risler, *August Weismann,* 1:245–246, Weismann to Dr. Otto Eiser, 18 August 1895.

61. Ibid., 1:249, Weismann to Dr. Otto Eiser, 13 October, 1895.

62. The only available source on Mina's illnesses and addiction are Weismann's letters. These, of course, present only one side of the picture. Particularly see ibid., 1:245–246, 249, 263, 269–270, 290–291, 357–359.

63. Ibid., 281–282, Weismann to Parker, 13 July 1897.

64. For Weismann's mixed views on Julius's wedding, see ibid., 366–367, 369–370.

65. August Weismann, "Ueber die Parthenogenesis der Bienen," *Anatomischer Anzeiger* 18 (1900): 492–499. A closely similar paper with the variant title, "Über die Dzierzonsche Theorie," was published by Weismann in *Die Biene* (Giessen). See also August Weismann, "In Erwiderung auf Ferdinand Dickel," *Anatomischer Anzeiger* 19 (1901): 108–110.

66. August Weismann, "Semons 'Mneme' und die 'Vererbung erworbener Eigenschaften,'" *Archiv für Rassen- und Gesellschafts* 31 (1906): 1–27.

67. August Weismann, "Vorwort," in Porfirii I. Bachmetjew, [Bakhmetev], *Experimentelle entomologische Studien* (Leipzig: Wilhelm Engelmann and Weismann, 1901); also "Eine hydrobiologische Einleitung," in Richard Woltereck's new *Internationale Revue der gesamten Hydrobiologie und Hydrographie* 1 (1908): 1–9.

68. August Weismann, "Versuche über Regeneration bei Tritonen," *Anatomischer Anzeiger* 22 (1903): 425–431. See also Churchill and Risler, *August Weismann,* 1:371, Weismann to Breinig, 4 September 1902.

69. It is not clear what experiments Weismann had in mind except for von Guaita's work with albino and Japanese waltzing mice. He made the same complaint to his old acquaintance Charles Otis Whitman, 7 September 1901 (ibid., 1:353–354). At that time, Weismann expressed envy of the American's success in breeding pigeons but commented that he himself was too busy at the moment with his *Vorträge* to make further hybridization attempts.

70. Weismann's letters to Gustav Fischer give some indication of the progress of the publication. Initially, Weismann was caught off guard as to when a new edition would be needed. It appears that the first edition was selling out more rapidly than he had anticipated. The second edition appeared in October 1904. Furthermore, in the middle of 1902, J. Arthur and Margaret Thomson had approached Weismann about an English edition; they timed their translation to take advantage of any changes in the second edition. The task also involved Weismann's negotiations with Fischer and his complete oversight of the English text, a duty he performed diligently. The task was completed by the end of the year.

71. Weismann, *Vorträge,* 2:49–50. William Bateson and E. R. Saunders, *Reports to the Evolution Committee of the Royal Society* (London: Harrison & Sons, 1902).

72. Weismann, *Vorträge,* 2:50.

73. The copybooks contain three letters from Weismann to Noorduyn: 6 April 1904; 18 April 1904; 22 October 1904. In one of the rare instances when a letter to Weismann has been preserved, Weismann's papers also contain Noorduyn's letter of 10 September 1903, which had clearly been set aside and marked by Weismann—most likely as he wrote a description of Noorduyn's experiments for the second edition of the *Vorträge*. This letter had been instigated because Noorduyn had been struck by Weismann's account of reversion in domesticated breeds of pigeons found in the first edition of the *Vorträge*.

74. Weismann, *Vorträge,* 2:47–48. See also C. L. W. Noorduyn, "Jets over Kleuren, Kleurverandering der Vogels en Paring von Varieteiten," *Album der Natur* (Haarlem: T. Jeink Willik, & Zoon, 1904).

75. Weismann, *Vorträge* (1902), 1:64; (1904) 2:49. It is interesting that Weismann would make such a strong distinction between what a botanist and a zoologist could do. This disciplinary divide never concerned Bateson.

76. Churchill and Risler, *August Weismann,* 1:460, Weismann to Iltis, 28 September 1906.

77. Ibid., 463, Weismann to Iltis, 19 December 1906.

78. Weismann was clearly responding to very recent research on tetrad formation. See *Vorträge* (1904), 2:37 and footnote. Montgomery had sent him reprints at the beginning of the year, which undoubtedly included "Some Observations and Considerations upon the Maturation Phenomena of the Germ Cells," *Biological Bulletin* 6 (1903–1904): 137–157. See Churchill and Risler, *August Weismann,* 1:413, Weismann to Montgomery, 29 March 1904. It is unclear when he studied Boveri's summary monograph *Ergebnisse über die Konstitution er chromatischen Substaz des Zellkerns* (Jena: Gustav Fischer, 1904), but he evidently knew much of its contents from earlier statements and Häcker's opposition to it.

79. Twenty-one colleagues and former students contributed scientific papers to this massive work (749 pages). His copybook contains explicit letters from Weismann to almost every author, describing the work of each and thanking him for the contribution in Weismann's honor. No letter to Häcker has been identified.

80. Valentin Häcker, "Bastardirung und Geschlechts-zellenbildung," *Zoologische Jahrbücher, Supplement Band VII* (1904): 161–256.

81. The contrasting theories can be seen in the illustration at the end of his text. In all fairness to Häcker, his paper was sent to the publishers in September 1903, before the publication of Boveri's important monograph of 1904. There is no indication that Häcker had yet studied Boveri's much shorter 1903 paper to the German Zoological Society.

82. Nettie Stevens's work on arbitrarily dividing the egg at the first cleavage stage also indicates the importance of the correct combination of chromosomes for future development.

83. Alexander Petrunkevitch, "Künstliche Parthenogenese," *Festschrift zum siebzigsten Geburtstage des Herrn Geheimen Raths Prof. Dr. August Weismann in Freiburg in Baden* (Jena: G. Fischer, 1904), 126.

84. Heinrich Ernst Ziegler, *Die embryonale Entwickelung von* Salmo salar (Freiburg i. Br: Rombach, 1882).

85. Heinrich Ernst Ziegler, *Die Vererbungslehre in der Biologie* (Jena: Gustav Fischer, 1905). The lecture in Wiesbaden was titled "Über den derzeitigen Stand der Vererbungslehre in der Biologie" (12 April 1905).

86. Theodor Boveri, *Ergebnisse über die Konstitution der chromatischen Substanz des Zellkerns* (Jena: Gustav Fischer, 1904).

87. Ziegler, *Die Vererbungslehre,* 21–23.

88. Ibid., 30.

89. This is a curious argument, but Ziegler makes it twice (ibid., 38, 50). He also argued that since every cell of the body contains a complete set of chromosomes, they all must be involved in producing the traits of any given cell. Both arguments reveal his unclear understanding of the new focus on transmission imposed by the Boveri-Sutton hypothesis.

90. Ibid., 51. Ziegler appears to resort to an embryological explanation for trait production. To this extent, he appears still to be thinking within the embryological framework embedded in Weismann's germ-plasm theory.

91. Churchill and Risler, *August Weismann,* 1:415, Weismann to Buttel-Reepen, 27 April 1904.

92. Ibid., 1:422, Weismann to Parker, 19 September 1904; 1:432–433, 9 March 1905 (commenting upon Bateson, in Bateson and Saunders, *Reports to the Evolution Committee,* 1902).

93. Churchill and Risler, *August Weismann,* 1:435, Weismann to Carl Correns, 21 March 1905.

94. Ibid., 475, Weismann to Lang, 30 August 1907. "Schön u. interessant" he called the experiments.

95. Ibid., 506, Weismann to Schallmeyer, 19 July 1910.

96. This manuscript is written in another hand, presumably by Schleip, but includes in Weismann's hand at the outset "von Schleip 1912." The entire manuscript consists of two parts, one of six pages and the other of thirteen. It is obvious that Weismann went through the entire text with care, for he also made many bold overwrites throughout and thanked Schleip in the preface to the new edition for his assistance in the areas covered by the manuscript.

97. *"Anteil der Eltern am Aufbau des Kindes.—Rückschlag."* A curious printer's mistake labeled the chapter inaccurately as chapter XXIV and the subsequent chapter as chapter XXIII. In comparing the numbering with the table of contents and the second edition, there can be little doubt that the numbering, not the chapters themselves, was reversed.

98. See Marsha L. Richmond, "Opportunities for Women in Early Genetics," *Nature Reviews Genetics* 8 (2007): 897–902.

99. Weismann, *Vorträge,* 2nd ed., 2:37.

100. Ibid., 3rd ed., 2:33.

101. Montgomery sent Weismann some of his papers—probably including his "Some Observations and Considerations upon the Maturation Phenomena of the Germ Cells" (1903). Boveri probably sent him his *Ergebnisse über die Konstitution*

der chromatischen Substanz des Zellkerns (Jena: Gustav Fischer, 1904), which Weismann read with enthusiasm. Churchill and Risler, *August Weismann,* 1:404–405, Weismann to Boveri, 18 February 1904. For a long time, Weismann had been aware of Hermann Henking's discovery of what he called the X-chromosome, because at least some of the work had been done in Weismann's institute.

102. Weismann is not always clear about the relationship between "Vollide," "Teilide," and chromosomes, perhaps because he treats lower organisms differently from higher in this regard. See particularly *Vorträge,* 3rd ed., 2:48, for statements equating chromosomes with "Teilide."

103. Ibid., 51.

104. On the controversy concerning Weismann and transverse division, see Frederick B. Churchill, "Rudolf Virchow and the Pathologist's Criteria for the Inheritance of Acquired Characteristics," *Journal of the History of Medicine* 31 (1976): 117–148. For a later perspective on the unimportance of the phenomenon, see Edmund Wilson, *The Cell in Development and Inheritance,* 3rd. ed. (New York: Macmillan, 1925).

105. With an understanding of tetrad formation came a recognition that synapsis and tetrads supplied a mechanism for guaranteeing that after two subsequent maturation divisions each gamete would receive one and only one chromosome of a homolog pair. At the same time, it became possible for Weismann to conceive of chromosomes as "partial ids" rather than as bearers of all the determinants of a complete organism. He continued to feel, however, that the number of Mendelizing traits must be limited to the number of diploid chromosomes. Boveri makes some of these points in *Ergebnisse* (50–78).

106. Weismann considered the chromatid segregation of the gametes as random, whereas he argued that recombinations due to fertilization were not.

107. *Vorträge* 3, II, 57–58, and the subsection on Rückschlag (58–64).

21. A Mechanical Model for Evolution

1. August Weismann, "Beiträge zur Kenntniss der ersten Entwicklungsvorgänge im Insektenei," in Jacob Festgabe, *Henle zum 4. April 1882 dargebracht von seinen Schülern* (Bonn: M. Cohen, 1882), 80–111.

2. "August Weismann zu seinem sechszigsten Geburtstage 17," *Zoologische Abhandlungen,* vol. 8 (January 1894).

3. *Bericht über die Feier des 70. Geburtstages von August Weismann am 17 Januar 1904 in Freiburg I. Breisgau* (Jena: Comite zur Stiftung der Weismann-Büste, 1904).

4. *Bericht* (Beilage III), 41–44. It is noteworthy that Weismann's old friend Ernst Haeckel was not among the contributors. The bust now stands in the building of the new zoological institute.

5. See Frederick B. Churchill and Helmut Risler, *August Weismann, Ausgewählte Briefe und Dokumente=Selected Letters and Documents,* 2 vols. (Freiburg: Universitätsbibliothek Freiburg, 1999), 2:537–606.

6. Weismann, *Bericht,* 32–33.

7. For Weismann's expression of gratitude to Spengel for his role in editing the volume, see Churchill and Risler, *August Weismann,* 1:418–419, Weismann to Spengel, 3 June 1904.

8. A complete list of the contributors and some of Weismann's letters of thanks appear ibid., 1:395–418.

9. Daniel L. Schacter, *Stranger Behind the Engram: Theories of Memory and the Psychology of Science* (Hillsdale, N.J.: Lawrence Erlbaum, 1982); see also Georg Uschmann, "Semon, Richard Wolfgang," *Dictionary of Scientific Biography,* 12:297–298; Jürg Schatzmann, *Richard Semon (1859–1918) und seine Mnemetheorie,* (Zürich: Juris Verlag, 1968).

10. Schacter, *Stranger,* 30–2, quotation from p. 31.

11. I had available the third German edition (Leipzig: Wilhelm Engelmann, 1911) and *The Mneme,* trans. Louis Simon (London: George Allen, 1921).

12. Schatzmann, *Richard Semon,* 68–69.

13. August Weismann, "Semons 'Mneme' und die 'Vererbung erworbener Eigenschaften,'" *Archiv für Rassen- und Gesellschafts-Biologie 3, ts-Biologie* 3 (1906): 1–27.

14. See Semon, *The Mneme,* 130–132, 230–235.

15. Semon also detailed Emil Fischer's experiments showing the multiple effects of temperature on the tiger moth (*Arctia caja*) and Paul Kammerer's early experiments on the fire salamander (*Salamandra maculosa*). These examples also supported the inheritance of acquired characters.

16. Weismann, "Semons 'Mneme,'" 3.

17. Ibid., 22.

18. Ibid., 27.

19. August Weismann, "Charles Darwin und sein Lebenswerk" (Jena: Gustav Fischer, 1909); also appearing from dictation as "Der Darwinismus in Vergangenheit und Zukunft; Festrede 12 Feb. 1909," *Allgemeine Zeitung* 20 (1909):168–171.

20. The full citation of the edited volume is August Weismann, "The Selection Theory," *Darwin and Modern Science: Essays in Commemoration of the Centenary of the Birth of Charles Darwin and of the Fiftieth Anniversary of the Publication of* The Origin of Species, ed. A. C. Seward (Cambridge: Cambridge University Press, 1909), 18–65. See also *Die Selektionstheorie, eine Untersuchung* (Jena: Gustav Fischer, 1909).

21. August Weismannn, "Über die Trutzstellung des Abendpfauenauges," *Naturwissenschaftliche Wochenschrift, N.F.,* 8 (1909): 271–726.

22. English translation appeared as August Weismann, "Charles Darwin," in *The Contemporary Review* (1909); reprinted in *Annual Report of the Smithsonian Institution* (Washington, D.C.: Government Printing Office, 1910), 431–452, quotations from pp. 443–444.

23. Weismann, "Selection Theory," 20–21.

24. Ibid., 25.

25. Ibid., 27.

26. Ibid., 39.

27. Helmut Risler, "August Weismanns Leben und Wirken nach Dokumenten aus seinem Nachlass," in *August Weismann (1834–1914) und die theoretische Biologie des 19. Jahrhunderts. Urkunden, Berichte und Analysen,* ed. Klaus Sander (Freiburg: Rombach Verlag, 1985), 38–40.

28. *Vorträge über Descendenztherorie gehalten an der Universität zu Freiburg im Breisgau* (Jena: Gustav Fischer, 1902, 1904, 1913); or *The Evolution Theory,* trans. J. Arthur Thomson and Margaret R. Thomson (London: Edward Arnold, 1904).

29. Weismann, *Evolution Theory,* 1:30.

30. Ibid., 55, Lecture 3.

31. Ibid., 56, Lecture 3.

32. Translation ibid., 198–199, Lecture 10; quotation from p. 198.

33. Ibid., 214, Lecture 11 (emphases mine).

34. Ibid., 251, Lecture 12.

35. Ibid., 2:49–50, Lecture 22.

36. Weismann did not distinguish between chromosomes and chromatids. They were both "Idanten."

37. Weismann, *Evolution Theory,* 2:39–41, Lecture 22, quotation from p. 41 (emphasis in the original).

38. Valentin Häcker, *Ueber das Schicksal der elterlichen und grosselterlichen Kernanteile: Morphologische Beiträge zum Ausbau der Vererbungslehre* (Jena: Gustav Fischer, 1902).

39. The mention of "affinities" between different idanten appears in a long footnote in *Vorträge,* 2:37 (emphasis in the original). The footnote also appears in the English edition, but the section was entirely rewritten in the third, final edition (1913). Any reference to "affinities" between idanten disappeared. Weismann comments on the confusion caused by the profusion of new facts being produced in this area of cytology by investigators, "vor allem Boveri und Häcker."

40. Helmut Risler, "Leben und fruhe deskriptive Veröffentlichungen von August Weismann," in Sander, *August Weismann (1834–1914),* 23–42. See particularly pp. 38–40, which include illustrations of the butterfly *Elymnias* and its model.

41. Ibid., 38. "Es kann also hier keine Rede davon sein, dass die Ursache der Artumwandlung in der direkten Einwirkung äusserer Umstände läge, es gibt nur die eine Möglichkeit, dass der Vorteil, den die Ähnlichkeit mit der immunen Art der

mimetischen gewährt, es ist, der ihr durchschnitt und im Laufe der Zeit immer wieder den Sieg verschaafft."

42. "Das Institut hält sich recht tapfer, gegen 30 Praktikanten u. in d. Vorlesu[n]g über Zoologie 300 Zuhörer." Churchill and Risler, *August Weismann,* 1:513, Weismann to Petrunkewitsch, 19 September 1911.

43. Ibid. "Aber ich werde sie nicht mehr lange haben, denn obgleich ich frei spreche ohne jede schriftliche Notiz, strengt es mir doch die Augen an." scheint es Zeit, die Last auf jünger Schultern abzuladen!"

44. Churchill and Risler, *August Weismann*, 1:513, Weismann to Petrunkewitsch, 19 September 1911; 2:737–740, Weismann to the Kultusminster, Franz Böhm.

Acknowledgments

This work has had a complicated history. It represents in part material collected after completing my dissertation and traveling by train in the summer of 1967 through what at the time was known as East Germany. My intent was to explore the libraries and to meet some of the scholars of the history of biology, about whom I had meager knowledge. I visited both Halle and Leipzig, where the respective institutes for history of medicine could provide an entrée to the likely scholars interested in the history of biology—above all, to Georg Uschmann and Reinhardt Mocek. Unfortunately, the former was unavailable, as I remember, on a summer vacation to the Baltic. Nevertheless, I was encouraged by scholars young and old at both universities to continue, and so I did. I remain in contact with a few of them even now. The trip also provided me with a mint copy of Weismann's *Das Keimplasma* for a bargain—some East German booksellers did not regard the advent of modern genetics as an important historical subject.

Traveling farther, I crossed the border back into West Germany and made my way to Freiburg, where I was surprised to find that the Institute for Zoology had just the previous week sponsored a lecture series on Weismann. The staff, headed by Prof. Bernhard Hassenstein, Direktor des Zoologischen Instituts, made arrangements for me to travel to Mainz, where Prof. Helmut Risler taught at the Allgemeine Institut für Zoologie. Prof. Risler had just described to his colleagues in Freiburg the historical materials he had inherited from his family, which contained many of Weismann's annual notebooks, seven Watt-style copybooks of letters Weismann had written, and smaller pocket books with hastily written, often illegible notes. Prof. Risler was pleased to receive me. He paraded forth his manuscript materials and indicated that he would be delighted if I were to spend more time in Mainz examining the collection. I promised to do this once I had earned my first leave of absence and the necessary grant—which I later did. The result would be an edition of selected Weismann letters and documents, most of which now reside at Freiburg Universität

(Frederick B. Churchill and Helmut Risler, *August Weismann, Ausgewählte Briefe und Dokumente* [*Selected Letters and Documents*], 2 vols., Universitätsbibliothek, Freiburg i. Br. 1999). My studies there led to my deeper interest in Weismann and his career.

It must be common, when an author has completed a substantial book with assistance from dozens of interested colleagues and a handful of concerned librarians, to thank all those who have contributed to that work, its format and contents. The specific gratitude I owe to those friends—as well as to the many others with whom I had casual interactions over the years, who provided their insights and special guidance—is nigh impossible to convey. Here, I will group those who most strongly influenced my academic development and will provide what appears to me a shamefully meager description of their involvement. Above all, I owe a great debt to my family and my schoolteachers of the distant past, who encouraged beyond all normal reason my academic career. My parents, Edward D. and Mary B. Churchill, were unflinching in their support from the very beginning of my pursuits; they understood me and pushed me gently from behind the scenes. My siblings, Mary C. Fischelis, Edward D. Churchill Jr., and Algernon C. Churchill, each published scholars in their own right, were always sympathetic and interested, providing critiques and suggestions. From the beginning, my formal education included unusual teachers and scholars such as E. Barton Chapin, Lillian Putnam, and James P. McCarthy of the Shady Hill School in Cambridge, Massachusetts, and George Van Santvoord, Richard C. Gurney, and Charles E. Berry of the Hotchkiss School in Lakeville, Connecticut. And I must not forget my two years of active service, mostly in Germany, at the lowest level of the U.S. Army, where I made friends and learned from a variety of servicemen and Landsmänner, who spontaneously, often unwittingly, helped me focus my life.

Later, Everett Mendelsohn and Ernst Mayr at the Harvard graduate school of Arts and Sciences befriended me and advanced my academic career. My professional colleagues, particularly Richard S. Westfall and Edward Grant at the Department of the History and Philosophy of Science at Indiana University in Bloomington, not only warmly supported me but provided me with outstanding models of what becoming a historian of science might entail. They strongly influenced my approach to our common profession: an emphasis on the contents of science that does not neglect its philosophical and institutional developments. Their argument was that the "internal intellectual events" of scientists provided the essential reasons for those scientists coming to historians' attention in the first place, an approach that remains controversial some forty years later. In addition, many graduate students—those whom I formally taught and those who unwittingly taught me—were instrumental in pushing me forward. I was fortunate to have both supportive and understanding colleagues throughout the history of science and medicine who focused on what was important in a life of teaching and scholarship. Saul Benison (at the time at Columbia

University), Helmut Risler at the University of Mainz, and William Coleman at the University of Wisconsin stand out in retrospect as the most important colleagues in my career-building stage. They and many others set the tone for serious scholarship that shaped my historical views.

In addition to formal and quasi-formal acknowledgments, most scholars benefit in special ways from their close contemporaries, be they teachers, colleagues, or students. Besides William Coleman, whom I have already mentioned, two other contemporaries stand out as critical to this work as it developed: Jane Maienschein and Paul L. Farber, who were students at Indiana University during their graduate student careers, continue to inform me and critique my work throughout. Jane, a longstanding professor in the School of Life Sciences at Arizona State University, and Paul, a recently retired professor from Oregon State University, have both candidly and forcefully commented on my work when it was in manuscript form. Their recommendations have been much appreciated, but it must be made clear that any failures in the final text are mine alone.

Finally, most historians owe a strong debt to the library systems and to the librarians who make our historical task possible. I consider myself fortunate to have worked at some fine institutions. Widener Library and the Library at the Museum of Comparative Zoology (now the Ernst Mayr Library), both at Harvard University, the library system at Indiana University with its diverse nineteenth- and twentieth-century collections, and the bibliothecas at both Mainz and Freiburg Universities in Germany were all, in their respective ways, essential for my work. Librarians Celestina Savonius-Wroth and Zachery Downey were particularly attentive and helpful.

Above all, I owe a deep gratitude to Sandra R. Churchill, who fully shared with me her life and family and served as an important force in encouraging my detailed research and writing. It is to her in particular I enthusiastically dedicate this book.

Index